Linear and Nonlinear Electron Transport in Solids

NATO ADVANCED STUDY INSTITUTES SERIES

A series of edited volumes comprising multifaceted studies of contemporary scientific issues by some of the best scientific minds in the world, assembled in cooperation with NATO Scientific Affairs Division.

Series B: Physics

RECENT VOLUMES IN THIS SERIES

Volume 10 – Progress in Electro-Optics
edited by Ezio Camatini

Volume 11 – Fluctuations, Instabilities, and Phase Transitions
edited by Tormod Riste

Volume 12 – Spectroscopy of the Excited State
edited by Baldassare Di Bartolo

Volume 13 – Weak and Electromagnetic Interactions at High Energies (Parts A and B)
edited by Maurice Lévy, Jean-Louis Basdevant, David Speiser, and Raymond Gastmans

Volume 14 – Physics of Nonmetallic Thin Films
edited by C.H.S. Dupuy and A. Cachard

Volume 15 – Nuclear and Particle Physics at Intermediate Energies
edited by J. B. Warren

Volume 16 – Electronic Structure and Reactivity of Metal Surfaces
edited by E. G. Derouane and A. A. Lucas

Volume 17 – Linear and Nonlinear Electron Transport in Solids
edited by J. T. Devreese and V. E. van Doren

Volume 18 – Photoionization and Other Probes of Many-Electron Interactions
edited by F. J. Wuilleumier

Volume 19 – Point Defects and Their Structure in Nonmetallic Solids
edited by B. Henderson and A. E. Hughes

Volume 20 – Physics of Structurally Disordered Solids
Edited by Shashanka S. Mitra

The series is published by an international board of publishers in conjunction with NATO Scientific Affairs Division

A	Life Sciences	Plenum Publishing Corporation
B	Physics	New York and London
C	Mathematical and Physical Sciences	D. Reidel Publishing Company Dordrecht and Boston
D	Behavioral and Social Sciences	Sijthoff International Publishing Company Leiden
E	Applied Sciences	Noordhoff International Publishing Leiden

Linear and Nonlinear Electron Transport in Solids

Edited by

J. T. Devreese
State University of Antwerp (RUCA) and UIA
Antwerp, Belgium

and

V. E. van Doren
State University of Antwerp (RUCA)
Antwerp, Belgium

SPRINGER SCIENCE+BUSINESS MEDIA, LLC

Library of Congress Cataloging in Publication Data

Nato Advanced Study Institute on Linear and Nonlinear Electronic Transport in Solids, Rijksuniversitair Centrum Antwerpen, 1975.
Linear and nonlinear electron transport in solids.

(NATO advanced study institutes series: Series B, physics; v. 17)
Includes indexes.
1. Electron transport–Addresses, essays, lectures. 2. Semiconductors–Addresses, essays, lectures. I. Devreese, Jozef T. II. Doren, V. E. van. III. Title. IV. Series.
QC176.8.E4N35 1975 537.6 76-15234

ISBN 978-1-4757-0877-6 ISBN 978-1-4757-0875-2 (eBook)
DOI 10.1007/978-1-4757-0875-2

Lectures presented at the NATO Advance Study Institute on Linear and Nonlinear Electronic Transport in Solids held at the State University of Antwerp (RUCA), Antwerp, Belgium, July 21-August 2, 1975

Originally published by Plenum Press, New York in 1976

Preface

The Advanced Study Institute on "Linear and Nonlinear Electronic Transport in Solids" was held at the Rijksuniversitair Centrum Antwerpen (R.U.C.A.) with its nice surroundings and excellent facilities, from July 21 till August 2, 1975. The Institute was sponsored by N.A.T.O. Co-sponsors were: Bell Telephone Mfg. Co. N. V. (Belgium), Esso Belgium N. V. (Belgium), Generale Bankmaatschappij (Belgium), Kredietbank (Belgium), Metallurgie Hoboken Overpelt N. V. (Belgium) and the National Science Foundation (USA). A total of 90 participants including the lecturers attended the Institute.

In recent years several important developments took place in the understanding of the electronic conduction in solids. Purely phenomenological approaches are now complemented by basic theories which provide a more exact description of the transport properties. The experimental studies produced also better results and raised several questions which need further studies for interpretation.

One of the aims of the summer school has been to develop a better understanding of the conduction problem both theoretically and experimentally, starting from an elementary level and proceeding to topics of increased complexity.

In the main the summer school dealt with general theories of transport phenomena in solids and the study of electronic properties of metals and semiconductors. As far as the general theory is concerned, the Boltzmann equation, the relaxation time concept and the inverse transport coefficients approach have been developed; these constitute the basic tools of the various investigations.

In the applications have been included a variety of topics covering the field of both crystalline and amorphous materials. The thermodynamical aspect of the electron gas in metals, of the electron-hole plasmas and of the charge carriers in liquid metals has been treated. The conductivity of small gap and disordered semiconductors was studied. The models of the polaron, and transport by hopping, have been used for the evaluation of transport quantities, including linear and nonlinear phenomena over a wide range of fields.

The transport properties of transition metal ion compounds, chalcogenide glasses and amorphous semiconductors were the main experimental results, presented together with their respective interpretations.

In addition to the lectures a variety of seminars, covering a wide range of topics, were given by members of the school which not only helped to supplement the programme, but also contributed to a more active participation.

The present volume covers the proceedings of the Summer School which are contained under the headings: General Theory, Metals and Semiconductors, Amorphous Semiconductors and Seminars.

I should like to thank the lecturers for their collaboration in preparing the manuscripts in time which enabled the participants to attend the lectures comfortably and more beneficially. The invited speaker for the opening session of the Institute was Professor Grosse of Physikalisches Institut der Rheinisch-Westfälischen Technischen Hochschule Aachen, whom I should like, taking this opportunity, to thank from this position once again. My thanks are also due to the members of the International Advisory Committee: Professors J. Bok, M. H. Cohen, J. Friedel, J. Hodby, S. Lundqvist, N. H. March, N. Mott, K. S. Singwi, and K. K. Thornber.

It gives me great pleasure to thank the Rector of the Rijksuniversitair Centrum Antwerpen (R.U.C.A.), Professor Dr. L. Massart, for his warm and continuous interest in the Institute, and for making available so promptly and effectively the university facilities.

I should also like to thank Mr. A. Drubbel, Chief Administrator of the University, and through him his staff, for their coordinated effort which so greatly contributed to the success of the Institute.

Finally, I wish to thank Drs. V. Van Doren and J. Van Royen, and Miss H. Evans for their dedication in carrying out the enormous secretarial task which enabled the smooth operation of the Summer School. My thanks are also due to Dr. F. Brosens, Dr. L. Lemmens, Drs. P. Van Camp and J. Van Royen for reading the proofs, and Miss H. Evans for preparing the author index.

Jozef T. Devreese
Professor of Theoretical Physics,
University of Antwerp (R.U.C.A. and U.I.A.)
Director of the Advanced Study Institute

March 1976

Contents

Part I
General Theory

LINEAR AND NONLINEAR RESPONSE THEORY WITH APPLICATIONS

David C. Langreth

Rutgers University

New Brunswick, New Jersey 08903, U.S.A.

I. INTRODUCTION

The basic perturbation theoretic method for dealing with non-equilibrium quantum statistical mechanical problems has been known for years. The theory discussed here evolved from the work of Schwinger and was developed further by Kadanoff and Baym[1]. In fact virtually all the theory discussed here is contained in Kadanoff's and Baym's book. Essentially the same theory was developed independently by Keldysh[2] who even anticipates a rather useful addendum by Craig[3].

My motivation for writing about such an old and well-established theory is twofold. First, it has become evident in talking with colleagues in condensed-matter and many-body theory that the earlier work is not as well known as perhaps it should be, even though it has been applied to problems in these fields for years. Secondly, it has become obvious that the time is ripe for its application to many problems in these fields.

Perhaps the reason that the use of non-equilibrium theory has been neglected is that as long as one is dealing only with linear response, there are well-known tricks which enable one to express the physical response function directly in terms of the time-ordered (at $T = 0$) or thermal response function (at $T \neq 0$); these latter functions are given by the familiar diagramatic expansions, so that in problems of this type the need to know non-equilibrium theory is "short-circuited". Nevertheless there are problems even in linear response where the non-equilibrium theory can be used to great advantage. It is, however, in non-linear response that the full non-equilibrium theory comes to the fore.

In the next section I develop the basic theory in a way which I hope will be useful for applications to condensed-matter physics. In section III I discuss one such application to linear response, and follow this by an application to non-linear response.

I assume that the reader has at least a vague familiarity with Green's function perturbation expansions of some type. For those who do not, I recommend the reading of Chapter 5 of Schrieffer's book[4] on superconductivity (or some similar treatment) before proceeding further in this article.

II. NON-EQUILIBRIUM THEORY

A. Path Formulation of Time Development

Consider a system which develops in time according to the Hamiltonian

$$\mathcal{H} = H + H' \tag{2.1}$$

where H is the time independent Hamiltonian of the "system" and H' represents a time dependent external field which drives the system out of thermal equilibrium. We assume that H' vanishes for times before t_o and that at this time the system is in thermal equilibrium at temperature $T \equiv (k_B\beta)^{-1}$; thus the density matrix at time t_o is

$$\rho_o = e^{-\beta H} / \mathrm{Tr}e^{-\beta H} \tag{2.2}$$

where we allow for an open system by measuring the single particle energies of each type of particle from their respective chemical potentials. In practice we will often let $t_o \to \infty$ and also suppose that H' is turned on adiabatically.

Generally, what one wishes to calculate is the average of a physical observable A at a time $t > t_o$, that is

$$\langle A \rangle = \mathrm{Tr}\rho_o A_{\mathcal{H}}(t) \tag{2.3}$$

where $A_{\mathcal{H}}(t)$ develops in the Heisenberg picture

$$i \frac{\partial A_{\mathcal{H}}(t)}{\partial t} = [A_{\mathcal{H}}(t), \mathcal{H}] \tag{2.4}$$

For the development of perturbation series it is, as usual, simpler to use operators which develop in the interaction picture according to

$$i \frac{\partial A_H(t)}{\partial t} = [A_H(t), H] . \tag{2.5}$$

The relationship between the two pictures can be expressed as

$$A_{\mathcal{H}}(t) = T_c \; e^{-i\int_c H'_H(\tau)d\tau} A_H(t) \tag{2.6}$$

where the integration contour c begins at t_o goes through t and then back to t_o. The symbol T_c indicates that all operators are to be ordered according to the positions of their time arguments on the contour c, with those occuring first on the right. That (2.6) satisfies (2.5) may be verified by direct differentiation; it trivally satisfies the initial condition $A_H(t_o) = A_{\mathcal{H}}(t_o)$. From now on I will use the interaction picture almost exclusively and will drop the subscript "H" from quantities developing in this picture.

The nice thing about (2.6) is that it is valid with very little restriction as to what contour c is used; the only feature that really matters is that it begins and ends at t_o and passes through t once. I implicitly assume that H' for complex τ is an analytic extension which coincides with the physical function H' on the real axis. In fact for this reason it is not even necessary to take t on the real axis because it can always be continued there. We shall use this freedom to deform the contour later to exhibit the connection between different formulations.

B. Correlation Functions and Green's Functions

We now define the two time correlation functions of the wave field operators ψ and $\psi^\dagger$, as these are related to the Green's functions determined by perturbation theoretic expansions:

$$g^<(t,t') = \langle \psi_H^\dagger(t')\ \psi_H(t)\rangle$$
$$g^>(t,t') = \langle \psi_H(t)\ \psi_H^\dagger(t')\rangle \qquad (2.7)$$

where ψ is assumed to have a column vector index α which denotes space, spin, etc. In terms of $g^>$ the one particle density matrix is simply $g^<_{\alpha\beta}(t,t)$. It is thus a simple matter to express for instance the number density $n(\vec{x},t)$ current density $\vec{j}(\vec{x},t)$, and spin density $\sigma(\vec{x},t)$ as

$$n(\vec{x},t) = \langle \vec{x}|g^<(t,t)|\vec{x}\rangle \qquad (2.8a)$$

$$\vec{j}(\vec{x},t) = (\vec{\nabla}_x - \vec{\nabla}_{x'})/2im\ \langle \vec{x}|g^<(t,t)|\vec{x}'\rangle|_{x=x'} \qquad (2.8b)$$

$$\vec{\sigma}(\vec{x},t) = \sum_{\mu\nu} \vec{\sigma}_{\mu\nu}\langle \vec{x}\ \nu|g^<(t,t)|\vec{x}\mu\rangle \qquad (2.8c)$$

where in (2.8a) and (2.8b) the spin components were not written and implicitly traced over; in (2.8c) $\vec{\sigma}$ represents the three Pauli matrices. One also notes that by introducing the sum $(\vec{R},T)$ and difference $(\vec{\xi},\tau)$ variables defined by

$$(\vec{R},T) = ((\vec{x} + \vec{x}')/2,\ (t + t')/2)$$
$$(\vec{\xi},\tau) = (\vec{x} - \vec{x}',\ t - t') \qquad (2.9)$$

and then Fourier transforming with respect to $(\vec{\xi},\tau)$ one can construct a generalized Wigner distribution function $g^<(\vec{p},p_o;\ \vec{R}T)$ where $(\vec{p},p_o)$ are the Fourier transform variables corresponding to $(\vec{\xi},\tau)$. Physically $g^<(\vec{p},p_o;\ \vec{R}T)$ can be thought of as the density of particles of momentum **p** and energy p_o at space-time point $(\vec{R},T)$, although because of the uncertainty principle this interpretation is not exact.

In terms of interaction representation operators $g^<$ (and similarly $g^>$) may be expressed as

$$g^<(t,t') = \langle [T_{c'}\ e^{-i\int_{c'} H'(\tau)d\tau}\ \psi^\dagger(t')]$$
$$\times\ [T_c e^{-i\int_c H'(\tau)d\tau}\ \psi(t)]\rangle \qquad (2.10)$$

where c and c' are contours which begin at t_o go through t or t' respectively and return to t_o. It is obviously more convenient to replace the two contours c and c' by a single contour C which starts at t_o goes through t and t' and then back to t_o. Note that the return of c to t_o after t is cancelled by the beginning of c' from t_o so that one does not need to require that C return to t_o between t and t'. We may write this all in compact notation convenient for manipulations if we define the path ordered Green's function

$$iG(t,t') = \langle T_c \; \psi_H(t) \; \psi_H^\dagger(t') \rangle \tag{2.11}$$

where T_c orders the operators ψ and $\psi^\dagger$ according to which time argument is further along the contour with "earlier" times to the right. [We will assume here and throughout this work that we are dealing with Fermion operators, so that the ordering $\psi^\dagger\psi$ receives an extra minus sign]. It is clear that G(t,t') has two analytic pieces

$$iG(t,t') = g^{>}(t,t') \tag{2.12a}$$

$$iG(t,t') = -g^{<}(t,t') \tag{2.12b}$$

where the first equality (2.12a) holds if t is further along C than t' and the second equality (2.12b) holding otherwise. In the interaction representation, then we use the argument following (2.10) to write (2.11) as

$$iG(t,t') = \langle T_C \; e^{-i\int_C H'(\tau)d\tau} \; \psi(t)\psi^\dagger(t') \rangle \tag{2.13}$$

where the order of all operators is determined by T_C.

C. Perturbation Expansion

One is now in a position to derive the diagrammatic perturbation theory for the non-equilibrium function G. There are two ways to proceed. The first is to apply the functional derivative method of Schwinger[5], which is described in Kadanoff and Baym[1]. The second is to let H contain only one body term, and include in H' all two (or more) body terms [we suppose the latter are turned on adiabatically] and directly expand the exponential in (2.13). In either case it is obvious that the diagrammatic expansion thus obtained is identical in form to the usual equilibrium expansions; the only difference is that all intermediate time integrations extend over the contour C, rather than from $-\infty$ to ∞ as they would in zero temperature equilibrium theory or from 0 to $-i\beta$ as in the usual form of finite temperature equilibrium theory. Because each reader will have his own way of keeping track of the i's and minus signs in the perturbation expansion [as well as in the definition of the Green's function itself], I will not give a set of diagrammatic rules and simply say: use your own rules; they will work here as well!

D. Linear Response Theory

As an example of the utility of this formalism, consider the linear density response of a system to an external potential $U(\tau)$, so that $H'(\tau) = U(\tau)n(\tau)$ with matrix multiplication implied in the space variables. The required density response is the change in $g^<(t,t')$ which is given by expanding the expotential in (2.13). Consequently we have

$$\delta g^<(t,t) = -i \int_C \langle T_C n(t)\ n(\tau)\rangle_o U(\tau) d\tau \qquad (2.14)$$

where $\langle\ldots\rangle_o$ represents the thermal equilibrium average. Notice that the path ordered correlation function $\langle T_c\ n(t)\ n(\tau)\rangle_o$ is precisely the quantity obtained from the diagrammatic theory described above. Since the physical t and τ are on the real axis (2.14) may be written

$$\delta g^<(t,t) = -i\int_{-\infty}^{\infty} \{\langle[n(t),\ n(\tau)]\rangle\theta(t-\tau)\}\ U(\tau)d\tau \qquad (2.15)$$

where θ is the unit step function which vanishes for negative arguments. The quantity in the curly brackets, i.e., the retarded commutator, is not the quantity given directly by the more usual diagrammatic formulations. For example, the quantity calculated in the usual zero temperature theory is the real axis time ordered product $\langle T\ n(t)\ n(\tau)\rangle$. Although there exists a simple relationship between this and the retarded commutator[6], it is not straightforwardly and simply generalizable to the case of non-linear response. To use a form of perturbation theory which gives directly the function one wishes to calculate is a real advantage.

E. Useful Identities[7]

Let me now derive a few identities which are useful in expressing products of Green's functions which occur in the various expansions in terms of integrals along the real axis (rather than along C). We first consider the multiplication of two Green's functions or related quantities "in series", that is let $D(t,t')$ be defined by

$$D(t,t') = \int_C d\bar{t}\ A(t,\bar{t})\ B(\bar{t},t') \qquad (2.16a)$$

or simply

$$D = AB \qquad (2.16b)$$

in matrix notation. For example, A and B might both be Green's functions (in a section of a diagram) connected by an external potential:

$$t \longleftarrow \bar{t} \longleftarrow t' \qquad (2.17)$$

As another example, A might be a self-energy Σ and B a Green's function, in a combination that occurs in Dyson's Equations

$$(G_o^{-1}-U)G = 1 + \Sigma G \qquad (2.18a)$$

$$G(G_o^{-1}-U) = 1 + G\Sigma \qquad (2.18b)$$

where G_o is the bare Green's function in the absence of many-body interactions (and also in the absence of any time-dependent vector or scalar potentials symbolized schematically by U). What we wish to do is to write expressions for the two analytic pieces of D (which we denote by $d^{\gtrless}$ in analogy with $g^{\gtrless}$) in terms of real axis integrals over $a^{\gtrless}$ and $b^{\gtrless}$. The most convenient choice of contour C for this purpose is one that has two loops C_1 and C_2, each of which begin and end at t_o. To calculate, for example, $d^<$ one takes t on C_1 and t' on C_2 so that

$$d^<(t,t') = \int_{C_1} d\tau\ A(t,\tau)\ b^<(\tau,t') + \int_{C_2} d\tau\ a^<(t,\tau)\ B(\tau,t') . \qquad (2.19)$$

This may easily be expressed if we define (for t,t' on the real axis) the retarded and advanced functions

$$ia_r(t,t') = [a^>(t,t') + a^<(t,t')]\theta(t-t') \qquad (2.20a)$$

$$-ia_a(t,t') = [a^>(t,t') + a^<(t,t')]\theta(t'-t) \qquad (2.20b)$$

Then in the limit $t_o \to \infty$, (2.19) becomes

$$d^<(t,t') = \int_{-\infty}^{\infty} d\tau [a_r(t,\tau) b^<(\tau,t')$$

$$+ a^<(t,\tau)\, b_a(\tau,t')]. \tag{2.21}$$

A similar relation can be written for $d^>$, which when combined with (2.21) can be written in matrix notation as

$$d^{\gtrless} = a_r b^{\gtrless} + a^{\gtrless}\, b_a. \tag{2.22}$$

This illustrates a general feature of our notation, which is that small letters represent matrix multiplication on the real axis, while capital letters represent matrix multiplication on the contour C. Since the advanced and retarded functions are basic to the theory it is useful to have rules for multiplying them as well. Application of (2.22) plus the definition (2.20) yields

$$d_r = a_r b_r \tag{2.23a}$$

and

$$d_a = a_a b_a \;. \tag{2.23b}$$

The use of (2.23) enables one to generalize (2.20). For example if we let

$$E \equiv ABC \tag{2.24}$$

then one has, on (2.22) and (2.23)

$$e^{\gtrless} = a_r b_r\, c^{\gtrless} + a_r b^{\gtrless} ca + a^{\gtrless} b_a c_a \,. \tag{2.25}$$

Of course (2.23) and 2.25) are trivially generalizable by induction to the product of n functions.

In a similar, but even more trivial manner, one can derive expressions for functions "in parallel" as in

$$F(t,t') = A(t,t')\; B(t',t) \tag{2.26}$$

For example, A and B might be the two Green's functions in random phase appromiation's "bubble"

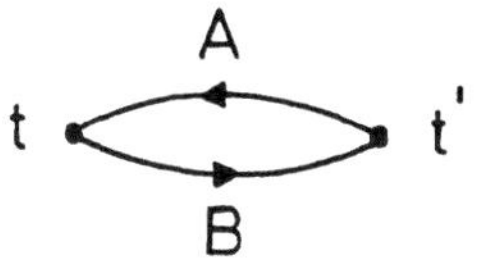

A

t ⟷ t'   (2.27)

B

The analogue of (2.22) and (2.23) are respectively

$$f^{\gtrless} = i\ a^{\gtrless}\ b^{\lessgtr} \qquad (2.28)$$

$$f_r = i[a_r b^< - a^< b_r] = i[a^> b_r - a_r b^>] \qquad (2.29)$$

and similarly for f_a. Note first that matrix multiplication in the time labels is not implied here, and second that the "parallel" product of two Fermion function is a boson function, so that there is an extra minus sign in the definition of $f^<$, and similarly an extra minus sign before $f^<$ in the definition of f_r and f_a. Needless to say these identities are easily generalized in an obvious way to the "parallel" products of more than two functions.

F. Transport Equations

I will now show how the Dyson Equations (2.18) or alternatively

$$G = G_o + G_o(U + \Sigma)G \qquad (2.29a)$$

$$G = G_o + G(U + \Sigma)G_o \qquad (2.29b)$$

are equivalent to the generalized Boltzmann transport equations first derived by Kadanoff[1]. We remind the reader that Σ is the sum of all the proper self-energy insertions due to the many-body interactions and U is the external time dependent potential which drives the system out of equilibrium. First we apply the identity (2.22) to both (2.18a) and (2.18b) yielding

$$(g_o^{-1} - U)g^{\gtrless} = \sigma_r\ g^{\gtrless} + \sigma^{\gtrless}\ g_a \qquad (2.30a)$$

$$g^{\gtrless}(g_o^{-1} - U) = g_r\sigma^{\gtrless} + g^{\gtrless}\ \sigma_a \qquad (2.30b)$$

where as before we use small letters to denote real axis time integrations. This can be written in a simple form if one makes the definitions

$$a = g^{>} + g^{<} = -i(g_a - g_r) \quad (2.31a)$$

$$\gamma = \sigma^{>} + \sigma^{<} = -i(\sigma_a - \sigma_r) \quad (2.31b)$$

and

$$g = (g_a + g_r)/2 \quad (2.32a)$$

$$\sigma = (\sigma_a + \sigma_r)/2 \; . \quad (2.32b)$$

In this non-equilibrium case g and ±a/2 play the same role as the real and imaginary parts of the advanced and retarded Green's function do in the equilibrium theory; σ and ± σ/2 have a similar relation to the self-energy; when Fourier transformed a is known as the spectral weight function and γ is related to the width of the state. Now if we subtract (2.30b) from (2.30a) and use (2.31) and (2.32) we obtain

$$(1/i)[g_o^{-1}-U-\sigma), g^{\gtrless}] - (1/i)[\sigma^{\gtrless}, g].$$
$$= - \{\gamma, g^{\gtrless}\}/2 + \{\sigma^{\gtrless}, a\}/2$$
$$= - \{\sigma^{\gtrless}, g^{\gtrless}\}/2 + \{\sigma^{\gtrless}, g^{\gtrless}\} \quad (2.33)$$

where the square brackets denote commutators and the curly brackets denote anticommutators. These are essentially the Generalized Boltzmann Equations of Kadanoff and Baym[1] except that we have not yet made the assumption that the external disturbances vary slowly with space or time.

Generally, it will be necessary only to consider one of the equations (2.33), say the one for $g^{<}$, because as is shown below, there exists a type of Dyson equation for each of g_a and g_r so that $g^{>}$ can be determined by (2.31) if $g^{<}$ is known. To determine these Dyson equations we use the identity (2.22) on (2.29), obtaining

$$g^{\gtrless} = g_o^{\gtrless} + g_{or} U g^{\gtrless} + g_{or}(\sigma_r g^{\gtrless} + \sigma^{\gtrless} g_a)$$
$$+ g_o^{\gtrless}(U + \sigma_a) g_a \quad (2.34)$$

where the definitions of $g_o^{\gtrless}$, g_{or}, and g_{oa} are the obvious ones. Upon adding the two equations (2.34) and using (2.31) we find

$$-g_a + g_{oa} + g_{oa}(U + \sigma_a)g_a$$
$$= -g_r + g_{or} + g_{or}(U + \sigma_r)g_r \quad . \tag{2.35}$$

Because of the definition of the retarded and advanced quantities (2.20) plus the identity (2.23), it is clear that depending on the time arguments, either the left or right-hand side of (2.35) is zero by definition. Therefore we must have

$$g_a = g_{oa} + g_{oa}(U + \sigma_a)g_a \ , \tag{2.36a}$$

and

$$g_r = g_{or} + g_{or}(U + \sigma_r)g_r \ . \tag{2.36b}$$

Therefore the retarded and advanced Green's function satisfy Dyson's equations as in equilibrium. From their solution one can obtain g and a for use in the transport equation (2.33) for $g^<$.

G. Gradient Expansion

The generalized Boltzmann equation of Kadanoff and Baym is obtained if one replaces the commutators in (2.33) by Poisson brackets and the anticommutators by simple products. Essentially the procedure used to derive this is to express all quantities in terms of sum and difference variables (2.9), and make a gradient expansion, by assuming all quantities are slowly varying functions of the sum variables. Finally, one Fourier transforms the equations with respect to the sum variables. Thus for example

$$g^<(\vec{x},t;\ \vec{x}',t') \rightarrow g^<(\vec{\xi}\ \tau;\ \vec{R},T)$$
$$\rightarrow g^<(\vec{p},p_o;\ \vec{R},\ T) \tag{2.37}$$

where $(\vec{p},\ p_o)$ is the 4-momentum conjugate to the difference variables $(\vec{\xi},\tau) \equiv (\vec{x}-\vec{x}',\ t-t')$. The algebraic details of the gradient expansion are tedious, and I refer the reader to appendix A of ref. (7). The result is that

$$\frac{1}{i}[a,b] \rightarrow \frac{\partial a}{\partial p_o}\frac{\partial b}{\partial T} - \frac{\partial a}{\partial T}\frac{\partial b}{\partial p_o} \tag{2.38a}$$
$$- \vec{\nabla}_p a \cdot \vec{\nabla}_R b + \vec{\nabla}_R a \cdot \vec{\nabla}_p b + \ldots$$

and

$$\{a,b\} \to ab + \ldots \qquad (2.38b)$$

In writing (2.38) we have assumed that the space and time variables were the only ones causing the functions a and b not to commute. This is obviously not true in a system with spin, and in this case there are still commutators with respect to the spin variables remaining on the right side of (2.38). It should be pointed out however, that if neither the external fields nor the interparticle potential couples to the spins, then the system can be treated as two coupled spinless systems, and (2.38) may be used.

I conclude this section by working out an example of the gradient expansion of great importance, that is the expansion for the inverse of the bare Green's function g_o^{-1} acting on an arbitrary function f. Since

$$g_o^{-1}(\vec{p},p_o;\vec{R},t) = p_o - \varepsilon_p , \qquad (2.39)$$

where ε_p is the single particle energy in the absence of interparticle interactions as a function of wave vector $\vec{p}$, one obtains from (2.38a)

$$\frac{1}{i}[g_o^{-1}, f] \to \left(\frac{\partial}{\partial T} + \vec{\nabla}_p\varepsilon_p\cdot\vec{\nabla}_R\right) f(\vec{p},p_o;\vec{R}T) \qquad (2.40)$$

The reader will note that the combination $-i[g_o^{-1}, g^<]$ occurs on the left side of the generalized Boltzmann equation (2.33), and that in deriving (2.40) we have generated two of the most important drift terms, which appear on the left side of the ordinary Boltzmann equations as well.

H. Choice of the Contour C.

I have already illustrated one choice of the contour C in deriving the identities (2.22) et seq. Actually, I have found those identities sufficient for most applications involving low orders of perturbation theory, and also for repeated scattering or t matrix types of theory. In such cases then one never has to worry about the problem. Some other properties can be gleaned simply from the property that the contour is closed and that it passes through t and t'. Nevertheless, it is sometimes useful to pick a specific contour, and here we mention several of the choices that occur in the literature.

Kadanoff-Baym Contour. The contour used extensively by Kadanoff and Baym[1] begins at t_o, goes through t and t' in proper order and returns to $t_o - i\beta$. The fact that it is not closed does not matter if one takes the limit $t_o \to -\infty$, so that the contribution from the closing segment would vanish. This contour has the advantage that the Green's functions, even in non-equilibrium have the periodicity property $G(t_o,t') = G(t_o - i\beta,t')$ and similarly for t'. Thus the contour is useful for making contact with the theory of equilibrium thermal Green's functions, which use such a contour, and to which this theory reduces in the absence of external fields.

Keldysh Contour. A contour very convenient for many purposes has been proposed by Keldysh[2]: this begins at $-\infty$ goes along the real axis to $+\infty$ and then back to $-\infty$ along the real axis again. Since the whole contour is to go through t and t' only once one must specify which of the two halves (label them 1 and 2) t and t' are respectively taken to be on. This leads one to a 2x2 matrix notation, where $G_{11}(t,t')$ implies t and t' both on 1, $G_{12}(t,t')$ implies t on 1 and t' on 2, and similarly for G_{21} and G_{22}. Clearly then $-iG_{12} = g^<$; $+iG_{21} = g^>$, while G_{11} is the ordinary time ordered (on the real axis)

$$iG_{11}(t,t') = \langle T\psi(t)\psi^\dagger(t')\rangle$$

$$= g^>(t,t')\theta(t-t') - g^<(t,t')\theta(t'-t) \tag{2.41}$$

while G_{22} is a reverse time ordered function

$$iG_{22}(t,t') = g^>(t,t')\theta(t'-t)-g^<(t,t')\theta(t-t') \tag{2.42}$$

It is common to regard the integrals on the return (2) part of the contour as running from $-\infty$ to ∞ instead of ∞ to $-\infty$, in which case one must remember to include a factor of (-1) in the matrix multiplication for points on this contour.

This contour is especially useful for taking the zero temperature limit in for a system in thermal equilibrium. In this case, as one knows, one can develop a closed theory for G_{11} alone, with time integrals only over 1, i.e. from $-\infty$ to ∞. As an example of how this happens consider a segment of a diagram where a particle is scattered by a one body potential:

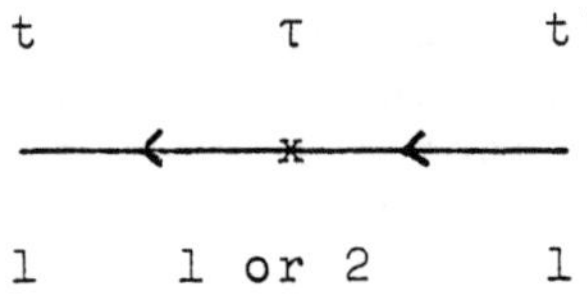

t τ t'

——<——×——<—— (2.43)

1 1 or 2 1

The contribution of this diagram is (aside from overall factors)

$$\int_{-\infty}^{\infty} d\tau V(\tau)[G_{11}(t,\tau)G_{11}(\tau,t') - G_{12}(t,\tau)G_{21}(\tau,t)] \tag{2.44}$$

where V is the potential represented by the cross. If V is time independent, then this may be Fourier transformed, so that the second term in (2.44) is

$$-VG_{12}(p_o)\,G_{21}(p_o) = -VG^{<}(p_o)G^{>}(p_o). \tag{2.45}$$

At zero temperature, all "greater than" quantities vanish for energies p_o less than the Fermi level, and all "lesser than" quantities vanish for p_o greater than the Fermi level. It is therefore clear that the 2nd term involving contour 2 vanishes, and that this result depends on energy conservation at the vertex, and hence upon V being time independent. A moments reflection shows that this is a general property that linked diagrams involving both contours 1 and 2 must vanish at zero temperature if there is energy conservation at all the vertices. Finally, we note that the unlinked parts of a diagram from contour 2 do not vanish, but that these are needed to cancel the unlinked parts from contour 1. The above serves to make contact with, and in some sense, derive the usual form of zero temperature perturbation theory.

I. Concluding Remarks

Before going to applications, I hope I have gotten at least one message across. This is that structurally and formally non-equilibrium theory is the same as equilibrium theory. Not only that, but if one gets used to thinking in terms of time integrals over an abstract contour C, then all the common versions of equilibrium theory merge into one as well.

III. APPLICATIONS

There have been a number of applications of this non-equilibrium theory over the years, and I will mention some typical ones which either I was involved in or which I find particularly interesting for some other reason. These should give the reader an idea how the theory can be applied in practice, but in no sense should such references be considered complete or even representative. Since the purpose of this article is pedagogy rather than review I say this without apology.

The applications seem to fall into two general classes, and I will give a worked, but simplified, example of each in sections A and B below. The first group consists in the derivation of transport equations suitable for determining linear response. This theory is particularly well suited for obtaining such transport equations, but since only linear response is involved, the results might also have been obtained by other well-known methods as mentioned earlier. In the second group of applications the quantity to be calculated is quadratic or higher order in the driving field. It is here in these non-linear applications where this theory has its most important contribution to make, as it enables such problems to be treated with the same formal ease as linear ones.

Of the applications in the first group let me mention three. The first[8] involved the use of (2.33) plus the gradient expansion (2.38) to derive a Boltzmann equation for electrons interacting with phonons in a metal; using a classic "theorem" by Migdal it was shown that one could even treat the case where the electron-phonon coupling was strong. A little later I used the same method to derive a Boltzmann equation for large polarons[9] (conduction electrons plus phonon dressing) in polar crystals; this equation was used to show how the mobility of the polaron varied with temperature at low temperatures and intermediate coupling strength. The most extensive application of the method which I am aware of was made by myself and John Wilkins[7,10], where we derived from first principles the coupled set of Bloch equations for the magnetic spin resonance and diffusion in a dilute magnetic alloy; we derived rigorous expansions for all the coefficients, including the Kondo effect, and settled a long standing argument about what is the effective magnetization toward which the conduction electron and localized spins relax.

In the second group of applications to non-equilibrium problems, a number of groups have developed theories of uv and x-ray photoemission and related spectroscopies, and in section B below we give a simplified example of this, similar to the theory of Chang and myself.[11] In photoemission one must compute the quadratic response of a material to the photon field so that one must compute a 3 point electronic correlation function. Yue and Doniach[12] have applied the theory to the theory of x-ray emission, as have McMullen and Bergersen[13] to Auger emission. Normally in, say, x-ray emission one thinks of starting with a prepared hole state which then decays; this is the approximation of a long-lived hole, and when treated theoretically it is only necessary to consider a 2 point correlation function. However, to the extent which the excitation has a finite lifetime, the initial excitation process affects the shape of the final spectrum; in this case one must calculate a 4 point electronic correlation function, and non-equilibrium techniques must be applied. Another similar and very fascinating application has recently been made to threshold effects in x-ray Raman spectroscopy by Nozières and Abrahams[14].

A. Boltzmann Equation

What I would like to do here is to give a simplified example of the types of uses of the theory which fall into the first group mentioned above, and I will use (2.33) to derive a simple Boltzmann equation. A perhaps more familiar way to do this is to use a correlation function or Kubo formula approach similar to (2.14). One finds that an infinite number of terms in perturbation theory must be summed, and that with appropriate approximation the Boltzmann equation is the equation obtained for the vertex part. In using (2.33) one avoids the necessity of worrying about infinite diagram summations and the Boltzmann equation automatically comes out in the desired form. In non-trivial examples, such as the spin resonance

Bloch equations mentioned earlier, this is a great advantage. However, one should note that the approximations, and hence the range of validity, of the two methods is slightly different: in the correlation function approach one assumes a weak but not necessarily slowly varying external disturbance; the use of (2.33) plus the gradient expansion (2.38) assumes a slowly varying, but not necessarily weak external disturbance.

I consider the simplest example I know of, that is a system consisting of electrons plus random impurities off which the electrons can scatter. In order that Eq. (2.33) for $g^<$ and (2.36) form a closed set, it is necessary to find an expression for the self-energy as a function of G. The perturbation expansion for this is well-known and aside from the trivial Hartree-like term the lowest order term in the perturbation series is

$$\Sigma(\vec{x},t;\vec{x}',t') = \quad \tag{3.1a}$$

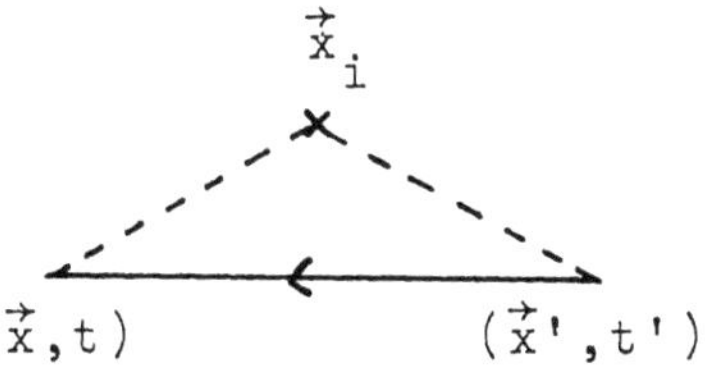

$\vec{x}_i$

$(\vec{x},t)$ $(\vec{x}',t')$

or

$$\Sigma(\vec{x},t;\vec{x}',t') = n_i \int d^3x_i\, v(\vec{x}-\vec{x}_i)\, v(\vec{x}_i-\vec{x}')\, G(\vec{x},t;\vec{x}',t') \tag{3.1b}$$

where n_i is the number of impurities per unit volume and $v(\vec{x}-\vec{x}_i)$ is the potential between an impurity at position $\vec{x}_i$ and an electron at $\vec{x}$. In more sophisticated approximations v is replaced by the single particle scattering matrix in expression similar to (3.1). One now writes (3.1) in terms of sum and difference variables (2.9) and Fourier transforms with respect to the difference variables, obtaining

$$\sigma^{\gtrless}(\vec{p},p_o;\vec{R}T) = n_i \int \frac{d^3p'}{(2\pi)^3}\, |v_{\vec{p}-\vec{p}'}|^2\, g^{\gtrless}(\vec{p}',p_o;\vec{R}_1T) \tag{3.1c}$$

where $v_{\vec{p}}$ is the three dimensional Fourier transform of $v(\vec{x})$, and where we have written the expression in terms of the two functions $\sigma^{\gtrless}$ which as before are the analytic pieces of Σ. Equations (3.2) have a simple physical interpretation. Note that $g^>(\vec{p},p_o)$ [from now on we supress the variables $\vec{R}T$ except where they are necessary for clarity] represents the density of unoccupied states of momentum $\vec{p}$ and energy p_o. Therefore Eq. (3.2) for $\sigma^>$ is simply the Born approximation for the rate of scattering out of state $(\vec{p},p_o)$ [given that it is initially occupied]; similarly $g^<(\vec{p},p_o)$ represents a density of occupied states, so that $\sigma^>(\vec{p},p_o)$ represents a scattering rate into $(\vec{p},p_o)$ given that it is initially unoccupied. Thus the right side of (2.33) for $g^<$ has an interpretation almost identical to that of the right side of the ordinary Boltzmann equation.

We rewrite (2.33) for $g^<$ as

$$(1/i)[(g_o{}^{-1}-U),g^<] + (1/i)[\sigma,g^<] - (1/i)[\sigma^<,g]$$

$$= -\gamma g^< + a\sigma^< = -\sigma^> g^< + \sigma^< g^>, \qquad (3.2)$$

where we have anticipated the gradient expansion (2.38b) in writing the right side as simple products. The quantity to the right of the last equality is in the form on which the physical interpretation of the last paragraph was based, while the quantity sandwiched between the equal signs is to be used in conjunction with (2.36) and (3.2) to form a closed set of equations. The most important role of the self-energy terms has thus already been illustrated - it is to provide scattering "in" and "out" of the beam and hence to drive the electrons toward thermal equilibrium. The terms on the left of (3.2) are the "drift" terms which tend to drive the system away from thermal equilibrium. There are also other effects of the self-energy terms; these are easily gleaned by examination of the Dyson equations (2.36) which determine the spectral weight function a. Within the context of the gradient expansion [see (2.38b)] these equations are solved by inspection, yielding [see (2.31)]

$$a(\vec{p},p_o) = -2\mathrm{Im}(p_o-\varepsilon_p-U-\sigma+i\gamma/2)^{-1}. \qquad (3.3)$$

Thus as in equilibrium the self-energy gives an energy shift σ and width γ to a spectrum which otherwise would be a delta function

$$a(\vec{p},p_o) \simeq 2\pi\ \delta(p_o-\varepsilon_p-U). \qquad (3.4)$$

We now make another approximation, which is that this shift and width are small, that is $\rho\gamma << 1$ and $\rho\sigma << 1$, where ρ is the electronic density of states; these inequalities may always be satisfied if the concentration of impurities is low enough. We therefore will use (3.4) for a.

To be consistent with a δ function approximation for a, we must also neglect the last two commutators on the left side of (3.2). To see this we note that these terms are small not only because σ is small, but also because they involve gradients of small slowly varying quantities, so that they are second order in small quantities [for example, $(\partial/\partial T)g^< \sim \omega g^< \sim \gamma g^<$ where ω is the frequency of the external disturbance; therefore the first commutator on the left, as well as the right is $\sim\gamma g^<$, while the second and third commutators on the left are $\sim\gamma^2\rho g^<$ and hence negligible by assumption]. That a is a δ function suggests the ansatz

$$g^<(\vec{p},p_o) = 2\pi\ \delta(p_o-\varepsilon_p-U)f(\vec{p}), \qquad (3.5)$$

Thus defining $f(\vec{p})$ which may be a function of $(\vec{R},T)$ as well [but because of the δ function it would be redundant to consider it a function of p_o].

Substitution of (3.5) in (3.2) and making use of (2.39) to move the δ function through the commutator in the first term on the left of (3.2) gives

$$a\frac{1}{i}[(g_o^{-1}-U),\ f(\vec{p})] \qquad (3.6)$$

which becomes upon making use of (2.40) for $[g_o^{-1},f]$ and (2.38a) for $[U,f]$

$$a\left(\frac{\partial}{\partial T} + \vec{\nabla}_p\varepsilon_p\cdot\vec{\nabla}_R + \vec{F}\cdot\vec{\nabla}_p\right)\ f(\vec{p}) \qquad (3.7)$$

where $\vec{F} \equiv -\vec{\nabla}_R U$ is the force, since we take U to be a scalar potential. Similarly, substitution of (3.1c), into the right side of (3.2), use of (3.4) and (2.31) yields

$$-a\int\frac{d^3p'}{(2\pi)^3}\,2\pi\,|v_{\vec{p}-\vec{p}'}|^2\,\delta(\varepsilon_p-\varepsilon_{p'})[f(\vec{p})-f(\vec{p}')]. \tag{3.8}$$

Finally, setting (3.7) equal to (3.8) and cancelling the factor of a from both sides gives the usual form of the Boltzmann equation for impurity scattering. The cancellation of a from both sides verifies that the ansatz (3.5) was correct. The function $f(\vec{p})$ has the usual distribution function meaning, and for example the current density $\vec{j}(\vec{R},T)$ is given by[15]

$$\vec{j}(\vec{R},T)=\int\frac{d^3p}{(2\pi)^3}\,\vec{\nabla}_p\varepsilon_p\;f(\vec{p}) \tag{3.9}$$

and the number density $n(\vec{R},T)$ by

$$n(\vec{R},T)=\int\frac{d^3p}{(2\pi)^3}\,f(\vec{p}) \tag{3.10}$$

as may be verified from (2.8).

In concluding this section, I should point out that the real utility of the method is in the non-trivial cases where all the commutators in (3.2) must be included. This is so in all the applications referenced at the beginning of this section.

B. X-ray Photoemission

In its simplest terms photoemission consists in applying an a.c. electric field and looking for a d.c. current response. We are thus dealing with quadratic response theory because (even though the exciting photon field is weak) in linear response the output frequency is the same as the input frequency. One therefore must calculate $g^<$, which in turn determines the desired momentum decomposition of the photocurrent, to second order in the photon field:

$$\delta g^< = \longleftarrow\bullet\longleftarrow\bullet\longleftarrow \tag{3.11a}$$

The above is the basic diagram for the 2nd order change $\delta g^<$ in the correlation function; the dots are proportional to $\vec{A}\cdot\vec{p}$, where $\vec{A}$ is the vector potential of the photon field which we take proportional to sin ωt, and $\vec{p}$ is the electronic momentum. Since according to (2.8) we need $g^{<}(t,t')$ at equal times $t = t'$, it is customary to twist (3.11a) around so the ends meet

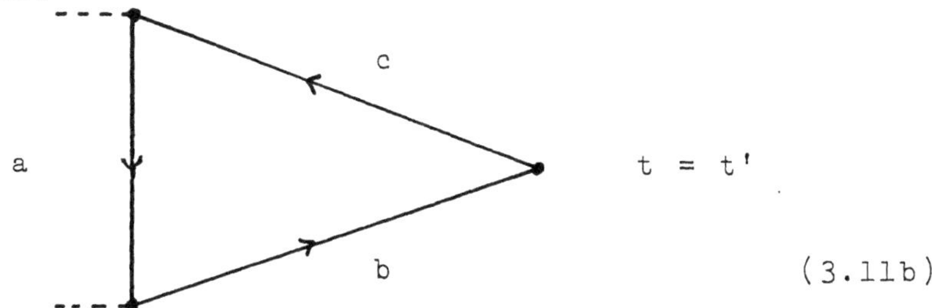

(3.11b)

where I have added the dashed photon lines to the left corners to distinguish them from the vertex at the right where the current is to be measured. Diagram (3.11b) has a simple Golden rule interpretation: the incident photon on the bottom of the diagram creates a hole (a) in a formerly occupied state and an electron (b) in a formerly unoccupied state; it is this electron which contributes to the photocurrent measured outside the sample at time t; the upper half of the diagram can be thought of as completing the square of the matrix element, as is required by the Golden rule.

To do the problem properly one must solve for all the Green's functions and related quantities in the presence of the surface; this is especially important for the functions b and c in (3.11b) because they originate inside the sample and end outside. We will content ourselves here, however, to use bulk Green's functions, and to calculate the momentum contribution to the photocurrent of electrons moving to the right, and originating at a point to the left of where they are measured. This is easy to calculate and is also a good approximation when the photoelectron's energy is high enough so that the scattering is highly peaked in the forward direction and that surface effects are small. In other words I will be using approximations appropriate for <u>x-ray photoemission</u>. I follow the treatment reported a long while ago for the admittance function for electron excited spectroscopies[16], and which was developed in detail by Chang and myself[17]. For a review of some of this work, including the surface effects[18] omitted here, see my recent article[11]. For a treatment of photoemission which fills in more of the details of the non-equilibrium theory I recommend Caroli[19], et al.

In the absence of many-body effects (3.11b) would be the only diagram there is. In the above approximation its evaluation is trivial, the change δG in G being given by

$$\delta G = GMGMG \tag{3.12}$$

where M is a matrix element proportional to $\vec{A}\cdot\vec{p}$ and whose time dependence is given by $\sin \omega t$. This is a case of Green's functions in series and is easily evaluated with identity (2.22), and then Fourier transformed. It is clear that if $\sin \omega t$ is written as $(e^{i\omega t}-e^{-i\omega t})/2i$ the d.c. contribution to the photocurrent comes from the cross terms between $e^{i\nu t}$ in one matrix element M and $e^{-i\nu t}$ in the other, where $\nu = \pm\omega$. Writing this term one gets [using (2.22)]

$$\begin{aligned}\delta g^<(p_o) &= g^<(p_o)\ M\ g_a(p_o-\nu)\ M\ g_a(p_o) \\ &+ g_r(p_o)\ M\ g^<(p_o-\nu)\ M\ g_a(p_o) \\ &+ g_r(p_o)\ M\ g_r(p_o-\nu)\ M\ g^<(p_o).\end{aligned} \tag{3.13}$$

One apparently has 6 terms all rather symmetric, but we note immediately that the only states which have asymptotic currents outside the sample are those above the vacuum level, and that these terms are unoccupied ($g^< = 0$). Therefore only the middle terms of (3.13) with $\omega = +\nu$ survives. For simplicity we take the initial state to be a deep atomic core state so that

$$g^<(p_o') = 2\pi\ \delta(p_o'-E) \tag{3.14}$$

where E is the energy of that state. Therefore (3.13) becomes

$$\delta g^<(\vec{p},p_o) = 2\pi|M_p|^2\ g_r(\vec{p},p_o)\ g_a(\vec{p},p_o)\ \delta(p_o-\omega-E). \tag{3.15}$$

We note that g_r and g_a are given by

$$g_r = g_a{}^* = (p_o-\varepsilon_p-\sigma+i\gamma/2)^{-1} \tag{3.16}$$

where σ and γ are the energy shift and width of the high energy states due to many-body interactions. Since I assume ω is large, this implies that σ and γ are small, which in turn means that

$$g_r g_a \rightarrow (2\pi/\gamma)\ \delta(\dot{p}_o - \varepsilon_p). \tag{3.17}$$

Substitution of (3.17) into (3.15) and integration over p_o (divided by 2π) to find $n_{\vec{p}}$ the number of electrons of momentum $\vec{p}$, gives

$$n_{\vec{p}} = 2\pi|M_p|^2\ \delta(E+\omega-\varepsilon_p)/\gamma \tag{3.18}$$

Eq. (3.18) states the physical fact that the population of some normally unoccupied state is expressed as a balance between its Golden rule rate of population due to the incident photons and its rate of depopulation γ due to the electron-electron interactions. Note that in all of this analysis we have implicitly assumed that the photon field does not vary in space, which is a good approximation so long as the skin depth is much greater than the mean free path, a situation always realized for x-rays on metals.

What about the electron-electron interaction? Since the electrons and holes involved directly in the photoemission process i.e. those represented schematically in (3.11b) interact with the electrons of the solid, there is the chance that excitations of the solid are created. These might be plasmons or other collective oscillations, electron-hole pairs, or something else, and are what produce the width γ in (3.16). I will not specify what they are, but assume they have a (Boson-like) propagator $D(\vec{q},q_o)$ and a coupling to electrons Vq. For a normal metal the spectral weight function for D [i.e. $D^>(q_o) = D^<(-q_o)$] has a sharp spike at the plasma frequency ω_p corresponding to the collective oscillations or plasmons and and a broad smeared out background due to the continuum of electron-hole excitations. I will simply represent the propagator for these excitations by a wiggly line. One then has to consider contributions to the photocurrent such as

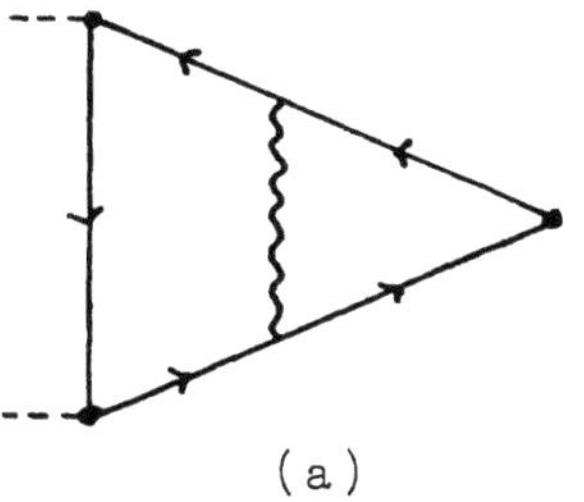
(a)

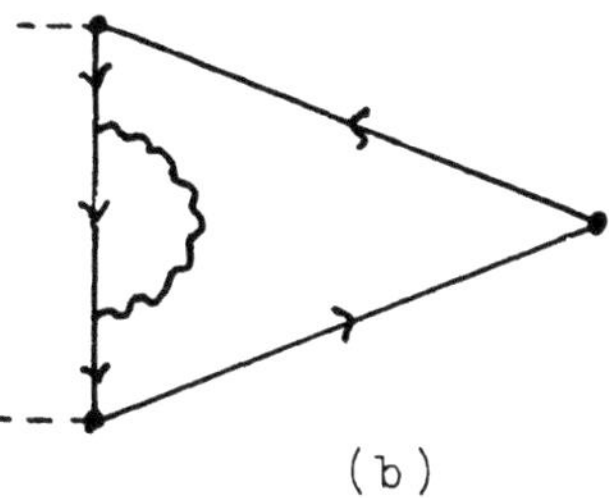
(b)

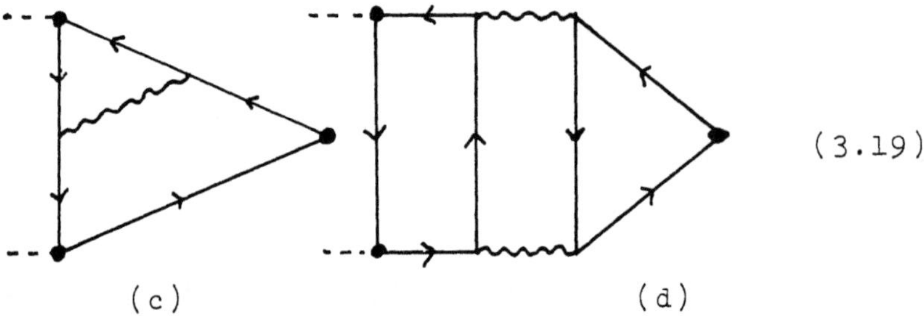

(3.19)

(c) (d)

Diagram (3.19a) represents an excitation emitted by the escaping photoelectron; (3.19b) represents the excitation emitted by the deep hole left behind, while (3.19c) is the quantum-mechanical interference term between the two. Diagram (3.19d) is a case where the signal is carried toward the surface for part of the time as an excitation of the solid rather than as an excited photoelectron. When the photon energy ω is high things simplify considerably. This is because the coupling of a fast escaping electron necessarily becomes weak [it is characterized by a dimensionless coupling constant $e^2/\hbar v$ where v is the photoelectron velocity; for 1.5 keV, $e^2/\hbar v \sim 0.1$]. Then (c) and (d) become small and may be neglected. Diagram (b) and others of its type are not small; their effect is to change the effective state density $g^<$ for the deep hole from a δ function to a complicated function with lots of structure. We will not consider these terms here because they do not relate to transport, but refer the reader to ref. (11). The diagram (3.19a) and ones like it relate to the escape of the photoelectron and what happens along the way, and I will discuss them below.

One might think that since the electron-electron interaction becomes small at high energies one could neglect terms like (3.19a). This is not true because the actual escape depth of the photoelectron is itself limited only by the interactions of this type (the skin depth is much greater). Therefore decreasing the strength of the interaction simply means that the observed photoelectrons come from a greater depth, and have proportionately a longer time to interact. Therefore, even for a very weak interaction (3.19a) is of the same order of magnitude as (3.11b).

Diagram (3.19a) is a series and parallel combination of Green's functions, written schematically as

$$\delta G = G V \underbrace{G M G M G}_{D} V G \tag{3.20}$$

and may most simply be evaluated using the generalization of (2.25) and (2.28). Using the same arguments that eliminated all but one term in (3.13), one can see by inspection that the analogue of (3.15) is

$$\delta g^{<}(\vec{p},p_o) = \int \frac{d^3q}{(2\pi)^3} \int \frac{dq_o}{2\pi} 2\pi |M_{\vec{p}-\vec{q}}|^2$$
$$\times |V_q|^2 \; g_r(\vec{p}-\vec{q},p_o-q_o) g_a(\vec{p}-\vec{q},p_o-q_o)$$
$$\times D^{>}(\vec{q},q_o) g_r(\vec{p},p_o) g_a(\vec{p},p_o) \delta(p_o-q_o-\omega-E). \tag{3.21}$$

It is now easy to see why this is of the same order as (3.11b), because both sets of Green's functions on the upper and lower legs of the ladder are forced by energy and momentum conservation to be the same; this implies through (3.17) that instead of one factor of γ in the denominator there are two:

$$\delta g^{<}(\vec{p},p_o) \to$$
$$\int \frac{d^3q}{(2\pi)^3} \int \frac{dq_o}{2\pi} 2\pi |M_{\vec{p}-\vec{q}}|^2 \; (2\pi/\gamma) \; \delta(p_o-q_o-\varepsilon_{p-q})$$
$$|V_q|^2 \; D^{>}(\vec{q},q_o)(2\pi/\gamma) \; \delta(p_o-\varepsilon_p) \; \delta(p_o-q_o-\omega-E), \tag{3.22}$$

Since

$$\gamma(\vec{p},p_o)$$
$$= \int \frac{d^3q}{(2\pi)^3} \int \frac{dq_o}{2\pi} |V_q|^2 D^{>}(\vec{q},q_o) 2\pi\delta(p_o-q_o-\varepsilon_{\vec{p}-\vec{q}}) \tag{3.23}$$

it is clear that no matter how small V, (3.22) is of the same order as (3.15).

It is thus necessary to sum the whole series of ladders

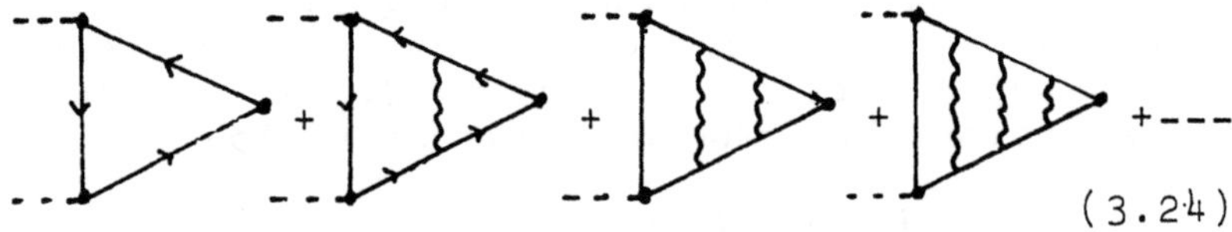

(3.24)

because clearly by the same argument all these diagrams have the same order of magnitude. On the other hand, other types of diagrams with interactions on the "escape legs" do indeed become negligible at high energies. After our experience with the first two diagrams of (3.24) it is easy to see that the sum (3.24) is given by

$$\delta g^{<}(P) = 2\pi |M_{\vec{p}}|^2 \; |g_r(P)|^2 \; \delta(p_o - \omega - E)$$

$$+ \int \frac{d^4Q}{(2\pi)^4} \; |V_q|^2 \; D^{>}(Q) |g_r(P)|^2 \; \delta g^{<}(P-Q) \tag{3.25}$$

where $P \equiv (\vec{p}, p_o)$ and $Q \equiv (\vec{q}, q_o)$. Upon making use of (3.17) and (3.23) this can be written in a very physical form (after multiplying both sides by γ)

$$r(P) = \int \frac{d^4Q}{(2\pi)^4} \; [R(P-Q \leftarrow P) \; \delta g^{<}(P)$$

$$- R(P \leftarrow P-Q) \; \delta g^{<}(P-Q)] \tag{3.26}$$

where r(P) is the rate (according to the Golden rule) at which the state $P \equiv (\vec{p}, p_o)$ is populated due to the photoeffect

$$r(P) = 2\pi |M_p|^2 \; 2\pi\delta(p_o - \varepsilon_p) \; \delta(p_o - \omega - E) \tag{3.27}$$

and $R(P-Q \leftarrow P)$ is the intrinsic Golden rule rate for the state P to be degraded to P-Q through the emission of an excitation of four-momentum Q:

$$R(P \leftarrow P-Q) = 2\pi |V_q|^2 D^{>}(Q) \; \delta(p_o - \varepsilon_p). \tag{3.28}$$

In physical terms (3.26) says that the rate of populating state P by the photoeffect must be equal to the rate at which it is depopulated by emission of excitations [first term on the right of (3.26)] minus the rate at which it is repopulated by the degradation of higher energy states [second term on the right of (3.26)]. One should note that $D^>$ is in its essentials given by

$$D^>(q,q_o) = 2\pi\ \delta(q_o-\Omega_{\vec{q}}) \tag{3.29}$$

where Ω is the energy of the excitation which may be a function of $\vec{q}$ as well as other summed over variables which convert the δ function into a function with width (lifetime effects could do this as well).

Eq. (3.26) may be simplified even further by noting that the solution has the form

$$\delta g^<(\vec{p},p_o) = 2\pi\ \delta(p_o-\varepsilon_p)\ n_{\vec{p}}\ . \tag{3.30}$$

Then (3.26) becomes

$$\gamma_{\vec{p}}\ n_{\vec{p}} = r_{\vec{p}} + \sum_{\vec{q}} R_{\vec{p}\leftarrow\vec{p}-\vec{q}}\ n_{\vec{p}-\vec{q}} \tag{3.31}$$

where

$$R_{\vec{p}\leftarrow\vec{p}-\vec{q}} = |V_q|^2\ D^>(\vec{q},\varepsilon_{\vec{p}}-\varepsilon_{\vec{p}-\vec{q}}) \tag{3.32a}$$

and

$$\gamma_{\vec{p}} = \sum_q |V_q|^2\ D^>(\vec{q},\varepsilon_{\vec{p}+\vec{q}}-\varepsilon_{\vec{p}}) \tag{3.32b}$$

$$r_p = 2\pi|M_{\vec{p}}|^2\ \delta(\varepsilon_{\vec{p}}-E+\omega). \tag{3.32c}$$

Equation (3.31) is the generalization of (3.18) and its physical meaning is obvious.

At high energies (3.31) may be greatly simplified, because the momentum $\vec{q}$ of any excitations is much smaller than the electronic momentum $\vec{p}$. Thus the scattering is highly peaked in the forward direction so that the only really active variable in (3.31) is the energy ε_p; the direction of $\vec{p}$ stays essentially the same through many emissions of excitations. For the same reason, it is a good approximation to take γ and r to be constants. Then (3.31) can be approximately rewritten as

$$\gamma n(\varepsilon) = \bar{r}\,\delta(\varepsilon-\omega-E) + \int d\varepsilon' R(\varepsilon \leftarrow \varepsilon') n(\varepsilon') \tag{3.33}$$

with slight redefinitions of n, r, and R. One can get a qualitative picture of the solution of (3.33) very simply. This is because the structure of R is dominated by plasmon emission; for illustrative purposes it is sufficient to take R to be of the form

$$R(\varepsilon \leftarrow \varepsilon') = \gamma^{1}\delta(\varepsilon+\omega_p-\varepsilon') + R^{pair}(\varepsilon,\varepsilon') \tag{3.34}$$

where ω_p is the energy of the plasmons and R^{pair} is a slowly varying function of ε and ε' representing the emission of electron-hole pairs. Substitution of (3.34) into (3.33) yields

$$\gamma n(\varepsilon) = \bar{r}\,\delta(\varepsilon-\omega-E) + \gamma^{1} n(\varepsilon+\omega_p) + S(\varepsilon) \tag{3.35}$$

where

$$S(\varepsilon) = \int d\varepsilon'\, R^{pair}(\varepsilon,\varepsilon')\, n(\varepsilon') \tag{3.36}$$

is also a slowly varying function. The solution of (3.35) thus has the form

$$n(\varepsilon) = \frac{\bar{r}}{\gamma} \sum_{m=0}^{\infty} \alpha^{m}\delta(\varepsilon-\omega-E+m\omega_p) + S'(\varepsilon) \tag{3.37}$$

where S' is a slowly varying background function and

$$\alpha = \gamma^{1}/\gamma\ , \tag{3.38}$$

that is α is the ratio between the rate of plasmon emission and the total decay rate. Thus we expect the x-ray photoemission spectrum to be dominated by a series of plasmon satellites whose strength falls off roughly exponentially with the order of the satellite. This is indeed born out experimentally, although it is less certain the extent to which these satellites originate from the effect discussed here.

Finally, note that Eq. (3.37) can be written as the convolution between the "bare" energy distribution $\bar{r}/\gamma\ \delta(\varepsilon-\omega-E)$ and an "escape" function, the singular part of which is given by

$$E(\nu) = \sum_{m=0}^{\infty} \alpha^m \delta(\nu - m\omega_p) \tag{3.39}$$

which gives the relative probability of escape with a particular energy loss ν. It is also clear that (3.39) can be turned around and used as an "admittance function" for core-hole spectroscopies which are triggered by an incident electron. It was in the latter connection in which I first derived[16,17] Eq. (3.39) using the method given above. Of course, it is also easy to derive (3.39) from simple probabilistic considerations. The real value of the theory is in the calculation of corrections to the simple intuitive theory, and the determination of when it is valid and when it is not. A more complete picture of the escape of a photoelectron requires the numerical solution of (3.31). At lower energies, the surface and other effects become important. The formalism developed here is capable of handling all these complications.

REFERENCES

1. L. P. Kadanoff and G. Baym, Quantum Statistical Mechanics, W. A. Benjamin, Inc., New York, 1962.

2. L. V. Keldysh, J. Exptl. Theoret. Phys. (U.S.S.R.) 47, 1515 (1964) [English Translation: Soviet Physics JETP 20, 1018 (1965)].

3. R. A. Craig, J. Math. Phys. 9, 605 (1968).

4. J. R. Schrieffer, Theory of Superconductivity, W. A. Benjamin, Inc., New York, 1964.

5. J. Schwinger, Proc. Nat'l. Acad. Sci. U.S. 37, 452 (1951).

6. See for example D. Pines, The Many-Body Problem, W. A. Benjamin, Inc., New York, 1961, p. 41, Eq. (2.49).

7. The derivation of the identities given here is taken from D. C. Langreth and J. W. Wilkins, Phys. Rev. 6, 3189 (1972).

8. R. E. Prange and L. P. Kadanoff 134, A566 (1964).

9. D. C. Langreth, Phys. Rev. 159, 717 (1967).

10. D. C. Langreth, D. L. Cowan and J. W. Wilkins, Solid State Commun. 6, 131 (1968); D. C. Langreth and J. W. Wilkins, Phys. Rev. 6, 3189 (1972); J. Sweer, D. C. Langreth and J. W. Wilkins, Phys. Rev. (to appear).

11. This work was reviewed by D. C. Langreth in Nobel Symposia - Medicine and Natural Sciences, Academic Press, New York and London, Vol. 24, p. 210, 1973.

12. J. T. Yue and S. Doniach, Phys. Rev. B8, 4578 (1973).

13. T. McMullen and B. Bergersen, Can. J. Phys. 50, 1002 (1972).

14. P. Nozières and E. Abrahams, Phys. Rev. B10, 3099 (1974).

15. For the band case where $\varepsilon_p \neq p^2/2m$ see D. C. Langreth, Phys. Rev. 148, 707 (1966).

16. D. C. Langreth, Phys. Rev. Lett. 26, 1229 (1971); see Eq. (8).

17. J. J. Chang and D. C. Langreth, Phys. Rev. B5, 3512 (1972).

18. J. J. Chang and D. C. Langreth, Bull. Am. Phys. Soc. 18, 55 (1973); Phys. Rev. B8, 4638 (1973).

19. C. Caroli, D. Lederer-Rozenblatt, B. Roulet and S. Saint-James, Phys. Rev. B8, 4552 (1973).

CONDUCTIVITY VIA NONEQUILIBRIUM STATISTICAL PHYSICS

G.J. Papadopoulos

University of Leeds, Leeds LS2 9JT, UK and E.S.I.S.,

Universitaire Instelling Antwerpen, Wilrijk, Belgium

I. INTRODUCTION

In these lectures we shall sketch the philosophy behind nonequilibrium Statistical Physics and illustrate the procedure using an exactly soluble model. We shall concentrate on the question of direct conductivity and mean energy.

The statistical properties of an independent thermodynamic system in equilibrium can be extracted from its density matrix, ρ^{eq}, which is fully determined by the system Hamiltonian H. It is assumed that we are considering a very large number of systems each described by the Hamiltonian H. What essentially the density matrix tells us is the fraction W_n of systems in the (stationary) state Ψ_n of the Hamiltonian H. Furthermore it is also tacitly assumed that these systems are not completely isolated from each other or the environment, for otherwise they would be in a position to have any desired density matrix. This can be easily seen from the fact that if we prepared the systems in various eigenstates these (without any disturbance) would have remained in the same eigenstates, and thus the equilibrium density matrix for a system could have been changed at will. However, this is not the case; an assembly of systems in thermodynamic equilibrium settles in a given distribution of states by itself. This occurs through slight environmental and intersystem interactions, which cause a given system to jump from a state it occupies to another one, but the fraction W_n of occupied states at any one time remains approximately constant, provided the thermodynamic variables are kept constant.

However, when we disturb our system externally the environmental and other slight interactions play a dominant role and manifest themselves in the non-equilibrium quantities in an effective manner, e.g. through a relaxation time.

In practice the situation under study involves the system of interest in whose nonequilibrium properties we are primarily interested, an energy reservoir to which the system of interest is permanently coupled, and an external system producing a variable interaction which can be switched at will on the system of interest.
Finally we have the environment whose influence is exerted on all three of the systems.

The figure below represents schematically the above state of affairs.

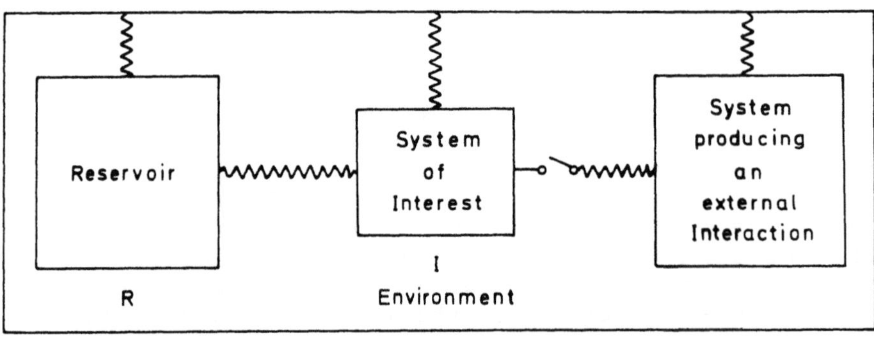

The above arrangement is realized for example in mobility studies of electrons in a polar crystal [1]. There the energy reservoir is played by the polarization field oscillators, which is coupled to the conduction electrons, forming the system of interest. The external system is responsible for producing the external field to which the electrons are subjected. The role of the environment is to establish a state of thermal equilibrium on the combined system made out of the reservoir and the system of interest, when these have been over a long time free from external disturbance.

We wish now to proceed towards formulating our problem in a mathematical fashion. If we denote collectively by $\vec{x}$ the degrees of freedom of our system of interest and by $\vec{X}$ the degrees of freedom of the reservoir the equilibrium density matrix of the combined system will be of the form :

$$\rho^{eq}(\vec{x},\vec{X}|\vec{x}',\vec{X}') = \Sigma W_n \Psi_n(\vec{x},\vec{X})\Psi_n(\vec{x}'\vec{X}') \qquad (1,1)$$

where the Ψ_n's represent the eigenfunctions of the combined reservoir - system of interest and W_n is the probability for the system to be in the state $\Psi_n(\vec{x},\vec{X})$. We have denoted by n collectively the quantum numbers in question.

At some time (say t=o) and onwards we apply the external interaction which, of course, will act on the system of interest with whose statistical response we are primarily concerned. The switching of the external interaction will give rise to a change in the pattern of the behaviour of the system of interest towards the reservoir and thus the state of equilibrium of the combined assembly will be disturbed. Initially most of the external interaction will be absorbed by the systems of interest but gradually they will be transmitting and partly receiving energy to and from the reservoir. This will go on for some time until a state of dynamic equilibrium, with regard to the energy exchange, is reached. The statistical content of these processes will be reflected in the nonequilibrium density matrix $\rho_1(x\ x';t)$ which we shall subsequently show how to obtain. This contains all thermodynamic properties pertaining to the system of interest at any one time.

As a first step towards obtaining the system of interest density matrix we work out the evolution of the density matrix of the combined system RI (= reservoir plus system of interest) under the influence of the external interaction but ignoring the slight environmental and other vanishingly small interactions responsible for the establishment of equilibrium in the assembly in the absence of external disturbances. To this end we argue as follows: The eigenstates Ψ_n , used to describe the equilibrium density matrix will go over to time-evolving wavefunctions $\Psi_n(\vec{x},\vec{X};t)$ obtained in accordance with the transformation:

$$\Psi_n(\vec{x},\vec{X};t) = \int K(\vec{x}\ \vec{X},t|\vec{x}'\vec{X}',o)\Psi_n(\vec{x}',\vec{X}')d\vec{x}'d\vec{X}' \quad (1.2)$$

where K is the propagator of the combined system RI involving the external interaction, U_{ext} . In the state of equilibrium we imagine the various RI - systems to be in eigenstates; the fraction of such systems in the state $\Psi_n(\vec{x},\vec{X})$ being W_n . With the application of the external field the set of all these eigenstates becomes the set of all the wave functions $\Psi_n(\vec{x},\vec{X};t)$ obtained through the transformation (1.2). Since two different eigenstates of the unperturbed system would not lead to the same time-evolving wave function (via the external interaction) the fraction of RI-systems with wave function $\Psi_n(\vec{x},\vec{X};t)$ at time t will remain W_n . Therefore the propagated density matrix will be given by

$$\rho(\vec{x}\vec{X}|\vec{x}'\vec{X}';t) = \Sigma W_n \Psi_n(\vec{x},\vec{X};t)\Psi_n(\vec{x}',\vec{X};t) \qquad (1.3)$$

Combining (1.1), (1.2) and (1.3) we obtain for the propagated RI-density matrix the expression :

$$\rho(\vec{x}\ \vec{X}|\vec{x}'\vec{X}';t) = \int K(\vec{x}\ \vec{X},t|\vec{x}_1\vec{X}_1,o)K^{\star}(\vec{x}'\vec{X}',t|\vec{x}_2\vec{X}_2,o)$$

$$x\rho^{eq}(\vec{x}_1,\vec{X}_1|\vec{x}_2,\vec{X}_2)d\vec{x}_1 d\vec{X}_1 d\vec{x}_2 d\vec{X}_2 \qquad (1.4)$$

It should be noted here that the propagated density matrix (1.4) cannot be correctly used as the nonequilibrium density matrix for the RI-systems, since the environmental and other interactions responsible for bringing about equilibrium have not entered the propagation procedure. However our prime concern lies in the thermodynamic properties pertaining to the assembly of the systems of interest, and in this connection the propagated density matrix, according to (1.4), is extremely good. In fact the larger the capacity of the reservoir the more effective is the screening of the system of interest from the outside world. In other words the reservoir forms effectively the rest of the universe for our system of interest.

For the derivation of the nonequilibrium density matrix for our system of interest we just have to equate the dashed coordinates $\vec{X}'$ of the reservoir with $\vec{X}$ in $\rho(\vec{x}\vec{X}|\vec{x}'\vec{X}';t)$ and integrate over all their space. The nonequilibrium density matrix for our system of interest is then given by :

$$\rho_i(\vec{x}|\vec{x}';t) = \int \rho(\vec{x}\vec{X}|\vec{x}'\vec{X};t)d\vec{X} \qquad (1.5)$$

$\rho_i(\vec{x}|\vec{x}';t)$ yields all thermodynamic information concerning our system of interest at any one time t .

Next we shall exemplify the above procedure using the model of harmonic oscillator assemblies, a mathematically tractable model, which received much attention as far as the statistical properties of one selected (test) particle are concerned. It has been shown, e.g., that under certain circumstances the test particle exhibits brownian motion [2].Furthermore, the statistical properties of the assembly as a whole in the presence of a magnetic field have been studied [3] . The problem has also been treated with quantum statistics [4] . The line of attack hitherto adopted usually concentrates on obtaining a Langevin equation for the test particle or on the evaluation of correlation functions[2,5] . Another way about is via nonequilibrium distribution functions, the evaluation of which received adequate attention in the classical regime [6] . Extensions to the quantum mechanical counterpart have been obtained under approximate conditions [7]. Here we shall present an exact derivation [8] .

The model, although essentially of a didactic character, can be also used for calculations of a more realistic nature. It could enable one to obtain effective quantities pertaining to a nonlinear system by averaging an appropriate equation involving the correct nonlinear structure, as in the case of the Thornber-Feynman theory [2] for the mean velocity of an electron in a polar crystal. We shall point out how to use this model for obtaining an effective relaxation time incorporating the nonlinear effects elsewhere.

II. NONEQUILIBRIUM DENSITY MATRIX FOR THE OSCILLATOR

We shall consider an assembly of charged oscillators, each of which is harmonically coupled to a harmonic lattice (or cluster) of neutral particles. The charged particles will form our system of interest, I, while the various clusters, to which they are bound, the reservoir, R . The charge-cluster systems will be assumed noninteracting among themselves. We can, therefore, describe the dynamics of the whole assembly by considering a representative charge-cluster system. To facilitate the analysis we shall tag the charge by o and the cluster particles by 1,2,...., N-1; we have N particles altogether in each representative charge-cluster.

The Lagrangian of a charged oscillator (if we imagine the coupling interaction cut off) will be :

$$L_1 = \frac{1}{2}m_o(\dot{\vec{x}}_o^{\,2} - \Omega^2\vec{x}_o^{\,2}) \tag{2.1}$$

Let us denote the N-particle Lagrangian for the combined system, made out of the charge and the associated cluster, by L_N and by L_{N-1} the bare cluster Lagrangian, i.e., without the coupling interaction. If we let U_{int} stand for the charge-cluster interaction we have :

$$L_N = L_1 + L_{N-1} - U_{int} \tag{2.2}$$

Let us now take for L_N the expression :

$$L_N = \sum_{j=o}^{N-1} \frac{1}{2}m_j\dot{\vec{x}}_j^{\,2} - \frac{1}{2}\sum_{i,j=o}^{N-1} \vec{x}_j^{\,T} A_{ij}\vec{x}_j \ , \tag{2.3}$$

with A_{ij} symmetric positive definite. In order that L_1 be part of (2.2) A_{oo} should be $A_{oo}=m_o\Omega^2$. L_{N-1} and U_{int} will be given by :

$$L_{N-1} = \sum_{j=1}^{N-1} \frac{1}{2} m_j \dot{\vec{x}}_j^2 - \frac{1}{2} \sum_{i,j=1}^{N-1} \vec{x}_j^T A_{ij} \vec{x}_j \tag{2.4}$$

$$U_{int} = \frac{1}{2} \sum_{j=1}^{N-1} (\vec{x}_o^T A_{oj} \vec{x}_j + \vec{x}_j A_{jo} \vec{x}_o) \tag{2.5}$$

It is through U_{int} that energy flows from the charged particle into the cluster particles and vice versa. The essential difference between this model Lagrangian and that of the polaron lies in the U_{int} which in the polaron case is highly nonlinear. We make no assumptions regarding the size of U_{int}. When an external field $\vec{E}(t)$, is applied on our system of interest $(U_{ext} = -e\vec{E}(t).x_o)$ the total representative system Lagrangian becomes :

$$L'_N = L_N + e\vec{E}(t).\vec{x}_o \tag{2.6}$$

In order to obtain the propagator K_N associated with the Lagrangian L'_N, given by (2.6), we make use of the normal coordinates which diagonalize the charge-cluster Lagrangian (2.3), and which does not involve the external interaction.

Let the diagonalizing transformation be :

$$\vec{q}_j = \sum_{k=o}^{N-1} C_{jk} \sqrt{m_k}\, \vec{x}_k \quad (j=0,1,2,\ldots,N-1) \tag{2.7}$$

where, as is well known, the matrix $\underline{C}$ is orthogonal $(\underline{C}^{-1} = \underline{C}^T)$. The inverse transformation then is :

$$\sqrt{m_j}\, \vec{x}_j = \sum_{k=o}^{N-1} C_{kj} \vec{q}_k \quad (j=0,1,2,\ldots,N-1) \tag{2.8}$$

Introducing this transformation into the Lagrangian (2.6), involving the external interaction, we obtain it in terms of the normal coordinates as :

$$L_N' = \sum_{j=o}^{N-1} \frac{1}{2} (\dot{\vec{q}}_j^{\,2} - \omega_j^2 \vec{q}_j^{\,2}) + (e/\sqrt{m_o}) \sum_{j=o}^{N-1} C_{jo} \vec{E}(t).\vec{q}_j \tag{2.9}$$

where ω_j are the eigenfrequencies associated with the normal coordinates. So, according to (2.9) our Lagrangian L_N' is decomposed into that of N independent forced harmonic oscillators ; the external force on the charge being now distributed among all the oscillators in the new frame of reference. The N-particle propagator in terms of normal coordinates will now be the product of the propagators for the corresponding oscillators. Utilizing from Feynman and Hibbs [9] the expression for the propagator of the forced harmonic oscillator, we obtain for our N-particle propagator the formula :

$$K_N(\vec{Q},t|\vec{Q}',0) = \prod_{j=o}^{N-1} \left(\frac{\omega_j}{2\pi i\hbar \sin\omega_j t}\right)^{3/2}$$

$$\times\exp\Big\{\frac{i}{\hbar} \sum_{j=o}^{N-1} \frac{\omega_j}{2\sin\omega_j t} [\vec{q}_j^{\,2} + \vec{q}_j'^{\,2})\cos\omega_j t - 2\vec{q}_j . \vec{q}_j']$$

$$+ \frac{i}{\hbar} \sum_{j=o}^{N-1} \frac{eC_{jo}}{\sqrt{m_o}\sin\omega_j t}$$

$$\times \int_0^t [\vec{q}_j' \sin\omega_j (t-\tau) + \vec{q}_j \sin\omega_j \tau] .\vec{E}(\tau)d\tau - \frac{i}{\hbar}\Phi(t)\Big\} \tag{2.10a}$$

where we have used $\vec{Q}$ to denote collectively the coordinates $(\vec{q}_o, \vec{q}_1, \dots, \vec{q}_{N-1})$. The time dependent function $\Phi(t)$ is given by :

$$\Phi(t) = \sum_{j=1}^{N-1} \frac{e^2 C_{jo}^2}{m_o \omega_j \sin\omega_j t} \int_o^t \int_o^\tau \sin\omega_j (t-\tau)\sin\omega_j \tau'$$

$$\times \vec{E}(\tau)\cdot\vec{E}(\tau')d\tau d\tau' \qquad (2.11b)$$

Let us now obtain the N-particle equilibrium density matrix ρ_N^{eq} ,(the equilibrium density matrix of the system RI) constituting the initial condition for the propagated N-particle density matrix. This, as is well known, in the regime of Boltzmann statistics can be obtained by suitably normalizing the N-particle propagator without the external field after replacing the time t by $-i\hbar\beta$ $(\beta=1/kT)$. In other words, utilizing K_N from (2.10a) we have :

$$\rho_N^{eq}(\vec{Q}|\vec{Q}') = \frac{1}{Z_N} [K_N(\vec{Q},-i\hbar\beta|\vec{Q}',o)]_{\vec{E}=0} \qquad (2.11a)$$

where Z_N is the N-particle partition function, given by :

$$Z_N = \int [K_N(\vec{Q},-i\hbar\beta|\vec{Q},o)]_{\vec{E}=o} \, d\vec{Q} \qquad (2.11b)$$

The resulting expression for the equilibrium density matrix, thus obtained, in terms of the $\vec{q}$ - coordinates is :

$$\rho_N^{eq}(\vec{Q}|\vec{Q}') = [\prod_{i=o}^{N-1} \frac{\omega_j}{\pi\hbar} \tanh(\tfrac{1}{2}\beta\hbar\omega_j)]^{3/2}$$

$$\times\exp\{-\sum_{j=o}^{N-1}\frac{\omega_j}{2\hbar\sinh(\beta\hbar\omega_j)}[(\vec{q}_j^{\,2}+\vec{q}_j^{\,\prime 2})\cosh(\beta\hbar\omega_j)$$

$$-2\vec{q}_j.\vec{q}_j^{\,\prime}]\} \quad (2.12)$$

Now, we make use of (1.4) for the propagation of the N-particle (RI representative system) density matrix but working in the $\vec{Q}$ - representation. We form $\rho_N^{eq}(\vec{Q}_1|\vec{Q}_2)$ from (2.12) and $K_N(\vec{Q}t|\vec{Q}_1o)$ and $K_N(\vec{Q}'t|\vec{Q}_2o)$ from (2.10a) which we insert in (1.4). We point out that the function $\Phi(t)$, appearing in K_N, will not enter our expression for the propagated density matrix $\rho_N(\vec{Q}|\vec{Q}';t)$, for in forming the product $K_N K_N^{\star}$ it will be cancelled out. Furthermore the integrations over $\vec{Q}_1$ and $\vec{Q}_2$ are gaussian and we obtain, after some rearrangements, the following expression for the propagated N-particle density matrix (i.e. the density matrix $\rho(\vec{x}\vec{X}|\vec{x}'\vec{X}';t)$ of sec. I) :

$$\rho_N(\vec{Q}|\vec{Q}';t) = [\prod_{j=o}^{N-1}\frac{\omega_j}{\pi\hbar}\tanh(\tfrac{1}{2}\beta\hbar\omega_j)]^{3/2}$$

$$\times\exp\Bigg(-\sum_{j=o}^{N-1}\frac{\omega_j}{2\hbar\sinh(\beta\hbar\omega_j)}$$

$$\times\{[(\vec{q}_j-\langle\vec{q}\rangle)^2+(\vec{q}_j^{\,\prime}-\langle\vec{q}_j\rangle)^2]\cosh(\beta\hbar\omega_j)$$

$$-2(\vec{q}_j-\langle\vec{q}_j\rangle).(\vec{q}_j^{\,\prime}-\langle\vec{q}_j\rangle)\}\Bigg)\exp[\frac{i}{\hbar}\sum_{j=o}^{N-1}\langle\vec{P}_j\rangle.(\vec{q}_j-\vec{q}_j^{\,\prime})]$$

(2.13a)

where :

$$\langle \vec{q}_j \rangle = (C_{jo}/\sqrt{m_o}\omega_j) \int_o^t \sin\omega_j(t-\tau) e\vec{E}(\tau) d\tau \qquad (2.13b)$$

$$\langle \vec{p}_j \rangle = (C_{jo}/\sqrt{m_o}) \int_o^t \cos\omega_j(t-\tau) e\vec{E}(\tau) d\tau \qquad (2.13b)$$

The quantities $\langle\vec{q}_j\rangle$, $\langle\vec{p}_j\rangle$ are the mean values of the position and momentum Heisenberg operators ; the averaging performed over the initial values of the operators with the aid of the density matrix ρ_N^{eq}, given by (2.12). Alternatively, they are the average values of the stationary operators $\vec{q}_j$, $\vec{p}_j=(-i\hbar\partial/\partial\vec{q}_j)$ with the averaging being done against the propagated density matrix $\rho_N(\vec{Q}|\vec{Q}';t)$, given by (2.13a) in accordance with the rules of the Schrödinger picture. One then can easily verify that $\vec{p}_j = d\langle\vec{q}_j\rangle/dt$. It is worth while noting that our propagated density matrix $\rho_N(\vec{Q}|\vec{Q}';t)$ is normalized since $\rho_N(\vec{Q}|\vec{Q}';o) = \rho_N^{eq}(\vec{Q}|\vec{Q}')$ is normalized and the propagation has been effected through a unitary transformation, the propagator K_N .

Denoting the averages $(\langle\vec{q}_o\rangle,\langle\vec{q}_1\rangle,\ldots,\langle\vec{q}_{N-1}\rangle)$ and $(\langle\vec{p}_o\rangle,\langle\vec{p}_1\rangle,\ldots,\langle\vec{p}_{N-1}\rangle)$ by $\langle\vec{Q}\rangle$ and $\langle\vec{P}\rangle$ respectively, we are able to write (2.13a) in the more transparent form:

$$\rho_N(\vec{Q}|\vec{Q}';t) = \rho_N^{eq}(\vec{Q}-\langle\vec{Q}\rangle|Q'-\langle\vec{Q}\rangle)\exp\left[\frac{i}{\hbar}\langle\vec{P}\rangle.(\vec{Q}-\vec{Q}')\right] \qquad (2.14)$$

The last factor on the r.h.s. of (2.14) is a plane wave carrying momentum equal to the average momentum of the system, while the other factor is the equilibrium density matrix referred to the system following the average motion. In other words, (2.14) is the N-particle drifted equilibrium density matrix.

In order to obtain the desired density matrix ρ_i for our system of interest, which in this case will be

$\rho_1(\vec{x}_0|\vec{x}_0';t)$, involving only the coordinates of the charge, we need go back to the direct space coordinates. Accordingly we express $\vec{Q}$ and $\vec{Q}'$ in $\rho_N(\vec{Q}|\vec{Q}';t)$, given by (2.13), in terms of the coordinates $(\vec{x}_0,\vec{x}_1,...,\vec{x}_{N-1})$ and $(\vec{x}_0',\vec{x}_1',...,\vec{x}_{N-1}')$ respectively via the transformation $\vec{Q}=\vec{Q}(\vec{x}_0,\vec{X})$ from (2.7) (where by $\vec{X}$ we have denoted $(\vec{x}_1,\vec{x}_2,...,\vec{x}_{N-1})$) and employ the formula :

$$\rho_1(\vec{x}_0|\vec{x}_0';t) = \mathcal{N}\int \rho_N(\vec{Q}(\vec{x}_0,\vec{X})|\vec{Q}(\vec{x}_0'\vec{X}))\prod_{j=1}^{N-1} d\vec{x}_j \qquad (2.15a)$$

where $\mathcal{N}$ is a constant taking care of the jacobian of the transformation and is evaluated by the normalization condition

$$\int \rho_1(\vec{x}_0|\vec{x}_0';t)\,d\vec{x}_0 = 1 \qquad (2.15b)$$

The integrations over $\vec{X}$ in (2.15a) are not so obvious, and for the details of carrying them out we invoke our ref. [8].

Furthermore, putting in the resulting expression from (2.15a,b) $m_0=m$ and $\vec{x}_0=\vec{x}$ (here $\vec{x}$ denotes the position vector of the charge) we have the following expression for the density matrix $\rho_i(=\rho_1)$ of our system of interest :

$$\rho_1(\vec{x}|\vec{x}';t) = (\frac{B}{2\pi})^{3/2}\exp\left[-\frac{1}{4}A(\vec{x}-\vec{x}')^2-\frac{1}{4}B(\vec{x}+\vec{x}'-2\vec{x}^{\star})^2\right]$$

$$\times \exp[\frac{i}{\hbar}\vec{p}^{\star}.(\vec{x}-\vec{x}')] \qquad (2.16a)$$

where the quantities A and B are positive and are given by :

$$A = \frac{m}{\hbar} \sum_{j=o}^{N-1} C_{jo}^2 \quad \omega_j \coth(\frac{1}{2} \beta\hbar\omega_j) \tag{2.16b}$$

$$B = \frac{m}{\hbar} \left[\sum_{j=o}^{N-1} \frac{C_{jo}^2}{\omega_j} \coth(\frac{1}{2} \beta\hbar\omega_j)\right]^{-1} \tag{2.16c}$$

and $\vec{x}^\star$, $\vec{p}^\star$ are the mean displacement and mean momentum of the charged particles, as will be shown, and are given by :

$$\vec{x}^\star = \sum_{j=o}^{N-1} \frac{C_{jo}}{\sqrt{m}} <\vec{q}> = \sum_{j=o}^{N-1} (C_{jo}^2/m\omega_j) \int_o^t \sin\omega_j(t-\tau)e\vec{E}(\tau)d\tau \tag{2.16d}$$

$$\vec{p}^\star = \sum_{j=o}^{N-1} C_{jo}\sqrt{m} <\vec{p}_j> = \sum_{j=o}^{N-1} C_{jo}^2 \int_o^t \cos\omega_j(t-\tau)e\vec{E}(\tau)d\tau \tag{2.16e}$$

The time dependence in the density matrix ρ_1 comes solely through the quantities $\vec{x}\star$ and $\vec{p}\star$.

Utilizing the formula

$$<F(\vec{x})> = \int [F(\vec{x})\rho_1(\vec{x}|\vec{x}';t)_{\vec{x}'=\vec{x}}]d\vec{x} \tag{2.17a}$$

for obtaining the average value of an operator $F(\vec{x})$, from the density matrix ρ_1, we find the following averages for the momentum and displacement of our charges at time t.

$$<\vec{p}> = \int [\frac{\hbar}{i} \frac{\partial}{\partial\vec{x}} \rho_1(\vec{x}|\vec{x}';t)]_{\vec{x}'=\vec{x}} d\vec{x} = \vec{p}\star \tag{2.17b}$$

$$<(\vec{x}-\vec{x}^{\star})> = 0 \quad \text{or} \quad <\vec{x}> = \vec{x}^{\star} \tag{2.17c}$$

Furthermore, we obtain for the mean kinetic energy per charged particle :

$$<-\frac{\hbar^2}{2m}\frac{\partial^2}{\partial\vec{x}^2}> = \frac{3}{4}\frac{\hbar^2}{m}A + \frac{\vec{p}^{\star 2}}{2m} \tag{2.17d}$$

The potential energy can be most easily obtained by finding initially the mean-square deviation of the displacement, which is $<(\vec{x}-\vec{x}^{\star})^2 = \frac{3}{2}B^{-1}$, and eventually obtain :

$$<\frac{1}{2}m\Omega^2\vec{x}^2> = \frac{3}{4}m\Omega^2B^{-1} + \frac{1}{2}m\Omega^2\vec{x}^{\star 2} \tag{2.17e}$$

$\vec{x}^{\star}, \vec{p}^{\star}$, A,B appearing in (2.17b) - (2.17e) are given by (2.16b) - (2.16e) . The first term on the r.h.s. of (2.17d) is the equilibrium kinetic energy per charged particle, while the first term of (2.17e) is the corresponding mean equilibrium potential energy. These quantities are correspondingly equal to the mean kinetic and the mean potential energy of the charged particle relative to the system following the average flow.

From (2.17d,e) we notice that the equilibrium kinetic energy and the equilibrium potential energy of our charged oscillator are not in general equal. This is due to the size of the reservoir-system of interest interaction, U_{int} hidden in the various ω_j's . If this interaction is too large the system of interest cannot be considered thermodynamically independent. In fact in such a case the role of the reservoir is more than that of getting the system of interest into a state of equilibrium described by the system bare Hamiltonian ;in addition it enters its equilibrium properties. However, for sufficiently small U_{int} the reservoir just steers the system of interest to equilibrium (in the absence of external forces) without in the end manifesting itself explicitely.

Next, we shall make use of the Schrödinger chain oscillator model to investigate the behavior of charged oscillators particularly in the limit of sufficiently small reservoir interactions.

III THE SCHROEDINGER CHAIN-OSCILLATOR MODEL

In this model the charge - cluster representative Lagrangian is :

$$L_N = \sum_{j=o}^{N-1} \frac{1}{2} m (\dot{\vec{x}}_j^2 - \Omega^2 \vec{x}_j^2) - \sum_{j=o}^{N-2} \frac{m}{2t_r^2} (\vec{x}_{j+1} - \vec{x}_j)^2 \tag{3.1}$$

where $\vec{x}_o = \vec{x}$.

Actually this is a variant of the Schroedinger model [10] in that in addition to the nearest neighbour interactions of the original model each particle in the chain undergoes the influence of an oscillator potential of freqnency Ω . This modification enables one to obtain the statistical properties of the quantum oscillator and also of the free particle by letting Ω , in the latter case, in the end go to zero. Furthermore, with an appropriate choice of Ω in terms of volume and temperature one can take account of the container effect [11]. The Lagrangian L_1 for the charged oscillator given by (2.1) takes the form

$$L_1 = \frac{1}{2} m(\dot{\vec{x}}^2 - \Omega^2 \vec{x}^2) \tag{3.1a}$$

while the interaction coupling the charge with the rest of the chain is given by :

$$U_{int} = \frac{m}{2t_r^2}(\vec{x}_1-\vec{x})^2 \qquad (3.1b)$$

t_r is a relaxation time, the inverse square of which characterises the strength of the interaction. We recall that the external interaction $[U_{ext}=-e\vec{E}(t)\cdot\vec{x}]$ affects directly only the charge, and is transmitted via U_{int} to the rest of the chain particles.

We wish now to obtain the density matrix for our charged particles of the Schrödinger model. As it appears from the formulae (2.22a) - (2.22e) , giving the density matrix of the charge in a general harmonic lattice, for the construction of this density matrix we need only know the transformation matrix (C_{jk}) from the direct-space coordinates $\{\vec{x}_j\}$ to the normal ones $\{\vec{q}_j\}$ and the frequency spectrum $\{\omega_j\}$.

The matrix elements of the transformation are [12]:

$$C_{jk} = \left(\frac{\varepsilon_j}{N}\right)^{1/2} \cos\left[\left(k+\frac{1}{2}\right)\pi\frac{j}{N}\right] \quad ; \; \varepsilon_j = \begin{cases} 2 & j\neq o \\ 1 & j=o \end{cases} \qquad (3.2a)$$

$$(j,k = 0,1,2,\ldots,N-1)$$

while the normal frequencies, in the present case, are given by :

$$\omega_j^2 = \Omega^2+(2/t_r)^2 \sin^2(\pi j/2N) \quad (j=0,1,2,\ldots,N-1) \qquad (3.2b)$$

The quantities A, B, $\vec{x}^{\star}$, $\vec{p}^{\star}$, given by formulae (2.16b)-(2.16e), can now be obtained and we are interested in their expressions for an infinite number of particles in the chain. By-passing the details of the limiting transition, which appear in ref. [8] we quote the results for the Schrödinger model :

$$A=\frac{4m}{\pi\hbar}\int_0^1 (1-u^2)^{1/2}[\Omega^2+(2u/t_r)^2]^{1/2}$$

$$\times \coth\{\tfrac{1}{2}\beta\hbar[\Omega^2+(\frac{2u}{t_r})^2]^{1/2}\}du \tag{3.3a}$$

$$B=\frac{m\pi}{4\hbar}\left(\int_0^1 \frac{(1-u^2)^{1/2}}{[\Omega^2+(2u/t_r)^2]^{1/2}} \coth \tfrac{1}{2}\beta\hbar[\Omega^2+(2u/t_r)^2]^{1/2}\}du\right)^{-1} \tag{3.3b}$$

$$\vec{x}^{\star}=\frac{4}{\pi}\int_0^1 du\frac{(1-u^2)^{1/2}}{[\Omega^2+(2u/t_r)^2]^{1/2}}\int_0^1 d\tau\sin\{[\Omega^2+(2u/t_r)^2]^{1/2}$$

$$\times(t-\tau)\}\frac{e\vec{E}}{m}(\tau) \tag{3.3c}$$

$$\vec{p}^{\star}=\frac{4}{\pi}\int_0^1 du(1-u^2)^{1/2}\int_0^t d\tau\cos\{[\Omega^2+(2u/t_r)^2](t-\tau)\}e\vec{E}(\tau) \tag{3.3d}$$

The above formulae are valid for any strength of interaction between the charges and the reservoir. As a result of a large size interaction the equilibrium properties of the charges (reflected in the quantities A and B) are not fully determined by the oscillator Hamiltonian $H_1=(\vec{p}^2/2m)+(m/2)^2\,\vec{x}^2$, but in addition reservoir characteristics come into play.

Let us now consider the case of sufficiently small (weak) coupling (t_r^{-1} small) so that the system of charges can be viewed as an independent thermodynamic system. Nevertheless, the coupling has to be of some size in order to enable the system of charges to steer itself to equilibrium, but sufficiently small as to justify its absence from its equilibrium density matrix, fully determined by the charge Hamiltonian H_1 .

Now, for sufficiently small t_r^{-1} the expressions A, B in (3.3a,b) can be replaced by the first term of their Taylor expressions, i.e. by putting $t_r^{-1}=o$. However, it should be emphasized that the influence of t_r does not altogether disappear since it is hidden in the dimensionless integration variable u. (see ref. [8]). The same thing cannot be done with the dynamical quantities $\vec{x}^\star$ and $\vec{p}^\star$.

Under the weak coupling conditions we have :

$$A=(m\Omega/\hbar)\coth(\tfrac{1}{2}\beta\hbar\Omega), \quad B=(m\Omega/\hbar)\tan(\tfrac{1}{2}\beta\hbar\Omega) \qquad (3.4)$$

Utilizing (2.16a) for the charged particle nonequilibrium density matrix with A,B from (3.6) and $\vec{x}^\star$, $\vec{p}^\star$ from (3.3c,d) we obtain the following density matrix for our charge :

$$\rho_1(\vec{x}|\vec{x}';t) = [\frac{m\Omega}{2\pi\hbar}\tanh(\tfrac{1}{2}\beta\hbar\Omega)]^{3/2}$$

$$\times \exp(-\frac{m\Omega}{2\hbar\sinh(\beta\hbar\Omega)}\{[(\vec{x}-\vec{x}^\star)^2+(\vec{x}'-\vec{x}^\star)^2]\cosh(\beta\hbar\Omega)$$

$$-2(\vec{x}-\vec{x}^\star).(\vec{x}'-\vec{x}^\star)\}\exp[\frac{i}{\hbar}p^\star.(\vec{x}-\vec{x}')] \qquad (3.5a)$$

This amounts to a drifted oscillator density matrix, and can be written in the more telling form :

$$\rho_1(\vec{x}|\vec{x}';t)=\rho_1^{eq}(\vec{x}-\vec{x}^\star|\vec{x}'-\vec{x}^\star)\exp[\frac{i}{\hbar}\vec{p}^\star.(\vec{x}-\vec{x}')] \tag{3.5b}$$

where $\rho_1^{eq}(\vec{x}|\vec{x}')$ is the single oscillator equilibrium density matrix. This is actually a brownian motion limit for the quantum oscillator.

We wish now to use the density matrix (3.5b) for obtaining several quantities of interest relating to our charged particle. We employ formula (2.17a) yielding the average value of an operator F of the charged particle, via its nonequilibrium density matrix. We have for the average momentum and displacement :

$$<\vec{p}> = \vec{p}^\star \tag{3.6a}$$

$$<\vec{x}> = \vec{x}^\star \tag{3.6b}$$

where $\vec{x}^\star$ and $\vec{p}^\star$ are given by (3.3c,d). Furthermore for the (bare) charged particle energy we find :

$$<-\frac{\hbar^2}{2}\frac{\partial^2}{\partial\vec{x}^2} + \frac{m}{2}\Omega^2\vec{x}^2 - e\vec{E}(t).\vec{x}>$$

$$= \frac{1}{2m}<\vec{p}>^2 + \frac{m}{2}\Omega^2<\vec{x}>^2 - e\vec{E}(t).<\vec{x}> + \frac{3}{2}\hbar\Omega\coth(\frac{1}{2}\beta\hbar\Omega) \tag{3.6c}$$

As is well known the last term in (3.6c) is the mean equilibrium oscillator energy.

Denoting by $H_1'(\vec{p},\vec{x})$ the Hamiltonian operator for our (bare) oscillator we cast (3.6c) in the more telling form :

$$\langle H_1'(\vec{p},\vec{x})\rangle = H_1'(\langle\vec{p}\rangle,\langle\vec{x}\rangle) + \langle H_1\rangle_{eq} \qquad (3.6d)$$

where $\langle H_1\rangle_{eq}$ is the mean equilibrium oscillator energy.

Finally we wish to specialize formulae (3.6a)-(3.6c) in the case of the free charge on which a constant electric field is applied from time $t=o$ onwards. To this end we make use of formulae (3.3c,d) with $\Omega=o$, $\vec{E}(t)=\vec{E}$ and obtain :

$$\langle\vec{P}\rangle_{free} = \frac{4}{\pi}\int_o^1 du(1-u^2)^{1/2}\int_o^t d\tau\ \cos[\frac{2u}{t_r}(t-\tau)]\, e\vec{E}$$

$$= et_r\vec{E}\int_o^t \frac{d\tau}{\tau}\, J_1(\frac{2\tau}{t_r}) \qquad (3.7a)$$

$$\langle\vec{x}\rangle_{free} = (e/m)t_r\vec{E}\int_o^t d\tau(\frac{t}{\tau}-1)J_1(\frac{2\tau}{t_r}) \qquad (3.7b)$$

where for the derivation of (3.7a,b) we have made use of ref. [13]. The time dependent integral on the r.h.s. of (3.7a) rises from zero to unity as t approaches infinity ; in fact it becomes 1 for t of the order of t_r and then oscillates about this value with ever-decreasing amplitude. Forgetting about the oscillations, we find that the average momentum of our charge obeys approximately the usual conductivity formula :

$$\langle\vec{p}\rangle_{free} = et_r\vec{E}\ [1-\exp(-t/t_r)] \qquad (3.8)$$

The average displacement can be obtained by integrating this over t. Furthermore, formula (3.6d) for the free case takes the form :

$$<-\frac{\hbar^2}{2m}\frac{\partial^2}{\partial\vec{x}^2} - e\vec{E}(t).\vec{x}>_{free} = \frac{1}{2m}<p>^2_{free} - e\vec{E}(t).<\vec{x}>_{free} + \frac{3}{2}kT \tag{3.9}$$

where $<\vec{p}>_{free}$ and $<\vec{x}>_{free}$ are obtained from (3.7a,b).

As a final remark, the average momentum and displacement predicted by this model do not involve quantum effects. However, quantum effects appear in the oscillator energy and this only through its equilibrium part. In contrast, the free - particle energy is completely classical, a consequence of the correspondence principle.

REFERENCES

1. Feynman, R.P., Hellwarth, R.W., Iddings, C.K. and Platzman, P.M., Phys. Rev. 127, 1004 (1962).
Thornber, K.K. and Feynman, R.P., Phys. Rev., B1, 4099 (1966).
Extensive work concerning the optical properties and conductivity within the polaron model has been carried out by Devreese and co-workers:
J. Devreese, J. De Sitter and M. Goovaerts, Phys. Rev. B, 5, 2367 (1972).
W. Huybrechts,J. De Sitter and J.T. Devreese, Solid State Comm. 13, 163 (1973).
L.F. Lemmens, J. De Sitter and J. T. Devreese, Phys. Rev. B, 8, 2717 (1973).
E. Kartheuser, R. Evrard, J.T. Devreese, Phys. Rev. Letters, 22, 94 (1969)
J.T. Devreese, J. Van Royen and L.F. Lemmens : Sum Rule for Optical Absorption (Pre-print).
J.T. Devreese, R.Evrard and E. Kartheuser: Self Consistent Equation of Motion Approach for Polarons (To appear in Phys. Rev. B).
Also the conductivity in the polaron problem has been treated with the Boltzmann equation by J.T. Devreese and R. Evrard: The momentum Distribution of Electrons in Polar Semiconductors for High Electric Field' (Pre-print).
Abstract: Bull. Am. Phys. Soc., EG5, 404 (1975).

2. Mazur, P. and Brown, E., Physica, 30 1973 (1964).
3. Mazur, P. and Siskens, Th.J., Physica, 47, 245 (1970).
4. Storer, R.G., J. Math. Phys., 12, 1296 (1971).
5. Deutch, J.M. and Silbey, R., Phys. Rev. A3, 2049 (1971).
   Ford, G.W., Kac, M. and Mazur, P., J. Math. Phys. 6, 505 (1965).
   Fujiwara, I., Hemmer, P.C. and Wergeland, H., Prog. Theor. Phys. Suppl. Nos 37 and 38, 149 (1966).
   Heurta, M.A. and Robertson, H.S., J. Stat. Phys., 3, 171 (1971).
   Mazur, P. and Møntroll, E., J. Math. Phys. 1, 70 (1960).
   Klein, G. and Prigogine, I., Physica 19, 1053 (1953).
   Prigogine, I., and Bingen, R., Physica 21, 299 (1955).
   For the electrical analogue of a harmonic chain lattice see: Stevens, K.W.H., Proc. Phys. Soc. 77, 515 (1961).
6. Rubin, R.J., J. Math. Phys., 1, 309 (1960); 2, 373 (1961).
   Turner, R.E., Physica 26, 274, (1960).
7. Ullersma P., Physica 32, 74, (1966).
8. Papadopoulos, G.J., Physica 74, 529 (1974).
9. Feynman, R.P. and Hibbs, A.R., Quantum Mechanics and Path Integrals, McGraw-Hill pp. 63-64. (New York,1965).
10. Schrödinger, E., Ann. Physik 44, 196 (1914).
11. Papadopoulos, G.J., J. Phys. A6, 1479 (1973).
12. Hemmer, P.C. and Wergeland, H.K., Norske Vindensk.Selsk, Forhandl., 30, 137 (1957).
13. Gradstein, I.S., and Ryzhik, I.M., Tables of Integrals, Series and Products, Academic Press, p. 419 (London,1965).

ASPECTS OF LINEAR AND NONLINEAR ELECTRONIC CONDUCTION IN DISSIPATIVE MEDIA

K. K. Thornber

Bell Laboratories, Incorporated

Murray Hill, New Jersey 07974, U.S.A.

ABSTRACT

We review a general approach to the calculation of electronic transport properties under steady-state but nonequilibrium conditions. Of particular interest is the velocity acquired by an electron in a finite electric field in a polar crystal as a (nonlinear) function of the applied electric field. The effect of finite magnetic fields applied in addition to the electric field is also considered. In solving the problem use is made of a dynamic self-consistency, the significance of which is indicated by an analysis of approximate solutions satisfying the free-energy sum rule. Emphasis throughout is on the choice of what one is calculating in addition to how one is to calculate.

I. INTRODUCTION

The purpose of these lecture notes is to discuss (in more detail than is usually permitted in a research paper) certain interesting and distinguishing features which enter into solutions of electronic transport problems using the Feynman operator calculus[1] and the Feynman path integral.[2-4] These methods have two very important advantages in treating transport properties of electrons interacting linearly with boson fields. First, using either method the boson (phonon) variables can be eliminated exactly. Second, using the path-integral method the problem can be treated regardless of the strength of the externally applied electric and magnetic fields. As a result, it is possible to discuss the behavior of the electron in terms of influence functionals which can be approximated so as to include all physical features of the system of interest. For example, results have been obtained for the self-energy,[5-7] impedance,[8-12] and field-velocity characteristic[13-15] of the Fröhlich polaron.[16] It has been possible to compute transport properties in cases when not only perturbation theory, but also the Boltzmann equation is inadequate.[8-15] Finite temperature, coupling constant, and electric and magnetic field strengths do not present insurmountable difficulties in this formalism; nonlinear dependences between applied fields and velocities can be obtained even in zero order.[13-15] Some numerical results have had direct impact on studies of certain electron devices.[13,14] In our discussion we shall explain how some of these interesting results come about.

Our discussion is divided into three parts. In Section II we review the general transport problem of interest. As this was the primary topic of a previous set of lectures,[15] our treatment here is confined to certain important physical features of the solution not adequately discussed before. In these lectures primary emphasis is given to the self-consistent solution,[10] Section III, and the equilibrium sum rules,[17-19] Section IV, which arise in connection with the transport problem. The former was only mentioned in passing in the previous lectures,[15] while the latter provide considerable insight into the adequacy of the Feynman-approximation solutions in general and the self-consistent solution in particular. As before,[15] the treatment is fully quantum-mechanical, and emphasis is on the choice of what one is calculating in addition to how one is to calculate. If we could solve these problems exactly, it would not matter what we calculated or how we went about it: we would always get the same answer for what we desired. Since, however, we calculate approximately, the choice of what to calculate and how to go about it is very important: we usually get different answers depending upon which choice is made. In this respect the self-consistent method[10] serves as an example of how one can construct a quantity to be calculated which yields with a bare minimum of mathematical effort significant physical results.

Much of the following is presented within the path-integral formalism. However, I have attempted to present the most basic concepts in such a manner that they can be understood with at most a minimum of familiarity with the path-integral approach. Furthermore, when calculations are carried out, sufficient detail is either presented or referred to so that few mathematical difficulties should be encountered. The emphasis here is on physics, especially internal consistency.

II. THE TRANSPORT PROBLEM

We wish to solve the following problem. Suppose we have an electron interacting with the phonon modes of a crystal. Since we are primarily interested in the effect of the phonon scattering, let us assume that in the undeformed lattice the electron would move as a free particle, possibly with an effective mass tensor. We are not interested in the problem of an electron weakly coupled to the phonons, in which the unperturbed electron states are better described by a band structure. Our primary interest lies in those problems where the amount of scattering by phonons can exceed that due to the fixed lattice.[14]

We now take the crystal and apply a spacially uniform and time invarient electric field $\underset{\sim}{E}$ and magnetic field $\underset{\sim}{H}$. We start the electron from rest and wait for the electron to acquire a steady-state velocity $\underset{\sim}{v}$. We seek the relation between $\underset{\sim}{v}$, $\underset{\sim}{E}$, and $\underset{\sim}{H}$ for arbitrary lattice temperature, electron-phonon coupling strengths and electric and magnetic field magnitudes. This relationship will be nonlinear in general. In various limits we can recover drift and Hall mobility, magnetoresistivities, and other interesting transport properties.

One's first impulse would be to attempt to calculate the expectation value of the velocity operator

$$\hat{\underset{\sim}{v}} = \hat{\dot{\underset{\sim}{x}}} = i[\hat{H},\hat{\underset{\sim}{x}}]/\hbar \tag{1}$$

where $\hat{H}$ is the Hamiltonian of the system. Such a calculation[13] has in fact been carried out for $\underset{\sim}{H} = 0$. Although results were found similar to those obtained more simply by methods discussed below, we had to work very hard for the following reason. At zero temperature, the system is in a well-defined state at all times and

$$v = \langle \hat{\underset{\sim}{v}} \rangle = \lim_{t\to\infty} \int \Psi_t^* i[\hat{H},\hat{\underset{\sim}{x}}]/\hbar \Psi_t \, d\underset{\sim}{x} \, d\underset{\sim}{Q} \tag{2}$$

where $\Psi_t = \Psi_t(\underset{\sim}{x},\underset{\sim}{Q})$, the wave function at time t of the system of electron ($\underset{\sim}{x}$) and lattice ($\underset{\sim}{Q}$) in the applied fields $\underset{\sim}{E}$ and $\underset{\sim}{H}$. For the H of interest here $\dot{\underset{\sim}{x}} = (\underset{\sim}{p}-q\underset{\sim}{A}/c)/\underset{\approx}{m}$, where $\underset{\sim}{H} = \underset{\sim}{\nabla}\times\underset{\sim}{A}$. Thus the $\dot{\underset{\sim}{x}}$ operator contains no specific features of the problem of interest. As a result, Ψ_t must contain all the features of the problem. Thus one must work very hard to ensure that Ψ_t is calculated accurately enough that meaningful results can be obtained from (2). At finite temperature, the system consists of an ensemble average of states each of which evolves quantum mechanically from an initial state. Under these circumstances

$$v = \langle \hat{v} \rangle = \lim_{t\to\infty} \mathrm{Tr}(\hat{\dot{x}}\rho_t) \tag{3}$$

where ρ_t is the density matrix of the entire system at time t. Again, since the operator $\dot{x}$ contains no information about the system, ρ_t must be sufficiently accurately calculated that it contain all features of the problem which are of interest. Again this is a difficult task.[13]

From the above it is clear that if we could determine $v(E,H)$ from the expectation value of an operator which itself contained some specific information about the problem, then we would no longer have to rely solely on ρ_t (or Ψ_t) to characterize the system. In particular, if the expectation value were insensitive to ρ_t, then the operator, which is known exactly, would carry the information desired. Let us see how we may attempt to achieve this situation for our transport problem.

We note two things. First, we are dealing with a steady-state problem, once unimportant transients decay. Second, in such a situation the electron is simultaneously acquiring momentum and energy from the applied electric field E and dissipating that momentum and energy to the phonons of the crystal. Thus, for example, qE, the external force on the electron, is in fact just the energy loss per unit distance suffered by an electron in a crystal under the influence of E and H and traveling with steady-state velocity v. Rather than regarding v as the dependent variable we may regard E as the dependent variable. This has one advantage in that $E(v,H)$ is single-valued in v whereas $v(E,H)$ is <u>double</u>-valued in v. Thus if we calculated $v(E,H)$, we would have to perform a separate calculation for each branch. It is also interesting to note that classically it is usually much easier to calculate the applied forces from the nature of the response than vice versa. Unfortunately, quantum mechanically the corresponding facility is <u>not</u> achieved (unless one passes to a classical limit). Nonetheless some advantage is gained.

Thus suppose we attempt to calculate $E(v,H)$. An operator expression involving qE, the energy loss/distance, or momentum loss/time, can be obtained from

$$d(m\hat{\dot{x}})/dt = m\hat{\ddot{x}} = i[\hat{H}, m\hat{\dot{x}}]/\hbar \tag{4}$$

the operator expression for the rate of change of the kinetic momentum of the electron. Since $\dot{x}$ includes $\hat{p}$ and $\hat{p}$ operates on the spacially varying potential in $\hat{H}$, we see that the operator $m\ddot{x}$ will contain some details of the specific problem of interest.

It also means that to proceed further, we must specify $\hat{H}$, the Hamiltonian describing the complete system of electron, crystal, interactions, and applied fields.

The most general Hamiltonian with which I shall work is the following

$$\hat{H} = \frac{1}{2}\,(\hat{p}-qA/c)\cdot(1/m)\cdot(\hat{p}-qA/c) - qE\cdot\hat{x} - qe_t\cdot\hat{x}$$
$$+ \sum_{n,k} \hbar\omega_{k,n} a^{\dagger}_{k,n} a_{k,n}$$
$$+ V^{-\frac{1}{2}} \sum_{n,k} \left(C_{k,n} a_{k,n} \exp(ik\cdot\hat{x}) + C^{*}_{k,n} a^{\dagger}_{k,n} \exp(-ik\cdot\hat{x})\right). \tag{5}$$

In this expression E is the applied electric field, $H = \nabla\times A$ is the applied magnetic field, e_t is a small, oscillatory, probe electric field, m is the (symmetric) effective-mass tensor of the fixed lattice, $\hbar\omega_{k,n}$ is the dispersion energy of phonons of type n and wave vector k, and C_{kn} is the amplitude for an electron to absorb a phonon of type n and wave vector k.

If we evaluate (4) for the rate of change of the kinetic momentum $m\dot{x}$, we obtain

$$m\hat{\ddot{x}} = i[\hat{H}, m\hat{\dot{x}}] = qE + qe_t + q\hat{\dot{x}} \times H/c - \sum_{n,k} k\hat{R}_{k,n} \tag{6}$$

where $\hat{R}_{k,n}$ is the operator for the net rate of phonon emission (emission less absorption) and is given by

$$\hat{R}_{k,n} = -i\left(C^{*}_{k,n} a^{\dagger}_{k,n} e^{-ik\cdot\hat{x}} - C_{k,n} a_{k,n} e^{ik\cdot\hat{x}}\right). \tag{7}$$

Equation (6) states, of course, that the rate of change of the kinetic momentum of the electron equals the applied electric and magnetic forces (which tend to increase the electron's momentum) minus the net rate at which momentum is lost to the lattice. In the steady-state limit, the expectation value of (6) can be written as

$$q\underset{\sim}{E} = - q\underset{\sim}{v} \times \underset{\sim}{H}/c + \sum_{n,\underset{\sim}{k}} \underset{\sim}{k} \langle \hat{R}_{\underset{\sim}{k},n} \rangle \tag{8}$$

where $\langle \dot{\underset{\sim}{x}} \rangle = \underset{\sim}{v}$, ($\langle \ddot{\underset{\sim}{x}} \rangle = 0$ in steady state). While $\langle \hat{R}_{\underset{\sim}{k},n} \rangle$ is in fact a function of $\underset{\sim}{E}$ and $\underset{\sim}{H}$, we shall see that we can write it as a function of $\underset{\sim}{v}$ and $\underset{\sim}{H}$. Hence (8) is an expression for $q\underset{\sim}{E}$, the applied electric force on the electron (the loss of energy per unit distance in steady state) as a function of $\underset{\sim}{v}$ and $\underset{\sim}{H}$.

Equation (8) offers a unique advantage over Eq. (3) for the calculation of transport properties. $\hat{R}_{\underset{\sim}{k},n}$ is unique to the problem of interest. This is a crucial feature. Unlike Eq. (3) in which all information specific to the problem had to be contained in the density matrix, in Eq. (8) the operator $\hat{R}_{\underset{\sim}{k},n}$ itself contains details of the electron-phonon interaction. Thus no longer need we rely solely on the density matrix; the presence of $\hat{R}_{\underset{\sim}{k},n}$ in the expectation value enables us to work with density matrices which contain less information than otherwise. In particular we shall see that significant results for transport properties can be obtained from Eq. (8) where the expectation values are calculated using a zero-order density matrix.

The actual calculation of the expectation value in (8) has been carried out in great detail elsewhere,[10,13-15] especially in Ref. 15 where very general initial conditions for the lattice were included. The final results were written in terms of the a.c. response of the electron in the dissipative medium to a small probe electric field superimposed on the finite electric and magnetic fields applied to the material. The nature of this response, including a dynamical, self-consistent approach for its calculation, is discussed in detail in Section III below. Certain features of the solution to (8), however, are worthy of brief mention here.

The calculation of (8) proceeds in two steps. First an initial phonon distribution is assumed and the phonon variables are eliminated exactly by any one of several means.[1-5,8,9,15] [The most realistic initial distribution is a thermal one, since the phonons in the material of interest will usually be initially in thermal equilibrium at some temperature. Of course, as the electron approaches its steady-state motion, both the electron and the lattice vibrations in its vicinity will become heated. This heating is included in the calculation since the entire interacting system is propagated in time according to the full hamiltonian (5).] Writing the desired expectation value in terms of path integrals over the electronic coordinates alone one finds that

$$\langle \hat{R}_{k,n} \rangle_{t_2} = \iint R_{k,n} e^{i\Phi_e} D(x)D(x') \tag{9a}$$

where $\iint D(x)D(x')$ is the path integral over x and x' of the electron between t_1 and t_2, and where

$$\Phi_e = \int_{t_1}^{t_2} dt \left(\frac{1}{2} \dot{x}_t \cdot m \cdot \dot{x}_t + \frac{1}{2} \dot{x}_t \cdot H \times x_t + f_t \cdot x_t \right)$$

$$- \int_{t_1}^{t_2} dt \left(\frac{1}{2} \dot{x}'_t \cdot m \cdot \dot{x}'_t + \frac{1}{2} \dot{x}'_t \cdot H \times x'_t + f'_t \cdot x'_t \right)$$

$$+ i \sum_{n,k} |C_{kn}|^2 \int_{t_1}^{t_2} dt \int_{t_1}^{t} dt' \Big[T_{\omega_{k,n}}(t-t') e^{-ik\cdot(x'_t - x'_{t'})}$$

$$+ T^*_{\omega_{k,n}}(t-t') e^{ik\cdot(x_t - x_{t'})} - T_{\omega_{k,n}}(t-t') e^{ik\cdot(x_t - x'_{t'})}$$

$$- T^*_{\omega_{k,n}}(t-t') e^{-ik\cdot(x'_t - x_{t'})} \Big] \tag{9b}$$

$$T_{\omega_{k,n}}(\tau) = e^{i\omega_{k,n}\tau}(1-e^{-\beta\omega_{k,n}})^{-1} + e^{-i\omega_{kn}\tau}(e^{\beta\omega_{kn}}-1)^{-1} \tag{9c}$$

with $f_t = f'_t = E$ when (9abc) is used in (8). [$R_{k,n}$ is also a functional in x_t and x'_t and is given explicitly in (20b) below.] Having (9abc) one can, by integrating out the electron's coordinates, complete the calculation of (8).

The influence functional Φ_e can be understood in the following way. The first term is the action integral of a free particle of mass m in a magnetic field H and electric field f_t. The second term in the sum over (n,k) arises from the electron's interaction at x_t with the disturbance it created earlier at $x_{t'}$. These two terms in the exponent, the free part and the interaction part, make up what is often referred to as the propagator of the wave function $\varphi(x)$ from t_1 to t_2. The second term in (9b) and the first term in the sum over $(n,\underset{\sim}{k})$ are the corresponding terms for the propagator of $\varphi^*(\underset{\sim}{x}')$, which also enters into the computation of the expectation value. In terms of phonon emission and absorption, these effects are shown schematically in Fig. 1. Curved lines which begin and end to the right of t_2 are associated with the propagator of $\varphi(\underset{\sim}{x})$, the motion $\underset{\sim}{x}_t$, while those which begin and end to the left of t_2 are associated with $\varphi^*(\underset{\sim}{x}')$. Such terms renormalize the electron's energy and mass and can be used to calculate the electron's self-energy.

Of greater interest, however, are the last two terms in the sum over $(n,\underset{\sim}{k})$ in (9b). These correspond to dissipation, as can be seen in Fig. 1. Note the curved lines which cross t_2. Their existence means that at t_2 the state of the system has changed. This in itself, of course, does not imply dissipation. However, we note that since energy can be conserved in the process of single phonon emission (absorption), one can have processes contributing to phonon lines crossing t_2 originating anywhere between t_1 and t_2, with comparable effects. Thus at t_2 one can have the system in any of a multitude of states of the same energy. The cross terms in $\underset{\sim}{x},\underset{\sim}{x}'$in (9b), therefore, are essential for the existence of dissipation. The reader is referred to van Hove[20] for an excellent discussion of this point.

One further aspect of (9b) is worth noting. If $\underset{\sim}{x}_t = \underset{\sim}{x}_t'$, then the four interaction terms cancel for arbitrary $\underset{\sim}{x}_t$. In the absence of fluctuations, $\underset{\sim}{x}_t$ and $\underset{\sim}{x}_t'$ would be equal to the classical (mean) path. Thus no fluctuation, no dissipation. This situation in fact arises for very energetic electrons which travel so fast that the lattice cannot respond to their motion. On the other hand, for the electrons of interest here, the dissipation and hence the fluctuations can be considerable. We consider the relation between dissipation and fluctuation in more detail in Section IV.

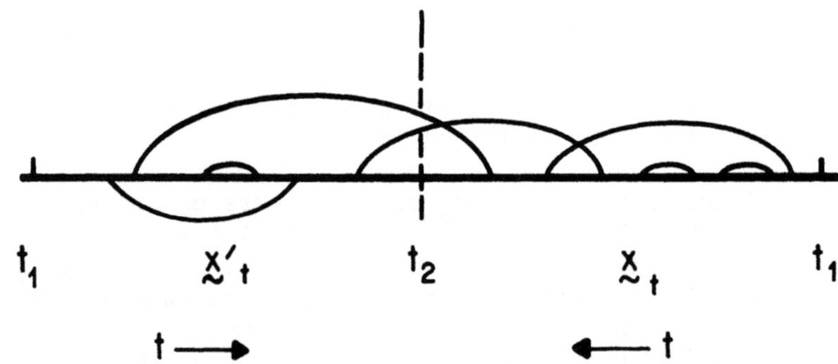

Figure 1. A contribution to renormalization and dissipation.

To summarize the first step, we have reduced a many-body problem to a one-body problem exactly. It is not a one-body problem in the usual sense, however. Ordinarily when a particle interacts with a potential (which may be a function of space and time), one can describe the motion of the particle with a one-particle Hamiltonian. This states essentially that the motion of the particle is governed by its motion at infinitesimally small previous times. It is clear from (9b) that the motion of the electron in the deformable lattice is governed by its motion at finite times in the past. Thus it is impossible to represent the behavior of the electron by a one-particle Hamiltonian. This is the price we pay in passing from a many-body problem which can be described by a Hamiltonian (albeit a many-body one) to a one-body problem which cannot.

The second step in calculating (8) is to carry out the integration over the electron variables $\underset{\sim}{x}_t, \underset{\sim}{x}_t'$. This we do not know how to do exactly. However, knowing the form of (9b), we can approximate Φ_e by a Φ_o which has a similar structure but for which we can carry out the integrations. We can then consider the problem as a perturbation in $(\Phi_e - \Phi_o)$. Before doing this, however, we first transform (9abc) to a frame drifting with the mean motion of the electron[10,13-15] ($\underset{\sim}{x}_t = \underset{\sim}{v}t + \underset{\sim}{y}_t$, $\underset{\sim}{x}_t' = \underset{\sim}{v}t + \underset{\sim}{y}_t'$) and write the path integrals in terms of $\underset{\sim}{y}_t$ and $\underset{\sim}{y}_t'$. Such a transformation can, of course, be done without approximation and tends to preserve the drifting nature of the electron when we finally approximate.

The details of actually carrying out the second step are available,[10,13-15] and need not be repeated there. The final result (8) has also been discussed in detail,[13-15] and (for $\underset{\sim}{H}=0$) numerical results have been presented for E versus $\underset{\sim}{v}$ for a variety of temperatures and coupling strengths.[13-15] One point, however, deserves special attention. Many interesting results can be obtained just from straightforward perturbation theory in the interaction energy. Therefore, what have we gained by our approach which might have been missed in perturbation theory. One feature which is not contained in perturbation theory in any simple form is that we calculate momentum and energy loss for arbitrary electron velocity. In perturbation theory in the interaction energy, the velocity has to exceed a certain critical velocity before the electron has enough energy and momentum to satisfy those conservation laws in order to emit a phonon and thereby lose energy and momentum to the medium. This crutial difference, this remarkable distinction between the lowest order results of these two approaches points to the obvious advantages of avoiding ordinary perturbation theory, which tries to make a dissipative state as a perturbation of a nondissipative one, for such problems.

What is more important for the present discussion is the nature of the approximation Φ_o' to Φ_e', where the prime refers to the influence functional being in the drifted frame. ($R_{\underset{\sim}{k},n}$ by contrast is never approximated.) Were we to let Φ_o' be just the influence functional of a free particle, we would clearly not be doing a good job of approximating the dissipation which we know to be present. If, on the other hand we let Φ_o' have the form

$$\Phi_o' = \int_{t_1}^{t_2} \left(\frac{1}{2}\,\dot{\underset{\sim}{y}}_t\cdot\underset{\approx}{m}\cdot\dot{\underset{\sim}{y}}_t + \frac{1}{2}\,\dot{\underset{\sim}{y}}_t\cdot\underset{\sim}{H}\times\underset{\sim}{y}_t + \underset{\sim}{f}_t\cdot\underset{\sim}{y}_t\right)dt$$

$$- \int_{t_1}^{t_2} \left(\frac{1}{2}\,\dot{\underset{\sim}{y}}_t'\cdot\underset{\approx}{m}\cdot\dot{\underset{\sim}{y}}_t' + \frac{1}{2}\,\dot{\underset{\sim}{y}}_t'\cdot\underset{\sim}{H}\times\underset{\sim}{y}_t' + \underset{\sim}{f}_t'\cdot\underset{\sim}{y}_t'\right)dt - i\int_{t_1}^{t_2}dt\int_{t_1}^{t_2}dt'$$

$$\times\Big[\left(\underset{\sim}{y}_t'-\underset{\sim}{y}_{t'}'\right)\cdot\underset{\approx}{G}^*(t-t')\cdot(\underset{\sim}{y}_t'-\underset{\sim}{y}_{t'}') + (\underset{\sim}{y}_t-\underset{\sim}{y}_{t'})\cdot\underset{\approx}{G}(t-t')\cdot(\underset{\sim}{y}_t-\underset{\sim}{y}_{t'})$$

$$- (\underset{\sim}{y}_t-\underset{\sim}{y}_{t'}')\cdot\underset{\approx}{G}^*(t-t')\cdot(\underset{\sim}{y}_t-\underset{\sim}{y}_{t'}') - (\underset{\sim}{y}_t'-\underset{\sim}{y}_{t'})\cdot\underset{\approx}{G}(t-t')\cdot(\underset{\sim}{y}_t'-\underset{\sim}{y}_{t'})\ , \tag{10}$$

then dissipation can be included even in lowest (zero) order. (In (10) $\underset{\sim}{f}_t$ and $\underset{\sim}{f}_t'$ no longer contain $\underset{\sim}{E}$ which gives rise to $\underset{\sim}{v}$.[10]) An influence functional of the form of Φ_o' would have arisen from eliminating the oscillator coordinates of an electron interacting with a distribution of oscillators, the interaction being linear in the coordinates of both the electron and the oscillators.

The problem of chosing the oscillator distribution $\underset{\approx}{G}(\Omega)$ to best use in (10) is the topic of the next section. In Refs. 5-9 and 11-14 $\underset{\approx}{G}$ was chosen to be a delta function at a frequency w with strength C, where w and C were chosen to minimize the ground state energy of the electron interacting with polar phonons at zero temperature. In Ref. 9, it was shown how for $\underset{\sim}{E}$=0, $\underset{\sim}{H}$=0, T=0, but α arbitrary, $\underset{\approx}{G}$ chosen to minimize the ground state energy could be related to the a.c. response of the electron to a small probe field calculated in perturbation theory. In Ref. 10, however, a dynamical self-consistency for the a.c. response was introduced which was valid for arbitrary $\underset{\sim}{E}$, $\underset{\sim}{H}$, T and α, the coupling constant. (It was also shown that for $\underset{\sim}{E}$=0, $\underset{\sim}{H}$=0 but now for arbitrary T and α, this self-consistency reduced to minimizing the free energy.) In Ref. 19, the subject of Section IV, it was found that using the $\underset{\approx}{G}(r)$ determined self-consistency one could calculate the a.c. response in a variety of ways and obtain the same answer each time. One has no a priori guarantee of this since the calculation is approximate. The determination of $\underset{\approx}{G}(\Omega)$ is clearly quite important if we wish to avoid expansions in $(\Phi_e'-\Phi_o')$ in calculating expectation values.

III. THE SELF-CONSISTENT SOLUTION

The self-consistent solution[10] for transport problems with electrons in dissipative media is based on the following physical idea. One applies static $\underset{\sim}{E}$ and $\underset{\sim}{H}$ fields, lets steady state be reached, hops on a coordinate system moving with the mean motion of the electron, probes the electron with a small electric field, and notes its linear response to the probe field. One then notes that this response can be calculated in two different ways. On the one hand one can calculate the linear response from the expectation value of $\hat{\underset{\sim}{x}}$ to lowest (zero) order in $(\Phi_e'-\Phi_o')$. The result is a very sensitive dependence of the response on the approximate Φ_o' chosen. On the other hand, one can calculate the linear response from (6) again to lowest (zero) order in $(\Phi_e'-\Phi_o')$. The result in this case is quite sensitive to $\hat{R}_{\underset{\sim}{k},n}$, that is to the exact interaction, and much less sensitive to Φ_o'. Requiring that the two results for the linear response be the same provides the self-consistency. This ensures that the Φ_o' chosen to calculate (8) will contain the proper spectral reactive and dissipative structure, and thus eliminates the ambiguity of which influence functional to use in path-integral treatments of electron transport.

While this self-consistent approach is applicable to many methods of calculation, we shall consider it only in the path-integral framework where it applies most naturally. Nonetheless, it is still rather complicated[10] when arbitrary electric and magnetic fields act on the electron. The reason for this is that in addition to having to match the trial oscillatory spectrum to that obtained from (8), one must also worry about the heating of the material in the vicinity of the electron. As this heating alters the actual scattering, so too will it alter the optimum oscillatory spectrum. In addition, the applied electric field introduces a distinction in the a.c. response parallel to from that perpendicular to this field. The magnetic field renders the a.c. response tensorial, a probe field in one direction giving rise to components of the response in perpendicular directions. It is, of course, desirable that our self-consistent approach does indeed include all these effects. On the other hand including so much all at once makes for a rather involved presentation.[10] For purposes of illustration, therefore, we shall discuss the self-consistency only for $\underset{\sim}{E}=0$, $\underset{\sim}{H}=0$, and an isotropic interaction, a problem of considerable interest in its own right.[8-12,17-19] It should then be easier to grasp the more general problem,[10] in which heating, anisotropy, etc. are included.

The problem we wish to solve is, therefore, as follows. If we can calculate the a.c. response to a small, probe electric field of an electron whose dynamical behavior is governed by (5) with $\underset{\sim}{E}=0$, $\underset{\sim}{A}=0$, then we in fact will have determined the absorption.[11]

To calculate the a.c. response, we must calculate the expectation value of $\hat{\underset{\sim}{x}}$. This we can do if we can calculate

$$A_e(f_t, f_t') \equiv \iint e^{i\Phi_e} D(\underset{\sim}{x})D(\underset{\sim}{x}') \tag{11}$$

where Φ_e is given by (9b), now with $\underset{\sim}{H}=0$, $\underset{\sim}{E}=0$, and $\underset{\approx}{m}=m\underset{\approx}{I}$. This follows, because if let $\underset{\sim}{f}_t=\underset{\sim}{e}_t+\underset{\sim}{\lambda}\delta(t-t_2)$ and if $\underset{\sim}{f}_t'=\underset{\sim}{e}_t$, then

$$\langle \hat{\underset{\sim}{x}} \rangle_{t_2} = -i \frac{\partial}{\partial \underset{\sim}{\lambda}} A_e(\underset{\sim}{e}_t+\underset{\sim}{\lambda}\delta(t-t_2), \underset{\sim}{e}_t)\Big|_{\underset{\sim}{\lambda}=0} \tag{12}$$

Expanding (12) to lowest order in $\underset{\sim}{e}_t$ will then yield the linear response.

We have already noted that we are unable to calculate (11) for the Φ_e of (9b). We have also noted that one approximate Φ_o of the form given in (10), with which we can calculate (11), might be a good starting point for our evaluation. Thus we can write (11) as

$$\begin{aligned} A_e(f_t, f_t') &= \iint e^{i\Phi_o} e^{i(\Phi_e-\Phi_o)} D(\underset{\sim}{x})D(\underset{\sim}{x}') \\ &= \sum_{n=0}^{\infty} \cdot \iint e^{i\Phi_o} [i(\Phi_e-\Phi_o)]^n/n! \; D(\underset{\sim}{x})D(\underset{\sim}{x}') \end{aligned} \tag{13}$$

(In this section we are concerned only with the n=0 term in this expansion; in Section IV we note the effect of including higher order terms.) Both Φ_e and Φ_o are of such form that if we can calculate

$$A_o(f_t, f_t') \equiv \iint e^{i\Phi_o} D(\underset{\sim}{x})D(\underset{\sim}{x}'), \tag{14}$$

then we can calculate each term in (13) as well.[15] The calculation of A_o is given in great detail in Refs. 10,13-15. The result is that

$$A_o(f_t, f_t') = \exp \int_{-\infty}^{\infty} dt \int_{-\infty}^{t} dt' (\underset{\sim}{f}_t - \underset{\sim}{f}_t') \cdot (L^*_{t-t'} \underset{\sim}{f}_{t'} - L_{t-t'} \underset{\sim}{f}_{t'}') \tag{15a}$$

where

$$L_\xi \equiv \int_{-\infty}^{\infty} \frac{dv}{2\pi i} \frac{4\pi i G(-v)}{Z_v \cdot Z_v^*} (-)e^{-iv\xi} \tag{15b}$$

$$Z_v \equiv -m(v+i\varepsilon)^2 - 4(v+i\varepsilon)^2 \int_{-\infty}^{\infty} d\Omega \frac{P}{\Omega} \frac{G(\Omega)}{\Omega^2-(v+i\varepsilon)^2} \tag{15c}$$

$$G(t) \equiv \int_{-\infty}^{\infty} d\Omega\, G(\Omega) e^{-i\Omega t} \tag{15d}$$

In evaluating (14) we have, of course, converted the Φ_0' of (10) to Φ_0 by noting that if $\underset{\sim}{E}=0$, $\underset{\sim}{v}$ (steady-state) $= 0$, and hence $\underset{\sim}{x}=\underset{\sim}{y}$, $\underset{\sim}{x}'=\underset{\sim}{y}'$.

Inserting (15) into (12) we find that

$$\langle \hat{\underset{\sim}{x}} \rangle_{t_2} = -i \int_{-\infty}^{t_2} dt\, (L^*_{t_2-t} - L_{t_2-t}) \underset{\sim}{e}_t \tag{16a}$$

Taking a Fourier transform with respect to t_2 and noting that all the poles of Z_v (Z_v^*) are in the lower (upper) half plane, it follows that[10]

$$\langle \hat{\underset{\sim}{x}} \rangle_\omega \equiv \int_{-\infty}^{\infty} dt\, \langle \hat{\underset{\sim}{x}} \rangle_t\, e^{+i\omega t}$$

$$= (1/Z_\omega) e_\omega \tag{16b}$$

where we have used (18) below, which follows directly from (15c). [Our result is linear in the probe field already, so that expanding (12) in e_ω is unnecessary. We can obtain nonlinear effects by including higher order terms in (13), but these are not of interest here.] If we define admittance (a.c. response) by

$$\langle \hat{x} \rangle_\omega \equiv Y_\omega e_\omega \tag{17a}$$

and impedance by

$$\langle \hat{x} \rangle_\omega \equiv (1/Z_\omega) e_\omega \tag{17b}$$

we note that because of (16b), what we have defined as Z_v in (15c) is in fact the impedance of the electron, and hence the same symbol is used. (Please note that FHIP and Ref. 11 use velocity response rather than position response in defining admittance and impedance.)

Before proceeding to the calculation of $\langle \hat{\underset{\sim}{x}} \rangle_\omega$ from (6), we should note how sensitive the $Y_\omega = 1/Z_\omega$ calculated above is to the oscillator distribution $G(\Omega)$ chosen for the calculation. Indeed from (15c) we find that

$$\mathrm{Im}\, Z_v = -2\pi[G(v)-G(-v)] \tag{18}$$

Clearly unless we have some criteria for determining G(v), (16) can yield little information about Y_ω.

Turning now to the next step, we rewrite (6) (setting q=1) in the form

$$\underset{\sim}{e}_{t_2} = m\langle\ddot{\hat{\underset{\sim}{x}}}\rangle_{t_2} + \sum_{n,\underset{\sim}{k}} |C_{\underset{\sim}{k},n}|^2 \underset{\sim}{k}\langle\hat{R}_{\underset{\sim}{k},n}\rangle_{t_2} \tag{19}$$

We then note that the Fourier transform of $\underset{\sim}{e}_t$ is just $\underset{\sim}{e}_\omega = Z_\omega\langle\hat{\underset{\sim}{x}}\rangle_\omega$, while that of $\langle\ddot{\underset{\sim}{x}}\rangle_t$ is just $-\omega^2\langle\hat{x}\rangle_\omega$. If now we can also calculate the Fourier transform of $\langle\hat{R}_{\underset{\sim}{k},n}\rangle_t$ to lowest order in $\langle\hat{\underset{\sim}{x}}\rangle_\omega$, the $\langle\hat{x}\rangle_\omega$ factor can be canceled and we shall have another expression for Z_ω, this time involving the actually interaction potential through $\hat{R}$. Elimination of the phonon variables proceeds as before, so that we obtain

$$\langle\hat{R}_{\underset{\sim}{k},n}\rangle_{t_2} = \iint R_{\underset{\sim}{k},n}\, e^{i\Phi_e} D(\underset{\sim}{x})D(\underset{\sim}{x}') \tag{20a}$$

where now in Φ_e one would set $\underset{\sim}{f}_t = \underset{\sim}{f}'_t = \underset{\sim}{e}_t$. In (20a) $R_{\underset{\sim}{k},n}$ is given by

$$R_{\underset{\sim}{k},n} = |C_{\underset{\sim}{k},n}|^2 \int_{t_1}^{t_2} dt \left(\frac{e^{-i\omega_{\underset{\sim}{k},n}(t_2-t)}\, e^{i\underset{\sim}{k}\cdot(\underset{\sim}{x}_{t_2}-\underset{\sim}{x}_t)}}{1-e^{-\beta\omega_{\underset{\sim}{k},n}}} - \frac{e^{i\omega_{\underset{\sim}{k},n}(t_2-t)}\, e^{-i\underset{\sim}{k}\cdot(\underset{\sim}{x}_{t_2}-\underset{\sim}{x}_t)}}{e^{\beta\omega_{\underset{\sim}{k},n}}-1} + \frac{e^{i\omega_{\underset{\sim}{k},n}(t_2-t)}\, e^{-i\underset{\sim}{k}\cdot(\underset{\sim}{x}_{t_2}-\underset{\sim}{x}'_t)}}{1-e^{-\beta\omega_{\underset{\sim}{k},n}}} - \frac{e^{-i\omega_{\underset{\sim}{k},n}(t_2-t)}\, e^{i\underset{\sim}{k}\cdot(\underset{\sim}{x}_{t_2}-\underset{\sim}{x}'_t)}}{e^{\beta\omega_{\underset{\sim}{k},n}}-1} \right) \tag{20b}$$

Let us evaluate the first term. The remaining three can be done by analogy.

We can calculate

$$B_1(t_2,t) = \iint e^{i\underset{\sim}{k}\cdot(\underset{\sim}{x}_{t_2}-\underset{\sim}{x}_t)} e^{i\Phi_e} D(\underset{\sim}{x})D(\underset{\sim}{x}')$$

if we first replace Φ_e by Φ_o and then in $A_o(f_\tau, f'_\tau)$ let $\underset{\sim}{f}_\tau = \underset{\sim}{e}_\tau + \underset{\sim}{k}\delta(\tau-t_2)-\underset{\sim}{k}\delta(\tau-t)$, and $\underset{\sim}{f}'_\tau = \underset{\sim}{e}_\tau$. Referring to (13), we are calculating (20) to zero order in $(\Phi_e-\Phi_o)$. Since $t_2>t$, we find that

$$B_1(t_2,t) = \exp\Big\{-k^2\bar{L}^*_{t_2-t} + \underset{\sim}{k}\cdot\int_{-\infty}^{t_2} dt'(L^*_{t_2-t'}-L_{t_2-t'})\underset{\sim}{e}_{t'} - \underset{\sim}{k}\cdot\int_{-\infty}^{t} dt'(L^*_{t-t'}-L_{t-t'})\underset{\sim}{e}_{t'}\Big\}$$

which using (16a) simplifies at once to

$$B_1(t_2,t) = \exp\Big\{-k^2\bar{L}^*_{t_2-t} + i\underset{\sim}{k}\cdot(\langle\hat{\underset{\sim}{x}}\rangle_{t_2} - \langle\hat{\underset{\sim}{x}}\rangle_t)\Big\}$$

or to

$$B_1(t_2,t) = i\underset{\sim}{k}\cdot\int_{-\infty}^{\infty}\frac{d\omega}{2\pi}\langle\hat{\underset{\sim}{x}}\rangle_\omega e^{-i\omega t_2}(1-e^{i\omega(t_2-t)})e^{-k^2\bar{L}^*_{t_2-t}}$$

when expanded to lowest order in $\langle\hat{\underset{\sim}{x}}\rangle_\omega$. Here $\bar{L}_\tau \equiv L_\tau - L_{0^+}$. Calculating the remaining three terms, inserting into (19), taking the Fourier transform and collecting terms, we finally obtain

$$Z_\omega = -m\omega^2 + \int_0^\infty d\xi(1-e^{i\omega\xi})\mathrm{Im}[S(\xi)] \tag{21a}$$

where

$$S(\xi) \equiv \sum_{n,\underset{\sim}{k}} |C_{\underset{\sim}{k},n}|^2(k^2/3)2T_{\omega_{\underset{\sim}{k},n}}(\xi)e^{-k^2\bar{L}(\xi)} \tag{21b}$$

Here now is a relation for Z_ω which is sensitive to the actual interaction $C_{\underset{\sim}{k},n}$ and somewhat insensitive to Φ_o and $G(\Omega)$ which determine $L(\xi)$.

Our self-consistency is now apparent. Calculating Z_ω from (21ab) depends upon knowing $L(\xi)$. On the other hand in order to calculate $L(\xi)$ according to (15a) we must know Z_ν. The Z_ω which satisfies (15b) and (21ab) is, therefore the self-consistent impedance of the electron in the dissipative media. Once $L(\xi)$ is known, we can also use it to calculate other expectation values with the help of (15a). In this manner we obtain not only a good expression for $Y_\omega = 1/Z_\omega$, but also an approximate Φ_o from which other quantities may be determined as well.

In our derivation for the determining equations of the self-consistent solution we have bypassed the oscillator distribution $G(\Omega)$. In the more general problem with applied $\underset{\sim}{E}$ and $\underset{\sim}{H}$ we cannot do this. For the zero field problem discussed above we can recover $G(\Omega)$ by comparing (21ab) and (15c). The result is that

$$G(\Omega) = \frac{1}{2} \sum_{n,\underset{\sim}{k}} |C_{\underset{\sim}{k},n}|^2 (k^2/3) \int_{-\infty}^{\infty} \frac{d\xi}{2\pi} e^{-i\Omega\xi} T_{\omega_{\underset{\sim}{k},n}}(\xi) e^{-k^2\bar{L}(\xi)} \qquad (22)$$

[Inserting (22) into (15c) will give (21ab).] If the path of the $d\xi$ integration is shifted from ξ to $\xi + i\beta/2$, it is easily seen that

$$G(-\Omega) = e^{-\beta\Omega} G(\Omega) \qquad (23)$$

in the light of which (18) becomes

$$\mathrm{Im}Z_v = -2\pi G(v)(1-e^{-\beta v}) = -\mathrm{Im}Z_{-v} \qquad (24)$$

From (24) it follows that (15b) can be rewritten as

$$L_\xi \equiv \int_{-\infty}^{\infty} \frac{dv}{2\pi i} \left(\frac{1}{Z_v} - \frac{1}{Z_v^*} \right) \frac{-e^{-iv\xi}}{e^{\beta v}-1} \qquad (15e)$$

for any $G(\Omega)$ satisfying (23). As we shall see in the next section, any $G(\Omega)$ which satisfies (23) can be used to predict fluctuations and dissipation which satisfy the fluctuation-dissipation theorem.[21,22] Here we note that our self-consistent solution will satisfy this important theorem, which must be satisfied <u>at each frequency</u> by the exact solution. As was shown in Ref. 10, the self-consistent solution for $\underset{\sim}{E}=0$, $\underset{\sim}{H}=0$ also minimizes the free energy at arbitrary temperature. In the next section we shall present a number of relations satisfied by the self-consistent solution and note how several other methods of calculating linear response yield the same $Y_\omega = 1/Z_\omega$ when the self-consistent $G(\Omega)$ is used to determine Φ_o.

One question which naturally comes to mind is what have we achieved over the Z_ω presented by FHIP or the corresponding Y_ω calculated in Ref. 11. In Ref. 10 an approximate, self-consistent solution was calculated for very low temperatures for the Fröhlich polaron model.[16] The results indicated that for sufficiently low temperatures one could see phonon fine structure in $\mathrm{Im}z_v$, that is the emission of one, two,... phonons. By contrast, in FHIP the structure in $\mathrm{Im}Z_v$ is solely due to the various electronic levels created in the electron's self-polarized well. What is needed to be checked is the extent to which the phonon fine structure which we calculated[10] is indeed superimposed onto the more gross electronic structure found by FHIP.

In the above we have limited ourselves to a path-integral treatment for zero fields. The extension to finite fields is more complicated but involves no additional concepts.[10] It is, moreover, much more interesting since heating and anisotropy effects enter in[10]. The extension to other methods of calculation should be possible: indeed, such extensions should be very rewarding. The most important feature of our self-consistency, however, is that it is dynamic: it is a frequency by frequency relation and not simply an average taken over frequency, as are many sum-rule relations. The significance of this feature will become more apparent in the next section.

IV. THE EQUILIBRIUM SUM RULES

In order to obtain an approximate solution to a difficult many-body problem, one usually must make several simplifications, the validity of which is quite difficult to ascertain directly. To obtain some qualitative idea of how adequate a specific solution may be, or to compare two different solutions, one often turns to sum rules which one knows are satisfied by the exact solution. For example, Lemmens, deSitter, and Devresse (LdSD) derived a ground-state theorem[18] for free polarons. This theorem relates the exact ground-state energy of the polaron to the exact absorption (dissipation) integrated over frequency. In order to test the accuracy of the Feynman one-oscillator approximation,[5,8] LdSD evaluated their theorem numerically using the ground-state energy[5] and the absorption[11] predicted using this model for zero temperature. Remarkably, agreement was found to within the errors of their numerical calculation. Such close agreement between these two quantities is unexpected, and the question arises as to whether this result is an independent check on the Feynman approach, or is in fact already contained in this approach.

In a subsequent paper[19] by the present author, it was demonstrated that the ground-state theorem of LdSD is in fact exactly satisfied when in place of the exact solution, one uses the Feynman one-oscillator solution. Indeed, it was shown that <u>any</u> solution based on a distribution of oscillators, <u>e.g.</u> (10), satisfies the ground-state theorem exactly, provided (a) that the dependence of the distribution on the electron-phonon coupling is chosen so as to minimize the ground-state energy, and (b) that the same approximate distribution is used to calculate both the energy and the absorption. Extending the ground-state theorem to arbitrary temperatures, a free-energy theorem for exact solutions was obtained. This latter theorem is also satisfied if in place of the exact solution one uses any approximate solution satisfying the two criteria given above, with this difference that for nonzero temperature one adjusts the distribution to obtain the lowest free energy.

The purpose of this section is to discuss these results in some detail. These results are significant for several reasons. On the one hand, they probably are the first nontrivial, approximate solutions to a nontrivial many-body problem that have been shown to satisfy exactly a nontrivial sum rule. On the other hand the fact that such a multiplicity of solutions satisfy the free-energy theorem suggests that this theorem or others like them are of little value in evaluating the accuracy of predicted absorption spectra. However, based on our analysis we can suggest an alternative criterion which compares at each frequency the absorption with the spectral density of the fluctuations in the electron's velocity. This criterion is closely related to the self-consistent approach discussed in the previous section. In the course of our discussion we shall consider a number of different ways to calculate absorption and comment on each. One remarkable result is that they all become equivalent when calculated with the self-consistent oscillator distribution.

We must first outline the ground-state[17,18] and free-energy[19] theorems. These relations as proven apply only to the exact solution of the physical problem of interest. We make use of (A) the fluctuation-dissipation (FD) theorem,[21,22] (B) the Feynman-Hellman theorem at arbitrary lattice temperature, and (C) scale transformations to derive three relations which can then be combined (D) to relate the free energy to the integrated absorption.

A. Fluctuation-Dissipation Theorem

If H_s is the Hamiltonian of a system at temperature T, then according to the FD theorem,[21,22] the real part of the Fourier transform of the autocorrelation function of the fluctuations in any observable v is related to the imaginary part of the spectrum of the generalized absorption function of the same observable by

$$\mathrm{Re}[K_r(\omega)] = \mathrm{Im}[X_r(\omega)]\frac{\hbar}{2}\,\frac{1+e^{-\beta\hbar\omega}}{1-e^{-\beta\hbar\omega}}, \tag{25a}$$

where $\beta=1/kT$, T being the temperature,

$$K_r(\omega) \equiv \int_0^\infty dt k_r(t) e^{i\omega t}, \tag{25b}$$

$$k_r(t) \equiv \tfrac{1}{2} \langle [\hat{r}(t), \hat{r}(0)]_+ \rangle, \tag{25c}$$

and

$$X_r(\omega) \equiv \int_{-\infty}^{\infty} dt\, x_r(t) e^{i\omega t}. \tag{25d}$$

In turn, $x_r(t)$ is the linear impulse response of r to a unit impulse "force" $\hat{f}$ of the system characterized by the Hamiltonian $H=H_s-f\hat{r}$. Letting $\hat{r}$ be the current operator in the ith direction $\hat{j}_i \equiv e\hat{p}_i/m$ (i=x, y, z), integrating Eq. (1) over all ω, and summing on i, it follows at once that

$$E_{kin} = \frac{3\hbar m}{2e^2} \int_{-\infty}^{\infty} \mathrm{Im}\chi_{jj}(\omega) \frac{1+e^{-\beta\hbar\omega}}{1-e^{-\beta\hbar\omega}}, \tag{26}$$

where $\chi_{jj}(\omega) \equiv X_r(\omega)$, $\hat{r}=\hat{j}$, and E_{kin} is the usual thermal-average expectation value of the kinetic energy of the electron when interacting with the polarizable lattice at temperature T:

$$E_{kin} \equiv \langle \hat{\underset{\sim}{p}} \cdot \hat{\underset{\sim}{p}}/2m \rangle$$

In the limit that $T \to 0$ $(\beta \to \infty)$ one obtains the corresponding relation on LdSD. [Note, $\mathrm{Im}\chi(\omega) = -\mathrm{Im}\chi(-\omega)$.] From the above derivation we may conclude only that Eq. (26) is valid for the exact E_{kin} and the exact $\mathrm{Im}\chi(\omega)$.

B. Feynman-Hellman Theorem for Arbitrary Temperature

For nonzero temperatures it is much more convenient to consider derivatives of the free energy F with respect to various parameters than to consider similar derivatives of the total energy E. If we define the free energy F in the usual way,

$$e^{-\beta F} \equiv \sum_n e^{-\beta E_n} = \sum_n \langle \varphi_n | e^{-\beta \hat{H}_s} | \varphi_n \rangle, \tag{27}$$

where the $E_n(\varphi_n)$ are the eigenvalues (eigenstates) of H_s, then it follows at once from the usual Feynman-Hellman theorem that

$$\frac{dF}{d\eta} = \left\langle \frac{\partial \hat{H}_s}{\partial \eta} \right\rangle , \tag{28}$$

where η is any parameter on which $\hat{H}_s$ depends. Following LdSD, we now let $\eta=\lambda=m^{-1}$, where m is the mass of the electron, and assume that m appears in the Hamiltonian only in the kinetic-energy term in the form $\hat{p}\cdot\hat{p}/2m$. It then follows from Eq. (28) that

$$\lambda \frac{dF}{d\lambda} = E_{kin.} \tag{29}$$

Again for zero temperature, F=E and one recovers the corresponding result in LdSD.

Another interesting relation can be derived from Eq. (28). In the polaron problem[16] the coupling between the electron and the lattice is proportional to $\alpha^{1/2}$, where α is dimensionless and is usually referred to as the coupling constant. Thus the electron-lattice interaction term in H_s is proportional to $\alpha^{1/2}$. Letting η in Eq. (8) equal α, it follows that

$$\alpha \frac{dF}{d\alpha} = \tfrac{1}{2} E_{inter}, \tag{30}$$

where E_{inter} is the thermal-average expectation value of the interaction energy. Once again, for zero temperature, we recover the corresponding result in LdSD. Again we realize that (10) as derived is valid only for the exact F and E_{inter}.

C. Scale Transformations

Since the Hamiltonian has the dimensions of energy, by factoring out an energy characteristic of the system, one can write the Hamiltonian as a dimensionless operator, which is in turn a function of dimensionless operators usually derived from x and p . (For the polaron problem the characteristic energy is chosen to be $\hbar\omega_L$, the energy of a single longitudinal-optical phonon.) Thus $\hat{H}_s=\hbar\omega_L\hat{h}_s$, where $\hat{h}_s$ is dimensionless, and where, as LdSD derive, $\hat{h}_s$ depends only on α. The expression for α is given by[16]

$$\alpha = \frac{1}{2}\frac{e^2}{\hbar\omega_L}\left(\frac{1}{\varepsilon_0} - \frac{1}{\varepsilon_s}\right)\left(\frac{2m\omega_L}{\hbar}\right)^{1/2} . \tag{31}$$

It follows that

$$\lambda \frac{dF}{d\lambda} = \lambda \frac{dF}{d\alpha}\frac{d\alpha}{d\lambda} = -\tfrac{1}{2}\alpha \frac{dF}{d\alpha}, \tag{32}$$

since from (31) $d\alpha/d\lambda = -\alpha/2\lambda$. Again at zero temperature F=E and the corresponding result in LdSD is recovered. Since the above scale transform can be performed whether one works with exact or only approximate eigenstates, Eq. (32) is also valid for the Feynman approximation.

D. Free-Energy Theorem for Polarons

If we now combine Eqs. (26) and (32) and integrate on α, we obtain

$$F(\alpha)-F(0) = -\frac{3m\hbar}{e^2}\int_0^{\alpha}\frac{d\alpha'}{\alpha'}\int_0^{\infty}\frac{d\omega}{\pi} \times \mathrm{Im}\chi_{jj}(\omega,\alpha')\,\frac{1+e^{-\beta\hbar\omega}}{1-e^{-\beta\hbar\omega}} \tag{33}$$

as a relation for the exact free energy $F(\alpha)$ as a function of the coupling α in terms of the exact absorption $\chi_{jj}(\omega,\alpha)$ for the polaron in a lattice at temperature T. In the zero-temperature limit (13) goes over into the ground-state theorem of LdSD. [F(0)=0 for zero temperature.] In what follows we shall show that our free-energy theorem (33) is valid in the Feynman approximation. In carrying out this program we shall prove that (30) is valid in the Feynman approximation as well.

The foregoing outline, (A) through (D), has of necessity been brief. More details can be found in Refs. 17-19. One reason for presenting the various steps leading up to (33), instead of just stating (33) and proceeding, is that in the course of our development we shall see how the self-consistent absorption satisfies not only (33) but also (25), (26), (29), (30), and (32), that is all the intermediate steps as well.

We now turn to a discussion of some of the various procedures available for calculating absorption. We must do this because, although the free-energy theorem tells us which relation the absorption must satisfy, neither it nor its derivation indicates how the absorption is to be calculated. The same comment applies, of course, to the free energy. However, in the case of the free energy a minimum principle exists[5] and indications are that the results of using it are very accurate.[23,24]

Below we consider six expressions for the admittance from which the absorption is readily obtained.[11] Included is the self-energy admittance[19] which, although it has received relatively little attention, will turn out to be the admittance which <u>exactly</u> satisfies (33) for <u>any</u> oscillatory distribution which minimizes the free energy.

Probably the most famous expression from which the admittance can be calculated is that due to Kubo, which we express as

$$Y_{t_2 t_1} \equiv i \langle [\hat{x}_{t_2}, \hat{x}_{t_1}] \rangle / \hbar \tag{34}$$

when $Y_{t_2 t_1}$ is the inverse Fourier transform of Y_ν. Given the exact wave functions of the system, (34) yields the exact admittance. Since the exact solution is seldom known, however, (34) must be evaluated using approximate solutions. The first three admittances discussed below are just such approximations. The fourth and fifth admittances are obtained as approximations to two other exact expressions from which the admittance can be calculated. These are respectively, the self-energy form derived in Appendix A of Ref. 19 and the conservation-of-momentum form which can be derived from (6), (9abc), and (11). The sixth admittance is just the reciprocal of the self-consistent impedance discussed in the last section. We now list these admittances.

1. Admittance in Zero Order (Y^a)

This is the simplest (and probably the least accurate) method of calculating the admittance. One approximates the exact propagator or influence function by one whose solution is known and calculates (34) to zero order in the difference, <u>e.g.</u> (13) with n=0 only. The calculation leading up to (16ab) is an example. In terms of the oscillator distribution, which we shall specify as some $G_a(\Omega)$, the zero order admittance $Y_\nu^a \equiv (Z_\nu^a)^{-1}$ is just

$$Y_\nu^a = \left(-m(\nu+i\varepsilon)^2 - 4 \int_{-\infty}^{\infty} d\Omega \, \frac{P}{\Omega} \, \frac{G_a(\Omega)(\nu+i\varepsilon)^2}{\Omega^2 - (\nu+i\varepsilon)^2} \right)^{-1} \tag{35}$$

This form is in general unsatisfactory. Since one is relying heavily on the accuracy of the form of the approximate solution, even though one may have chosen the parameters of the solution to minimize the free energy, one has little guarantee of high accuracy for the resulting absorption spectrum.

2. Admittance Perturbation Expansion (Y^1)

One expects to do better if, instead of going merely to zero order as above, one expands the admittance in a power series expansion about the approximate solution one is using. Thus, if instead of stopping with the zero-order solution, one expands (34) to first order, then one obtains for the inverse Fourier transform of the first-order admittance Y^1_v,

$$Y^1_{t_2 t_1} = Y^a_{t_2 t_1} + \left.\frac{\delta Y_{t_2 t_1}}{\delta S_{t_3 t_4}}\right|_a (S^e_{t_3 t_4} - S^a_{t_3 t_4}) \tag{36}$$

where integration over repeated times (t_3, t_4) is understood. Here $S_{tt'}$ is defined by

$$\Phi \equiv \int_{-\infty}^{\infty} dt \int_{-\infty}^{\infty} dt' \; S_{tt'}$$

where Φ is the influence functional, Φ_e of (9abc) for the exact case and Φ_a equals the Φ_o of (10) with $\underset{\sim}{y}=0$ ($\underset{\sim}{x}=\underset{\sim}{y}$, $\underset{\sim}{x}'=\underset{\sim}{y}'$) and $G(\Omega)=G_a(\Omega)$. Higher functional derivatives give rise to higher orders in perturbation theory equivalent to the expansion shown in (13). (This method of expansion is not limited to path-integral applications.) The second and third terms in (35) can be calculated in a manner analogous to our calculation of (20ab). One obtains

$$Y^1_v = Y^a_v - Y^a_v Z^F_v Y^a_v + Y^a_v \tag{37}$$

where Z^F, the "Feynman impedance" calculated in FHIP for the special case in which $G_a(\Omega)$ represents a single oscillator [see the discussion following (10) above], is discussed below. Although one expects Y^1 to be more accurate than Y^a since the former includes the actual electron-phonon interaction in S^e explicitly (via Z^F), Y^1, like Y^a, remains rather sensitive to the choice of S^a through the explicit dependence of Y^1 on Y^a.

3. Impedance Perturbation Expansion (Y^F)

To circumvent the above difficulty FHIP noted that if one focused attention on the reciprocal of Y_v, the impedance Z_v, then the poles of Y_v, which show resonance structure, appear in the numerator of Z_v. Hence an expansion of Z_v might be expected to do a better job of reproducing the resonance in Y_v than an expansion of Y_v as above. If we expand Z_v to first order, we obtain

$$Z^1_{t_2t_1} = Z^a_{t_2t_1} + \left.\frac{\delta Z_{t_2t_1}}{\delta S_{t_3t_4}}\right|_a (S^e_{t_3t_4} - S^a_{t_3t_4})$$

$$= Z^a_{t_2t_1} - Z^a_{t_2t_5} \left.\frac{\delta Y_{t_5t_6}}{\delta S_{t_3t_4}}\right|_a Z^a_{t_6t_1} (S^e_{t_3t_4} - S^a_{t_3t_4}), \tag{38}$$

from which it follows that

$$Z^1_v = Z^a_v - Z^a_v(-Y^a_v Z^F_v Y^a_v + Y^a_v) Z^a_v$$

$$= Z^F_v \text{ and } Y^F_v \equiv 1/Z^F_v. \tag{39}$$

[Note that $Z^1 \neq (Y^1)^{-1}$.] This expression can be expected to be superior to (37) for two reasons. First it does not depend explicitly on Z^a_v (or Y^a_v). In fact, the primary dependence of Z_v on v is governed by S^e, the exact generalized Lagangian of the problem. The second reason is that if S^e is itself derived from a distribution of oscillators, then Z_v given in (20) is exact no matter what distribution is chosen for $G_a(\Omega)$. We shall call Z^F_v the Feynman impedance and note that it is obtained from the first-order expansion of the impedance in powers of the difference between the actual and the approximate S's. By contrast Z^a is the impedance associated with the approximate S^a. For an example of Z^F_v derived for a one-oscillator approximation for S^a, the reader is referred to FHIP, Eqs. (41) and (35). [The method used in FHIP differs only in appearance and not in substance from the expansion in (38).] Reference to an explicit example of Z^F_v will be deferred to subsection 5 below.

In the above three subsections we have discussed three different approximate admittances derived from the basic Kubo expression. We turn now to three different approximate admittances derived from other exact expressions.

4. Self-Energy Admittance (Y^{se})

A fourth expression from which Z_v and hence Y_v can be obtained is derived in Ref. 19, Appendix A. It states that

$$Z^{se}_v = Z^0_v \left(1 - \frac{i}{\hbar} F_v\right)^{-1}, \tag{40a}$$

where

$$Z_\nu^0 = -m(\nu+i\varepsilon)^2 \tag{40b}$$

is the free-particle impedance, and where

$$F_\nu = \int_0^\infty dt \langle[(\nabla_x \hat{H}_{int})_t, (\hat{x})_0]\rangle e^{i\nu t}. \tag{40c}$$

The quantity H_{int} is the interaction Hamiltonian between the electron and the rest of the system. If one evaluates F_ν for the polaron problem (using, for example, the methods of FHIP or of Ref. 15), then one obtains for the self-energy admittance, the expression

$$Y_\nu^{se} \equiv 1/Z_\nu^{se} = [1-(Z_\nu^F-Z_\nu^0)/Z_\nu^a]/Z_\nu^0. \tag{41}$$

The initials se stand for self-energy, a label arising from the form of (40a). Here Z_ν^F and Z_ν^a are as defined above. Expression (41) clearly differs from (37) and (39). Its good features include the fact that it is a nontrivial expression which involves Z_ν^F, and, therefore, the actual potential, but is derived by calculating only to zero order, and the fact that it is Y_ν^{se} which exactly satisfies the free-energy-theorem sum rule [Eq. (33)] for any oscillator distribution adjusted so as to minimize the free energy. We shall come back to this point when we discuss the numerical calculations presented by LdSD in which (39) rather than (41) was used in (33). A bad feature of (41) is that its resonance structure is partially determined by Z_ν^a, which in turn depends very sensitively on $G_a(\Omega)$.

5. Conservation-of-Momentum Admittance Y^{cm}

The center-of-momentum impedance $Z_\nu^{cm} \equiv (Y_\nu^{cm})^{-1}$ was in fact introduced and discussed in the previous section in connection with the self-consistent solution. Z_ν^{cm} is given explicitly by (21ab) with $L(\xi)$ from (15c) in which the Z_ν shown is Z_ν^a as given by (35) inverted. Alternatively Z_ν^{cm} is Z_ν^a when the latter is calculated according to (35) inverted with $G_a(\Omega)$ replaced by (22), which we shall refer to as $G_{cm}(\Omega)$. Although we shall not go through the details, it is straightforward to show that for arbitrary $G_a(\Omega)$, one which need not even minimize the energy or satisfy (23), $Z_\nu^F = Z_\nu^{cm} : Y_\nu^F = Y_\nu^{cm}$. Y^{cm} thus possesses the same advantages and disadvantages as Y^F. [Of course, for influence functionals of a more general form than (10), Z^F and Z^{cm} need <u>not</u> be equal.]

6. Self-Consistent Admittance (Y^{sc})

In the previous section we derived the equations for the self-consistent impedance $Z_V^{sc} \equiv (Y_V^{sc})^{-1}$. In terms of the above this amounts to using Z_V^{sc} instead of Z_V^a to calculate $L(\xi)$ in (15b), and requiring the resulting Z_V in (21ab) to be the same Z_V^{sc}. For $G_a(\Omega)$ chosen to be the oscillator distribution $G_{sc}(\Omega)$ which yields self-consistency, it follows that $Z_V^{sc} = Z_V^a$, and $Z_V^a = Z_V^{cm}$. From these and $Z_V^{cm} = Z^F$, it follows that

$$Y_V^a = Y_V^1 = Y_V^F = Y_V^{se} = Y_V^{cm} = Y_V^{sc} \tag{42}$$

where self-consistency is satisfied. [Y_V^1 is included because from (37), $Y^F = Y^a$ implies $Y^1 = Y^a$; Y_V^{se} is included because from (41), $Z^F = Z^a$ implies $Z^{se} = Z^a$. Relation (42) is valid even in the presence of static, uniform $\underset{\sim}{E}$ and $\underset{\sim}{H}$ fields.] Thus while self-consistency equates Y^a with $\tilde{Y}^{cm}$, in fact Y^1 and Y^{se} are included as well in (42). What has <u>not</u> been proven, however, is that any method of calculating the admittance (under self-consistency) will give the same result as (42) or that self-consistency gives the exact admittance.

In the above six subsections we have outlined six different ways of calculating the admittance. While for any $G_a(\Omega)$, $Y^{cm} = Y^F$, even for any specific $G_a(\Omega)$, <u>e.g.</u>, one, two... oscillators, etc., varied so as to minimize the free energy, in general Y^a, Y^1, Y^F, and Y^{se} will be different. One, therefore, seeks criteria with which to choose among these or other methods. Self-consistency is one such criterion, which in fact makes all these methods equivalent. Another is satisfying sum rules, to which we now turn.

We now ask the question, what conditions are sufficient for an approximate solution [derived from a $G_a(\Omega)$] to satisfy (25), (26), (29), (30) and (33). What we shall find may be summarized as follows:

[1] The fluctuation-dissipation theorem (25) for $\hat{r}=\hat{x}$ and hence for $\hat{r}=\hat{j}$, and therefore (26) as well, are satisfied by any approximate solution derived from a $G(\Omega)$ satisfying (23), so long as the fluctuations, kinetic energy, and linear response are calculated to zero order. One is, therefore, here calculating Y^a.

[2] The standard relation between free-energy and interaction energy (30) is satisfied if the free energy is calculated according to Feynman[5], and the interaction energy is calculated to zero order using any approximate solution derived from a $G(\Omega)$ which has been varied so as to minimize the free energy.

[3] The free-energy theorem (33) is satisfied if the free energy is calculated according to Feynman[5] and the absorption is determined from the self-energy admittance calculated to zero order using any approximate solution derived from a $G(\Omega)$ which has been varied so as to minimize the free energy.

[4] The self-consistent solution satisfies the fluctuation-dissipation theorem (25) and its integral over frequency (26), the interaction-energy theorem (30), the kinetic energy-theorem (29), (32) (as do all solutions whatever), and, therefore, the free-energy theorem (33). Of course, (33) is seen to be satisfied at once since $Y^{sc} = Y^{se}$, and Y^{se} satisfies (33) from the statement [3] above. But what is most significant is that the self-consistent solution is the only approximate solution which satisfies each of (25), (26), (30), (29), (32) as well as (33). The other solutions which satisfy (33) need not satisfy the others.

As the detailed calculations involved in the above are somewhat subtle, we shall merely outline their derivations here. In fact, since the details involve path-integral calculations while the basic features are more general, emphasis is best placed on the latter. Further details can be found in Ref. 19.

Two types of arguments enter into our discussion. On the one hand is the fluctuation-dissipation theorem satisfied for each frequency. On the other hand are other relations involving quantities integrated over frequency. For the latter, the starting point is the free-energy.

To derive the FD theorem, we first calculate to zero-order

$$2k_x(t) \equiv \langle x_t x_0 + x_t' x_0' \rangle , \quad t > 0$$

The first term follows from

$$- \frac{\partial}{\partial \lambda_1} \frac{\partial}{\partial \lambda_2} A_0(\lambda_1 \delta(\tau - t) + \lambda_2 \delta(\tau), 0) \Big|_{\lambda_1 = \lambda_2 = 0}$$

while the second follows from

$$+ \frac{\partial}{\partial \lambda_1} \frac{\partial}{\partial \lambda_2} A_0(\lambda_1 \delta(\tau - t), \ \lambda_2 \delta(\tau)) \Big|_{\lambda_1 = \lambda_2 = 0}$$

where A_0 is defined in (14). Thus using (15e)

$$\begin{aligned} 2k_x(t) &= -(L_t^* + L_t) \\ &= \int_{-\infty}^{\infty} \frac{dv}{2\pi i} \left(\frac{1}{Z_v^a} - \frac{1}{Z_v^{a*}} \right) \frac{1 + e^{-\beta v}}{1 - e^{-\beta v}} \, e^{-ivt} \end{aligned}$$

Multiplying by $e^{i\omega t}$, integrating over t as in (25b), and taking the real part, it follows at once that

$$\mathrm{Re}K_x(\omega) = \mathrm{Im}Y^a_\omega \frac{1}{2}\frac{1+e^{-\beta\omega}}{1-e^{-\beta\omega}} \tag{43}$$

where we have let $\hbar=1$. A similar derivation can be given for $\hat{r}=\dot{x}=\hat{p}_x/m=\hat{j}_x/e$, which is actually what is desired. Alternatively, one may note that

$$K_{\dot{x}}(\omega) = +\omega^2 K_x(\omega),\ X_{\dot{x}}(\omega) = +\omega^2 X_x(\omega) \tag{44}$$

Equations (43) and (44) imply that (25a), and therefore that (26), are valid whenever (23) is satisfied.

Turning now to the expressions involving free energy, (29), (30) and (33), suffice it to note that the free energy can be written as a function of the mass m and the coupling constant α and as a functional of $\bar{Z}^a(\omega_\ell)$ where $\bar{Z}^a(-i(v+i\varepsilon)) \equiv Z^a_v$ with $\omega_\ell = 2\pi\ell/\beta$, $\ell=1,2,\ldots$, and of $\Sigma(\eta)$ which is in term a functional of $\bar{Z}^a(\omega_\ell)$:

$$F \equiv F(m,\alpha,\bar{Z}^a,\Sigma) \tag{45}$$

The essence of our derivation is to note that the impedance Z^a_v (15c) depends on the parameters ξ_i which characterize the trial oscillator distribution $G_a(\Omega)$, and that the parameters ξ_i take values ξ^o_i determined for each i by

$$\left.\frac{\partial F}{\partial \bar{Z}^a}\frac{\partial \bar{Z}^a}{\partial \xi_i}\right|_{\xi^o_i} = 0 \tag{46}$$

in order to minimize the free energy. Thus

$$\frac{dF}{d\eta} = \frac{\partial F}{\partial \eta} + \frac{\partial F}{\partial \bar{Z}^a}\frac{\partial \bar{Z}^a}{\partial \eta} + \left.\sum_i \frac{\partial F}{\partial \bar{Z}^a}\frac{\partial \bar{Z}^a}{\partial \xi_i}\frac{d\xi_i}{d\eta}\right|_{\xi_i=\xi^o_i}$$

$$= \frac{\partial F}{\partial \eta} + \frac{\partial F}{\partial \bar{Z}^a}\frac{\partial \bar{Z}^a}{\partial \eta} \tag{47}$$

where the second equality follows from (46). Thus, to calculate $dF/d\eta$ one need only differentiate F with respect to its <u>explicit</u> dependence on η.

Using (47) it is straightforward to obtain (29) ($\eta=\alpha$), and (33) ($\eta=\lambda$) if in the latter the absorption is calculated from the self-energy admittance. For the self-consistent impedance Z^{sc}, (30) is also obtained. However, (30) need not be satisfied in order for (33) to be derived from (47) for $\eta=\lambda$. For the interested reader, further details can be found in Ref. 19.

Perhaps the most interesting result of the above was the fact that any Feynman-approximation solution exactly satisfies the free-energy theorem (33) when one derives the absorption from the self-energy admittance. Since we noted how sensitive the self-energy admittance is to the choice of the form of the oscillator distribution used to simulate the actual electron-lattice interaction, we might expect that the sum rule (33) may not be very sensitive to the calculated absorption. This expectation is borne out when we examine the numerical results reported by LdSD.

As a test of the breadth of the Feynman one-oscillator model[5] for calculating the ground-state energy (F for T=0) and the FHIP derivation of the impedance using the same one-oscillator model, LdSD compared numerical calculations of the left- and right-hand sides of (33). The absorption[11] was calculated based on the Feynman[8] impedance Z^F. To the accuracy of the calculations, the two sides were found to be equal.

On the basis of the results of the preceding section we would not have expected such accuracy, since we found that (33) was exactly satisfied if the absorption was calculated from the self-energy impedance. What the numerical results of LdSD tell us is that to numerical accuracy

$$\int_0^\infty \frac{d\omega}{2\pi} \, \mathrm{Im}\chi_{jj}(\omega)\Big|_{\text{self-energy}} = \int_0^\infty \frac{d\omega}{2\pi} \, \mathrm{Im}\chi_{jj}(\omega)\Big|_{\text{FHIP}} . \tag{48}$$

(We have been unable to determine whether this expression is exact or not exact.) But irrespective of Eq. (48),

$$\mathrm{Im}\chi_{jj}(\omega)\Big|_{\text{self-energy}} \neq \mathrm{Im}\chi_{jj}(\omega)\Big|_{\text{FHIP}} \tag{49}$$

for nearly all values of ω. The most striking departure can be seen as follows. From Eq. (41) for the self-energy admittance we note that some of the resonance structure in the absorption will be due to zeroes of Z^a_v. For the Feynman[5] one-oscillator model

$$Z^a_\omega = -\omega^2 \frac{v^2-\omega^2}{\omega^2-\omega^2} , \tag{50}$$

which has a zero at ω=v. Thus we expect a sharp piece of structure in the absorption at ω=v for all α. The absorption as derived from the FHIP impedance shows quite different structure.[11] (See in particular Ref. 11, Figs. 1-5.) For $\alpha \leq 5$ there is no sharp piece of structure; for $\alpha \geq 6$ the sharp structure, the relaxed excited state (RES), is shifted to about ω=v-1. Nonetheless, to numerical accuracy, the quantities which enter the sum rule (48) are equal.

The above represents an interesting dilemma. On the one hand one expects the FHIP impedance to be superior to the self-energy impedance. Again this is because the former is far less sensitive to the details of the form of the oscillator distribution $G_a(\Omega)$ and is exact in the case where the electron does indeed interact linearly with the oscillators of the boson field. On the other hand, the absorption derived from the self-energy approach satisfies the ground-state-theorem sum rule exactly. Thus no matter how closely this sum rule is satisfied numerically by the absorption calculated by Devreese et al.,[11] if satisfying the sum rule is to be used as a criterion for the accuracy of the absorption, the above results would favor the self-energy form of the absorption over the Devreese[11] form. Since physically this is clearly not the case,[8,10,11] we can only conclude that, however adequate the ground-state theorem may be for testing the internal consistency of an approximation which is being used to calculate energy and/or integrated absorption, it is of little use for testing the accuracy of the frequency dependence of the absorption. If we desire a criterion for assessing the accuracy of the absorption, frequency by frequency, it would seem more appropriate to make use of a scheme which compares two functions of frequency, rather than two integrals over such functions. The fact that the self-consistent solution renders the self-energy and FHIP solutions equal, hence resolving the above difficulty, suggests that more attention might be paid to this solution. For the present let us also note that if one derives the absorption from $3/2\ Y_v^a$ (rather than from Y_v^a, as one might be tempted to try) for the Feynman one-oscillator model at zero temperature, one satisfies (33) to order α for $\alpha \ll 1$, and does very well for $\alpha < 7$. The author is unaware of the significance of this result.

As a final example, consider the Thomas-Reich-Kuhn sum rule discussed by Pines.[25] It can be stated in the following form. If N is the number of electrons present,

$$i\left\langle\left[\dot{\hat{\rho}}_{\underset{\sim}{k}}(t),\hat{\rho}^{\dagger}_{\underset{\sim}{k}}(0)\right]\right\rangle/N\Big|_{t=0^+} = k^2/m \tag{51}$$

where $\hat{\rho}_{\underset{\sim}{k}}$ is the $\underset{\sim}{k}$ component of the electron-density operator.

$[\hat{\rho}_{\underset{\sim}{k}} \equiv \Sigma_j \exp(-i\underset{\sim}{k}\cdot\underset{\sim}{r}_j)]$ This sum rule, the left-hand side being equivalent to an integral over frequency (energy), is satisfied by the exact solution for every $\underset{\sim}{k}$. Despite this nontrivial feature, (51) is nonetheless satisfied by any approximate solution generated by (10) ($\underset{\sim}{v}$=0) no matter what one chooses for $G(\tau)$; (51) is satisfied for any oscillatory distribution whatever. That this is so amounts to nothing more than noting that such solutions satisfy the Heisenberg uncertainty principle. [Pines[25] discusses the interacting electron gas. Putting such a gas in a dissipative media does not alter (51). Our above statement applies whether or not our approximation includes electron-electron interactions. Pines[25] seems to imply that such sum rules are useful. Our results indicate the contrary.]

V. THE SELF-CONSISTENT APPROACH REVISITED

In Section III a dynamical self-consistent procedure was formulated for calculating the impedance of an electron interacting with an arbitrary boson field at arbitrary temperature. In the absence of external fields, the self-consistent solution minimizes the free energy for the lattice at the temperature of interest. The method is capable of including even the fine structure of individual phonon emission and absorption.

In the context of this discussion, a potentially very powerful test of any absorption spectrum is whether it satisfies the FD theorem for each frequency. Of course, as we have seen, if we use a harmonic-oscillator-distribution influence functional to calculate both the kinetic energy and the (Y_v^a) admittance, then we shall satisfy this theorem, so long as the oscillator distribution satisfies (23). If, on the other hand, we use for the admittance the reciprocal of the conservation-of-momentum impedance, then the only impedance (or, equivalently, the only absorption spectrum, or the only oscillator distribution influence functional) which will satisfy the FD theorem at every frequency is the dynamical self-consistent impedance. Thus in the absence of applied magnetic and electric fields, the self-consistent approach may be viewed as providing a solution which satisfies a frequency-by-frequency criterion which must be satisfied by the exact solution. This alternative method clearly provides a more sensitive procedure with which to judge the accuracy of calculated absorption spectra. By requiring equality between two functions for each value of their argument, the method easily distinguishes between spectra, each of which may satisfy the free-energy theorem very closely.

VI. CONCLUSION

In this discussion we have focused on the spectral structures generated by approximate influence functionals similar to those used in calculating linear and nonlinear transport properties of electrons in dissipative media. We concentrated primarily on self-consistent solutions for zero applied fields, although this method has been developed for finite fields as well.[10] We then discussed six ways of calculating impedance in the context of equilibrium sum rules. We found that while any Feynman-approximation solution satisfies the free-energy theorem, our self-consistent impedance is the <u>only</u> Feynman-approximation solution which satisfies our alternative criterion. The sensitivity of this criterion results from the fact that one is comparing the value of two functions for each frequency rather than merely the value of two numbers as with sum rules.

REFERENCES

1. R. P. Feynman, Phys. Rev. 84, 108 (1951).
2. R. P. Feynman, Rev. Mod. Phys. 20, 367 (1948).
3. R. P. Feynman and A. R. Hibbs, Quantum Mechanics and Path Integrals, New York: McGraw-Hill, 1965.
4. R. P. Feynman, Statistical Mechanics, A Set of Lectures, Reading, Mass.: W. A. Benjamin, 1972.
5. R. P. Feynman, Phys. Rev. 97, 660 (1955).
6. P. M. Platzman, Phys. Rev. 125, 1961 (1962).
7. R. W. Hellwarth and P. M. Platzman, Phys. Rev. 128, 1599 (1962).
8. R. P. Feynman, R. W. Hellwarth, C. K. Iddings, and P. M. Platzman, Phys. Rev. 127, 1004 (1962), hereafter referred to as FHIP.
9. P. M. Platzman, in Polarons and Excitons, C. G. Kuper and G. D. Whitfield, eds, New York: Plenum, 1963.
10. K. K. Thornber, Phys. Rev. B3, 1929 (1971); B4, 675E (1971).
11. J. T. Devreese, J. deSitter, and M. Goovarts, Phys. Rev. B5, 2367 (1972).
12. J. T. Devreese, in Polarons in Ionic Crystals and Polar Semiconductors, J. T. Devreese, ed., Amsterdam: North-Holland, 1972.
13. K. K. Thornber, Ph.D. Thesis, California Institute of Technology, 1966 (unpublished).
14. K. K. Thornber and R. P. Feynman, Phys. Rev. B1, 4099 (1970); B4, 674E (1971).
15. K. K. Thornber, in Ref. 12.
16. M. Fröhlich, Advan. Phys. 3, 325 (1954).
17. L. F. Lemmens and J. T. Devreese, Solid State Commun. 12, 1067 (1973).

18. L. F. Lemmens, J. deSitter, and J. T. Devreese, Phys. Rev. B8, 2717 (1973), hereafter referred to as LdSD.
19. K. K. Thornber, Phys. Rev. B9, 3489 (1974). Note that the second integral in (C14) should be preceded by a plus sign (+) and not by a minus (-) sign as given.
20. L. van Hove, Physica 21, 517 (1955).
21. H. B. Callen and T. A. Welton, Phys. Rev. 83, 34 (1951).
22. K. K. Thornber, B.S.T.J. 53, 1041 (1974).
23. J. Devreese and R. Evrard, Phys. Lett. 11, 278 (1966).
24. J. T. Marshall and L. R. Mills, Phys. Rev. B2, 3143 (1970).
25. D. Pines, Elementary Excitations in Solids, New York: Benjamin (1964) pp. 130-136.

[illegible]

ON THE ELECTRONIC TRANSPORT IN POLAR SOLIDS¶

J. Devreese* and R. Evrard**

*Institute for Applied Mathematics
Rijksuniversitair Centrum Antwerpen and
Departement Natuurkunde, Universitaire
Instelling Antwerpen, Wilrijk, Belgium
**Institut de Physique, Université de Liège,
Belgium

I. Introduction to polarons [1], [2]

An electron (or hole) moving through a polar crystal polarizes its neighbourhood (see fig.1). This polarization modifies the physical properties of the electron. The electron together with the self induced polarization has been considered as one entity and was called a polaron. The differences between the polaron and a band electron are seen via
1) a change in the groundstate energy of the system (see fig.2). The dashed lines in fig.2 show typical valence and conduction bands without lattice polarization. If the lattice polarization is taken into account the band gap decreases.
2) a change in the effective mass of the system : the self induced polarization contributes to the inertia of the polaron. Therefore the polaron mass is always larger than the band mass of the electron.
3) a typical behaviour of the mobility. The mobility of polarons, its temperature and field dependence are determined by electron-phonon interaction. This gives rise to drastic changes from a band picture. In I.2. it will be seen that a radius can be associated to the

¶ This work was performed in the framework of the project E.S.I.S. on electronic structure in solids (joint project between the University of Antwerpen and the University of Liège).

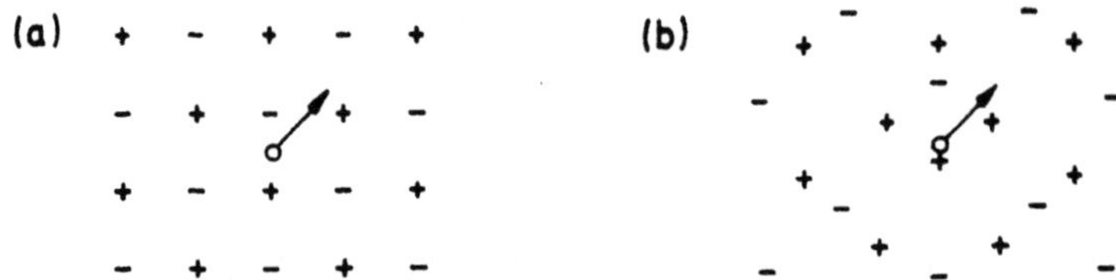

Fig.1. An electron moving through an ionic crystal: a) unpolarized medium; b) polarized medium.

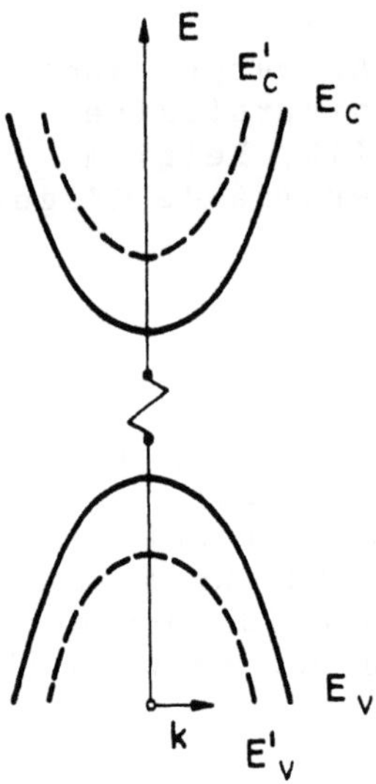

Fig.2. Valence and conduction band in a polar crystal with polaron effects (solid curves) and without polaron effects (dashed curves). The energy shifts due to the lattice polarization can range from less than a millivolt (InSb) to 0.1 eV (KBr).

polaron; it is a measure for the radius of a sphere which contains e.g. half the polarization energy induced by the electron.

I.1. Dynamic view

It is not sufficient to treat the selfinduced polarization as an electrostatic phenomenon. One should analyze the lattice vibrations as quanta, the phonons, and the interaction between the electron and the phonons has to be taken into account to describe the polaron. For non piezoelectric crystals it turns out that the dominant interaction is between the electron and the long wavelength longitudinal phonons. Indeed if the crystal is not piezoelectric the dipolar field induced by the acoustical modes is

negligible compared to that due to the longitudinal modes. (As the divergence of the transverse polarization is zero 'at least if the polaron radius is large enough').
The dominant self induced polarization is caused by the L.O. phonons. If the wavelength of the L.O. phonon is too small compared to the polaron radius the electrical fields due to the self induced dipoles will tend to average out leaving zero net effect on the phonon. Consequently the long wavelength L.O. phonons are responsible for the dominant polaron effects.

I.2. The polaron radius

Consider the L.O. phonon field with frequency ω_o interacting with an electron. Denote by Δv the quadratic mean square deviation of the electron velocity. If the electron-phonon interaction is weak, the electron can travel a distance

$$\Delta x \approx \frac{\Delta v}{\omega_o} \tag{1}$$

during a time ω_o^{-1}, characteristic for the lattice period.
Because it is the distance within which the electron can be localized using the phonon field as measuring device. From the uncertainty relations it follows

$$\Delta p \; \Delta x = \frac{m}{\omega_o} \Delta v \; \Delta v \approx \hbar \tag{2}$$

$$\Delta v \sim \sqrt{\frac{\hbar\omega_o}{m}}$$

$$\Delta x \sim \sqrt{\frac{\hbar}{m\omega_o}}$$

At weak coupling Δx is a measure of the polaron radius r_p. To be consistent the polaron radius r_p must be considerably larger than the lattice parameter a.

Indeed our previous arguments are based on a continuum approximation :

$$r_p \gg a \tag{3}$$

Experimental evaluation [3] of the polaron radius, the details of which are not discussed here, lead to the following typical values :

Alkali-halides : $r_p \approx 10$ Å
Silver-halides : $r_p \approx 20$ Å
II-VI, III-V semiconductors : $r_p \approx 100$ Å
The continuum approximation is not always very well satisfied in solids. This is the case for transition metal oxydes (NiO, CaO, MnO), in UO_2, NbO_2, etc... For those solids the "small polaron" concept has often been used.

In the perovskites it seems that some intermediate region between large and small polarons is realized.

I.3. Introduction of the coupling constant

To understand the structure of the coupling constant we follow an argument due to Fröhlich [1]. First consider the case of strong electron lattice interaction. The electron is then localized and can to a first approximation be considered as a static charge distribution with radius ℓ_1 (see fig.3).

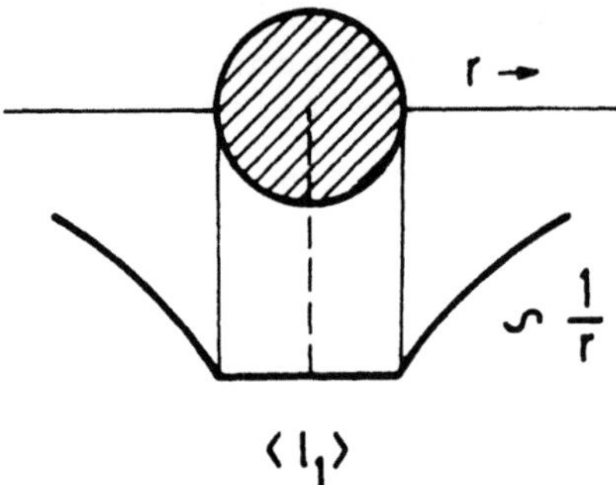

Fig.3 : Strong coupling model for the potential seen by the electron in a polaron.

The medium is now characterized by an average dielectric constant $\bar{\varepsilon}$. The binding energy U_1 of the electron is easily obtained by minimizing the total energy with respect to ℓ_1

$$U_1 = - \frac{m e^4}{8 \hbar^2 \bar{\varepsilon}^2 \pi^2} \tag{4}$$

For weak coupling one can neglect the potential energy of the electron. Taking the polaron radius from (2) $r_p = \frac{\hbar}{m\omega_o}$ the binding energy U_2 is

$$U_2 = -\frac{e^2}{\bar{\varepsilon}\, r_p} = \frac{e^2}{\bar{\varepsilon}} \cdot \sqrt{\frac{m\omega_o}{2\pi\hbar}} \tag{5}$$

Note that

$$-\frac{U_1}{\hbar\omega_o} = \frac{1}{8\pi^2} \frac{m\, e^4}{\bar{\varepsilon}^2\, \hbar^3\, \omega_o} = -\frac{1}{4\pi} \left(\frac{U_2}{\hbar\omega_o}\right)^2 \tag{6}$$

It is convenient, following the conventions of field theory, to write the self energy at weak coupling as $U_2 = -\alpha\hbar\omega_o$. Therefore the so called Fröhlich polaron coupling constant is

$$\alpha = \frac{1}{\sqrt{2}} \frac{e^2}{\bar{\varepsilon}} \sqrt{\frac{m}{\hbar^3 \omega_o}} \equiv \frac{e^2}{\hbar c} \sqrt{\frac{m\, c^2}{2\, \hbar\, \omega_o}}\; \frac{1}{\bar{\varepsilon}} \tag{7}$$

From experimental studies, involving cyclotron resonance and mobility measurements, one infers that

for alkali halides $\alpha \approx 3$

for II-VI semiconductors $\alpha \approx 10^{-2}$

for III-V semiconductors $\alpha \approx 10^{-1}$

A few examples of experimentally determined coupling constants are now given (for a more detailed list see ref.[4]).

KBr	3.05
KCl	3.44
KI	2.50
RbCl	3.81
RbI	3.16
CsI	3.67
GaAs	0.068
GaP	0.2
CdS	0.527

For the "average" dielectric constant $\bar{\varepsilon}$ one shows that

$$\frac{1}{\bar{\varepsilon}} = \frac{1}{\varepsilon_\infty} - \frac{1}{\varepsilon_o}$$

where ε_∞ is the electronic dielectric constant

ε_o is the static dielectric constant of the solid.

The difference $\frac{1}{\varepsilon_\infty} - \frac{1}{\varepsilon_o}$ arises because the ionic vibrations occur in the infrared and the electrons in the shells can follow the conduction electron adiabatically. In determining the coupling constant α it is necessary to perform two independent measurements because m, the band mass, enters the theory. One has not been able to measure the band mass directly. Such an experiment would require a "quenching" of the polarization. To avoid this difficulty one considers the theoretical expression for m^* the polaron mass and μ the polaron mobility as functions of α and m. The measurement of μ [1], [2] at a certain temperature provides a theoretical estimate for m; from this m together with ε_∞, ε_o, ω_o one then calculates α and the polaron mass m^*. This polaron mass m^* is then compared with a cyclotron resonance experiment.

I.4. Summary of some polaron properties

In this paragraph we list some physical properties of polarons. The groundstate energy of polarons is given by the following expressions :

$$\frac{\Delta E}{\hbar\omega_o} = -\alpha - 0.01592\,\alpha^2 \qquad \alpha \to 0$$

$$\frac{\Delta E}{\hbar\omega_o} = -.106\,\alpha^2 - 2.83 \qquad \alpha \to \infty \tag{9}$$

The effective mass is

$$\frac{m^*}{m} = 1 + \frac{\alpha}{6} \qquad \alpha \to 0$$

$$\frac{m^*}{m} \doteq \frac{16\,\alpha^4}{81\,\pi^2} \qquad \alpha \to \infty \tag{10}$$

Apart from the static properties, like groundstate energy and effective mass, the dynamical response properties are of great interest.
The mobility will be the subject of the next paragraphs. Optical properties have also been studied extensively. For weak coupling the dominant scattering process in absorption is given in fig.4.

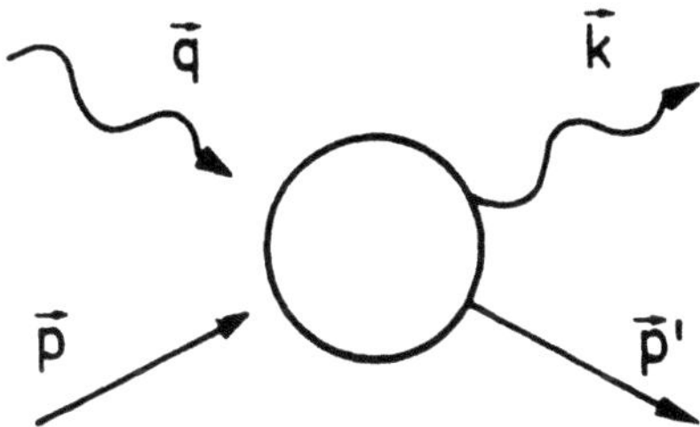

Fig.4 : Dominant weak coupling contribution to polaron mobility. An electron with momentum $\vec{p}$ interacts with a phonon of momentum $\vec{q}$. In the final state the electron has momentum $\vec{p}'$ the phonon has momentum $\vec{k}$.

This leads to an optical absorption coefficient

$$\Gamma(\Omega) = \frac{2 e^2 \omega_o^2}{3 \varepsilon_o \, c n \, m \, \Omega^3} \, \alpha \sqrt{\frac{\Omega}{\omega_o} - 1} \qquad (11)$$

Ω is the frequency of the incident radiation
ε_o is the permittivity of free space
n^o is the refractive index of the medium
c is the velocity of light

Corrections to (11) have been studied in ref.[5]. In fig.5 the optical absorption coefficient for CdO is shown and compared with theory, following ref.[6]. The main contribution for $\Omega > \omega_o$ ($\lambda < 20$ μm) is due to the polaron corrections eq.(11). This comparison provides good evidence for the occurence of polaron absorption.

For stronger coupling the possibility of a relaxed internal excited state of the polaron has to be considered. The idea [7] is that the most stable excited state of the polaron occurs in a new potential (fig.6) because the lattice readapts to the changed electronic distribution. For $\alpha = 5$ the resulting absorption [5] is shown in fig.7.

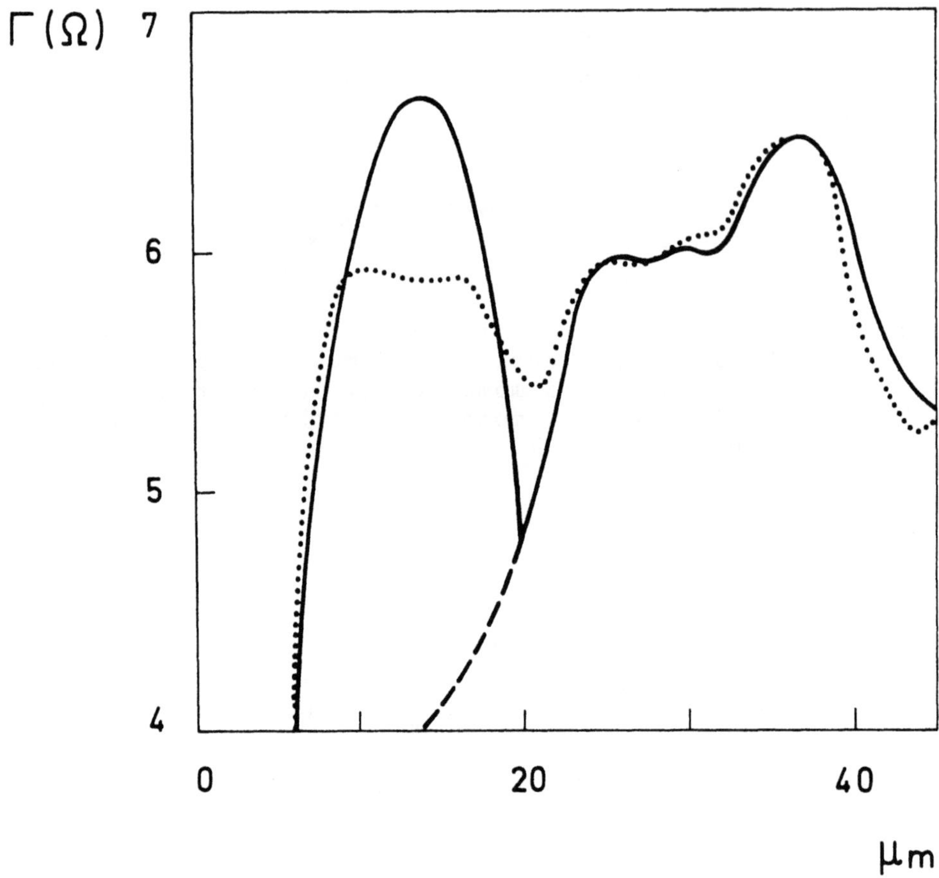

Fig.5 : Infrared optical absorption in CdO as measured by Finkenrath and Uhle and Waidelich (ref.[6]). The absorption for λ < 20 μm is predominantly due to polaron effects.

Some evidence exists [8] that infrared optical absorption in TiO_2, $SrTiO_3$ is associated with transitions to a R.E.S. polaron state, although the continuum approximation does not seem to provide accurate description.

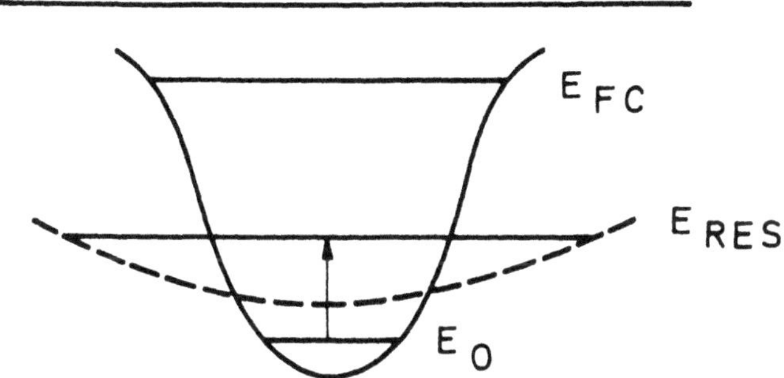

Fig.6 : Scheme of strong coupling potentials for polarons. In the self induced potential belonging to the groundstate, the polaron can be in the groundstate (E_o), and in Franck-Condon type excited states (e.g. E_{FC}). In the final state the lattice can also relax (dashed parabola) and the polaron goes to the relaxed excited state (RES).

I.5. The Fröhlich Hamiltonian and Lagrangian for polaron

A complete derivation of Hamiltonian and Lagrangian is rather lengthy.
For the sake of convenience we will follow here a simple derivation of the structure of the Lagrangian given by Feynman [9] . One wants to find the force felt by an electron and produced by a lattice deformation of wave number k. The medium is treated as a continuum. The polarization $\vec{P}$ is written as a longitudinal wave

$$\vec{P} = \frac{\vec{k}}{k} a_k e^{i\vec{k}.\vec{r}} \tag{12}$$

the charge density then is

$$\rho = \vec{\nabla}.\vec{P} = k a_k e^{i\vec{k}.\vec{r}} = \Delta V \tag{13}$$

where V is the potential.
If now q_k is the amplitude of the longitudinal running wave with wave number k then the <u>polarization a_k</u> is <u>proportional to q_k</u>.
The interaction between the phonon (polarization wave) and the electron is then proportional to the sum over all $\vec{k}$ values of the contributions $\frac{q_k}{k} e^{i\vec{k}.\vec{r}}$.

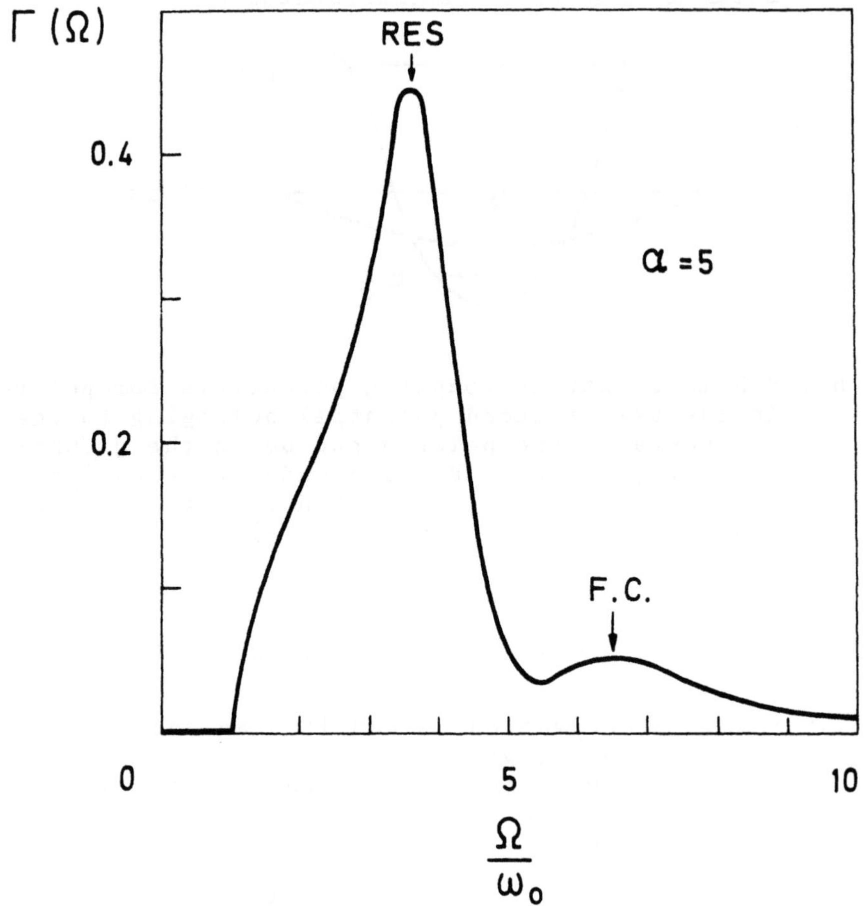

Fig.7 : Optical absorption by free polarons for $\alpha \approx 5$ as obtained in ref.[5].

Although a more detailed argument is needed to obtain the factor in front of the interaction term, it can also be obtained by imposing $E_0 = -\alpha\hbar\omega_0$ and using α as defined by (6).
The Lagrangian of the interacting system consisting of the electron and the phonon field then becomes :

$$L = \frac{m}{2} \dot{\vec{r}}^2 + \sum_{\vec{k}} \frac{1}{2} (\dot{q}_k^2 - \omega_o^2 q_k^2) + e \sqrt{\frac{4\pi}{V}} \omega_o \left(\frac{1}{\varepsilon_\infty} - \frac{1}{\varepsilon_o}\right) \sum_{\vec{k}} \frac{q_k}{k} e^{i\vec{k}.\vec{r}} \quad (14)$$

The first term is the kinetic energy of the electron, the second is the energy of the free phonon field, the third is the electron-phonon interaction as derived above. Introducing standing waves

$$q_k = -i \left(\frac{\hbar}{\omega_o}\right)^{1/2} \frac{a_{\vec{k}} - a_{\vec{k}}^+}{\sqrt{2}}$$
$$\pi_k = -(\hbar\omega_o)^{1/2} \frac{a_{-\vec{k}} + a_{\vec{k}}^+}{\sqrt{2}} \quad (15)$$

one obtains the celebrated Fröhlich polaron Hamiltonian

$$H = \frac{p^2}{2m} + \sum_k \hbar\omega_o a_{\vec{k}}^+ a_{\vec{k}} + \sum_k (V_k a_{\vec{k}} e^{i\vec{k}.\vec{r}} + V_k^* a_{\vec{k}}^+ e^{-i\vec{k}.\vec{r}})$$
$$V_k = \frac{ie}{k} \left(\frac{2\pi}{V}\right)^{1/2} (\hbar\omega_o)^{1/2} \left(\frac{1}{\varepsilon_\infty} - \frac{1}{\varepsilon_o}\right)^{1/2} \quad (16)$$

The Hamiltonian (16) and the Lagrangian (14) have been used extensively in theoretical work relating to polarons.

II. Polaron transport theory, with emphasis on the path integral approach.

II.1. Mobility with linear response and using path integrals.

As the subject of the present school is on linear and nonlinear electronic transport in solids we will now digress on the transport properties of polarons. Because much of the transport theory presented at the present school is formulated using the path integral formalism (lectures of Prof.Papadopoulos [10] and Dr. Thornber) and because two introductory lectures have been provided on path integrals (by Prof.Papadopoulos [12] and by one of us (JD) [13]) we will concentrate on the physics of polaron transport as resulting from path integral work relating to the Feynman model for polarons. The present remarks will be rather introductory in nature and also overlap with Dr. Thornber will be avoided. The linear response properties in the Feynman model have been studied by Feynman, Hellwarth, Iddings, Platzman(FHIP)[14]. Their trial action describes the polaron as a system of 2 particles (electron and a particle with mass described by an action S_o which "simulates" the phonon field). The variational principle for path integrals involves S the expectation value of the rigorous action (via a path integral) and therefore the physics of the problem as contained in the Lagrangian (14) is described much more completely than could be expected from the simple model described by S_o only. It would be quite misleading to think of the Feynman model for the polaron as being decribed by S_o itself. E.g. if S_o is taken without an interaction term (just the kinetic energy of the electron) one obtains $E_o = -\alpha\hbar\omega_o$ for the selfenergy from the variational principle. This is because S, the expectation value of the true action, which contains also the interaction, occurs in the application of the variational principle. The problem faced by FHIP is now to describe the response of polarons as accurately as the groundstate properties. In an introductory lecture on path integrals[13] was shown how the mobility can be expressed in terms of the path integral. In the case of linear response, the formulation given in FHIP is general of course the action Φ of eq.(51) in ref [13] cannot be handled exactly.).

Two crucial approximations are now made by FHIP :

1. The Coulomb-type terms in equation (50) ref.[13] are replaced by a quadratic approximation $\Phi \to \Phi_o$ e.g. for the third term in equation (51) there comes :

$$-iC \int_{-\infty}^{+\infty} dt \int_{-\infty}^{+\infty} ds\, [\vec{X}(t)-\vec{X}(s)]^2\, [e^{-iW(t-s)} + 2\, P(\beta\omega)\cos(t-s)] \qquad (17)$$

Although this is the same type of approximation as that used for the groundstate properties of the polaron, it is more difficult to justify this approximation for excited states which are important now.

2. $G(\Omega)$ is approximated as a geometric series : first g is developed as follows :

$$g = \iint e^{i\Phi}\mathcal{D}X\mathcal{D}X' = \iint e^{i\Phi_o}\mathcal{D}X\mathcal{D}X' + \iint e^{i\Phi_o}(\Phi-\Phi_o)\mathcal{D}X\mathcal{D}X \qquad (18)$$

It takes a considerable effort to calculate the two path integrals in the rhs of equation (18) and it

seems unrealistic to try actual evaluation of higher order terms.
The expansion [18] gives for $G(\Omega)$ (see eq.(49) of ref. [13])

$$G(\Omega) = G_o(\Omega) + G_1(\Omega) \tag{19}$$

However, this expression for $G(\Omega)$ does not provide any resonances and FHIP conjecture that the expression

$$\Omega Z(\Omega) \simeq \frac{1}{G_o(\Omega)} - \frac{G_1(\Omega)}{G_o^2(\Omega)} \tag{20}$$

is far more accurate and is the one to be used. This then amounts to taking a geometrical series for $G(\Omega)$. There is no real justification for it in the FHIP approximation except that for simplified cases (particle bound to a fixed point) equation (20) works. In ref.[15] it is shown that (20) also follows from a self consistent treatment of the polaron equations of motion. We omit any further details and write the results for the impedance with the FHIP approximation :

$$-i\Omega Z(\Omega) = \Omega^2 - \chi(\Omega) \tag{21}$$

with

$$\chi(\Omega) = \int_o^\infty [1 - e^{i\Omega u}] \text{ Im } S(u)\, du \tag{22}$$

and

$$S(u) = \frac{2\alpha}{3\sqrt{\pi}} \{[D(u)]^{-3/2} (e^{iu} + 2P(\beta) \cos u)\} \tag{23}$$

$$D(u) = \frac{w^2}{v^2} [\frac{v^2-w^2}{vw^2} (1-e^{ivu}+4P(\beta v)\sin^2(\tfrac{1}{2}vu)-iu+\frac{u^2}{\beta})] \tag{24}$$

The mobility obtained in this way is

$$\mu_{FHIP} = \frac{e}{2m\omega\alpha} \frac{3}{2} kT (\frac{w}{v})^3 \exp(\frac{v^2-w^2}{w^2 v}) \frac{e^{\hbar\omega_o/kT}}{\hbar\omega_o} \tag{25}$$

The standard weak coupling result for the polaron mobility is

$$\mu_o = \frac{e}{2m\omega\alpha} e^{\hbar\omega_o/kT} \tag{26}$$

(26) can be obtained using the Boltzmann equation.

There is a factor $\frac{3}{2}$ kT in the FHIP theory which seems absent in all others. Although in practice this factor does not seem to be of much importance it is certainly intriguing. It has been suggested that this factor 3/2 kT is due to the approximations inherent to the Boltzmann equation. However (26) can also be obtained from the Kubo formula [16] and this makes it more difficult to understand the problem.

It may be noted, in passing, that the weak coupling expansion for the mobility as obtained by Langreth [16] by diagrammatic techniques is

$$\mu = \mu_o \left(1 - \frac{\alpha}{6}\right)$$

For (25) one finds $\mu = \mu_o (1 - \alpha)$.
It has been shown [17] that the FHIP approximation satisfies exactly the sum rules for the first and second moment of the optical absorption. This, of course, is a very reassuring point as far as transport properties are concerned, because the sum rules inform us about static limits like mobility. See Dr.Thornber's lectures in this context. The FHIP approximation predicts a mobility which for large α, increases with α. This is contrary to the result which we obtain using ref.[15] at weak coupling. The result of FHIP, as pointed out by Platzman, is exactly the same as a rigorous solution of the Boltzmann equation if a nonzero frequency of the external field is considered.
We will not enter the discussion of other polaron mobility calculations. Relevant references are given in ref.[2]. It suffices to state that eq.(25) has been quite useful in comparing (linear response) polaron theory to experiment. Especially if the coupling constant is not longer very small (alkali halides, silver halides)(25) is appropriate as it is based on the Feynman model for polarons which is rather accurate for all α.

II.2. Nonlinear transport by polarons

Thornber and Feynman [18], [19] have provided a powerful treatment of polaron mobility devised for all fields, couplings, and electrical field strengths. This treatment, which again is in the framework of path integrals, is not limited to the Boltzmann approximation. It is in a sense the extension of FHIP to nonlinear response. This work will be further discussed at this school by Dr. K.Thornber [11].

III. Momentum distribution for polarons

III.1. Introduction

Most papers dealing with transport properties of polarons give direct approaches to the electrical mobility. To our knowledge, this is the first attempt to obtain the momentum distribution of polarons as a function of the temperature and electric field.

The reason of our interest in the distribution function is twofold. First, it gives detailed informations on the range (in temperature and electric strength) of validity of the approximations made in the calculations of the polaron mobility. Second, physicists working on electronic transport properties of polar semiconductors are interested in the effects of the interaction between the charge carriers and the polar lattice modes on the electron distribution. Most often they developed their own approach to this problem, independently from polaron theory and, on the other hand, there has been little effort from the specialists of polaron theory to provide the semiconductor physicists with predictions susceptible of application to their problems. It is hoped that the present lecture will help in bringing these two points of view closer to each other.

Regarding this latter aspect, let us immediately point out that in practice, scattering by longitudinal optical phonons of large wavelength is not the only contribution to the resistivity of polar semiconductors. Often, it even does not play the predominant role. Scattering by ionized impurities or acoustic phonons are important as well. Therefore, from the practical point of view the polaron must be considered as an ideal model which can serve as a guide to understand the effects of the polar coupling with the lattice vibrations, except probably in strongly ionic crystals (like the silver-halides, the alkali halides, etc...) or in the compounds where the concept of small polaron is required to explain the electrical properties.

III.2. Boltzmann equation for polarons at weak interaction strength

The polaron momentum distribution depends on three different parameters : the coupling constant α, the temperature T and the electric field E. In what follows only the case of weak coupling will be considered, so that perturbation theory can be used. More precisely, we will here adopt the point of view of semiconductor physicists, i.e. suppose that the electron behaves freely except during the time of a collision which produces transitions between these free electron states. Therefore, the Born expression for the transition probability (or, if you prefer, for the differential cross-section) is used throughout these lectures.

This means that the change in the effective mass with the polaron velocity is neglected. For weak coupling strengths, like in the III-V or II-VI compounds, it can be hoped that the polaron effective mass does not differ too much from the electron band mass and, therefore, that the approximation made here is valid. However, it is not clear that this is still true when the polaron kinetic energy becomes of the order of the phonon energy $\hbar\omega$. However, the conclusions of the works devoted to the study of this type of polaron-phonon resonance do not seem sufficiently well established to take this effect into account in the present discussion.

With this approximation, Boltzmann's equation becomes

$$\frac{\partial f(\vec{p})}{\partial p_z} eE = \int f(\vec{p}\,')\Pi(\vec{p}\,'\to\vec{p})d^3p' - f(\vec{p}) \int \Pi\,(\vec{p}\to\vec{p}\,')d^3p' \tag{27}$$

where $f(\vec{p})$ is the momentum distribution and $\Pi(\vec{p}\to\vec{p}\,')$ is the probability density for electron transitions from momentum $\vec{p}$ to momentum $\vec{p}\,'$; e denotes the electron charge, E the strength of the electric field and the z-direction in the p-space is taken along the electric force.

At first order of perturbation, two processes contribute to the transition probability. The first one is absorption of a phonon (cf. fig.8a) and the second one is emission of a phonon. Obviously, the absorption of a phonon is possible only at non zero temperature and the emission only when the electron kinetic energy is larger that the phonon energy. One easily obtains (20)

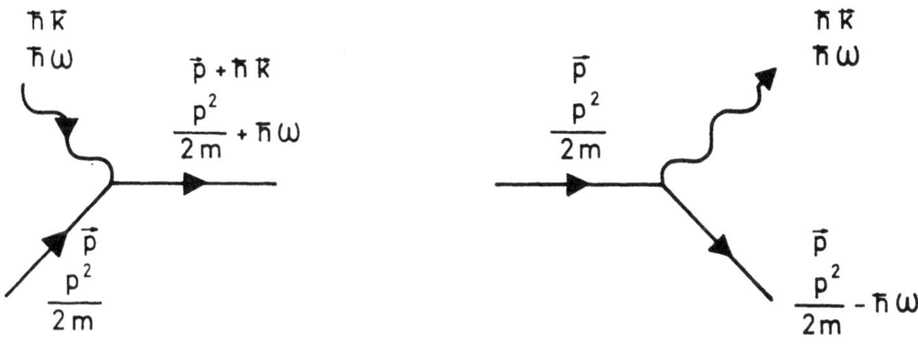

Fig.8a Fig.8b

$$\Pi_e(\vec{p}\rightarrow\vec{p}\,') = \frac{\alpha}{\pi\sqrt{2}} \frac{\delta(p^2/2 - 1 - p'^2/2)}{|\vec{p} - \vec{p}\,'|^2} \tag{28}$$

$$\Pi_e(\vec{p}) = \int \Pi_e(\vec{p}\rightarrow\vec{p}\,')d^3p'$$

$$= 2\ \alpha\ \frac{\sqrt{2}}{p} \cosh^{-1}\left(\frac{p}{\sqrt{2}}\right) \tag{29}$$

$$\Pi_a(\vec{p}\rightarrow\vec{p}\,') = \frac{\alpha}{\pi\sqrt{2}}\ \overline{N}\ \frac{\delta(p'^2/2 - p^2/2 - 1)}{|\vec{p} - \vec{p}\,'|^2} \tag{30}$$

$$\Pi_a(\vec{p}) = \int \Pi_a(\vec{p}\rightarrow\vec{p}\,')d^3p'$$

$$= 2\ \alpha\ \overline{N}\ \frac{\sqrt{2}}{p}\ sh^{-1}\left(\frac{p}{\sqrt{2}}\right) \tag{31}$$

From now on, units such that $\hbar = m = \omega = 1$ will be used. In these relations the subscripts e and a denote emission and absorption respectively, and $\overline{N}$ is the average number of thermal phonons

$$\overline{N} = (e^{\beta\hbar\omega} - 1)^{-1}$$

with

$$\beta = 1/kT$$

where k is Boltzmann's constant.

If one considers states of the total system, electron plus phonons, rather than states of the electron alone, the transition occurs between states containing one phonon and electron with energy $p^2/2m$ and states containing no phonon and electron with energy $\hbar\omega + p^2/2m$. The transition probabilities between these states with absorption and with emission of the phonon are equal, as predicted by the principle of detailed balance. This transition probability is

$$\Pi = \frac{\alpha}{\pi\sqrt{2}} |\vec{p} - \vec{p}'|^{-2}$$

It is interesting to note that the detailed balance principle does not apply to states of the electron alone, (see expressions of Π_e (29) and Π_a (31)) but to states of the total system.

Notice that absorption of a phonon increases the electron energy by $\hbar\omega$. Emission lowers the energy by the same amount. This leads to divide the p-space in spheres of radii such that $p^2/2m = n\hbar\omega$ where n is a positive integer. The intersection by a plane containing p_z gives circles as shown on fig.9 where $p_\perp$ represents the component of $\vec{p}$ along the direction perpendicular to the electric field. With our units the radii of these concentric circles are $\sqrt{2n}$, i.e. $\sqrt{2}, 2, \sqrt{6}, 2\sqrt{2}, \ldots$

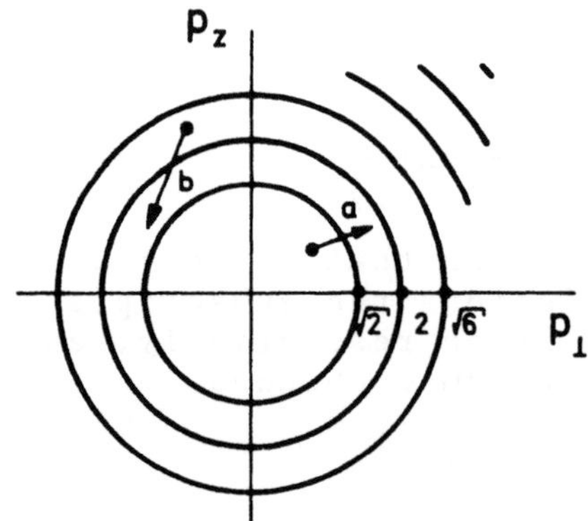

Fig.9 : a) absorption process
b) emission process

In conclusion for weak coupling strengths, transitions occur from a circular ring to an adjacent one.

III.3. The equilibrium distribution function

First, consider the situation at zero applied electric field and "low" temperatures ($kT < \hbar\omega$). This condition of low temperature does not constitute a strong restriction since $kT = \hbar\omega$ corresponds in most crystals to at least 300° K.

The longitudinal optical phonons do not allow the electrons to relax their energy. Indeed, when electrons initially inside the first circle (called here the initial circle and corresponding to $p = \sqrt{2}$) absorb a phonon and are brought into the region beyond this circle, they eventually fall back inside the first region in emitting a phonon. As the dispersion of the phonons is neglected in the Fröhlich model of polarons, the electrons gain and subsequently lose the same energy $\hbar\omega$. Therefore the collisions with the longitudinal optical phonons modify the direction of the momentum $\vec{p}$, but not the kinetic energy $p^2/2m$: the electrons on a circle of radius $p < \sqrt{2}$ remain on this circle. Therefore, the shape of the distribution inside the critical circle is not determined by the interactions between the electrons and the longitudinal optical phonons. A system of non interacting polarons is not ergodic since the polarons cannot exchange energy when their kinetic energy is lower than $\hbar\omega$.

On the contrary, the distribution outside the critical circle is governed by the collisions with longitudinal optical phonons. In a sense, it is the image of the distribution inside the circle, image formed by the transitions with absorption of phonons.

Therefore, the population outside the circle is proportional to $\bar{N}$, the number of thermal phonons which is small since it is assumed that $kT < \hbar\omega$. Thus the population at energies higher than $2\hbar\omega$ is of the order of $\bar{N}^2$ and can be neglected. With this approximation, the Boltzmann equation for the region outside the critical circle under equilibrium conditions is

$$\int f_1(\vec{p}\,')\ \Pi_a(\vec{p}\,' \rightarrow \vec{p})d^3p' - f_2(\vec{p})\ \Pi_e(\vec{p}) = 0 \quad (32)$$

where the subscripts 1 and 2 refer to inside, respectively outside, the first circle.

In the absence of electric field or other external perturbation, the distribution is isotropic. Integrating in spherical coordinates gives

$$f_2(\vec{p}) = \alpha\overline{N}\sqrt{2}\,\Pi_e^{-1}(\vec{p})\int_{\sqrt{2}}^{\infty} dp'p'^2 f_1(p')\delta\left(\frac{p'^2}{2} - \frac{p^2}{2} + 1\right)$$

$$\int_{-1}^{1} \frac{du}{p'^2 + p^2 - 2pp'u}$$

where u is the cosine of the angle between $\vec{p}$ and $\vec{p}'$. With the expression (29) of Π_e, one obtains :

$$f_2(\vec{p}) = \overline{N}\, f_1\,(\sqrt{p^2-2})$$

At low temperature

$$\overline{N} = e^{-\beta}$$

so that

$$f_2(\vec{p}) = f_1\,(\sqrt{p^2-2})\; e^{-\beta} \qquad (33)$$

If, due to effects not included in Fröhlich's model of polarons, the electrons are in thermal equilibrium with the lattice, i.e.

$$f_1(\vec{p}) = \exp\,(-\beta p^2/2)$$

then the solution at all energies is a Maxwell-Boltzmann distribution since

$$f_2(\vec{p}) = e^{-\beta}\; e^{-\beta(p^2/2-1)}$$

$$= e^{-\beta p^2/2}$$

The Maxwell-Boltzmann distribution remains a solution of Boltzmann's equation even at higher temperature, at least for thermal equilibrium with the lattice. This has been discussed by Platzman [21].

On the contrary, if the electrons are not in thermal equilibrium with the lattice, the distribution is no longer Maxwellian. For example, if the electrons are heated by some external perturbation up to temperatures above the lattice temperature, the distribution outside the circle is depressed by the transitions with emission of phonons.

Consider the case where the bottom of the conduction band is suddenly populated (e.g. by photo-ionization of donor centers) in a range of energies lower than kT, the evolution towards thermal equilibrium can take a relatively long time, if the energy relaxation time is long as in the case of InSb for instance. During this time, the transitions with absorption of phonons give extra wings to the distribution outside the critical circle, in accordance with (33).

The distribution inside the circle obeys the following Boltzmann equation

$$-f_1(\vec{p})\ \Pi_a(\vec{p}) + \int d^3p' f_2(\vec{p}')\ \Pi_e(\vec{p}'\to\vec{p}) = 0 \qquad (34)$$

It is straightforward to show that this equation is identical to (32) obtained for outside the circle.

III.4. Distribution in the ohmic regime at low temperature

First consider the region outside the critical circle. As pointed out previously, the absorption of phonons in this region is negligible compared to the emission of phonons. Indeed the former is proportional to $\overline{N}$ whereas the latter is practically temperature independent. The Boltzmann equation outside the circle is then (cf.equation (27)) :

$$\frac{\partial f_2(\vec{p})}{\partial p_z} eE = \int f_1(\vec{p}')\ \Pi_a(\vec{p}'\to\vec{p})d^3p' - f_2(\vec{p})\ \Pi_e(\vec{p})$$

All the terms in this expression behave like $\overline{N}$. At weak electric field (ohmic conditions) the left hand side is small compared to each term of the right-hand side. Therefore it can be taken equal to zero, so that the behaviour of the system outside the critical circle is governed by the same equation as in the absence of an external electric field :

$$\int f_1(\vec{p}')\ \Pi_a(\vec{p}'\to\vec{p})d^3p' - f_e(\vec{p})\ \Pi_e(\vec{p}) = 0 \qquad (35)$$

The physical reason for this approximation is the following : Consider an electron appearing in the region outside the critical circle (due to a transition from inside with absorption of a phonon). As the electric field is vanishingly weak, the drift it suffers before going back to the region inside the circle (with emission of a phonon) is negligible, so that the distribution is simply given by the balance (35) between the appearance and disappearance of electrons in the region under consideration.

In the ohmic regime, the change in the distribution due to the electric field is a small perturbation. Let us write

$$f(\vec{p}) = g(\vec{p}) + h(\vec{p})$$

where $g(\vec{p})$ is the unperturbed distribution and $h(\vec{p})$ the perturbation. At first order in the electric field, equation (35) gives

$$h_2(\vec{p}) = \Pi_e^{-1}(\vec{p}) \int h_1(\vec{p}\,') \, \Pi_a(\vec{p}\,' \to \vec{p}) d^3p' \qquad (36)$$

The field acting in the z-direction, the angular dependence of $h_1(p')$ is of the type

$$h_1(\vec{p}\,') = \ell_1(p') \cos\theta'$$

where p' is the modulus of the p'-vector and $\cos\theta'$ its angle with the z-axis. Therefore, the integral in the right-hand side of (36) is of the type

$$J = \int_0^\infty dp' p'^2 \ell_1(p') \delta\left(\frac{p'^2}{2} - \left(\frac{p^2}{2} - 1\right)\right) \int_{-1}^{1} \frac{\cos\theta' \, d\theta'}{p^2 + p'^2 - 2pp' \cos\Theta} \qquad (37)$$

where Θ is the angle between $\vec{p}$ and $\vec{p}\,'$.

At low temperature, the distribution $\ell_1(p')$ is different from zero only for small values of p'. Because of the energy conservation, p remains then close to the critical value $\sqrt{2}$. Therefore $p'^2 - 2pp'\cos\Theta$ is negligible compared to p^2 in the denominator of (37) and the integral J tends to zero faster than the factor $\cosh^{-1}(p/\sqrt{2})$ coming from $\Pi_e(p)$ in the denominator of (36).

In conclusion, the perturbation $h_2(\vec{p})$ due to the electric field on the distribution outside the critical circle is negligible :

$$f_2(\vec{p}) \simeq g_2(\vec{p}) \qquad (38)$$

In fact, this is valid only at very low temperature. It would be interesting to study the effect of this contribution at moderate temperature.

With this approximation (38), the Boltzmann equation (restricted to terms linear in the electric field) for inside the circle becomes

$$eE \frac{\partial g_1(\vec{p})}{\partial p_z} = - \Pi_a(\vec{p})\; g_1(\vec{p}) - \Pi_a(\vec{p})\; h_1(\vec{p}) + \int d^3p'\; \Pi_e(\vec{p}'\to\vec{p})\; g_2(\vec{p}') \qquad (39)$$

But the unperturbed distribution $g(\vec{p})$ is the solution of equation (34) valid when $E = 0$. Therefore the first and the last terms of the right hand side of (39) cancel each other. This gives :

$$eE \frac{\partial g_1(\vec{p})}{\partial p_z} = - \Pi_a(\vec{p})\; h_1(\vec{p})$$

At low temperature, the distribution is restricted to a region close to the origin. In this region $\Pi_a(\vec{p})$ is nearly constant, so that

$$eE \frac{\partial g_1(\vec{p})}{\partial p_z} = - \Pi_a(o)\; h_1(\vec{p})$$

and therefore

$$f_1(\vec{p}) = g_1(\vec{p}) - eE\; \Pi_a^{-1}(o)\; \frac{\partial g_1(\vec{p})}{\partial p_z}$$

If the electric field is weak enough this reduces to

$$f_1(\vec{p}) = g_1(\vec{p} - e\vec{E}\; \Pi_a^{-1}(o))$$

Inside the critical circle, the distribution perturbed by a weak electric field, in first approximation is the unperturbed distribution shifted in the direction of the electric force by $eE\Pi_a^{-1}(o)$ with

$$\Pi_a(o) = 2\;\alpha\;\overline{N}$$

The average velocity is obviously

$$\langle v\rangle = \int d^3p\; p_z\; f(\vec{p}) / \int d^3p\; f(\vec{p})$$

At low temperature the contribution from outside the critical radius is negligible, so that

$$\langle v\rangle = \int d^3p\; p_z\; g_1(\vec{p} - e\vec{E}\Pi_a^{-1}(o))\{\int d^3p\, g_1(\vec{p}-e\vec{E}\Pi_a^{-1}(o))\}^{-1}$$

A substitution, replacing p_z by $p_z + eE\Pi_a^{-1}(o)$, gives immediately

$$\langle v \rangle = \frac{eE}{\Pi_a(o)} = \frac{eE}{2\alpha \bar{N}}$$

Therefore, with the approximation used here and after reintroducing quantities with dimensions, the mobility is

$$\mu = \frac{e}{2\,\alpha\, m\, \bar{N}\, \omega}$$

which is the result at first order of a relaxation time approach of the problem.

III.5. The momentum distribution at zero temperature in the presence of an electric field

More than twenty years ago, Shockley [22] predicted a deviation from Ohm's law for the electric conductivity in polar semiconductors at low temperature and high electric field. The reason given by Shockley for this non-ohmic behaviour is the following.

Before the field is applied, the electrons (or holes in p-type semiconductors) have very low energy. Then they are accelerated by the external field. At relatively high field and for good samples at low temperature and low electron density, the probability that the electrons are scattered (by imperfections, thermal phonons or e-e collissions) is very small. Thus the speed of most of the electrons reaches the critical value which corresponds to the kinetic energy becoming equal to the energy $\hbar\omega$ of a longitudinal optical (LO) phonon.

In a polar semiconductor, the electrons interact rather strongly with the LO phonons and when the critical speed is reached they sooner or later emit a real phonon in a kind of Čerenkov effect and lose almost all their kinetic energy. Then the process described above repeats itself again and again so that the electron velocity depends on time according to a kind of "saw tooth" law. With this law, the mean velocity is about half the critical speed and depends very little on the electric field. This leads to the prediction of a saturation in the curve giving the electric current versus the field.

Obviously the situation in an actual semiconductor is more complicated and other effects are then important, like scattering by thermal phonons or imperfections, e-e interactions, non-parabolicity of the band, intervalley or interband transitions,etc. However,a quantitative understanding of the electron distribution for Shockley's model seems an important prerequisite for the treatment of more realistic situations.

During the phase of constant acceleration, the electrons spend equal time in intervals of equal length along p_z in the momentum space (the axes are oriented so that Oz lies in the direction of the force $e\vec{E}$ acting on the charge carrier). If the probability of emission were infinite at the critical speed, the electron distribution in a perfect lattice at zero temperature and neglecting the e-e interaction would be the following : all the electrons have $p_x = p_y = 0$ and along p_z the distribution is constant between zero and the critical value $p_{z,c}$ (such that $\frac{p_{z,c}^2}{2m} = \hbar\omega$). It is zero elsewhere.

This model of distribution seems too rough to give a satisfactory description of actual situations, even in first approximation. Indeed, for values of temperature and electric field encountered in practice, the probability of finding electrons with speed beyond the critical value is far from negligible. Therefore, a more precise description of the electron momentum distribution for Shockley's model is needed.

From the practical point of view the most successful approach to calculate the distribution function of hot electrons in (polar) semiconductors and to solve the Boltzmann equation has been the use of numerical methods : a Monte Carlo method [23] and an Iterative method [24] , [25] , have been introduced for this purpose (for a review see e.g. ref.[26] . Analytical solutions have only been obtained with quite restrictive approximations and further development seems desirable [26] . The purpose of the present chapter is to develop an analytical method for the evaluation of the distribution function of hot electrons interacting with LO-phonons at T = 0. This leads to an expression of the momentum distribution involving a single integral which is evaluated numerically [27] . The study of the effects of the temperature are in progress.

Here it is assumed that the distribution function is zero for $p^2 > 4$ (corresponding to $\frac{p^2}{2m} > 2\hbar\omega$). Thus the emission of phonons brings electrons always from outside to inside the critical circle. Therefore, for the region outside the critical circle, at T = 0 the first term on the right-hand side of Boltzmann's equation (27) is identically zero, so that

$$\frac{\partial f_2(p_\perp,p_z)}{\partial p_z}eE = -f_2(p_\perp,p_z)\int \Pi_e(\vec{p}\to\vec{p}\,')d^3p' \qquad p > \sqrt{2} \quad (40)$$

On the contrary, the second term is missing when $p < \sqrt{2}$ and

$$\frac{\partial f_1(p_\perp,p_z)}{\partial p_z}eE = \int f_2(p_\perp',p_z')\,\Pi_e(\vec{p}\,'\to\vec{p})d^3p' \qquad p < \sqrt{2} \quad (41)$$

Moreover, in the latter case, the electrons come from outside the circle : $f_2(p_\perp',p_z')$ in the integral of (41) is the distribution for $p' > \sqrt{2}$, i.e. the solution of (40).

With these considerations, the procedure for solving the problem becomes evident. First solve (40) which is a first-order differential equation and then introduce the solution into (41) and integrate to get the complete solution.

With the expression (29) of the rate of transition, (40) becomes

$$\frac{\partial f_2(p_\perp,p_z)}{\partial p_z}eE = -f_2(p_\perp,p_z)\,\frac{2\alpha\sqrt{2}}{p}\cosh^{-1}\left(\frac{p}{\sqrt{2}}\right) \qquad p > \sqrt{2} \quad (42)$$

It is quite natural to take the critical value as the initial point for integrating (42). Then one has

$$f_2(p_\perp,p_z) = f_o(p_\perp,p_z)\exp\left\{-\frac{2\alpha\sqrt{2}}{eE}\int_{\sqrt{2-p_\perp^2}}^{p_z} dp_z\,\frac{\cosh^{-1}\left(\frac{p}{\sqrt{2}}\right)}{p}\right\} \qquad p > \sqrt{2}$$

where $f_o(p_\perp,p_z)$ is the distribution measured on the critical circle, i.e.

$$f_o(p_\perp,p_z) = f(p_\perp,\sqrt{2-p^2})$$

Taking $p = \sqrt{p_\perp^2 + p_z^2}$ as a new variable of integration gives

$$f_2(p_\perp, p_z) = f_o(p_\perp, p_z)\exp\{- \frac{2\alpha\sqrt{2}}{eE} \int_{\sqrt{2}}^{\sqrt{p_\perp^2+p_z^2}} dp \frac{\cosh^{-1}(\frac{p}{\sqrt{2}})}{\sqrt{p^2 - p_\perp^2}}$$

$$p > \sqrt{2} \qquad (43)$$

It does not seem possible to perform the integration in (43) analytically. However, remembering the rough model for the electron distribution described at the end of the introduction, it is expected that $f_2(p_\perp, p_z)$ goes rapidly to zero when $p > \sqrt{2}$. As a consequence, the physically interesting interval of integration is small and the variable of integration p does not differ much from the critical value $\sqrt{2}$. The square root in (43) can then be expanded in powers around $p = \sqrt{2}$. At first order, this gives for the distribution outside the critical circle

$$f_2(p_\perp, p_z) = f_o(p_\perp, p_z)\ e^{-g(p)} \qquad p > \sqrt{2} \qquad (44)$$

with

$$g(p) = \frac{2\alpha}{eE}\{p \cosh^{-1}(\frac{p}{\sqrt{2}}) - \sqrt{p^2-2}\} \qquad (45)$$

and

$$p = \sqrt{p_\perp^2 + p_z^2}$$

Introducing (44) into (41) and using (28) gives the following equation for the distribution inside the critical circle :

$$eE \frac{\partial f_1(p_\perp, p_z)}{\partial p_z} = \frac{\alpha}{\pi\sqrt{2}} \int \frac{f_o(p_\perp', p_z')e^{-g(p)}}{|\vec{p} - \vec{p}'|^2} \delta(\frac{p'^2}{2} - \frac{p^2}{2} - 1)d^3p' \qquad p < \sqrt{2}$$

In polar co-ordinates (p', θ, φ), this relation becomes :

$$eE \frac{\partial f_1(p_\perp, p_z)}{\partial p_z} = \frac{\alpha}{\pi\sqrt{2}} \frac{1}{\sqrt{p^2 + 2}} \int_0^{2\pi} d\varphi \int^{\pi} \sin\theta\ d\theta \int_{\sqrt{2}}^{\infty} dp'$$

$$\frac{p'^2\ f_o(p_\perp', p_z')\ e^{-g(p')}}{p'^2 + p^2 - 2\ pp' \cos\theta} \times \delta(p' - \sqrt{p^2+2}) \quad p < \sqrt{2}$$

where θ is the angle between $\vec{p}$ and $\vec{p}\,'$. As $f_o(p_\perp', p_z')$ is measured on the critical circle ($p = \sqrt{2}$), it can be taken outside the integral over p', leading to

$$eE \frac{\partial f_1(p_\perp, p_z)}{\partial p_z} = \frac{\alpha}{2\pi\sqrt{2}} \sqrt{p^2+2} \int_0^{2\pi} d\varphi \int_0^{\pi} d\theta$$

$$\frac{\sin\theta\, f_o(p_\perp', p_z')\, e^{-g(\sqrt{p^2+2})}}{p^2 + 1 - p\sqrt{p^2+2}\cos\theta} \qquad p < \sqrt{2} \quad (46)$$

The exponential $\exp[-g(\sqrt{p^2+2})]$ falls off rapidly when the argument of g becomes larger than $\sqrt{2}$. This means that the main contribution to the integral in (46) comes from the region close to $p = 0$. Therefore the denominator in this integral can be expanded in powers of p. At first order, this gives

$$eE \frac{\partial f_1(p_\perp, p_z)}{\partial p_z} = \frac{\alpha}{\sqrt{2}} \eta\, e^{-g(\sqrt{p^2+2})} \sqrt{p^2+2} \qquad p < \sqrt{2}$$

and, after integration,

$$f_1(p_\perp, p_z) = f_1(p_\perp, p_z^o) + \frac{\alpha}{\sqrt{2}} \frac{\eta}{eE} \int_{p_z^o}^{p_z} e^{-g(\sqrt{p^2+2})} \sqrt{p^2+2}\, dp_z$$

$$p < \sqrt{2} \qquad (47)$$

where g(x) is defined by eq.(45) and

$$\eta = \int_0^{\pi} \sin\theta\, f_o(p_\perp', p_z')\, d\theta$$

is essentially the integral of the distribution taken on the critical circle.

Far enough in the negative direction of p_z, the distribution is practically zero. Therefore, if one chooses p_z^o negative and large enough, the expression (47) for the distribution inside the circle becomes

$$f_1(p_\perp, p_z) = \frac{\alpha}{\sqrt{2}} \frac{\eta}{eE} \int_{p_z^o}^{p_z} e^{-g(\sqrt{p^2+2})} \sqrt{p^2+2}\, dp_z \qquad p < 2 \quad (48)$$

Moreover, Boltzmann's equation (together with initial conditions) gives the distribution except for a constant factor which must be determined by a normalization condition. Thus the constant factors in (48) can be dropped, leading to the final expression

$$f_1(p_\perp, p_z) = \int_{p_z}^{p_z^o} e^{-g(\sqrt{p^2+2})} \sqrt{p^2+2}\, dp_z \qquad p < \sqrt{2} \quad (49)$$

It does not seem possible to perform the remaining integration analytically.

The last step is to ensure the continuity of the distribution on the circle. This requires that values obtained from (49) for $p_z = \sqrt{2 - p^2}$ be used as initial values for $f_o(p, p_z)$ in (44).

The results of the computation of $f(p_\perp, p_z)$ are shown in figs.10 to 15. The electron phonon coupling constant is chosen as $\alpha = 0.02$, the value for InSb. The values for the parameter $2\alpha\sqrt{2}/eE$ (called AEM in the figures) are 500 (figs.10,11,12) and 50 (figs.13, 14,15). For InSb this corresponds approximately to electric field strenghts of 12 V/cm and 120 V/cm.

In figs.10,11,13,14 the two dimensional surfaces for $f(p_\perp, p_z)$ are displayed. In figs.12 and 15 contours of constant $f(p_\perp, p_z)$ are shown.

It is seen that the electron distribution is far from Maxwellian in the direction of the electric field (p_z - direction). The distribution function drops sharply at the critical circle $p = \sqrt{2}$. It has a constant value over a substantial range of p_z-values (corresponding to the classical region of constant acceleration) and a rather broad "tail" around the origin.

The decrease of $f(p_\perp, p_z)$ at the critical circle, due to emission of phonons by a Čerenkov like effect, becomes less sharp for increasing electrical fields. The acceleration being larger, the electron travels a larger average distance in the p-space before emitting a phonon.

A remarkable feature of the distribution function is an essential singularity occurring at E = 0 and $p > \sqrt{2}$. It is seen most clearly from eq.(43) which is a rigorous solution of the Boltzmann equation (27) with probabilities defined according to eq.(28). This essential singularity also occurs in the integrand of (48) which defines $f(p_\perp, p_z)$ for $p > \sqrt{2}$. The physical meaning of this essential singularity is not completely clear but it might be related to the fact that for

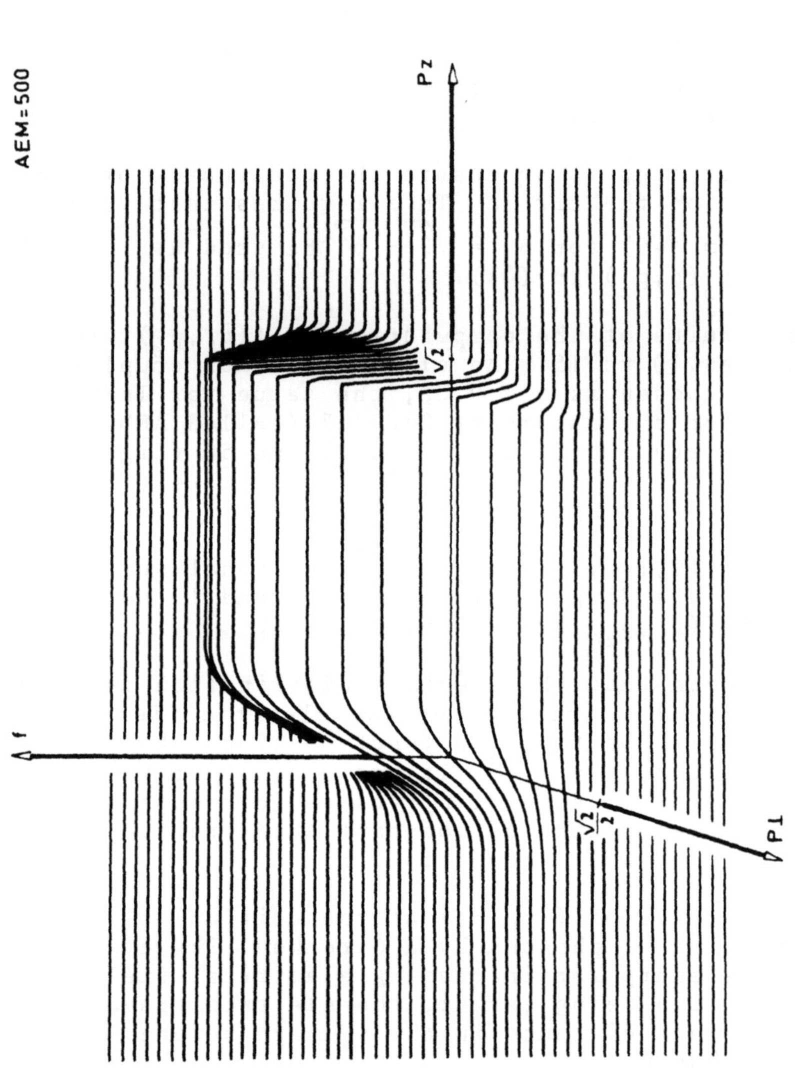

Fig.10: Momentum distribution function $f(P_\perp, P_z)$ for polarons in an external electric field in the P_z-direction at $T = 0$. $E = 12$ V/cm and $\alpha = 0.02$ (InSb) have been choosen.

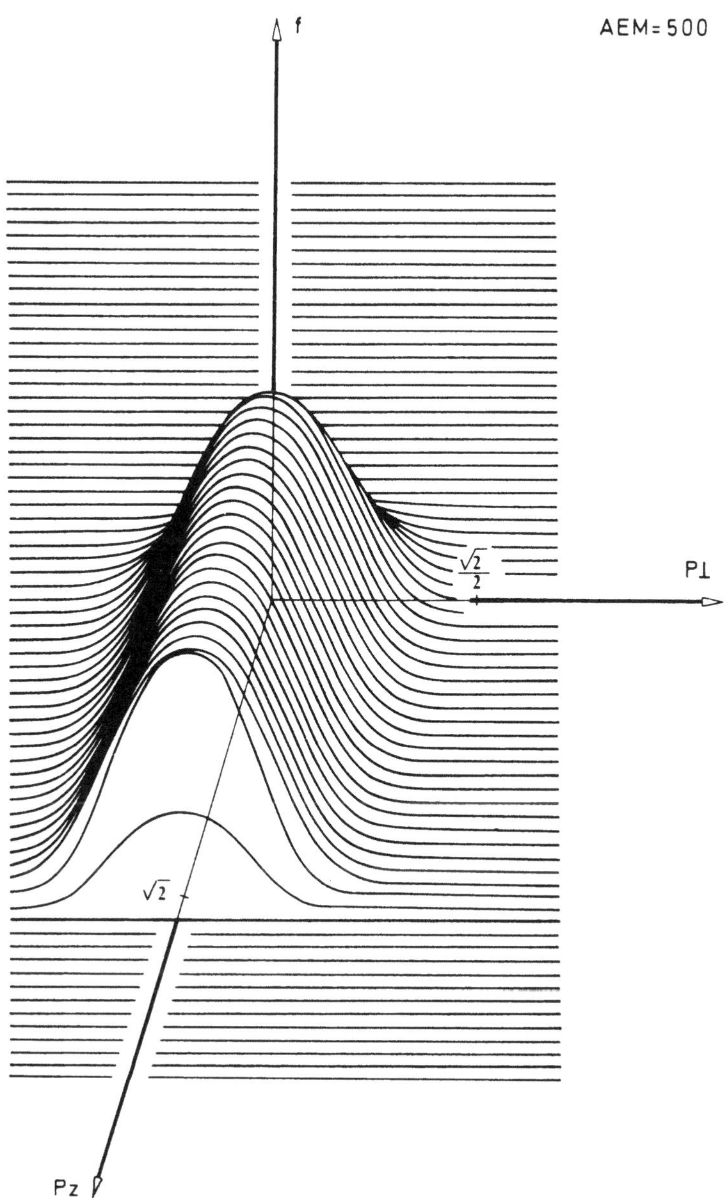

Fig.11 : Same as fig.10, seen under a different angle.

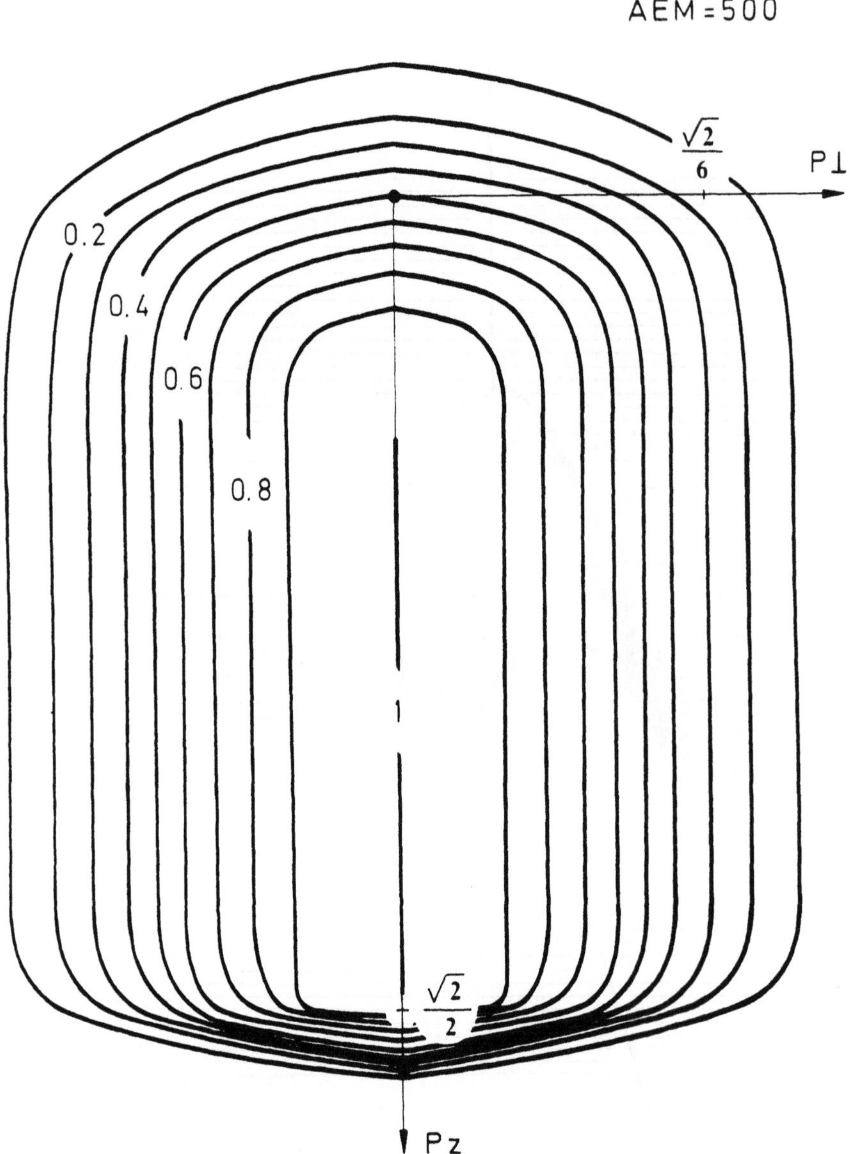

Fig.12 : Contour of constant $f(P_z, P_\perp)$ for the parameters of fig.10.

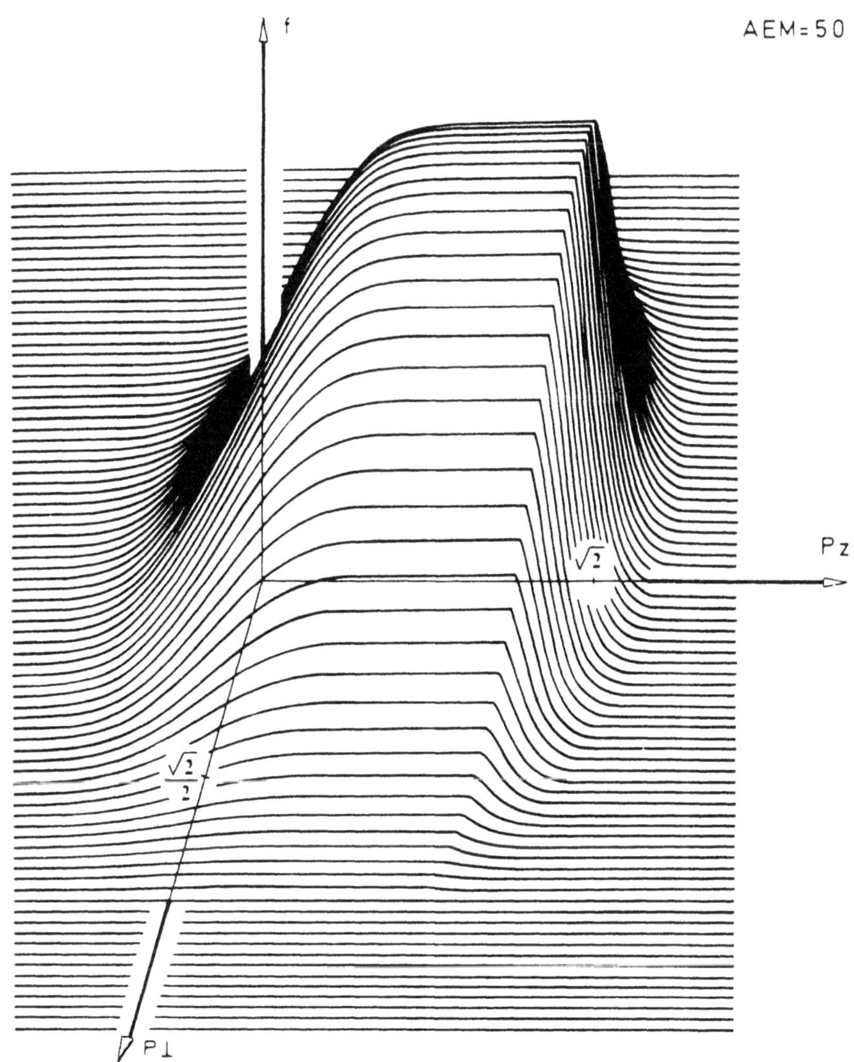

Fig.13 : Momentum distribution function $f(P_\perp, P_z)$ for polarons in an external electric field in the P_z-direction at T = 0. E = 120 V/cm and α = 0.02 (InSb) have been choosen.

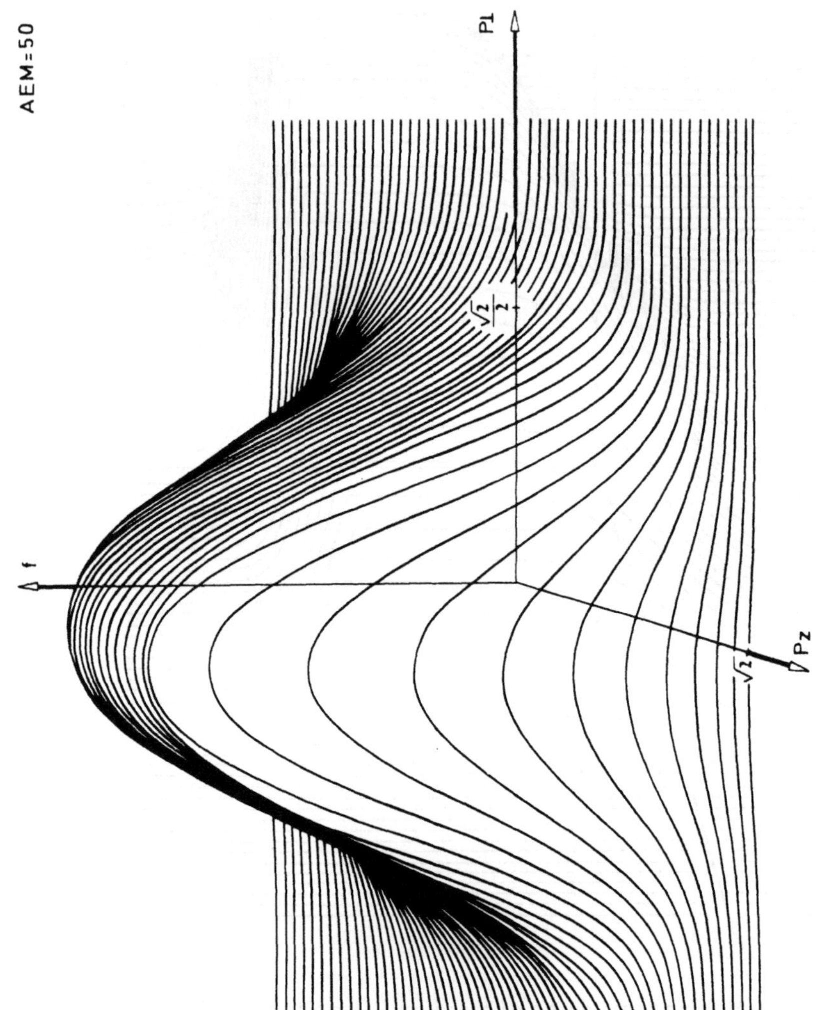

Fig.14 : Same as fig.11, but E = 120 V/cm.

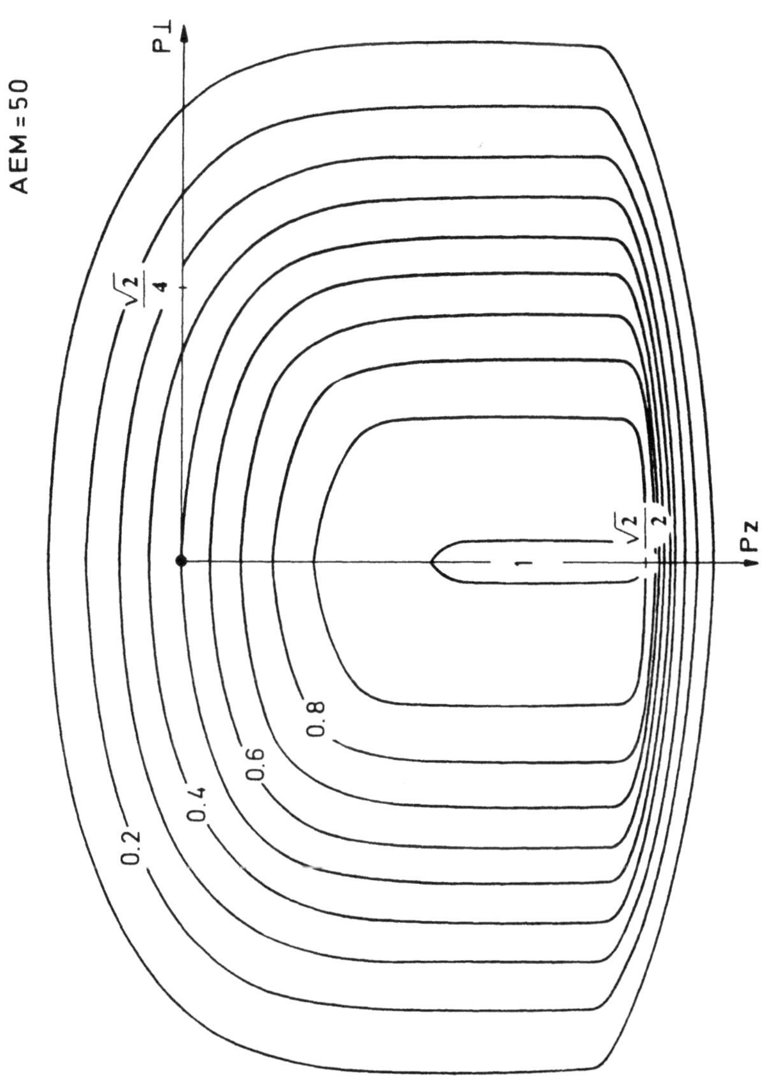

Fig.15 : Contour of constant $f(P_z,P_\perp)$ for the parameters of fig.13.

$E = 0$ the Maxwell distribution is a solution of eq.(27) while for any $E \neq 0$ the distribution is asymmetric and not connected to the solution at $E = 0$ in a continuous way.

Because of the assumption that $f(p_\perp, p_z) = 0$ for $p^2 > 4$, the present approximation violates the conservation of the number of electrons : those electrons which reach the region $p^2 > 4$ are not treated. This approximation becomes less satisfactory with increasing electrical field.

A steep decrease of the distribution function $f(p_\perp, p_z)$ at the cut-off (due to LO phonon emission) had also been obtained from an Iterative method calculation [28]. Before a meaningful comparison with such numerical calculations can be made, however, our method should be extended to finite temperatures [27] and also other scattering mechanisms should be included. In particular the continuous transition from the Maxwell distribution at $E = 0$ to the asymmetric distribution at $E \neq 0$, obtained with the Iterative method but absent here, seems to be related to other scattering mechanisms than those involving optical phonons. It should be noted that the "tail" of the distribution and the "cut-off" cover roughly the same p^2 interval. The p_z-intervals for cut-off and tail are - of course - different.

IV. Conclusions

In these lectures, we have discussed the role played by polaron effects in the electrical resistivity of polar semiconductors. Among other results, it has been shown that it is possible to develop analytic formalisms to study the momentum distribution of polarons in the ohmic as well as non ohmic regimes.

These lectures, we hope, will contribute to give a better understanding of the effects of the coupling with the polar modes of the lattice on the momentum distribution of electrons and on their mobility. An important contribution is that most of the results reported here have been obtained in analytical forms. This should allow to treat the polar coupling in an analytical way even in more realistic formalisms where other scattering mechanisms are taken into account. In this way, exceedingly involved and heavy numerical computations could be avoided, saving important computation time.

REFERENCES

For the review part of these lectures no complete list of references is given. They can be found in other review works :

[1] C.Kuper and G.Whitfield (editors), Polarons and Excitons, Oliver and Boyd (1963)

[2] J.T.Devreese (editor), Polarons in Ionic Crystals and Polar Semiconductors, North Holland (1972)

[3] See. e.g. F.C.Brown in ref.1

[4] E.Karthauser in ref.2

[5] J.Devreese, J.De Sitter, M.Goovaerts, Phys.Rev.B5 2367 (1972)

[6] H.Finkenrath, N.Uhle and W.Waidelich, Solid State Comm.7, 11 (1969)

[7] R.Evrard, E.Karthauser and J.Devreese, Phys.Rev. Letters 22, 94-97 (1969)

[8] W.Huybrechts, J.Devreese, to be published Solid State Communications

[9] R.P.Feynman and H.Hibbs, Quantum Mechanics and Path Integrals, McGraw Hill Book Co., New York (1965)

[10] G.Papadopoulos, in this volume

[11] K.Thornber, in this volume

[12] G.Papadopoulos, in this volume

[13] J.Devreese, in this volume

[14] R.P.Feynman, R.Hellwarth, C.Iddings, P.Platzman, Phys.Rev. 128, 1599 (1962)

[15] J.Devreese, R.Evrard and E.Karthauser, to be published, Phys.Rev. B (1975)

[16] D.Langreth, Phys.Rev., 137, A760 (1965)

[17] L.F.Lemmens, J.De Sitter, J.T.Devreese, Phys. Rev.B8 2717 (1972)

[18] K.K.Thornber and R.P.Feynman, Phys.Rev.B1 4099 (1970)

[19] K.K.Thornber, Phys.Rev.B3, 1929 (1971)

[20] R.P.Feynman, Statistical Mechanics : A set of Lectures, Benjamin (1972)

[21] P.M.Platzman in "Polarons and Exitons", edited by C.G.Kuper and G.D.Whitfield, Oliver and Boyd, 1963.

[22] W.Shockley, Bell Syst. Tech. J. (US) 30, 990 (1951)

[23] T.Kurosawa, J.Phys.Soc.Japan, Suppl.21, 424 (1966). 31 668 (1971)

[24] H.F.Budd, J.Phys.Soc.Japan, Suppl.21 420 (1966); Phys.Rev.158, 798 (1967)

[25] W.Fawcett, A.D.Boardman, S.Swain, J.Phys.Chem. Solids 31, 1963 (1970)

[26] G.Bauer in "Springer Tracts in Modern Physics", vol.74 (1974), Springer-Verlag, New York

[27] J.Devreese, R.Evrard, E.Karthauser, to be published

[28] D.Kranzer, H.Hillbrand, H.Pötzl, O.Zimmerl, Acta Phys.Austriana 35, 110 (1972)

THEORY AND APPLICATION OF INVERSE TRANSPORT COEFFICIENTS

N.H. March

Physics Department, Imperial College
Prince Consort Road
London SW7 2BZ, England

ABSTRACT

The ideas behind the theory of inverse transport coefficients are outlined by considering two classical examples :

(a) Self-diffusion in a classical liquid. Here the diffusion constant D is related to a friction constant ζ. In turn, ζ is connected with the force-force correlation function.

(b) The electrical resistivity of a classical plasma.

The quantum-mechanical generalization of the force-force correlation function is then introduced. This correlation function is shown explicitly to lead to the correct electrical resistivity of metals and metallic alloys for :

(c) Dilute impurity scattering, both in the Born approximation and for arbitrary phase shifts.

(d) Weak scattering (Ziman) theory of liquid metals Approximations applicable under strong scattering conditions are also outlined.

(e) The Kondo effect arising from conduction electron scattering from a localized magnetic moment.

1. INTRODUCTION

In these lectures, we shall tackle transport problems via the theory of inverse transport coefficients. It should be said immediately that a good deal of controversy still exists at the foundations of such a theory. References to this controversy are included at the end of these notes, but in these lectures we shall adapt a pragmatic approach and focus a lot of attention on the evaluation of the quantum-mechanical formula for the electrical resistivity of metals and metallic alloys. However, it will be instructive to begin with two classical examples. These will help us to understand the structure of the quantum-mechanical formula.

2. DIFFUSION AND FORCE-FORCE CORRELATION FUNCTION

Let us begin with the usual model of Brownian motion. A macroscopic body of mass M moving through a continuous fluid experiences, according to Stokes' law, a damping force proportional to its velocity. Langevin's equation then describes the motion, if it is assumed that the macroscopic body, with an average behaviour governed by Stokes' law, has a superimposed erratic motion caused by a fluctuating or stochastic force $\vec{X}(t)$ say. Then the Langevin equation is simply

$$\frac{M d\vec{v}}{dt} + M\zeta\vec{v}(t) = \vec{X}(t) \tag{2.1}$$

According to Stokes' law, ζ is given in terms of the shear viscosity of the liquid and the radius a of the macroscopic particle by

$$\zeta = 6\pi\eta a \tag{2.2}$$

From eqn (2.1) it is straightforward to construct the velocity auto-correlation function $\langle v(0).v(t)\rangle$ as

$$\langle\vec{v}(0).\vec{v}(t)\rangle = \frac{3}{\beta M}\exp(-\zeta|t|) \;:\; \beta = \frac{1}{k_B T} \tag{2.3}$$

Hence the macroscopic self-diffusion constant D is given by

$$D = \frac{1}{3}\int_0^\infty dt\langle\vec{v}(0).\vec{v}(t)\rangle = (M\beta\zeta)^{-1} \tag{2.4}$$

This equation shows us that D is inversely proportional to the friction constant ζ. But if we have a charged fluid, then the Einstein relation yields for the electrical conductivity σ

$$\sigma = \frac{ne^2 D}{k_B T} \tag{2.5}$$

n being the number density of carriers with charge e. Hence the electrical resistivity $R = 1/\sigma$ is given by

$$R = \frac{k_B T}{ne^2 D} = \frac{M\zeta}{ne^2} \tag{2.6}$$

If we compare this with the conductivity formula

$$\sigma = \frac{ne^2 \tau}{M} \tag{2.7}$$

it is clear that a calculation of the friction constant ζ is quite equivalent to finding the inverse of the relaxation time τ.

2.1. Kirkwood's Formula for Friction Coefficient

If we adopt the initial condition

$$v(t=0) = v_0 \tag{2.8}$$

then we can solve the Langevin equation (2.1) explicitly to obtain

$$v(t) = v_0 e^{-\zeta t} + e^{-\zeta t} \int_0^t e^{\zeta s} X(s) ds \tag{2.9}$$

Squaring eqn (2.9) and averaging we obtain

$$\langle v^2(t) \rangle = v_0^2 e^{-2\zeta t} + e^{-2\zeta t} \int_0^t \int_0^t e^{\zeta(s+s')} \langle \vec{X}(s).\vec{X}(s') \rangle \, ds \, ds' \tag{2.10}$$

Evaluating the integral and taking the limit $t \to \infty$ in eqn (2.10) yields since

$$\lim_{t\to\infty} \langle v^2(t) \rangle = 3k_B T \tag{2.11}$$

$$\zeta = (k_B T)^{-1} \int_0^\infty ds \langle \vec{X}(0).\vec{X}(s) \rangle e^{-\zeta s} \tag{2.12}$$

If, as seeems to be the case, $\langle \vec{X}(0).\vec{X}(s) \rangle$ has a sharp maximum and falls to zero in a time $\tau_1 \ll \zeta$, then

$$\zeta = (k_BT)^{-1} \int_0^\infty ds \langle \vec{X}(0).\vec{X}(s) \rangle \tag{2.13}$$

However, though this is a correct formula, it seems much more useful in practice to express ζ in terms of the (total) force-force correlation function <FF>, F being dp/dt. This can again be worked out in the Brownian motion example to obtain

$$\langle \vec{F}(t).\vec{F}(t+\tau) \rangle = \langle \vec{X}(t).\vec{X}(t+\tau) \rangle - 3\zeta^2 k_B T m e^{-\zeta t}$$

Then we get finally

$$\zeta = \frac{1}{3k_BT} \int_0^\tau \langle \vec{F}(t).\vec{F}(t+s) \rangle ds \quad , \quad \tau > \tau_1 \tag{2.14}$$

where as remarked τ_1 is the time in which $\langle \vec{X}(0).\vec{X}(s) \rangle$ falls to zero.

There is a difficulty in general (actually not with Brownian motion) with formula (2.14) if τ is allowed to go to infinity as

$$\int_0^\infty \langle \vec{F}(t).\vec{F}(t+s) \rangle ds = 0 \tag{2.15}$$

However, provided τ in equation (2.14) is greater than τ_1 and $\ll \frac{1}{\zeta}$, the expression (2.14) is essentially independent of τ.

2.2. Practical Utility of Force-force Correlation Function

The above discussion shows that great care needs to be exercised when working with time-dependent force-force correlation functions. In the quantum-mechanical case, the same remarks obtain when working with Heisenberg operators. For later purposes, we remark that in quantum mechanics we can avoid these difficulties by working with the Schrodinger picture. Then, as we shall discuss, by using 'scattering' rather than 'box' boundary conditions, it appears possible to circumvent the difficulties associated with a result like eqn (2.8).

Returning briefly to eqn (2.14), it would appear entirely feasible by means of molecular dynamical computations to get useful results for the friction constant ζ. However, it is not clear that such an approach would have merit over the usual procedures of calculating D directly from the velocity auto-correlation function via eqn (2.4).

In this connection, it is of some interest to remark that a route to the calculation of η exists via the frequency spectrum $g(\omega)$ of the liquid, defined in terms of the velocity autocorrelation function by

$$g(\omega) = \frac{1}{3\pi} \int_0^\infty dt \langle \vec{v}(0).\vec{v}(t) \rangle \exp(i\omega t) \tag{2.16}$$

As shown by Gaskell and March (1970), it follows from the $t^{-3/2}$ long time tail of the velocity autocorrelation function that the second term in the small ω expansion of $g(\omega) \propto \omega^{1/2}$ and explicitly

$$g(\omega) = \frac{D}{\pi} + d_1\omega^{\frac{1}{2}} + d_2\omega + \ldots \tag{2.17}$$

where

$$d_1 = -(2\pi)^{\frac{1}{2}} \frac{2}{3\rho} [4\pi(D + \frac{\eta}{\rho m})]^{-\frac{3}{2}} \frac{k_B T}{m\pi} \tag{2.18}$$

Though the long-time tail of the velocity autocorrelation function was, in fact, discovered by Alder during machine computations on hard spheres, it is in fact difficult to compute quantitatively. Thus, for the purpose of estimating η, a route via equations like (2.14) and (2.2) might prove advantageous.

Finally, in connection with η, it is worth remarking that the customary correlation function expression for η can be transformed into

$$\eta = \frac{1}{k_B T} \int\int \langle \tau_{xy}(\vec{r}t)\tau_{xy}(00) \rangle d\vec{r}\, dt \tag{2.19}$$

where τ_{xy} is an off-diagonal element of the stress tensor, but we shall not go into further details here.

3. ELECTRICAL RESISTIVITY OF AN IONIZED PLASMA

After Kirkwood's work, Edwards (1965) using quite different methods discussed electrical resistivity and we want briefly to summarize his work on a classical plasma before turning to the quantum-mechanical case. First, we summarize how the conductivity can, in principle at least, be calculated for a classical Lorentz gas. Here the electrons are scattered off random centres, described by scattering potentials

$\phi(\vec{r})$, the total potential Φ being given by

$$\Phi(\vec{r}) = \sum_\alpha \phi(\vec{r}-\vec{R}_\alpha) \tag{3.1}$$

where the scattering centres are evidently at positions $\vec{R}_\alpha$. The force F acting on an electron is given by

$$\vec{F} = -\nabla\Phi = \sum_\alpha \vec{F}(\vec{r}-\vec{R}_\alpha) \tag{3.2}$$

The equilibrium distribution function f_e is given in terms of the Helmholtz free energy F_n and the Hamiltonian H by

$$f_e = \exp\left(\frac{F_n - H}{k_B T}\right) \tag{3.3}$$

with

$$H = \frac{1}{2} mv^2 + \sum_\alpha \phi(\vec{r}-\vec{R}_\alpha) \tag{3.4}$$

3.1. Electrical Conductivity

The conventional argument is then to write down the Liouville equation in the presence of an applied electric field $\vec{\varepsilon}$, to calculate the perturbed distribution function f. Thus

$$\frac{\partial f}{\partial t} + \vec{v}.\frac{\partial f}{\partial \vec{r}} + \sum_\alpha \frac{1}{m} \vec{F}(\vec{r}-\vec{R}_\alpha).\frac{\partial f}{\partial \vec{v}} + \frac{e\vec{\varepsilon}}{m}\frac{\partial f}{\partial \vec{v}} = 0 \tag{3.5}$$

Writing

$$f = f_e + f_p \tag{3.6}$$

and denoting the Liouville operator by $\mathcal{L}$ we readily find

$$\mathcal{L} f_p = \frac{-e}{m} \vec{\varepsilon}.\vec{v}\, f_e \tag{3.7}$$

If the electric field is turned on at time t', then a formal solution may be written in terms of the Green function G of the Liouville equation as

$$f_p = -\int_{t'}^{t} G(\vec{r}\ \vec{r}''\ \vec{v}\ \vec{v}''\ t\ t'')\ \frac{e\vec{\varepsilon}.\vec{v}''}{k_B T}\ f_e(\vec{r}''\vec{v}'')d\vec{r}''d\vec{v}''dt'' \tag{3.8}$$

where

$$\mathcal{L}G = \delta(\vec{v}-\vec{v}')\delta(\vec{r}-\vec{r}')\delta(t-t') \tag{3.9}$$

The current $\vec{j}$ is constructed in the usual way as

$$\vec{j}(\vec{r}t) = \int e\vec{v}f d\vec{v} = \int e\vec{v}f_p d\vec{v}$$

$$= \int e\vec{v}G \frac{\vec{v}''.\vec{\varepsilon}}{k_BT} f_e(\vec{r}''\vec{v}'')d\vec{r}''d\vec{v}''dt'' \tag{3.10}$$

or if we write as usual

$$j_\mu = \sigma_{\mu\nu}\,\varepsilon_\nu \tag{3.11}$$

then

$$\sigma_{\mu\nu} = \frac{e^2}{k_BT}\int v_\mu G v_\nu'' f_e'' d\vec{v}''d\vec{r}''dt''d\vec{v} \tag{3.12}$$

It is clear that to follow this route one needs essentially complete knowledge of the classical motion (i.e. G). Moreover, G cannot be expanded in an elementary way in powers of the scattering potentials, since in the absence of the scatterers the electron is accelerated and the integral for σ diverges. Thus, at very least, one must sum sub-series of the expansion in $F(\vec{r}-\vec{R}_\alpha)$ up to the level of the customary collision-time approximation.

Actually Edwards and Sanderson (1961) have pushed the calculation of the conductivity through, using a Green function in which the spatial coordinates are integrated out. We summarize their approach in Appendix 1. We merely note here that, if the spatially averaged Green function G tells us the probability that, in equilibrium, one finds a particle at time t' with a velocity $\vec{u}'$ and that subsequently the <u>same</u> particle has velocity $\vec{u}$ at time t. Then in the Lorentz gas with static scatterers, they construct the two-particle Green function which refers now to the velocities of two particles as

$$G_{12} = G_1G_2 \tag{3.13}$$

After lengthy calculation outlined in Appendix 1, by analyzing G_1 into eigenfunctions, they find the perturbed distribution function as

$$f_p = [Ne(\vec{\varepsilon}.\vec{u})u^3/(2\alpha_0 k_B T)] f_0 \tag{3.14}$$

where the quantity α_0 is related to the Debye length λ by

$$\alpha_0 \cong \frac{-2\pi Ne^4}{m^2} \ln(2\lambda/r_0) \tag{3.15}$$

The conductivity σ comes out as

$$\sigma = 2(2k_B T/\pi)^{\frac{3}{2}}(e^2 m^{\frac{1}{2}} \ln(2\lambda/r_0))^{-1} \tag{3.16}$$

The physical situation in the plasma is that the conductivity can be related to the rate at which a single particle, moving through the plasma, loses momentum, and how the momentum lost by such a particle is, in part, transferred to other particles in such a way that it still contributes to the flow of current.

The essence of the problem lies in knowing the probability G_1 that a particle with velocity $\vec{u}'$ at t' will have velocity $\vec{u}$ at t. Similarly, the probability G_{12} for a pair of particles is of central importance. The great simplification in a plasma is that, as in eqn (3.13), G_{12} is the product $G_1 G_2$ to a good approximation, because of the predominance of weak collisions.

Any acceptable theory must regain the result (3.16) and the merit of the argument of Edwards (1965) was that for this classical problem he could do so by direct calculation of the electrical resistivity. We shall content ourselves below to giving his derivation of the resistivity formula; the interested reader can refer to his paper for the calculation of eqn (3.16) from his basic formula.

3.2. Resistivity for Classical Lorentz Gas

Consider again the Liouville equation (3.5), with $\vec{\varepsilon}$ switched on at time t'. The mean velocity $\vec{V}(t)$ will start at zero and tend to some limiting value. However, the velocity distribution function for the steady state will not only describe the mean drift but will be distorted in shape from the mean, i.e.

$$f(\vec{r}\ \vec{v}\ \vec{t}) \neq f_e(\vec{r},\ \vec{v}-\vec{V}(t),\ t) \tag{3.17}$$

Thus, we shall write

$$f = f_e(\vec{v}-\vec{V}(t')) + f_2 \equiv f_1 + f_2 \tag{3.18}$$

Since we know that the entire mean flow is contained in the first term, we have

$$\int \vec{v} f_2 d\vec{v} = 0 \tag{3.19}$$

At $t = t'$, $f = f_1$ and therefore since

$$\mathcal{L}(f_1 + f_2) = 0 \tag{3.20}$$

we can write

$$f_2 = -\int_{t'}^{t} \int\int G\mathcal{L} f_1 d\vec{r}'' d\vec{v}'' dt'' \tag{3.21}$$

At this point, the hydrodynamic equations stemming from Liouville's equation are invoked. To obtain the electrical conductivity, the density N_e of the system can be taken as uniform. Taking the first moment of v we find from

$$\int d\vec{v}\ \vec{v}\ [\frac{\partial}{\partial t} + \vec{v}.\frac{\partial}{\partial \vec{r}} + \frac{1}{m}\{\sum_\alpha \vec{F}(\vec{r}-\vec{R}_\alpha) + e\vec{\varepsilon}\}.\frac{\partial}{\partial \vec{v}}] f = 0 \tag{3.22}$$

$$\frac{\partial \vec{V}}{\partial t} - \frac{e\vec{\varepsilon}}{m} - \frac{1}{mN_e}\int \sum_\alpha \vec{F}(\vec{r}-\vec{R}_\alpha) f d\vec{v} = 0 \tag{3.23}$$

since

$$\int d\vec{v}\ \vec{v}\ \frac{\partial f}{\partial \vec{r}} = \frac{\partial}{\partial \vec{r}} \vec{V}(t) = 0 \tag{3.24}$$

and

$$\int d\vec{v}\ \vec{v}\ \frac{e\vec{\varepsilon}}{m}\frac{\partial f}{\partial \vec{v}} = \frac{-e\vec{\varepsilon}}{m} \tag{3.25}$$

on integrating by parts.

Average over Positions of Random Scatterers

At this point, we perform a random ensemble average on eqn (3.23). Then we get writing F for total force $\sum_\alpha \vec{F}$

$$\frac{\partial}{\partial t}(m\vec{V}) - e\vec{\varepsilon} - \frac{1}{N_e}\langle \vec{F} f\rangle = 0 \tag{3.26}$$

Clearly we must have $\langle F\ f_1\rangle = 0$ and hence $\langle F\ f\rangle = \langle F\ f_2\rangle$ Now we can write

$$f_1^{-1}\mathcal{L}f_2 = \frac{e}{k_BT}\vec{\varepsilon}\cdot(\vec{v}-\vec{V}) - \frac{1}{k_BT}\sum_{\alpha}\vec{v}.\vec{F}(\vec{r}-\vec{R}_\alpha) + \frac{m}{2k_BT}\frac{\partial\vec{V}}{\partial t}.(\vec{v}-\vec{V}) \qquad (3.27)$$

so that

$$f_2 = \frac{-e^2}{k_BT}\int_{t'}^{t}\iint G(\vec{r}\ \vec{r}''\ \vec{v}\ \vec{v}''\ \vec{t}\ \vec{t}'')\vec{\varepsilon}(t'') .\{\vec{v}''-\vec{V}(t'')\}f_1''d\vec{v}''d\vec{r}''dt$$
$$+ \int_{t'}^{t}\iint G(\vec{r}\ \vec{r}''\ \vec{v}\ \vec{v}''\ t\ t'')\vec{F}(\vec{r}'').\vec{V}(t'')f_1''d\vec{v}''d\vec{r}''dt''$$
$$\frac{-m}{k_BT}\int_{t'}^{t}\iint G(\vec{r}\ \vec{r}''\ \vec{v}\ \vec{v}''\ t\ t'')\{\vec{v}''-\vec{V}(t'')\}.\dot{\vec{V}}(t'')f_1''d\vec{v}''d\vec{r}''d\vec{t}'' \qquad (3.28)$$

In an obvious notation, with

$$v^* = v-V \qquad (3.29)$$

we have then

$$\frac{\partial\vec{V}}{\partial t} - \frac{e}{m}\vec{\varepsilon} + \frac{1}{mN_e}\langle\vec{F}G\ \vec{v}^*.\vec{\varepsilon}f_1\rangle + \frac{m}{2k_BTN_e}\langle\vec{F}G\ \vec{v}^*.\dot{\vec{V}}f_1\rangle - \frac{1}{mN_e}\langle\vec{F}G\vec{F}.\vec{V}f_1\rangle = 0 \qquad (3.30)$$

As discussed by Edwards and Sanderson (1961), G can be decomposed into a bilinear combination of eigenfunctions, each of which decays with an appropriate rate, except the equilibrium distribution which does not enter the integrals in eqn (3.30).

This implies that, if the electric field is gradually increased to its final value in a time very much longer than the collision time, $\dot{V}$ can be dropped. After a sufficient time $\vec{V}$ and $\vec{\varepsilon}$ can be taken outside the integrals to obtain

$$e\varepsilon_\alpha - \frac{e}{k_BT}\varepsilon_\beta\langle F_\alpha G v_\beta^* f_1\rangle = \langle F_\alpha G F_\beta f_1\rangle V_\beta \qquad (3.31)$$

In the isotropic case (e.g. a classical one-component plasma) this gives with

$$\vec{J} = eN_e\vec{V} \qquad (3.32)$$

$$\vec{\varepsilon} = \left\{ \frac{\frac{1}{3} \langle FGFf_1 \rangle}{1-(1/3k_BT)\langle FGv^* f_1 \rangle} \right\} \frac{\vec{J}}{N_e} \tag{3.33}$$

yielding

$$R = \frac{\frac{1}{3} \langle G\ f_1 \rangle}{1-(1/3k_BT)\langle FGv^* f_1 \rangle\ N_e} \tag{3.34}$$

As remarked above, Edwards has shown that eqn (3.16) is regained from this formula, but we shall not give the details here. We stress that an essential ingredient in this result is again a force-force correlation function in the numerator of eqn (3.34)¶ .

4. QUANTUM-MECHANICAL FORMULA FOR RESISTIVITY

In his 1965 paper, Edwards also gave a quantum-mechanical version of the expression (3.34). Rousseau, Stoddart and March (1972) calculated R by a different approach using a method nearer to Greenwood's derivation of the Kubo-Greenwood formula. However, each of these derivations have been criticised, and clearly are far from rigorous. A lot of authors have subsequently tried to work with 'fluctuating forces' (e.g. Ballentine and Heaney) but it seems to us that, in this way, the calculational merit of the inverse transport approach is sacrificed (compare, for example, the use in the classical case of eqns (2.13) and (2.14) for the friction constant).

Because of these difficulties we shall not go over the original derivations here. However, one way of seeing the form of the resistivity R is to consider the relation

$$R = \pi \langle \vec{E}.\vec{E} \rangle / T \tag{4.1}$$

given by Landau and Lifshitz (1960). Here T is the absolute temperature and $\vec{E}$ is the fluctuating electric field in the conductor. If one assumes, by analogy with the classical discussion of section 2, that one can replace $\vec{E}$ by the force per unit charge acting on an

¶ There is a lot of argument about the presence of the denominator in eqn (3.34). We merely remark here that any deviation of the denominator from unity does <u>not</u> contribute to the result (3.16).

electron, then by interpreting the brackets < > as in the Kubo formula

$$\sigma = T\langle\vec{J}.\vec{J}\rangle/\pi \qquad (4.2)$$

then it can be shown that (cf Rousseau, Stoddart and March, 1972)

$$R = \frac{-2\pi}{2\Omega\rho_e^2}\int_0^\infty dE\,\frac{\partial f(E)}{\partial E}\int d\vec{r}_1 d\vec{r}_2 \nabla_1 V(\vec{r}_1).\nabla_2 V(\vec{r}_2)\,|\rho^1(\vec{r}_1\vec{r}_2E)|^2 \qquad (4.3)$$

where V is the total scattering potential while ρ^1 is the energy derivative of the Dirac density matrix defined by

$$\rho^1(\vec{r}_1\vec{r}_2E) = \sum_n \psi_n^*(\vec{r}_1)\psi_n(\vec{r}_2)\delta(E-E_n) \qquad (4.4)$$

the ψ_n's and E_n's being the eigenfunctions and eigenvalues of V.

As we shall discuss further below, Silver and McGill (1974) point out that this same force-force correlation function formula for resistivity is obtained using a procedure proposed by Kubo for evaluating the inverse of the relaxation time τ in equation (2.7). We outline the argument, because of the obvious interest in the fundamental status of eqn (4.3), in Appendix 2.

5. RESISTIVITY OF DILUTE ALLOYS

To gain confidence in the use of equation (4.3) we turn immediately to discuss the simplest possible theory for the excess resistivity ΔR of a dilute alloy. By dilute we mean simply that regime in which ΔR is simply proportional to the concentration c of the impurities. We can therefore consider the scattering of completely degenerate conduction electrons, represented by plane waves $e^{i\vec{k}.\vec{r}}$, off a single spherical potential V(r).

5.1. Conventional Theory

The resistivity due to impurities in a dilute metallic solution was first calculated by Huang (1948), for a strong spherical scattering potential.

Following Mott (1936), who worked out the Born approximation, the resistivity increment ΔR is found to be

$$\Delta R = \frac{\hbar k_f}{e^2} cA \qquad (5.1)$$

where

$$A = \int_0^{\pi} (1-\cos\theta) I(\theta) 2\pi \sin\theta d\theta \qquad (5.2)$$

As Huang showed, using

$$I(\theta) = |f(\theta)|^2 \qquad (5.3)$$

where the scattering amplitude is given as usual by

$$f(\theta) = \frac{1}{2ik_f} \sum_{l=0}^{\infty} (2l+1)[\exp(2i\eta_l - 1] P_l(\cos\theta) \qquad (5.4)$$

$\eta_l(k_f)$ being the phase shifts for scattering at the Fermi surface, the integration over angles in eqn (5.2) can be completed to yield

$$\Delta R = \frac{2h}{e^2 k_f} \sum_{l=0}^{\infty} [(2l+1)\sin^2\eta_l - 2l\sin\eta_l \sin\eta_{l-1} \cos(\eta_{l-l} - \eta_l)] \qquad (5.5)$$

An alternative form can be written

$$\Delta R = \frac{4\pi}{k_f^2} \sum_{l=1}^{\infty} l \sin^2(\eta_{l-l} - \eta_l) \qquad (5.6)$$

Born Approximation

The Born approximation replaces $\exp(2i\eta_l)-1$ by $2i\eta_l$ and η_l is given in terms of the (weak) spherical scattering potential by

$$\eta_l(k) = -\pi \int_0^{\infty} V(r) J^2_{l+\frac{1}{2}}(kr) r\, dr \qquad (5.7)$$

Then we find

$$f(\theta) = 2\int_0^{\infty} \frac{\sin Kr}{Kr} V(r) r^2 dr \qquad (5.8)$$

where $K = 2k_f \sin\frac{\theta}{2}$ as usual.

If we take the Thomas-Fermi screened potential for an excess charge Ze in a Fermi gas, namely

$$V(r) = -\frac{Ze^2}{r}\exp(-qr) \;:\; q^2 = \frac{4k_f}{\pi a_0} \tag{5.9}$$

then we get from eqn (5.8)

$$f(\theta) = \frac{-2Zq^2}{K^2(q^2+K^2)} \tag{5.10}$$

Inserting this into the formula for the resistivity change ΔR yields (Mott 1936)

$$\Delta R = \frac{q\pi^2 Z^2 c}{2r_s^2 k_f^6}\left[\ln\left(1+\frac{1}{y}\right) - \frac{1}{1+y}\right] \tag{5.11}$$

with $y = q^2/4k_f^2$. The initial slope $\frac{\partial \Delta R}{\partial c}$ is therefore proportional to Z^2, in agreement with the law of Linde (1939) for alloys of noble metals with Z>0 (e.g. Zn, Ga, Ge in Cu).

5.2. Inverse Transport Theory

Let us start from the formula (4.3) in the completely degenerate limit and in Born approximation. Since (see eqn (5.11) above), this means calculating ΔR to order V^2 only, we can evidently insert in eqn (4.3) the free electron result for the Dirac density matrix, i.e.

$$\Delta R = \frac{2\pi}{3\Omega\rho_e^3}\int d\vec{r}_1 d\vec{r}_2 \nabla_1 V(\vec{r}_1).\nabla_2 V(\vec{r}_2)\,|\rho_0'(\vec{r}_1\;\vec{r}_2\;E_f)|^2 \tag{5.12}$$

E_f being of course the Fermi energy.

But the Dirac density matrix for free electrons in volume w is immediately

$$\rho_0(\vec{r}_1\vec{r}_2E_f) = \sum_{|k|<k_f}\frac{1}{w}e^{i\vec{k}.(\vec{r}_1-\vec{r}_2)} = \frac{k_f^2}{\pi^2}\,\frac{j_1(k_f|\vec{r}_1-\vec{r}_2|)}{k_f|\vec{r}_1-\vec{r}_2|} \;:\; E_f = \frac{k_f^2}{2} \tag{5.13}$$

where j_1 is the first-order spherical Bessel function

$$j_1(x) = \frac{\sin x - x\cos x}{x^2} \tag{5.14}$$

Using the result (5.13) for general energy E, differentiating with respect to E to form $\rho_0'(\vec{r}_1\vec{r}_2E_f)$, it is a straightforward, if slightly tedious, matter to perform the angular integrations in eqn (5.12) for a spherically symmetric potential. Completing the integration over r_1 and r_2 for the screened Coulomb potential (5.9), the result (5.11) is regained. Thus, the force-force correlation function formula certainly leads back to the well established result for weak scattering by spherically symmetrical potentials. It is also of interest that in the non-degenerate limit when $f(E)\alpha\exp(-\frac{E}{k_BT})$ we also regain the Brooks-Herring formula for ionized impurity scattering, q^{-1} in(5.9) now being the Debye length.

Strong Spherical Scatterers

As was pointed out by Rousseau, Stoddart and March (1972), the formula (4.3) can be rewritten in terms of the total T matrix of the scattering assembly as

$$R = -\frac{1}{3(2\pi)^5\Omega\rho_e^2}\int_0^\infty dEf'(E)\int d\vec{k}_1 d\vec{k}_2(\vec{k}_1-\vec{k}_2)^2$$
$$\times|T(\vec{k}_1\vec{k}_2)|^2\delta(E-k_1^2)\delta(E-k_2^2) \tag{5.15}$$

We shall not reproduce the derivation here as it is given in full by Rousseau, Stoddart and March (1973). However, for single-centre scattering, we can write the T matrix in terms of the phase shifts, and by this route we can regain the usual result (5.6) for strong scattering.

However, just as we built up the Dirac density matrix (5.13) for free electrons from plane waves, so we can build up the full density matrix for a single scatterer from the radial wave functions R_l (rE) and the sphericalharmonics $Y_{lm}(\theta,\phi)$.

Following March and Murray (1960) we can decompose $\frac{\partial\rho}{\partial E}(\vec{r}\,\vec{r}'E)$ into l components $\frac{\partial\rho_l(rr'E)}{\partial E}$ which, as shown by Rousseau (1971; see also Harris, 1972) takes the form

$$\frac{\partial\rho_l(rr'E)}{E} = \frac{\sqrt{E}}{\pi}\cos^2\eta_l(E)R_l(rE)R_l(r'E) \tag{5.16}$$

The angular integrations in the force-force correlation function can again be completed. Furthermore, with the (trivially different) normalization

$$R_l = j_l\cos\eta_l - n_l\sin\eta_l \tag{5.17}$$

j and n being the usual spherical Bessel and Neumann functions, Gaspari and Gyorffy (1972) have shown that

$$\int_0^\infty dr\; r^2 R_l \frac{dV}{dr} R_{l+1} = \sin(\eta_{l+1} - \eta_l) \tag{5.18}$$

To get this nice result, the following steps are involved :

(i) Differentiate the Schrodinger equation for $u_l = rR_l$, multiply by u_{l+1} and integrate.

(ii) Repeat for u_{l+1} and add the two results.

(iii) After integrating by parts, the remaining integrals are evaluated again using the Schrodinger equation.

This yields

$$\int_0^a dr\; r^2 R_l \frac{dV}{dr} R_{l+1}$$

$$= \frac{1}{2}[u_l u''_{l+1} - 2u'_l u'_{l+1} + u''_l u_{l+1}]_0^a$$

$$+ [(l+1)/a][u_l u'_{l+1} - u'_l u_{l+1}]_0^a \tag{5.19}$$

For the limit a greater than the range of the potential, or in the case we are interested in, $a\to\infty$, this gives $\sin(\eta_{l+1}-\eta_l)$.

Using this result, we readily regain the full Faxen-Holtsmark theory, and the resistivity result (4.10).

Thus, the force-force correlation function formula gives back the well established results for impurity scattering in a dilute metallic alloy. We shall turn next to discuss the application to liquid metals.

6. RESISTIVITY OF LIQUID METALS

In the weak scattering theory of liquid metals we start out from the general assumption

$$V(\vec{r}) = \sum_{\vec{l}} v(\vec{r}-\vec{l}) \tag{6.1}$$

for scatterers on sites $\vec{l}$, and calculate the probability $P_{\vec{k}\vec{k}'}$ of scattering from $\vec{k}$ to $\vec{k}'$ via the golden rule. We get

$$P_{\vec{k}\vec{k}'} \propto |v(\vec{k}-\vec{k}')|^2 \; |e^{i(\vec{k}-\vec{k}').\vec{l}}|^2 \tag{6.2}$$

and we now ensemble average to replace $|e^{i(\vec{k}-\vec{k}').\vec{l}}|^2$ in favour of the liquid structure factor i.e.

$$P_{\vec{k}\vec{k}'} \propto |v(K)|^2 \; S(K) \quad : K = |\vec{k}-\vec{k}'| \tag{6.3}$$

This is essentially I (θ) in eqn (5.2) and we find the usual Ziman formula

$$R = \frac{\rho_i}{48\pi^3\rho_e^3} \int_0^{2k_f} dK K^3 S(K)\,|v(K)|^2 \tag{6.4}$$

From the T matrix formula (5.15), it is simple to show for weak scattering that

$$<|T(k_1k_2)|^2> = NS(|\vec{k}_1-\vec{k}_2|)\,|v(\vec{k}_1-\vec{k}_2)|^2 \tag{6.5}$$

and the weak scattering result is regained from eqn (5.15) which is equivalent to the force-force correlation function result (4.3).

6.1. Strong Scattering Theory

Different approximations have been proposed for the resistivity of a liquid metal when the scattering is strong. None of these is completely satisfactory as regards its theoretical foundations.

Thus Rousseau, Stoddart and March (1973) propose to replace the formula (6.5) for weak scattering by

$$<|T(k_1k_2)|^2> = NS(|\vec{k}_1-\vec{k}_2|)\,|\tau(\vec{k}_1\vec{k}_2)|^2 \tag{6.6}$$

where $\tau(\vec{k}_1\vec{k}_2)$ is related to the single-centre t matrix through

$$\tau(\vec{k}_1\vec{k}_2) = t(\vec{k}_1\vec{k}_2) + \frac{1}{(2\pi)^3}\int d\vec{k}\; t(\vec{k}_1\vec{k})g(\vec{k}_1-\vec{k})\, G_0^+(\vec{k})\tau(\vec{k}_1\vec{k}_2) \tag{6.7}$$

where g(k) is the Fourier transform of the radial distribution function g(r). Here $G_0^+(\vec{k})$ is simply

$$G_0^+(\vec{k}) = (E-k^2+i\varepsilon)^{-1} \tag{6.8}$$

Preliminary calculations in which eqn (6.7) was solved for liquid Ca have been reported by Rousseau, Stoddart and March (1973). It should be stressed that the Ziman pseudopotential is here being transcended by the use of a structure-dependent t matrix. The Bristol group have used the single-centre t matrix rather than the weak pseudopotential in extensive calculations (see Evans et al 1973 and other references there).

7. KONDO EFFECT

As an example of a more sophisticated kind than considered so far in these lectures, we shall now use the force-force correlation function expression to calculate the resistivity of a free-electron metal in the presence of local moments (cf Silver and Mc Gill 1974). In Appendix 2 we record the appropriate generalization of the force-force correlation function required to treat this problem.

We follow Kondo in taking the conduction electron-spin interaction to have the form

$$V_{\alpha\beta}(\vec{x}t) = J\,\vec{\sigma}_{\alpha\beta}\cdot\sum_{\alpha}\vec{S}_{\alpha}(t)\delta(\vec{x}-\vec{R}_{\alpha}) \tag{7.1}$$

where $\vec{\sigma}_{\alpha\beta}$ denote the Pauli spin matrices, $\vec{S}_{\alpha}(t)$ is the time-dependent spin, while $\vec{R}_{\alpha}$ denotes the position of the αth spin.

The approach adopted is to develop the resistivity formula as a perturbation expansion in J, the strength of the conduction electron-spin interaction. Though the interesting term for the resistance minimum is the J^3 contribution, we shall consider the J^2 term in some detail and then merely indicate how the J^3 term can be got, before giving the result to this order.

To evaluate the resistivity formula (A2.5) we need the quantity

$$\int d\vec{x}\ \psi^{+}_{\beta}(\vec{x}t)\nabla_{\mu}V_{\beta\alpha}(\vec{x}t)\psi_{\alpha}(\vec{x}t)$$

It is convenient in calculating this to define

$$S^{i}(\vec{k}t) = \sum_{\alpha} \exp(-i\vec{k}.\vec{R}_{\alpha})S^{i}_{\alpha}(t) \qquad (7.2)$$

Then we can express the above matrix element of the force as

$$\int dx\ \psi^{+}_{\beta}(\vec{x}t)\nabla_{\mu}V_{\beta\alpha}(\vec{x}t)\psi_{\alpha}(\vec{x}t)$$

$$= J\int \frac{d\vec{k}_1}{(2\pi)^3}\frac{d\vec{k}_2}{(2\pi)^3}\, i(\vec{k}_1-\vec{k}_2)_{\mu} \times \sum_{i} a^{+}_{\vec{k}_1\beta}(t)a_{\vec{k}_2\alpha}(t) \times S^{i}(\vec{k}_1-\vec{k}_2 t) \qquad (7.3)$$

the a^{+} and a being creation and annihilation operators related to the ψ's in the usual way :

$$\psi^{+}_{\alpha}(\vec{x}t) = \int \frac{d\vec{k}_1}{(2\pi)^3}\exp(-i\vec{k}_1.\vec{x})a^{\pm}_{\vec{k}_1\alpha}(t)$$

$$\psi_{\beta}(\vec{x}t) = \int \frac{d\vec{k}_2}{(2\pi)^3}\exp(-i\vec{k}_2.\vec{x})a_{\vec{k}_2\beta}(t) \qquad (7.4)$$

Due to the fact that we are using a perturbation expansion in J, we also have

$$a^{+}_{\vec{k}\beta}(t) = a^{+}_{k\beta}\exp(i\varepsilon_{\vec{k}}t/\hbar);\ a_{\vec{k}\beta}(t) = a_{\vec{k}\beta}\exp(-i\varepsilon_{\vec{k}}t/\hbar) \qquad (7.5)$$

with $\varepsilon_{\vec{k}} = h^2k^2/2m$. Matrix elements of the form $Tr\{\hat{\rho}_a a^{+}a\ldots a^{+}a\}$ can be evaluated by procedures set out, for example, by Fetter and Walecka (1971). Then the contribution to the resistivity R_{μ} of order J^2 is given by

$$R^{(2)}_{\mu} = \frac{1}{e^2n^2\Omega}\frac{J^2}{k_BT}\int_0^{\infty}dt\ \frac{d\vec{k}_1}{(2\pi)^3}\frac{d\vec{k}_2}{(2\pi)^3}\frac{d\vec{k}_3}{(2\pi)^3}\frac{d\vec{k}_4}{(2\pi)^3}$$

$$[i(\vec{k}_1-\vec{k}_2)_{\mu}\times i(\vec{k}_3-\vec{k}_4)_{\mu}] \times \sum_{\alpha\beta\gamma\delta}\sum_{ij}\sigma^{i}_{\beta\alpha}\sigma^{i}_{\gamma\delta}e^{-\varepsilon t}e^{i(\varepsilon_{\vec{k}_1}-\varepsilon_{\vec{k}_2})t/h}$$

$$Tr[\rho^{0}_{G}\ S^{i}(\vec{k}_1-\vec{k}_2,t) \times S^{j}(\vec{k}_3-\vec{k}_4,0)]$$

$$\times (2\pi)^3\delta(\vec{k}_1-\vec{k}_4)\delta_{\beta\delta}(2\pi)^3\delta(\vec{k}_2-\vec{k}_3)\delta_{\alpha\gamma}n^{0}_{1}(1-n^{0}_{3}) \qquad (7.6)$$

Here n_1^0 is the Fermi function corresponding to momentum 1 etc. while ρ_G^0 is the grand canonical statistical operator in the absence of a conduction electron-spin interaction. Defining a second-order spin correlation function by

$$S^{(2)}(\vec{k}\omega) = \sum_i \int dt\ e^{i\omega t} \mathrm{Tr}[\rho_G^0 S^i(\vec{k}t) S^i(-\vec{k}\ 0)] \qquad (7.7)$$

we obtain

$$R_\mu^{(2)} = \frac{2J^2}{e^2 n^2 \Omega k_B T} \int \frac{d\omega}{2\pi} \int \frac{d\vec{k}_1}{(2\pi)^3} \frac{d\vec{k}_3}{(2\pi)^3} \frac{(\vec{k}_1-\vec{k}_3)_\mu (\vec{k}_3-\vec{k}_1)_\mu}{i(\varepsilon_{\vec{k}_1} - \varepsilon_{\vec{k}_3} - \hbar\omega)/\hbar - \delta}$$
$$\times\ S^{(2)}(\vec{k}_1-\vec{k}_3, \omega)\ n_1^0(1-n_3^0) \qquad (7.8)$$

Just as in the weak (2nd order) scattering theory of liquid metals, the liquid pair function entered (cf eqn 6.4), so the second-order contribution to the resistivity here involves a second-order spin correlation function. It turns out that, in general, the nth order contribution to the resistivity is determined by an nth-order spin-correlation function.

In Appendix 3, we outline the parallel argument to obtain the third-order term. This is essential in the Kondo problem, as the contributions leading to the resistance minimum come from third and higher orders.

7.1. Resistance Minimum for a Single Local Moment

We now specialize the argument to the case of a single spin. Then we can write from eqn (7.2)

$$S^i(\vec{k}t) = e^{-i\vec{k}\cdot\vec{r}_\alpha} S_\alpha^i \qquad (7.9)$$

and one has in the equation for S_3 (cf eqn A3.3)

$$-i\varepsilon^{ijk} \mathrm{Tr}[\rho_G^0 S^i(\vec{k}_1 t_1) S^j(\vec{k}_2 t_2) S^k(\vec{k}_3 t_3)] = S(S+1) \qquad (7.10)$$

when $\vec{k}_1+\vec{k}_2+\vec{k}_3 = 0$ and S is the spin of the local moment. Therefore

$$S^{(3)}(\vec{k}_1\vec{k}_3\omega_1\omega_3) = S(S+1)2\pi\delta(\omega_1)2\pi\delta(\omega_3) \qquad (7.11)$$

for a single moment.

Performing the angular integrations in $R_\mu^{(3)}$ leads then to

$$R_\mu^{(3)} = \frac{2S(S+1)}{3e^2n^2\Omega}\frac{J^3(-i/\hbar)}{k_BT(2\pi)^9} \times \int k_1^2dk_1k_3^2dk_3k_5^2dk_5(k_1^2+k_3^2+k_5^2)$$

$$\times \Big(\frac{(-1)n_1^0(1-n_5^0)n_3^0}{[i(\varepsilon_1-\varepsilon_5)/\hbar-\varepsilon][i(\varepsilon_5-\varepsilon_3)/\hbar+\varepsilon]}$$

$$+ \frac{(-1)n_5^0(1-n_1^0)(1-n_3^0)}{[i(\varepsilon_5-\varepsilon_1)/\hbar-\varepsilon][i(\varepsilon_3-\varepsilon_5)/\hbar+\varepsilon]}\Big) \tag{7.12}$$

We now use the relation that for a function $f(k_1)$ invariant under $k_1\to-k_1$ and $k_5\to 0$

$$\int_0^\infty \frac{dk_1f(k_1)}{i(\varepsilon_1-\varepsilon_5)/\hbar-\varepsilon} - \int_0^\infty \frac{dk_1f(k_1)}{i(\varepsilon_5-\varepsilon_1)/\hbar-\varepsilon} = \frac{-\pi\,\mathrm{Im}f(k_5)}{\hbar k_5} \tag{7.13}$$

Then we get

$$R_\mu^{(3)} = \frac{-2S(S+1)J^3\pi m}{3e^2n^2\Omega(2\pi)^9\hbar^2k_BT}\int k_5^2dk_5k_3^2dk_3$$

$$\times (2k_5^2+k_3^2)n_5^0(1-n_5^0)\frac{2n_3^0-1}{\varepsilon_5-\varepsilon_3} \tag{7.14}$$

Qualitative Form in Low Temperature Limit

Consider the form

$$I(T) = \frac{1}{k_BT}\int d\varepsilon\; n^0(1-n^0)f(\varepsilon) \tag{7.15}$$

where $f(\varepsilon)$ is a smoothly varying function of ε. At low temperatures, the limits in the integral can be taken as $-\infty$ to ∞ without appreciable error and then

$$I(T) = \int_{-\infty}^{\infty}dx\,\frac{1}{e^x+1}\,\frac{1}{1+e^{-x}}\,f(kTx+\mu) \tag{7.16}$$

where μ is the Fermi energy. At low temperatures we then find

$$I(T) \simeq f(\mu) \tag{7.17}$$

Using this in eqn (7.14) for $R_\mu^{(3)}$ one finds

$$R_\mu^{(3)} \propto - J^3 \int d\varepsilon_3 \frac{(2n_3^0-1)g(\varepsilon_3)}{\mu-\varepsilon_3} \qquad (7.18)$$

where $g(\varepsilon_3)$ is a positive, smoothly varying, function of ε_3. This is exactly the form obtained by Kondo in his third-order calculation. It is easily shown that

$$R_\mu^{(3)} \propto + J^3 \ln k_B T \qquad (7.19)$$

Adding this to the phonon contribution which clearly decreases with decreasing T, then when J is negative (antiferromagnetic) the resistance minimum follows. The divergence at T = 0K is removed by the higher-order terms in the perturbation expansion of R.

Resistance Minima in Amorphous Ferromagnets

Silver and McGill emphasize that it is not possible to neglect the interaction between spins in alloys with appreciable concentrations of ferromagnetic elements and which tend to magnetically order as the temperature is lowered. Amorphous ferromagnetic alloys continue to exhibit resistance minima well below the magnetic ordering temperature. They enquire what properties of the spin correlation functions are needed in order that the sum of the spin and phonon contributions yield a minimum in the resistance at some temperature. Specifically they assume that the Kondo mechanism continues to hold, and that, in third order

$$R_\mu^{(3)} \doteq A(T)\ln kT + B(T) \qquad (7.20)$$

where A(T) and B(T) are both well behaved in the low temperature limit.

Using the resistivity formalism discussed above, and studying the triple-spin correlation function as $T\to 0$, Silver and McGill conclude that the only way A(T) can remain non-zero in the low temperature limit is for the triple spin correlation function to have the form

$$S^{(3)}(k_1 k_3 \omega_1 \omega_3) = S(k_1 k_3)\delta(\omega_1)\delta(\omega_3) \qquad (7.21)$$

+ part analytic in ω_1 and ω_3.

We must refer the reader to their paper for the argument. Because of the sum rules relating $S^{(3)}$ and $S^{(2)}$, this implies that

$$S^{(2)}(k\omega) = S(k)\delta(\omega) + \text{part analytic in } \omega.$$

The physical meaning of this statement is that the alloy must have a finite density of zero frequency excitations in order that the coefficient of the ln kT term shall remain non-zero in the low temperature limit. Thus, unless some other mechanism than the conduction electron-spin interaction is operating, it appears that the spin excitation spectra of amorphous ferromagnets exhibiting resistivity minima must have a finite density of (near) zero frequency excitations.

ACKNOWLEDGMENTS

During the preparation of these lectures, Dr. Barry McCoy has given me very generous help with the classical problems discussed here and I am most grateful to him. Dr.Peter Schofield also made valuable comments relating to diffusion in classical liquids.

APPENDIX 1

Electrical Conductivity of Ionized Plasma

Edwards and Sanderson (1961) work with the Green function G, which is the probability that, in equilibrium, one finds a particle at time t' with a velocity u' and that subsequently the same particle has velocity $\vec{u}$ at time t. Note that the spatial coordinates are integrated out.

In the Lorentz gas, with static scatterers, the two-particle Green function which is similarly referring to the velocities of two particles is given by

$$G_{12} \doteq G_1 G_2 \tag{A1.1}$$

The defining equation for G_1 is

$$\left(\frac{\partial}{\partial t} + \frac{\alpha_0}{u^2}\frac{\partial^2}{\partial\Omega^2}\right)G_1(\vec{u}\ u'tt') = \delta(\vec{u}-\vec{u}')\delta(t-t') \tag{A1.2}$$

where

$$\alpha_0 = -\frac{2\pi Ne^4}{m^2}\int_0^{2/r_0}\frac{kdk}{(k^2+\lambda^{-2})}$$

$$= -\frac{2\pi Ne^4}{m^2}\ \ln(2\lambda/r_0+1)$$

$$\simeq -\frac{2\pi Ne^4}{m^2}\ln(2\lambda/r_0) \tag{A1.3}$$

One can write

$$G_1 = G_1^1\delta(|u|-|u'|) \tag{A1.4}$$

and as

$$\delta(\vec{u}-\vec{u}') = \frac{1}{v^2}\delta(\cos\theta-\cos\theta')\delta(\phi-\phi')\delta(|\vec{u}|-|\vec{u}'|) \tag{A1.5}$$

the equation for G_1^1 becomes

$$\left(\frac{\partial}{\partial t}+\frac{\alpha_0}{u^3}\frac{\partial^2}{\partial\Omega^2}\right)G_1^1 = \frac{1}{u^2}\,\delta(\cos\theta-\cos\theta')\delta(\phi-\phi') \tag{A1.6}$$

This is solved in terms of spherical harmonics $Y_{1n}(\theta,\phi)$ which are eigenfunctions of the differential operator $\partial^2/\partial\Omega^2$.

$$G_1^1 = \frac{1}{u^2}\,\Sigma\,\exp[-\alpha_0(t-t')1(1+1)/u^3]$$
$$Y_{1n}(\theta,\phi)Y^*_{1n}(\theta',\phi')\ H\ (t-t') \tag{A1.7}$$

$\sigma_{\mu\nu}$, which is clearly isotropic, is given by

$$\sigma_{\mu\nu} = \sigma\delta_{\mu\nu} \tag{A1.8}$$

where, when the electric field is switched on at $t = 0$,

$$\sigma = \frac{Ne^2}{k_BT}\left(\frac{m}{2\pi k_BT}\right)^{3/2}\int_0^t \cos\theta\cos\theta'\ \Sigma\ \frac{1}{u^2}\ Y_{1n}Y^*_{1n}u^2$$
$$\times\ \exp[-\alpha_0(t-t')1(1+1)/u^3-mu^2/2k_BT]\,u^2du \tag{A1.9}$$

Only $1 = 1$, $m = 0$ contribute and as $t\to\infty$, σ tends to

$$\sigma = \frac{2Ne^2}{3k_BT}\left(\frac{m}{2\pi k_BT}\right)^{3/2}\int_0^\infty\int_0^\infty \exp(-2\alpha_0t/u^3)dt$$
$$\times\ 2\pi u^4du\ \exp(-mu^2/2k_BT)$$

$$= \frac{2Ne^2\pi}{3\alpha_0k_BT}\left(\frac{m}{2\pi k_BT}\right)^{3/2}\int_0^\infty u^7\exp(-mu^2/2k_BT)du$$

$$= 2(2k_BT/\pi)^{3/2}(e^2m^{1/2}\ln(2\lambda/r_0))^{-1} \tag{A1.10}$$

The distribution function is given by $f = f_0 + f_1$

$$= f_0 + \frac{Ne}{k_BT}\left(\frac{m}{2\pi k_BT}\right)^{3/2} \int_0^t \sum_{ln} Y_{ln} Y^*_{ln} \exp(-\alpha_0 l(l+1)(t-t')/u^3)$$
$$\times\ u^{-2}\delta(|\vec{u}|-|\vec{u}'|)\vec{\varepsilon}.\vec{u}' \exp(-mu^2/2k_BT)\,d\vec{u}$$

$$= f_0 + Ne(\vec{\varepsilon}.\vec{u})u^3/(2\alpha_0 k_BT)f_0 \qquad (A1.11)$$

when t-t' tends to infinity.

APPENDIX 2

Connection of Resistivity Formula (4.3) with Kubo's Evaluation of Inverse of Relaxation Time

As Silver and McGill (1974) emphasize, Kubo has proposed what he considers to be an approximate procedure to evaluate the relaxation time in the conductivity formula (2.7). Thus he writes

$$\frac{1}{\tau} = \lim_{\eta\to 0^+} \frac{m}{n} \int_\eta^\infty \frac{\partial^2}{\partial t^2} I_\mu(t)\,dt \qquad (A2.1)$$

Here $I_\mu(t)$ is, in its most elementary form

$$I_\mu(t) = (n/m)e^{-t/\tau}\theta(t) \qquad (A2.2)$$

and comes in via a variant of the Kubo formula. The above form leads to a frequency dependent conductivity

$$\sigma_\mu(\omega) = ne^2\tau_\mu/m(1-i\omega\tau_\mu) \qquad (A2.3)$$

leading back to the usual relation for $\omega\to 0$.

In the Kubo formula

$$\sigma_{\mu\nu}(\omega) = -\frac{ie^2}{\hbar\Omega}\int d\vec{x}\,d\vec{x}'d(t-t')e^{i\omega(t-t')}\theta(t-t')$$
$$\times\ \mathrm{Tr}\{\rho_G[\hat{X}_\nu(\vec{x}'t'),\ J_\mu(\vec{x},t)]\} \qquad (A2.4)$$

the central problem is to calculate $I_\mu(t)$ which is the diagonal form of

$$I_{\mu\nu}(t-t') = -\frac{i\theta(t-t')}{\hbar\omega}\int d\vec{x}\,d\vec{x}'$$
$$\times\ \mathrm{Tr}\{\rho_G[\hat{X}_\nu(\vec{x}',t'),\ J_\mu(\vec{x},t)]\} \qquad (A2.5)$$

One then obtains for the diagonal resistivity R_μ the result (cf Silver and Mc Gill, 1974)

$$R_\mu = \lim_{\eta\to 0^+} \frac{m^2}{n^2 e^2 \Omega} \frac{1}{k_B T} \int_\eta^\infty dt \int d\vec{x}\, d\vec{x}'$$

$$\mathrm{Tr}[\rho_G \frac{\partial J_\mu(\vec{x},t)}{\partial t} \frac{\partial J_\mu(\vec{x}',0)}{\partial t}] \qquad \text{(A2.6)}$$

For a Hamiltonian of the form

$$H_{el} = \sum_\lambda \int d\vec{x}\, \psi^*_\lambda(x)\left(\frac{-\hbar^2\nabla^2}{2m}\right)\psi_\lambda(x)$$

$$+ \sum_{\alpha\beta} \int d\vec{x}\, \psi^+_\alpha(x) V_{\alpha\beta}(x)\psi_\beta(x) \qquad \text{(A2.7)}$$

one has

$$\frac{\partial J_\mu(\dot{x},t)}{\partial t} = -\frac{1}{m}\sum_{\alpha\beta} \psi^*_\beta(xt)\nabla_\mu V_{\alpha\beta}(xt)\psi_\alpha(xt) \qquad \text{(A2.8)}$$

and thus

$$R_\mu = \lim_{\eta\to 0^+} \frac{1}{n^2 e^2 \Omega} \frac{1}{k_B T} \int_\eta^\infty dt \int d\vec{x}\, d\vec{x}'$$

$$\times\ \mathrm{Tr}\{\rho_G \sum_{\alpha\beta} \psi^*_3(xt)\nabla_\mu V_{\beta\alpha}(xt)\psi_\alpha(xt)$$

$$\times \sum_{\gamma\delta} \psi^+_\gamma(x'0)\nabla_\mu V_{\gamma\delta}(x',0)\psi_\delta(x'0)\} \qquad \text{(A2.9)}$$

This formula contains the result (4.3) of Rousseau, Stoddart and March (1972). Eqn (A2.5) is, however, somewhat more general and, in particular, the Fermi factors are not explicitly identified. Such an identification is possible if the potential becomes spin and time independent.

The generalized result (A2.5), due to Silver and McGill, was used by them to discuss the Kondo effect (see section 7).

APPENDIX 3

Third-order Term in Kondo Problem

In the main text, we calculated the resistivity contribution to order J^2. The contributions leading to the resistance minimum come from third and higher orders. In calculating the J^3 contribution, one must bear in mind the cancellation of disconnected parts and remove such pieces as they arise. Then one can write

$$R_\mu^{(3)} = \frac{1}{e^2 n^2 \Omega} \frac{1}{k_B T} \int_0^\infty dt \int d\vec{x} \int d\vec{x}' \int_0^\infty dt_1 \int d\vec{x}_1 \left(-\frac{i}{\hbar}\right) \mathrm{Tr}[\, \rho_G^0 T$$

$$\left(\sum_{\alpha\beta} \psi_\beta^+(\vec{x},t) \nabla_\mu V_{\beta\alpha}(\vec{x},t) \psi_\alpha(\vec{x},t) + \sum_{\gamma\delta} \psi_\gamma^+(\vec{x}',0) \nabla_\mu V_{\gamma\delta}(\vec{x}',0)\right.$$

$$\left.\psi_\delta(\vec{x}',0) \sum_{\varepsilon\phi} \psi_\varepsilon^+(\vec{x}_1 t_1) V_{\varepsilon\phi}(\vec{x}_1 t_1) \psi_\phi(\vec{x}_1, t_1)\right)]_{\text{connected}} \tag{A3.1}$$

T as usual means that the operators inside the trace should be time-ordered. In momentum space

$$R_\mu^{(3)} = \frac{1}{e^2 n^2 \Omega} \frac{J^3}{k_B T} \left(-\frac{i}{\hbar}\right) \int_0^\infty dt e^{-\varepsilon t} \int_{-\infty}^\infty dt' e^{-\varepsilon |t'|}$$

$$\int \frac{d\vec{k}_1}{(2\pi)^3} \frac{d\vec{k}_2}{(2\pi)^3} \frac{d\vec{k}_3}{(2\pi)^3} \frac{d\vec{k}_4}{(2\pi)^3} \frac{d\vec{k}_5}{(2\pi)^3} \frac{d\vec{k}_6}{(2\pi)^3} i(\vec{k}_1 - \vec{k}_2)_\mu i(\vec{k}_3 - \vec{k}_4)_\mu$$

$$\times \sum_{\alpha\beta\gamma\delta\varepsilon\phi} \mathrm{Tr}\{\rho_G^0 T[\, (\sum_i a^+_{\vec{k}\beta}(t) \sigma^i_{\beta\alpha} a_{\vec{k}_2\alpha}(t)\ S^i(\vec{k}_1 - \vec{k}_2, t))$$

$$\times (\sum_j a^+_{\vec{k}_3,\gamma}(0) \sigma^j_{\gamma\delta} a_{\vec{k}_4,\delta}(0) S^i(\vec{k}_3 - \vec{k}_4, 0)$$

$$\times \sum_k a^+_{\vec{k}_5,\varepsilon}(t') \sigma^k_{\varepsilon\phi} a_{\vec{k}_6,\phi}(t') S^i(\vec{k}_5 - \vec{k}_6, t))]\}_{\text{connected}} \tag{A3.2}$$

The subscript 'connected' now means that in the reduction of this expression, all terms with $\vec{k}_1 = \vec{k}_2$, $\vec{k}_3 = \vec{k}_4$ or $\vec{k}_5 = \vec{k}_6$ are to be removed.

The evaluation of this third-order result is lengthy but fairly straightforward. It is useful to express the result in terms of a triple-spin correlation function. Thus define

$$i\varepsilon^{ijk}\mathrm{Tr}[\rho_G^0 s^i(\vec{k}_1\vec{t}_1)s^j(\vec{k}_2\vec{t}_2)s^k(\vec{k}_3\vec{t}_3)]$$

$$\equiv \int \frac{d\omega_1}{2\pi}\frac{d\omega_3}{2\pi} e^{-i\omega_1(t_1-t_2)} e^{i\omega_3(t_3-t_2)} S^{(3)}(\vec{k}_1\vec{k}_3\omega_1\omega_3) \tag{A3.3}$$

Two momentum labels are enough because the only correlation functions to be considered are those for which $\vec{k}_1 + \vec{k}_2 + \vec{k}_3 = 0$. The triple correlations can be shown to satisfy

$$[S^{(3)}(\vec{k}_1\vec{k}_3\omega_1\omega_3)]^+ = S^{(3)}(-\vec{k}_3,\vec{k}_1,\omega_3,\omega_1) \tag{A3.4}$$

Of course $S^{(2)}(\vec{k}\omega)$ and $S^{(3)}(\vec{k}_1\vec{k}_3\omega_1\omega_3)$ are not independent. In particular there are sum rules

$$\int \frac{d\omega_1}{2\pi} S^{(3)}(\vec{k}_1\vec{k}_3\omega_1\omega_3) = S^{(2)}(-\vec{k}_3,\omega_3) \tag{A3.5}$$

and

$$\int \frac{d\omega_3}{2\pi} S^{(3)}(\vec{k}_1\vec{k}_3\omega_1\omega_3) = S^{(2)}(\vec{k}_1\omega_1) \tag{A3.6}$$

Then the desired expression for the third-order term in the resistivity is

$$R_\mu^{(3)} = \frac{2J^3}{e^2n^2k_BT\Omega}\left(\frac{-i}{\hbar}\right)\int\frac{d\vec{k}_1}{(2\pi)^3}\frac{d\vec{k}_3}{(2\pi)^3}\frac{d\vec{k}_5}{(2\pi)^3}\int\frac{d\omega_1}{2\pi}\int\frac{d\omega_3}{2\pi}$$

$$\times [i(\vec{k}_1-\vec{k}_3)_\mu\, i(\vec{k}_3-\vec{k}_1)_\mu + i(\vec{k}_3-\vec{k}_5)_\mu\, i(\vec{k}_5-\vec{k}_1)_\mu]$$

$$\times S^{(3)}\ (\vec{k}_1-\vec{k}_5,\vec{k}_5-\vec{k}_3,\omega_1\omega_3)$$

$$\left(\frac{(-1)n_1^0(1-n_5^0)n_3^0}{[i(\varepsilon_1-\varepsilon_5-\hbar\omega_1)/\hbar-\varepsilon][i(\varepsilon_5-\varepsilon_3+\hbar\omega_3)/\hbar+\varepsilon]}\right.$$

$$\left.+\frac{(-1)n_5^0(1-n_1^0)(1-n_3^0)}{[i(\varepsilon_5-\varepsilon_1-\hbar\omega_1)/\hbar-\varepsilon][i(\varepsilon_3-\varepsilon_5+\hbar\omega_3)/\hbar+\omega]}\right)$$

REFERENCES

L.Ballentine and R.Heaney, 1974, J.Phys. C7, 1985

W.G.Chambers, 1973, J.Phys. C6, 2586

S.F.Edwards, 1965, Proc.Phys.Soc.Lond. 86, 977

S.F.Edwards and J.J.Sanderson, 1961, Phil.Mag.6, 71

R.Evans, B.L.Gyorffy, N.Szabo and J.M.Ziman, 1973, Proc.Conf. The Properties of Liquid Metals Ed. S. Takeuchi, (Taylor and Francis, London)

A.L.Fetter and J.D.Walecka, 1971, Quantum theory of many-particle systems (McGraw-Hill, San Francisco)

T.Gaskell and N.H.March, 1970, Phys.Lett. 33A, 460

G.D.Gaspari and B.L.Gyorffy, 1972, Phys.Rev.Lett. 28, 801

H.G.Ghassib, R.Gilbert and G.J.Morgan, 1973, J.Phys. C6, 184

R.Harris, 1972, J.Phys.C5, L56

K.Huang, 1948, Proc.Phys.Soc.60, 161

W.Jones, 1974, J.Phys. C7, 1974

1974, J.Phys. C7, 3357

J.O.Linde, 1939, Dissertation, Stockholm

N.H.March and A.M.Murray, 1960, Phys.Rev. 120, 830

N.F.Mott, 1936, Proc. Camb.Phil.Soc. 32, 281

J.S.Rousseau, 1971, J.Phys. C4, L351

J.S.Rousseau, J.C.Stoddart and N.H. March, 1972, J.Phys. C5, L175, 1973, Proc.Conf. The Properties of Liquid Metals Ed. S.Takeuchi (Taylor and Francis) London p.249

R.N.Silver and T.C.McGill, 1974, Phys.Rev. B9, 272

N.Szabo, 1972, J.Phys. C5, L241

1973, J.Phys. C6, L437

Part II
Metals and Semiconductors

INTERACTING ELECTRON GAS IN METALS ★

K.S. Singwi

Physics Department, Northwestern University

Evanston, Illinois, U.S.A.

In these lectures I propose to discuss the behavior of an interacting electron gas in the "Jellium" model of a metal. Results, which we shall obtain on the basis of this model, are indeed applicable to simple metals. In this connection, I would derive an expression for the wave number and frequency dependent dielectric function of the electron gas, the latter, as you perhaps know, plays an important role in all transport phenomena in metals and in the calculation of other interesting physical properties. In deriving this expression, I shall use the equation of motion method rather than the more fashionable diagrammatic technique. Using the latter technique, although very powerful and suggestive, not much progress has been made since the important work of Hubbard [1]. I shall closely follow the approach as developed by Singwi <u>et al</u> [2]. We shall also attempt to establish the connection between our work and that of Hohenberg, Kohn and Sham [3,4]. As occasion arises, I shall point out some of the unsolved problems.

★ A course of lectures given at the NATO Advanced Study Institute on "Linear and Nonlinear Electronic Transport in Solids", Rijksuniversitair Centrum Antwerpen, Belgium, July 21 - August 2, 1975.

PRELIMINARIES

Let us begin by reviewing some results of linear response theory as applied to a system of electrons of mean density n, moving in a neutralizing background of a static, uniform, positive charge distribution (Jellium). Such an idealized model of a metal is not very far from truth, at least we assume so, for simple metals. Consider the response of the system to an arbitrarily weak applied external potential $V_{ext}(\vec{r},t)$ which couples to the density fluctuations in the system. In a linear response regime, the induced density change is formally given by

$$\langle \delta n(\vec{r},t)\rangle = \int d\vec{r}\,'\, dt'\, \chi(\vec{r},\vec{r}\,',t,t')\, V_{ext}(r',t'), \qquad (1.1)$$

where $\chi(\vec{r},\vec{r}\,',t,t')$ is the density-density response function, which is entirely determined by the properties of the system in the absence of any external potential. For a system which is translationally invariant in space and time, χ depends only on the difference of the space-time coordinates. The Fourier transform of Eqn.(1.1) is

$$\langle n(\vec{q},\omega)\rangle = \chi(\vec{q},\omega)\, V_{ext}(\vec{q},\omega)\ . \qquad (1.2)$$

By definition

$$\chi(\vec{r}-\vec{r}\,',t-t') = -i\theta(t-t')\,\langle [\rho(\vec{r},t),\ \rho(\vec{r}\,',t')]\rangle \qquad (1.3)$$

where $\theta(t)$ is the step function =1 for $t > 0$, and =0 for $t < 0$. Since the response is causal, $\chi(\vec{q},\omega)$

is analytic in the upper half of the complex ω-plane.

The dielectric function $\varepsilon(\vec{q},\omega)$ is defined through the relation

$$V_t(\vec{q},\omega) = \frac{V_{ext}(\vec{q},\omega)}{\varepsilon(\vec{q},\omega)} , \tag{1.4}$$

where V_t is the potential felt by a test charge. This potential is

$$V_t(\vec{q},\omega) = V_{ext}(\vec{q},\omega) + \frac{4\pi e^2}{q^2} \langle\rho(\vec{q},\omega)\rangle . \tag{1.5}$$

Equations (1.2), (1.4) and (1.5) give

$$\frac{1}{\varepsilon(\vec{q},\omega)} = 1 + \frac{4\pi e^2}{q^2} \chi(\vec{q},\omega) . \tag{1.6}$$

Let us define a new function $\overline{\chi}(\vec{q},\omega)$ through

$$\langle\rho(\vec{q},\omega)\rangle = \overline{\chi}(\vec{q},\omega)\, V_t(\vec{q},\omega) , \tag{1.7}$$

then, it follows from Eqn. (1.4), (1.5) and (1.7) that

$$\varepsilon(\vec{q},\omega) = 1 - \frac{4\pi e^2}{q^2}\,\bar{\chi}(\vec{q},\omega)\ . \tag{1.8}$$

$\bar{\chi}(\vec{q},\omega)$ is also referred to as the proper (or irreducible) polarizability.

There exists the following exact relation between $\chi(\vec{q},\omega)$ and the dynamic form factor $S(\vec{q},\omega)$:

$$\chi(\vec{q},\omega) = \frac{1}{\hbar}\int_{-\infty}^{+\infty} d\omega'\,[\,S(\vec{q},\omega)-S(\vec{q},-\omega')]\,\frac{1}{\omega-\omega'+i\eta+}\ . \tag{1.9}$$

$S(\vec{q},\omega)$ is the Fourier transform of the pair-correlation function $g(\vec{r},t)$. The latter is defined by

$$g(\vec{r},t) = \frac{1}{n}\int d\vec{r}\,'\,\langle\rho(\vec{r}\,'-\vec{r},0)\rho(\vec{r}\,',t)\rangle\ , \tag{1.10}$$

where n is the number density of particles.

From (1.9), it follows that

$$\mathrm{Im}\chi(\vec{q},\omega) = -\frac{\pi}{\hbar}[\,S(\vec{q},\omega) - S(\vec{q},-\omega)] \tag{1.11}$$

From the principle of detailed balance, we have

$$S(\vec{q},-\omega) = e^{-\hbar\omega/k_BT} S(\vec{q},\omega) \quad . \tag{1.12}$$

The imaginary part of $\chi(\vec{q},\omega)$ is thus directly related to the scattering properties of the system. It is $S(\vec{q},\omega)$ which is measured directly by inelastic scattering experiments. Equation (1.11) implies that $\mathrm{Im}\chi(\vec{q},\omega)$ is an odd function of ω .

From Eqns.(1.6), (1.11) and (1.12), it follows that

$$S(\vec{q}) = -\frac{\hbar q^2}{4\pi^2 ne^2}\int d\omega\,(1-e^{-\hbar\omega/k_BT})\ \mathrm{Im}\,\frac{1}{\varepsilon(\vec{q},\omega)} \ , \tag{1.13a}$$

where $S(\vec{q})$, the static structure factor, is defined by

$$S(\vec{q}) = \frac{1}{n}\int_{-\infty}^{+\infty} d\omega\ S(\vec{q},\omega) \quad . \tag{1.14}$$

For $T=0$, Eq. (1.13a) simplifies to

$$S(\vec{q}) = -\frac{\hbar q^2}{4\pi^2 ne^2}\int_0^{\infty} d\omega\ \mathrm{Im}\,\frac{1}{\varepsilon(\vec{q},\omega)} \tag{1.13b}$$

The density-density response function and therefore the dielectric function satisfies a number of sum rules:

$$\lim_{\omega\to\infty} \chi(q,\omega) = \frac{nq^2}{m\omega^2} , \tag{1.15a}$$

which gives

$$\lim_{\omega\to\infty} \frac{1}{\varepsilon(\vec{q},\omega)} = 1 + \frac{\omega_p^2}{\omega^2} , \tag{1.15b}$$

where $\omega_p = (\frac{4\pi ne^2}{m})^{1/2}$ is the classical plasma frequency.

Also

$$\lim_{q\to 0} \frac{1}{\varepsilon(\vec{q},0)} = (1 - \frac{4\pi n^2 e^2}{q^2} K) , \tag{1.16}$$

K is the isothermal compressibility .

I. APPROXIMATE TREATMENT OF $\varepsilon(\vec{q},\omega)$

Let us introduce the average self-consistent potential due to the polarization of the medium in the form

$$V_{Pol.}(\vec{q},\omega) = \psi(\vec{q},\omega) \langle\rho(\vec{q},\omega)\rangle , \tag{1.17}$$

and write

$$\langle\rho(\vec{q},\omega)\rangle = \chi_{sc}(\vec{q},\omega)[V_{ext}(\vec{q},\omega) + V_{Pol}(\vec{q},\omega)] . \tag{1.18}$$

Equation (1.18) dfines a new response function χ_{sc}, which describes the response of the system to the external potential and the polarization potential. From the above equations, it follows that

$$\chi(\vec{q},\omega) = \frac{\chi_{sc}(\vec{q},\omega)}{1-\Psi(\vec{q},\omega)\chi_{sc}(\vec{q},\omega)} , \tag{1.19}$$

which on using Eqn. (1.6) yields

$$\varepsilon(\vec{q},\omega) = 1 - \left(\frac{4\pi e^2}{q^2}\right) \frac{\chi_{sc}(\vec{q},\omega)}{1-\left(\Psi(\vec{q},\omega) - \frac{4\pi e^2}{q^2}\right)\chi_{sc}(\vec{q},\omega)} \tag{1.20}$$

The functions χ_{sc} and Ψ are not known a <u>priori</u>. We try to make various approximations for these functions.

I.a.Random-Phase Approximation

Here we replace $\chi_{sc}(\vec{q},\omega)$ by $\chi_{o}(\vec{q},\omega)$ i.e. by the free particle response as given by Lindhard and $\chi(q,\omega)$ by $\frac{4\pi e^2}{q^2}$, then Eqn. (1.20) simplifies to

$$\varepsilon_{RPA}(\vec{q},\omega) = 1 - \frac{4\pi e^2}{q^2}\chi_{o}(\vec{q},\omega) . \tag{1.21}$$

In the limit $\omega=0$ and $\vec{q}\to 0$, (1.21) reduces to the well-known Thomas-Fermi approximation

$$\lim_{\vec{q}\to 0} \varepsilon(\vec{q},0) = 1 + \frac{q^2_{FT}}{q^2} , \tag{1.22}$$

where $q_{FT} = (\frac{4q_F}{\pi a_0})^{1/2}$, a_0 being the Bohr radius.

From the way we have arrived at Eq. (1.21) it is clear that in RPA , the inter-particle interaction is taken into account only via the average coulomb potential of the polarization charges. No allowance has been made for exchange and correlation effects. Besides, there is no reason to believe that the electrons will respond to the potential as free particles. Also by comparing Eqn.(1.22) with Eqn. (1.16), it can easily be verified that the RPA corresponds to the free-electron value for the compressibility, which obviously is not right. We also find that, if we calculate the structure factor $S(q)$ and then the static pair-correlation function $g(r)$ using Eqn.(1.21) in the fluctuation-dissipation theorem (Eqn. (1.13)), $g(r)$ turns out to be negative [2] for small inter-particle separation. This is manifestly absurd.

I.b. Hubbard Approximation

This corresponds to replacing in Eqn. (1.20) χ_{sc} by χ_0 and putting

$$\Psi(q,\omega) = \frac{4\pi e^2}{q^2} (1 - \frac{1}{2} \frac{q^2}{q^2+q_F^2}) . \tag{1.23}$$

The correction term in the local field Ψ is such that in the limit $q\to 0$, one recovers the RPA result and in the limit $q\to\infty$ this term is just $\frac{1}{2}$. Hubbard's modification takes care of exchange correlations very well but neglects coulomb correlations between electrons. It is a considerable improvement over the RPA . It still gives a negative pair-correlation function and violates the compressibility sum rule. The compressibility sum rule requires that the compressibility of the system as evaluated by differentiating the total energy twice with respect to volume should be the same as obtained by using Eq.(1.16).

I.c. STLS Approximation

In the approximation of Singwi et al. [2] χ_{sc} is replaced by χ_0 and the function $\Psi(\vec{q},\omega)$ is given by

$$\Psi(\vec{q},\omega) = \frac{4\pi e^2}{q^2} \left[1 + \frac{1}{n}\int \frac{d\vec{q}\,'}{(2\pi)^3} \frac{\vec{q}\cdot\vec{q}\,'}{q'^2} (S(\vec{q}-\vec{q}\,')-1)\right] \tag{1.24}$$

One interesting feature of the above expression is that it contains $S(q)$ i.e. the instantaneous local structure of the electron liquid around a given particle. Being a physical quantity, $S(q)$ has both exchange and coulomb correlations built in it. In fact, if for $S(q)$ we use the expression for the noninteracting electron gas, i.e., the Pauli hole in Eq. (1.24), we immediately recover [2] for Ψ the expression as given by Hubbard. In the scheme of Singwi et al $S(q)$ is determined in a self-consistent manner as is obvious through the set of equations (1.13b), (1.20) and (1.24). Detailed numerical calculations have been done [2] and it is found taht the pair distribution $g(r)$ is positive for values of density $r_s \leq 4$ and very slightly negative for $r_s \geq 5$. The compressibility sum rule is violated, but this defect has been rectified in a later modification of the theory by Vashishta and Singwi [5].

Even the modified version of the STLS theory is an approximate one. In the next lecture, I shall attempt to show that it is exact in the static situation i.e. $\omega=0$ and in the long wave lenght limit. We shall also establish a correspondence between this theory and that of Hohenberg, Kohn and Sham [3-4].

II. MEAN FIELD APPROACH [6] ($\omega=0$ AND $q\to 0$)

Consider a homogeneous electron gas with a uniform positive background which is perturbed by a weak potential varying slowly in space and time. The exact quantum-mechanical equation of motion for the density $\langle n(\vec{r},t)\rangle$ is [2]

$$\frac{\partial^2}{\partial t^2}\langle n(\vec{r},t)\rangle - \frac{1}{m}\nabla_\alpha\nabla_\beta\,\pi_{\alpha\beta}(\vec{r},t) - \frac{1}{m}\int d\vec{r}\,'\nabla_\alpha$$

$$[\nabla_\alpha v(\vec{r}-\vec{r}\,')\langle n(\vec{r},t)n(\vec{r}\,'t)\rangle] = \frac{1}{m}\nabla_\alpha[\langle n(\vec{r},t)\rangle\nabla_\alpha V_{ext}(\vec{r},t)] , \tag{1.11}$$

where ∇_α denotes differentiation with respect to the αth cartesian compoment of $\vec{r}$ and the usual convention of summation over repeated indices is used. The angular brackets denote the ground-state expectation value. The kinetic tensor $\pi_{\alpha\beta}(\vec{r},t)$ is defined by

$$\pi_{\alpha\beta}(\vec{r},t) = \sum_\sigma \int d\vec{p}\,\frac{p_\alpha p_\beta}{m}\,f_\sigma(\vec{r},\vec{p},t) , \tag{11.2}$$

where $\vec{p}$ is the momentum, f_σ is the Wigner phase-space distribution function, and σ is the spin index. $v(r)$ is the coulomb interaction and V_{ext} is the external perturbing potential.

If $V_{ext}(\vec{r},t)$ varies slowly in space and time, we may justifiably assume that $\langle n(\vec{r},t)n(\vec{r}\,',t)$ and $\pi_{\alpha\beta}(\vec{r},t)$ have relaxed to their local equilibrium values. In that case, these will depend only on the value of the local density. Therefore

$$<n(\vec{r},t)n(\vec{r}',t)> \;=\; <n(\vec{r},t)><n(\vec{r}',t)>g(\vec{r}-\vec{r}';<n>) \tag{11.3}$$

and

$$\pi_{\alpha\beta}(r,t) = \frac{2}{3}\,\delta_{\alpha\beta}t(<n>) \;, \tag{11.4}$$

where $t(<n>)$ is the local value of the kinetic energy density. Equation (11.3) is the definition of the pair correlation function, which now depends on $(\vec{r},\vec{r}')$ and the local density $<n>$. We are concerned here with the deivation

$$<\bar{n}(\vec{r},t)> = <n(\vec{r},t)> - n$$

of the density from its unperturbed value n. We have from Eqn. (11.3)

$$\delta<n(\vec{r},t)n(\vec{r},t)>=n(1+\frac{n}{2}\,\frac{\partial}{\partial n})g(\vec{r}-\vec{r}';n)$$

$$\times\;[\,<\bar{n}(\vec{r},t)>-<n(\vec{r}',t)>] \;, \tag{11.5}$$

where we have written

$$\delta g(\vec{r}-\vec{r}';\langle n\rangle)=\frac{1}{2}\,\frac{\partial g(\vec{r}-\vec{r}',n)}{\partial n}[\langle\bar{n}(r,t)\rangle+\langle\bar{n}(r',t)\rangle]\,. \tag{11.6}$$

Substituting Eqs. (11.5) and (11.4) in Eq. (11.1) we have after linearization, the following equation of motion for $\langle\bar{n}(\vec{r},t)\rangle$:

$$\frac{\partial^2}{\partial t^2}\langle\bar{n}(\vec{r},t)\rangle-\frac{2}{3m}\left(\frac{\partial t}{\partial n}\right)\nabla^2\langle\bar{n}(\vec{r},t)\rangle-\frac{n}{m}\nabla_\alpha\left\{\int d\vec{r}'\nabla_\alpha v(\vec{r}-\vec{r}')\right.$$

$$\left.\cdot\left[\left(1+\frac{n}{2}\frac{\partial}{\partial n}\right)g(\vec{r}-\vec{r}';n)\right]\langle\bar{n}(\vec{r},t)\rangle\right\}=\frac{n}{m}\nabla^2V_{ext}(\vec{r},t)\,. \tag{11.7}$$

Defining an effective potential through

$$\nabla\vec{V}_{eff}(\vec{r},t)=\nabla V_H(\vec{r},t)$$

$$+\int d\vec{r}'\vec{\nabla}v(\vec{r}-\vec{r}')\left[\left(1+\frac{n}{2}\frac{\partial}{\partial n}\right)(g(\vec{r}-\vec{r}')-1)\right]\langle\bar{n}(\vec{r};t)\rangle\,, \tag{11.8}$$

where V_H is the Hartree potential, Eq. (11.7) can be written in a compact form as

$$\frac{\partial^2}{\partial t^2}\langle\bar{n}(\vec{r},t)\rangle-\frac{2}{3m}\left(\frac{\partial t}{\partial n}\right)\nabla^2\langle\bar{n}(\vec{r},t)\rangle=\frac{n}{m}\nabla^2V_{eff}(\vec{r},t)\,. \tag{11.9}$$

Expanding

$$<\bar{n}(\vec{r}\,',t)> = <\bar{n}(\vec{r},t)> + (\vec{r}-\vec{r}\,').\ \nabla<\bar{n}(\vec{r},t)> + \ldots \tag{11.10}$$

and using it in Eq. (11.7) we have

$$\frac{\partial^2<\bar{n}(\vec{r},t)>}{\partial t^2} - \frac{1}{m}\left[\frac{2}{3}\left(\frac{\partial t}{\partial n}\right) - \frac{n}{3}\int \vec{r}.\vec{\nabla}v(r)\times\left(1+\frac{n}{2}\frac{\partial}{\partial n}\right)\right.$$

$$\left.[g(\vec{r},n)-1]\,d\vec{r}\right] \times \nabla^2<\bar{n}(\vec{r},t)> = \frac{n}{m}\nabla^2 V_H(\vec{r},t) \quad . \tag{11.11a}$$

In the absence of V_{ext}, the RHS of the above equation vanishes and one has the familiar wave equation for the density fluctuation. It then follows that

$$\frac{\partial p}{\partial n} = \frac{2}{3}\left(\frac{\partial t}{\partial n}\right) - \frac{n}{3}\int r\frac{dv(r)}{dr}\ .\ \left(1 + \frac{n}{2}\frac{\partial}{\partial n}\right)[g(r,n)-1]\ d\vec{r}$$

$$= \frac{\partial}{\partial n}\left\{\frac{2}{3}t(n) - \frac{n^2}{6}\int r\frac{dv(r)}{dr}\cdot[g(r,n)-1]\ d\vec{r}\right\}, \tag{11.12}$$

where p is the static pressure and the compressibility is given by $(n\frac{\partial p}{\partial n})^{-1}$. The quantity within the curly brackets is the pressure as obtained from the virial theorem $(3P\Omega=2T+V)$. For a static potential, Eq. (11.11a) on using Eq. (11.12) becomes

$$\nabla^2 \left[\frac{\partial p}{\partial n} \langle \bar{n}(r) \rangle + n\, V_H(r) \right] = 0$$

or

$$\frac{1}{n} \frac{\partial p}{\partial n} \langle \bar{n}(r) \rangle + V_H(r) = 0 \; . \tag{11.11b}$$

$t(n)$ is the kinetic energy density of an interacting electron gas and is, therefore, not the same as that for a non-interacting gas. There exists the following interesting relation [6] between the kinetic-energy density and $g(r)$:

$$\frac{5}{3}t(n) - n\frac{\partial t(n)}{\partial n} = \frac{n^2}{2} \int v(r) \left(1 + n\frac{\partial}{\partial n}\right)[g(r,n)-1] \; d\vec{r}$$

$$+ \frac{n^2}{6} \int r\frac{dv(r)}{dr} [g(r,n)-1] \; d\vec{r} \; , \tag{11.13}$$

which has the solution

$$t(n)-t_o(n) = -\frac{n^2}{2} \int v(r) \, [g(r,n)-1] \; d\vec{r}$$

$$- \frac{n^{5/3}}{6} \int v(r)d\vec{r} \int_{\infty}^{n} n'^{-2/3} \, [g(r,n')-1] \, dn' \tag{11.14}$$

where $t_0(n)$ is the kinetic energy density in the non-interacting case. Equation (11.14) on differentiation yields

$$\frac{2}{3n}\frac{\partial}{\partial n}\left(t(n)-t_o(n)\right) = -\frac{7}{9}\int v(r)\,[g(\vec{r},n)-1]\,d\vec{r}$$

$$-\frac{n}{3}\int v(r)\frac{\partial g(\vec{r},n)}{\partial n}d\vec{r} - \frac{5}{27}n^{-1/3}\int v(r)d\vec{r}\int_{\infty}^{n} n'^{-2/3}[g(\vec{r},n') -1]\,dn' . \tag{11.15}$$

If we choose to include the interacting part of the kinetic energy in the effective potential V_{eff} in (11.8) it can be written as

$$\tilde{V}_{eff}(\vec{r},t) = V_H(\vec{r},t) + [-\frac{1}{3}\int r\frac{dv(r)}{dr} \cdot (1+\frac{n}{2}\frac{\partial}{\partial n})(g(\vec{r},n)-1)\,d\vec{r}$$

$$+\frac{2}{3n}(\frac{\partial t(n)}{\partial n} - \frac{\partial t_o(n)}{\partial n})] <\overline{n}(\vec{r},t)> . \tag{11.16}$$

Equation (11.16) is an exact equation for the effective potential in the equation

$$<n(\vec{q},\omega)> = \chi_o(\vec{q},\omega)\; V_{eff}(\vec{q},\omega)$$

only in the limit $\omega=0$ and $q\to 0$. Even in this limit we already see that (11.16) is fairly complicated when we note that the last term in the square brackets is given by (11.15). In the approximation of Vashishta and Singwi [5] (hereafter referred to as VS) this term was replaced by

$$-\frac{n}{18}\int \frac{dv(r)}{dr}\ \frac{\partial}{\partial n}\,[g(\vec{r},n)]\ d\vec{r}$$

thus giving for the effective potential the following simple form :

$$\tilde{V}_{eff}(\vec{r},t) = V_H(\vec{r},t) - \frac{1}{3}\int r\frac{dv(r)}{dr}(1 + an\frac{\partial}{\partial n})$$

$$[g(r,n)-1]\,d\vec{r}\ \langle\bar{n}(\vec{r},t)\rangle\ , \qquad (11.17)$$

where $a = 2/3$

With this choice of "a"VS found that the compressibility sum rule was very well satisfied. Later on, Vaishya and Gupta [7] checked this approximation using the self-consistent pair correlation function of VS and found it to be exceedingly accurate for r_s values of interest .

Comments :

In the VS theory the expression for the dielectric function $\varepsilon(\vec{q},\omega)$ is given by Eq. (1.20) with χ_{sc} replaced by χ_o , i.e.,

$$\varepsilon_{VS}(\vec{q},\omega) = 1 - (\frac{4\pi e^2}{q^2})\ \frac{\chi_o(\vec{q},\omega)}{1+G(\vec{q})\ \frac{4\pi e^2}{q^2}\ \chi_o(\vec{q},\omega)} \quad ,(11.18)$$

where we have written

$$\Psi_{VS}(\vec{q}) = \frac{4\pi e^2}{q^2}\,(1-G(\vec{q}))\ , \qquad (11.19a)$$

where

$$G(\vec{q}) = -\frac{1}{n}\int \frac{\vec{q}.\vec{q}'}{q'^2}\,\frac{d\vec{q}'}{(2\pi)^3}\ \left(1 + an\frac{\partial}{\partial n}\right)[\,S(\vec{q}-\vec{q}')-1]\ . \qquad (11.19b)$$

Now in the limit $q\to 0$, it can easily be demonstrated that $\Psi_{VS}(q)$ gives the same effective potential as (11.17). Thus we can assert that only in the static limit and $q\to 0$, ε_{VS} is exact. ε_{VS} has, however, been used as if it were valid for all q's and ω's. We must bear this fact in mind. The use of ε_{VS} in phenomena involving frequencies much less than plasma frequencies and all wave numbers (e.g. lattice dynamics in simple metals) has been found to be very satisfactory. The deficiencies of the theory of Singwi *et al*. are now beginning to surface as a result of the measurements on $S(\vec{q},\omega)$ in metals through inelastic scattering experiments. I shall comment on this later in my lectures.

III. THE KOHN-SHAM APPROACH

Kohn and Sham [4] are concerned only with the static situation. They write the ground state energy in the form :

$$E[\langle n\rangle] = T_o + \frac{1}{2}\int d\vec{r}d\vec{r}'v(\vec{r}-\vec{r}')\langle n(\vec{r})\rangle\langle n(\vec{r}')\rangle + \int d\vec{r}\langle n(\vec{r})\rangle$$

$$V_{ext}(\vec{r}) + E_{xc}[\langle n\rangle]\ ,\quad \ldots \qquad (111.1)$$

where

$$E_{xc}[\langle n\rangle] = \frac{1}{2}\iint d\vec{r}d\vec{r}'v(\vec{r}-\vec{r}')\ \langle n(\vec{r})n(\vec{r}')\rangle_c + T(\langle n\rangle) - T_o(\langle n\rangle) \tag{111.2}$$

and the suffix c has the usual meaning of the correlated part of $\langle n(\vec{r})h(\vec{r}')\rangle$.

The functional $E[n]$ has the following property : Its minimum value is the exact ground state energy of the system and is attained when $n(r)$ is the correct ground state density .

Kohn and Sham also write Eq.(111.2) in the form

$$E_{xc}[\langle n\rangle] = E_{xc}[n] + \frac{1}{2}\iint d\vec{r}d\vec{r}'\ K_{xc}(\vec{r}-\vec{r}';n)\langle\bar{n}(\vec{r})\rangle\langle\bar{n}(\vec{r}')\rangle \tag{111.3}$$

Equation 111.3 in the long wavelength limit becomes

$$E_{xc}[\langle n\rangle] = E_{xc}[n] + \frac{1}{2}K_{xc}(n)\int d\vec{r}(\langle\bar{n}(r)\rangle)^2\ , \tag{111.4}$$

where

$$K_{xc}(n) = \int K_{xc}(r,n)d\vec{r}\ .$$

Similarly

$$T_o[\langle n\rangle] = T_o(n) + \frac{1}{2}\int d\vec{r}d\vec{r}'\ K_o(\vec{r}-\vec{r}',n)\langle\bar{n}(\vec{r})\bar{n}(\vec{r}')\rangle \tag{111.5}$$

and $$T_o[\langle n\rangle] = T_o(n) + \frac{1}{2}K_o(n)\int d\vec{r}(\langle\bar{n}(\vec{r})\rangle)^2 \tag{111.6}$$

in the same limit.

Minimization of E of Eq. (111.1) leads to the following basic equation of Hohenberg-Kohn-Sham :

$$\frac{\delta T_o[\langle n\rangle]}{\delta\langle n(\vec{r})\rangle} + \frac{\delta E_{xc}[\langle n\rangle]}{\delta\langle n(\vec{r})\rangle} + V_H(\vec{r}) = 0\ , \tag{111.7}$$

which on using Eqs. (111.3) and (111.5) becomes

$$\int d\vec{r}'[K_o(\vec{r}-\vec{r}',n) + K_{xc}(\vec{r}-\vec{r}',n)]\langle\bar{n}(\vec{r}')\rangle + V_H(\vec{r}) = 0\ . \tag{111.8}$$

For a slowly varying disturbance, this becomes

$$[K_o(n) + K_{xc}(n)]\langle\bar{n}(\vec{r})\rangle + V_H(\vec{r}) = 0 \tag{111.9}$$

where

$$K_o(n) = \frac{1}{V} \frac{\partial^2 T_o(n)}{\partial n^2} \tag{111.10}$$

and

$$K_{xc}(n) = \frac{1}{V} \frac{\partial^2 K_{xc}(n)}{\partial n^2} \tag{111.11}$$

Now comparing Eq.(111.9) with the corresponding Eq. (11.11b) of the mean field theory we have

$$K_o(n) = \frac{2}{3n} \left(\frac{\partial t_o}{\partial n}\right) \tag{111.12}$$

and $$K_{xc}(n) = \frac{2}{3n}\left[\left(\frac{\partial t}{\partial n}\right) - \left(\frac{\partial t_o}{\partial n}\right)\right] - \frac{1}{3}\int r\frac{dv(r)}{dr}\left(1+\frac{n}{2}\frac{\partial}{\partial n}\right)[g(\vec{r},n)-1]\,d\vec{r} \tag{111.13}$$

Equation (111.2) can be written as

$$E_{xc}[\langle n\rangle] = \frac{1}{2} Vn^2\int v(r)[g(\vec{r},n)-1]\,d\vec{r} + V(t(n)-t_o(n)) \tag{111.14}$$

The exchange and correlation part of the chemical potential, $\mu_{xc}(n)$ is

$$\mu_{xc}(n) = \frac{1}{V}\frac{\partial E_{xc}(n)}{\partial n}$$

$$= n\int v(r)(1+\frac{n}{2}\frac{\partial}{\partial n})(g(\vec{r},n)-1)d\vec{r} + \frac{\partial t(n)}{\partial n} - \frac{\partial t_o(n)}{\partial n} \tag{111.15}$$

and

$$K_{xc}(n) \equiv \frac{\partial \mu_{xc}}{\partial n}$$

$$= \int v(r)[1 + 2n\frac{\partial}{\partial n} + \frac{n^2}{2}\frac{\partial^2}{\partial n^2}](g(\vec{r},n)-1)d\vec{r}$$

$$+ (\frac{\partial^2 t(n)}{\partial n^2} - \frac{\partial^2 t_o(n)}{\partial n^2}) \tag{111.16}$$

Now differentiating Eq. (11.15) with respect to n we have

$$\frac{2}{3}(\frac{\partial t}{\partial n}) - \frac{n}{3}\int r\frac{dv(r)}{dr}(1 + \frac{n}{2}\frac{\partial}{\partial n})[g(\vec{r},n)-1]d\vec{r}$$

$$= n[\frac{\partial^2 t}{\partial n^2} + \int v(r)(1 + 2n\frac{\partial}{\partial n}+\frac{n^2}{2}\frac{\partial^2}{\partial n^2})[g(\vec{r},n)-1]d\vec{r} \tag{111.17}$$

Remembering that

$$\frac{2}{3}\frac{\partial t_o}{\partial n} = n\frac{\partial^2 t_o}{\partial n^2} \tag{111.18}$$

and using Eq. (111.17) in Eq. (111.16) we have $K_{xc}(n)$ which is exactly the same as $K_{xc}(n)$ of Eq. (111.13) of the mean field theory.

We have thus been able to show that in the limit of long wave length and static disturbance (i.e. $\omega=0$) $K_{xc}(n)$ as obtained from the mean field approach of VS is identical with that obtained from the Kohn-Sham approach. In the procedure of VS, $K_{xc}(n)$ has been taken as

$$K_{xc}(n) = -\frac{1}{3}\int r\frac{dv(r)}{dr}\left(1 + an\frac{\partial}{\partial n}\right)[g(\vec{r},n)-]\,d\vec{r} \tag{111.19}$$

with $a = 2/3$. As mentioned before, Eq. (111.19) is for all practical purposes as good as the exact one. The VS procedure has the computational advantage in that only first density derivatives appear.

Static Dielectric Function

In the mean field approach the density response function is given by

$$\chi(\vec{q},\omega) = \frac{\chi_o(\vec{q},\omega)}{1-v(q)[1-G(\vec{q})]\chi_o(\vec{q},\omega)}\quad , \tag{111.20}$$

where $\chi_0(\vec{q},\omega)$ is the usual Lindhard function, and the function $G(q)$ in the VS scheme is

$$G(\vec{q}) = (1+\frac{2}{3}n\frac{\partial}{\partial n})\times(-\frac{1}{n}\int\frac{d\vec{q}\,'}{(2\pi)^3}\frac{\vec{q}\cdot\vec{q}\,'}{q'^2}(S(\vec{q}-\vec{q}\,',n)-1) \tag{111.21}$$

It is important to realize that all we know about the function $G(q)$ is that it is exact in the limit $q\to 0$. In the limit $q\to\infty$, it tends to $(1+\frac{2}{3}n\frac{\partial}{\partial n})(1-g(0))$, where $g(0)$ is the value of the pair correlation function at $r=0$. The exact value in this limit, as shown by Niklasson,[8] is $\frac{2}{3}(1-g(0))$. At present we have no knowledge of the precise form of $G(q)$ for intermediate values of q . One cannot rule out the possibility [9] that the true $G(q)$ has a maximum for some intermediate value of q.

The induced density is

$$\langle n(\vec{q},\omega)\rangle = \chi(\vec{q},\omega)V_{ext}(\vec{q},\omega) \quad , \tag{111.22}$$

which on using Eq. (111.20) can be written as

$$[-\frac{1}{\chi_0(\vec{q},\omega)} - v(q)G(\vec{q})]\langle\bar{n}(\vec{q},\omega)\rangle + V_H(\vec{q},\omega) = 0 \tag{111.23}$$

Comparing this equation for $\omega=0$ with the Fourier transform of Eq.(111.8) we have the following identification:

$$K_0(q) = -\frac{1}{\chi_0(q,0)} \tag{111.24}$$

and

$$K_{xc}(q) = - v(q)G(q) \qquad (111.25)$$

The function $K_{xc}(q)$ is, as we shall presently see, an important function and is related to the "local field" correction $G(q)$ through Eq. (111.25). As an illustration let us consider the exchange contribution to $K_{xc}(0)$. In this case [5]

$$G(q) \underset{q \to 0}{\rightarrow} (1 + \frac{n}{2}\frac{\partial}{\partial n}) \frac{3}{8} (\frac{q}{q_F})^2$$

$$= \frac{1}{4} (\frac{q}{q_F})^2 \qquad . \qquad (111.26)$$

Therefore,

$$\lim_{q \to 0} K_x(q) = - \frac{4\pi e^2}{q^2} \frac{1}{4} \frac{q^2}{q_F{}^2}$$

$$= - \pi e^2 (3\pi^2 n)^{-2/3}$$

Hence, the exchange potential is

$$\mu_x(n) = \int_0^n K_x(n')dn'$$

$$= - \frac{e^2}{\pi} (3\pi^2 n)^{1/3} \qquad (111.27)$$

which is, as expected, the same as the Kohn-Sham exchange.

Gradient Correction

For a slightly inhomogeneous electron gas, Kohn and Sham suggested expanding $K_{xc}(q)$ in powers of q. This suggestion has formed the basis for the calculations of surface energy of metals. For a slowly varying density $(|\nabla n/n| < K_F, K_{FT})$, we can write

$$E_{xc}[n] = \int d\vec{r}\ \varepsilon_{xc}(n)\langle n(\vec{r})\rangle + \frac{1}{2}\int d\vec{r}\ g_{xc}^{(2)}(n)\,|\langle \vec{\nabla} n(r)\rangle|^2 + .., \tag{111.28}$$

where $\varepsilon_{xc}(n)$ is the exchange and correlation energy per electron in the homogeneous electron gas with density n. The quantity $g_{xc}^{(2)}(n)$ can be shown [4] to be the coefficient of the q^2 term in the q expansion of $K_{xc}(q)$. According to Eq. (111.25), $g_{xc}^{(2)}$ is, therefore, the q^2 coefficient in the expansion of $-v(q)G(q)$. In using Eq. (111.28) for practical calculations of surface energy, one should bear in mind that it has only lowest order gradient expansion terms. The condition $|\nabla n/n| < K_{FT}$ is not satisfied in the surface region. Besides, the coefficient $g_{xc}^{(2)}(n)$ is known exactly only in the limit $r_s \ll 1$ and is therefore unlikely to be meaningful at realistic metallic densities in the bulk metal, let alone in the tenuous tail of the surface density distribution.

In the scheme of Vashishta and Singwi, $G(q)$ is known for all electron densities although approximately. Let us examine the small q expansion of $G(q)$ of the VS theory. We shall do this separately in two cases : (a) exchange only and (b) exchange and correlation together.

(a) Exchange (H-F Approximation):

In the HF approximation the structure factor $S(q)$ is known and hence $G_{HF}(q)$ can be evaluated using Eq. (111.21) and remembering that we have the factor 1/2 instead of 2/3. The result is

$$G_{HF}(q) = (1 + \frac{n}{2}\frac{\partial}{\partial n})\ [\frac{3}{8}(\frac{q}{q_F})^2 - \frac{31}{400}(\frac{q}{q_F})^4 + \frac{6}{5}(\frac{q}{2q_F})^4 \ln(\frac{q}{2q_F}) + \ldots] \quad , \qquad (111.29)$$

where we have ignored terms of order q^6 and higher. This then yields

$$-v(q)G_{HF}(q) \equiv K_x(q)$$

$$= -\frac{\pi e^2}{(3\pi^2 n)^{2/3}} + \frac{23}{150}\frac{\pi e^2}{(3\pi^2 n)^{4/3}} q^2 - \frac{1}{10}\frac{\pi e^2}{(3\pi^2 n)^{4/3}} q^2 \ln(\frac{q}{2q_F}) \quad . \qquad (111.30)$$

We see from Eq. (111.30) that $g_x^{(2)}$ has a logarithmic singularity for $q \to 0$, implying that in the VS scheme no gradient expansion exists. In the exact calculation of Sham [9] to order e^2 , only three graphs in the perturbation expansion contribute to $K_x(q)$. Each one of these graphs gives a singular contribution of the same kind as above, but they exactly cancel out in the resultant contribution. This subtle cancellation is not present in the Hartree-Fock of VS . Sham's result is

$$K_x(q) = -\frac{\pi e^2}{(3\pi^2 n)^{2/3}} - \frac{21}{216}\frac{\pi e^2}{(3\pi^2 n)^{4/3}} q^2 \quad , \qquad (111.31)$$

which can be compared with Eq.(111.30) . It is seen that in Eq.(111.30) we not only have an extra singular term but the sign of the q^2 term is wrong. Recently Geldart et al. [10] have shown that the Hartree-Fock result including all orders in e^2, is not analytic for $q \to 0$. This would again imply that the gradient expansion does not exist in the H-F case.

Very recently Gupta [11] has used the following expression for $G(q)$ as given by Ichimaru [12] for the classical electron plasma :

$$G(\vec{q}) = -\frac{1}{n}\int \frac{\vec{q}\cdot\vec{q}\,'}{q'^2} S(\vec{q}\,')[S(\vec{q}\,'-\vec{q})-1]\,d\vec{q}\,' \quad , \qquad (111.32)$$

and has evaluated $K_x(q)$ by taking $S(q)$ in (111.32) to be that given in the HF approximation. Gupta's result is

$$K_x(q) = -\frac{36}{35}\frac{\pi e^2}{(3\pi^2 n)^{2/3}} + \frac{4}{75}\frac{\pi e^2}{(3\pi^2 n)^{4/3}}\, q^2 \ . \qquad (111.33)$$

Remarkably enough there is now no singular term but the sign of the q^2 term is still wrong. $G(q)$ of Eq.(111.32) takes three particle correlations into account in an approximate manner. Since the first term in Eq.(111.33) is very nearly the same as the exact one, it indicates that long rango three-particle correlations are reasonably well taken care of. The absence of the singular term in (111.33) has, therefore, something to do with taking three particle correlations into account. What connetion, if any, exists between Sham's calculation and that based on $G(q)$ of Eq.(111.32) is not at all clear.

(b) Exchange and Correlation :

The essence of the VS scheme and that of the earlier work is that it is self-consistent in the sense that the pair correlation function entering in G(q) is the same as one obtains from the resulting response function. It is, therefore, not legitimate to separate exchange and correlation. In what follows we shall consider both exchange and correlation.

It is easy to show that

$$G(q) = (1+\frac{2}{3} n \frac{\partial}{\partial n})\{ - q^2 \frac{1}{6\pi^2 n} \int_0^\infty dk(S(k)-1) - q^4 \frac{1}{5\pi^2 n} \int_0^\infty \frac{dk}{k^2} (S(k)-1) + O(q^6)\} \qquad (111.34)$$

Now

$$S(k) = \frac{\hbar k^2}{2m\hbar\omega_p} , \qquad (111.35)$$

Lim k→0

so that

$$\int_0^\infty \frac{dk}{k^2} (S(k)-1) = \int_0^\delta \frac{dk}{k^2} (S(k)-1) + \int_\delta^\infty \frac{dk}{k^2} (S(k)-1),$$

which on using (111.35) becomes

$$\int_0^\infty \frac{dk}{k^2}(S(k)-1) = \frac{\hbar\delta}{2m\omega_p} + \int_\delta^\infty \frac{dk}{k^2} (S(k)-1) - \int_0^\delta \frac{dk}{k^2} \qquad (111.36)$$

In Eq. (111.36) the second term is finite but the last term diverges. One can show that the coefficients of q^6 and all higher order terms also diverge. It implies that G(q) of VS is not an analytic function of q at q=0. Thus in a strict mathematical sense the coefficient of the q^2 term in the expansion of $K_{xc}(q)$ does not exist in the VS theory. We believe this to be the weakness of the theory.

VS have done a fully self-consistent calculation of G(q) for different values of r_s and have found that this function can be well represented for $q \leq 2q_F$ by the following analytic expression

$$G(q) = A(1 - e^{-B(q/q_F)^2}) \quad , \tag{111.37}$$

the parameters A and B are very gentle functions of r_s and have been tabulated by these authors. From their numerical computations, we conclude that the above mentioned singular behavior would exist only in the immediate vicinity of q=0 and should, therefore, not pause any problem in extracting the slope of the function $K_{xc}(q)$ in the region of $q/q_F \ll 1$. In reality, it is indeed so. Using Eq. (111.37) we find that

$$g_{xc}^{(2)}(n) = \frac{2}{3} AB^2 \{ (\frac{e^2}{\pi}) \frac{n^{-4/3}}{(3\pi^2)^{1/3}} \} \tag{111.38}$$

Using the tabulated values of A and B , one finds that the factor AB^2 varies from 0.064 for r_s=1 to 0.055 for r_s=6 . For $r_s \ll 1$, it extrapolates to the value 0.066 .

There is no other calculation available for $g_{xc}^{(2)}(n)$ for the range of metallic densities with which the above values can be compared. Such a comparison, however, is possible for $r_s \ll 1$. Ma and Brueckner [13] find that the correlation correction in the local density approximation can be written as

$$\Delta E_c[n] = \int d\vec{r}\ B_c(n)\,|\nabla\langle n(r)\rangle|^2, \qquad (111.39a)$$

where

$$B_c(n) \equiv \frac{1}{2}\, g_c^{(2)}(n) = 0.00847\ n^{-4/3}\ \mathrm{Ry}\ . \qquad (111.39b)$$

To order e^2, Sham's estimate [9] for the corresponding exchange contribution is

$$g_x^{(2)}(n) = -0.00667\ n^{-4/3}\ \mathrm{Ry}\ . \qquad (111.40)$$

The resulting exchange and correlation contribution is then

$$g_{xc}^{(2)}(n) = 0.01027\ n^{-4/3}\ \mathrm{Ry}\ . \qquad (111.41)$$

Our estimate for $r_s \ll 1$ is

$$g_{xc}^{(2)}(n) = 0.0136\ n^{-4/3}\ \mathrm{Ry}\ . \qquad (111.42)$$

We now have estimates trough Eq. (111.38) of $g_{xc}^{(2)}(n)$ in the metallic density range, and it would be interesting if one could check them through some independent means.

IV. CONCLUDING REMARKS

From our foregoing discussion, it is clear that the VS dielectric response function is exact in the static situation and in the long wavelength limit. One is, however, interested in the behavior of $\varepsilon(\vec{q},\omega)$ for all values of $\vec{q}$ and ω. Let us recall here that in the very first version of the theory of Singwi *et al.* [2], the primary aim was to take account of short-range correlations which reflect the behavior of fluctuations whose wavelength is of the order of inverse Fermi wave vector. In this version of the theory one was able to remedy the bad behavior of the pair correlation function for small interparticle separation which was present in the earlier theories. Recent calculations [14-15] based on a diagrammatic analysis of the perturbation theory in which an infinite sum of electron-electron ladder diagrams are included, have yielded results for the pair correlation function which are very similar to those of Singwi *et al*. The modification of the theory due to VS gives good compressibility while at the same time retaining the good behavior of the pair correlation function. The use of the VS dielectric function in the calculation of phonon dispersion curves in simple metals has yielded excellent results. All this seems to indicate that the VS dielectric function is satisfactory for treating phenomena involving frequencies much less than plasma frequency and perhaps for any value of wave number.

How good the VS dielectric function is in the region of large q and ω ? The answer to this question is provided by recent inelastic x-ray and electron scattering experiments in simple metals. Of particular interest are the experiments of Platzman and Eisenberger [16]. Their experiments have revealed some very interesting features of the dynamical structure factor $S(\vec{q},\omega)$ of the electron liquid, which is simply related to $\text{Im}\,\frac{1}{\varepsilon(q,\omega)}$ through Eq. (1.11). These features are : (1) contrary to what one might expect in the momentumtransfer region $q_F < q < 2q_F$, where q_F is the Fermi momentum, the excitation spectrum is comprised of two parts; one part is broad and corresponds to particle-hole excitations and the other is relatively sharper and has been interpreted as the continuation of the plasmon excitation in the particle-hole continuum. For $q < 2q_F$, the latter has larger strength than the former, and at $q \approx 2q_F$, there occurs a switching over of the two strengths. For $q > 2q_F$, one observes only a single broad peak corresponding to free-particle excitations. (2) Surprisingly enough, the plasmon peak in the entire region $q_c < q < 2q_F$, where q_c is the critical wave vector at which the plasmon joins the particle-hole continuum, shows no dispersion. Platzman and Eisenberger have even observed a small negative dispersion in the region $q_F < q < 2q_F$.

A very significant observation of Platzman and Eisenberger is that the above-mentioned features are commo common to such diverse systems as C, Be, and Al thereby indicating that these features are not the result of simple one-electron properties of these metals, but are indeed the manifestations of the dynamical behavior of the electron gas. It is this aspect that renders the problem so interesting from a many-body theoretical point of view.

It has recently been pointed out by Kalia and Mukhopadhyay [17], among others, that the observed $S(\vec{q},\omega)$ in the region $q_c < q < 2q_F$ cannot be understood in terms of the random-phase approximation (RPA). Nor can it be understood in terms of the dielectric response function of Vashishta and Singwi.

Obviously, something has gone wrong with the VS theory in this region of frequency and wave number. In my own view two important effects, which now make themselves feel, have been ignored in the VS theory : One is the dynamics of the correlation hole or in other words ω dependence of the local field factor $G(q)$; and the other is the finite life time of electrons and holes or the modification of $\chi_0(q,\omega)$. How one is to achieve this in a consistent manner is not very clear at present. Very recently, Mukhopadhyay, Kalia and Singwi [18] have tried to tackle this problem in a somewhat adhoc fashion. What they have done is to add to the free particle energies $\varepsilon(\vec{p})$ occuring in the energy denominators of the Lindhard function $\chi_0(q,\omega)$ a term $i/\tau(p)$, where $\tau^{-1}(p)$ is given by the imaginary part of the self-energy $\Sigma(\vec{p},\varepsilon(p))$. The latter has been calculated in the lowest-order approximation and evaluated at the free particle energy $\varepsilon(\vec{p})$. The real part of $\Sigma(\vec{p},\varepsilon(p))$ has been ignored for the reason that part of it is already contained in some average fashion in the local field factor $G(q)$. This procedure helps them to preserve the f-sum rule. In this way Mukhopadhyay *et al.* have been able to account for the above mentioned observed features of $S(\vec{q},\omega)$. However, one very disquieting feature of this modification of the VS theory is that it violates the equation of continuity. A more systematic calculation which does not have this defect is now warranted.

All what I have said so far in my lectures had almost nothing to do with non-linear phenomena to which, if not most, some attention will be given at this School. If I had time, I would have liked to talk about non-linear screening. I would, therefore, conclude these lectures by just making a few cursory remarks and for details I would suggest you to look into the original papers. I have only attempted in these lectures to prepare the ground work for those of you who are interested in non-linear screening.

Two very important and interesting examples of non-linear screening are provided one by a positron and the other by a proton in a metal. Both these entities produce a very strong perturbation when introduced in an electron gas because of their strong coulomb field. The positron, which is much less massive than a proton, is a weaker perturbation because of its recoil ; but inspite of this the positron annihilation rate in metals which is proportional to the electron density at the position of the positron, is substantially larger than what is given by the lowest order perturbation theory. This fact clearly demonstrates the failure of the linear theory. Sjölander and Stott [19] have generalized the theory of Singwi <u>et al.</u> [2] to a two component plasma consisting of electrons and positrons, and then by letting the density of positrons go to zero, they obtained a simple integral equation for the Fourier transform of the polarization charge around the positron. Numerical results of these authors for the annihilation rates for densities $r_s \leq 4$ are in excellent agreement with experiments. For values of $r_s \geq 5$, the theory of Stott and Sjölander seems to diverge. Later on, Bhattacharyya and Singwi [20] using the VS form of the local field $G(q)$ with an adjustable parameter "a" obtained a non-linear integral equation for the Fourier transform of the polarization charge around a positron. Their calculated annihilation rates are in good agreement with experiment over the entire metallic density range. These authors performed a similar calculation of the screening charge around a proton for $r_s=2$. The results are very similar to those of Popovic and Stott [21] and Almbladh and Von Barth [22], who in their calculation used the density functional formalism of Hohenberg-Kohn and Sham. The latter calculations are based on a firmly founded theoretical basis, whereas the calculations of Bhattacharyya and Singwi are open to objections for the meaning of the parameter "a" in their theory is not clear. It seems to me that it is now necessary to improve upon the VS theory so as to take into account short-range threep particle correlations. The long-range three-particle correlations are there because of the presence of the density derivative term in G(q).

The work of Sjölander and Stott can be generalized to a multi-component plasma in a straight forward manner. A very interesting application of this generalization has recently been made in a series of papers by Vashishta et al. [23] to calculate the ground state energy of the electron-hole liquid in Ge, Si and in highly strained Ge [111] and Si [100]. The calculated properties such as the binding energy of the electron-hole liquid are in very good agreement with experiment. This again, in my view, points out the importance of taking multiple scattering into account between all components of the plasma (non-linear effects).

My thanks are due to Dr. A.K. Gupta for reading through the manuscript.

REFERENCES

I have made no attempt to give a complete list of references. Interested readers should be able to find them in the references given below.

For a comprehensive discussion of the theory of Fermi liquids, see text : D. Pines and P. Nozières : The theory of Quantum Liquids, Vol. 1 (Benjamin, New York, 1966) .

1. J. Hubbard, Proc. Roy. Soc. (London) A243, 336 (1957).
2. K.S. Singwi, M.P. Tosi, R.H. Land and A. Sjölander, Phys. Rev. 176, 589 (1968).
3. P. Hohenberg and W. Kohn, Phys. Rev. 136, B864 (1964).
4. W. Kohn and L. Sham, Phys. Rev. 140, A1133 (1965).
5. P. Vashishta and K.S. Singwi, Phys. Rev. B6, 875 (1972).
6. G. Niklasson, A. Sjölander and K.S. Singwi, Phys. Rev. B11, 113 (1975).
7. J.S. Vaishya and A.K. Gupta, Phys. Rev. B7, 4300 (1973).
8. G. Niklasson, Phys. Rev. B10, 3052 (1974).
9. L. Sham, in Computational Methods in Band Theory, edited by P.M. Marcus, J.F. Janak and A.R. Williams (Plenum, New York, 1971) p. 458.
10. D.J.W. Geldart, M. Rasolt and C.O. Almbladh, Solid State Communication 16, 243 (1975).
11. A.K. Gupta, Private communication.
12. S. Ichimaru, Phys. Rev. A2, 494 (1970).
13. S. Ma and K.A. Brueckner, Phys. Rev. 165, 18 (1968).
14. H. Yasuhara, J. Phys. Soc., Jpn 36, 361 (1974).
15. G. Brown and D. Lowy, Phys. Rev. (to be published).
16. P.M. Platzman and P. Eisenberger, Phys. Rev. Lett. 33, 152 (1974), and previous references cited therein. See also P. Zacharias, J. Phys. C 7, 126 (1974).
17. R.K. Kalia and G. Mukhopadhyaya, Solid State Communication, 15, 1243 (1974).
18. G. Mukhopadhyay, R.K. Kalia and K.S. Singwi, Phys. Rev. Lett. 34, 950 (1975).
19. A. Sjölander and M.J. Stott, Phys. Rev. B5, 2109(1972).
20. P. Bhattacharyya and K.S. Singwi, Phys. Rev. Lett. 29, 22 (1972).
21. Z.D. Popovic and M.J. Stott, Phys. Rev. Lett. 33, 1164 (1974).
22. C.O Almbladh and U. Von Barth (to be published).

23. P. Vashishta, P. Bhattacharyya and K.S. Singwi, Nuovo Cimento 23B, 172 (1974). See also Phys. Rev. Lett. 30, 1248 (1973), Phys. Rev. B10, 5108 (1974), and Phys. Rev. Lett. 33, 911 (1974).

EFFECTS OF CORRELATIONS ON THE CONDUCTIVITY OF ELECTRON-HOLE PLASMAS

N. Tzoar and P. M. Platzman

City College of the City University of New York
New York, New York 10031 and
Bell Laboratories, Murray Hill, New Jersey 07974

Calculations of the high frequency conductivity for electron-hole system are presented. The treatment rests on the Kubo's formula for the conductivity and the temperature dependent Green's function technique. The frequency and r_s dependence of the resistivity are calculated for small electron-hole mass ratio. The effects of dynamical correlations are discussed.

I. INTRODUCTION

An extensive study of the interaction of electromagnetic waves with quantum and classical plasmas were carried out more than a decade ago.[1] Here the high frequency conductivity of systems composed of negatively and positively charged particles was obtained. In these calculations electron and ions were treated on an equal footing, i.e. no approximation in the formulation were made taking into account the massive nature of the ions. Thereafter in the computation of the frequency dependence of the conductivity one always took the limit of infinite ion mass as a realistic case for known plasma systems.

Recently[2] much interest has been generated in the condensed "electron-hole" droplets in indirect band gap semiconductors such as Ge and Si. Here the life time of the electron-hole pair is of the order of microseconds. It is accepted that for high levels of ionization the electron-hole pairs will condense to create a plasma fluid, which in general occupies a volume characterized by a radius of 1-3 μm. The densities of the droplets are larger than 10^{18} particles/cc, thus its plasma frequency is in the infrared.

To decouple correlation effects from surface effects we need an infinite-extent system. In the case of the condensed electron-hole plasma this can be realized provided the radius of the droplet will be 1-2 mm. Here one group has reported[3] the existence of macroscopic droplets of the above-quoted size in strained samples.

In these situations we proceed to investigate new properties which should be studied in such systems. One which comes to mind are dynamical correlation effects in the transport and optical properties. We note that the condensed electron-hole plasmas constitute an interesting system in which the ratio of the heavy to light mass of the charged particles is 1-2 or 1-3. Here dynamical correlation effects will be much larger than in the case of massive ions. Also the frequency dependent conductivity will be qualitatively different from plasmas with massive ions as we shall see. We shall next study the dynamical correlation effects on the conductivity for condensed electron-hole plasmas, using an infinite homogeneous, isotropic electron-hole gas model. The conductivity will be affected by the anisotropy and the mild inhomogeneousity quantitatively but not qualitatively. Nevertheless it is an essential approximation to make due to the complexity of the problem.

In section II we present the general formulation of the problem and in section III we evaluate the expression for the conductivity. Here we follow closely the work of Ron and Tzoar of ref. 1. In section IV calculations of the resistivity are presented and section V is reserved for a discussion.

II. GENERAL FORMULATION OF THE PROBLEM

We start from the general expression of the long wavelength conductivity for a system of charged particles as given by Kubo[4]

$$\sigma_{\mu\nu}(\omega) = \frac{1}{V}\int_0^\infty d\tau e^{i\omega\tau}\int_0^\beta d\lambda \langle j_\mu(\tau - i\hbar\lambda) j_\nu(0)\rangle \tag{1}$$

where ω is the frequency of the electromagnetic wave,

$$j_\mu(t) = e^{iHt/\hbar} j_\mu e^{-iHt/\hbar} \tag{2}$$

is the current operator in the Heisenberg representation, and the average of an operator $\mathcal{O}$ is given by

$$\langle \mathcal{O} \rangle = \mathrm{Tr}\left\{ e^{\beta\left(\Omega + \sum_s \mu_s N_s - H\right)} \mathcal{O} \right\} . \tag{3}$$

In Eqs. (2) and (3), H represents the total Hamiltonian of the system, Ω is defined by

$$e^{-\beta\Omega} = \mathrm{Tr}\left\{ e^{\beta\left(\sum_s \mu_s N_s - H\right)} \right\}$$

where μ_s and N_s are, respectively, the chemical potential and the number operator of the s species in the system, and β the inverse of the temperature in energy units. In order to render Eq. (1) in a more convenient form we intergrate it by parts and obtain:

$$\sigma_{\mu\nu}(\omega) = \sigma^{(0)}_{\mu\nu}(\omega) + \sigma^{(1)}_{\mu\nu}(\omega) \tag{4}$$

where

$$\begin{aligned}\sigma^{(0)}_{\mu\nu}(\omega) &= -\frac{1}{i\omega}\frac{1}{V}\int_0^\beta d\lambda \langle j_\mu(-i\hbar\lambda) j_\nu(0) \rangle \\ &= i\,\frac{1}{4\pi\omega}\sum_s \omega_s^2\,\delta_{\mu\nu} \\ &= i\sigma_0(\omega)\delta_{\mu\nu} \quad .\end{aligned} \tag{5}$$

Here $\omega_s^2 = (4\pi e_s^2 n_s/m_s)$ is the plasma frequency, e_s is the charge, n_s is the average particle density and m_s is the particle mass of the s species, and

$$\sigma^{(1)}_{\mu\nu}(\omega) = \frac{1}{\hbar\omega}\frac{1}{V}\int_0^\infty d\tau\, e^{i\omega\tau} \langle [j_\mu(\tau),\ j_\nu(0)] \rangle . \tag{6}$$

In Eq. (6) [,] denotes the commutator. In order to facilitate the temperature-dependent Green's function formalism[5] we define a function

$$Y_{\mu\nu}(z) = \frac{1}{\hbar}\int_{-\infty}^{+\infty} \frac{d\omega'}{\omega'-z}\,(1-e^{-\hbar\omega'\beta})\Phi_{\mu\nu}(\omega') \tag{7}$$

which is analytic in the entire z plan, except for a cut on the real axis. $\Phi_{\mu\nu}(\omega)$ is real and given by

$$\Phi_{\mu\nu}(\omega) = \frac{1}{V}\sum_{n,n'} \exp\left[\beta\left(\Omega + \sum_{s} \mu_s N_s - E_n\right)\right]$$

$$\langle n|j_\mu(0)|n'\rangle\langle n'|j_\nu(0)|n\rangle\,\delta\left(\omega - \frac{E_{n'}-E_n}{\hbar}\right) \tag{8}$$

where n and n′ represent eigenstates of the Hamiltonian and the number operator with

$$H|n\rangle = E_n|n\rangle;\; N_s|n\rangle = N_s^{(n)}|n\rangle \tag{9}$$

and $N_s^{(n)} = N_s^{(n')}$ in Eq. (9) due to the fact that j_μ commutes with the number operator. If we represent explicitly the average in Eq. (6) as a sum over states and use the fact that $N_s^{(n)} = N_s^{(n')}$ we obtain

$$\sigma^{(1)}_{\mu\nu}(\omega) = \frac{1}{i\omega}\,Y^{+}_{\mu\nu}(\omega) \tag{10}$$

where for any function f(z) in the complex z plane, we denote by

$$f^{\pm}(\omega) = \lim_{z\to\omega\pm i\epsilon} f(z);\; \epsilon \to 0_+ \;. \tag{11}$$

In order to obtain the function $Y_{\mu\nu}(z)$ of Eq. (6) we define a Green's function

$$M_{\mu\nu}(u) = \frac{1}{V}\langle T\{j_\mu(u) j_\nu(0)\}\rangle .$$

$$-\beta < u < \beta , \tag{12}$$

where T is the Dyson ordering operator and

$$j_\mu(u) = e^{uH} j_\mu e^{-uH} . \tag{13}$$

By expressing our Green's function in term of the sum over states, one can show that $M_{\mu\nu}(u)$ is a periodic function i.e. $M_{\mu\nu}(u+\beta) = M_{\mu\nu}(u)$. Thus its "Fourier Transform" with respect to u

$$M_{\mu\nu}(n) = \int_0^\beta du\ e^{i2\pi nu/\beta}\ M_{\mu\nu}(u)$$

$$n = 0,\ \pm 1,\ \pm 2,\ \ldots \tag{14}$$

is given by

$$M_{\mu\nu}(n) = \frac{1}{\hbar}\int \frac{d\omega'}{\omega' - i2\pi n/\beta\hbar}\,(1-e^{-\beta\hbar\omega'})\Phi_{\mu\nu}(\omega') . \tag{15}$$

If we now compare Eqs. (15) and (7) we realize that $Y_{\mu\nu}(z)$ is the analytical continuation of $M_{\mu\nu}(n)$ from the infinite set of points $i2\pi n/\hbar\beta$ $(n > 0)$ on the position imaginary axis of z to the entire upper half-plane of z. We thus conclude that our problem reduces to the calculations of $M_{\mu\nu}(n)$ which will be evaluated using the well-known Green's function diagram technique.

III. EVALUATION OF THE CONDUCTIVITY

We turn now to the calculation of M(n) using perturbation expansion, and then resuming all diagrams (terms) which contribute to the conductivity for quantum (classical) plasma, when the number of particles in the Bohr (Debye) sphere is large, the frequency is high compared to the collision frequency, and the wavelength of the incident field is large compared to the Bohr (Debye) radius. Thus, in resuming the diagrams (terms) of the perturbation expansion, we consider processes proportional to the number of particles, N, as finite and include them to all orders while those processes which are not proportional to N are treated as small.[6]

We write now our Green's function in the second quantization formalism in the interaction representation as

$$M_{\mu\nu}(u) = \frac{\hbar^2}{V} \sum_{s,s'} \frac{e_s e_{s'}}{m_s m_{s'}} \sum_{\vec{p}\vec{p}'} p_\mu p_\nu \langle U(\beta) \rangle_0^{-1}$$

$$\langle T[a_p^\dagger(s,u)a_p(s,u)a_{p'}^\dagger(s',0)a_{p'}(s',0)U(\beta)]\rangle_0 \tag{16}$$

where use has been made of the second quantization representation of the current operator

$$\vec{j} = \hbar \sum_s \frac{e_s}{m_s} \sum_{\vec{p}} \vec{p}\, a_p^\dagger(s) a_p(s) \;. \tag{17}$$

The symbol $\langle\ \rangle_0$ corresponds to the average defined in Eq. (3) for noninteracting particles,

$$U(\beta) = \exp\left\{-\int_0^\beta du\ H_I(u)\right\} \tag{18}$$

and

$$H_I = \frac{1}{2V} \sum_{s,s'} \sum_{\vec{p},\vec{p}',\vec{k}} \frac{4\pi e_s e_{s'}}{k^2} a^\dagger_{\vec{p}+\vec{k}}(s) a^\dagger_{\vec{p}'-\vec{k}}(s') a_{\vec{p}'}(s') a_{\vec{p}}(s) \tag{19}$$

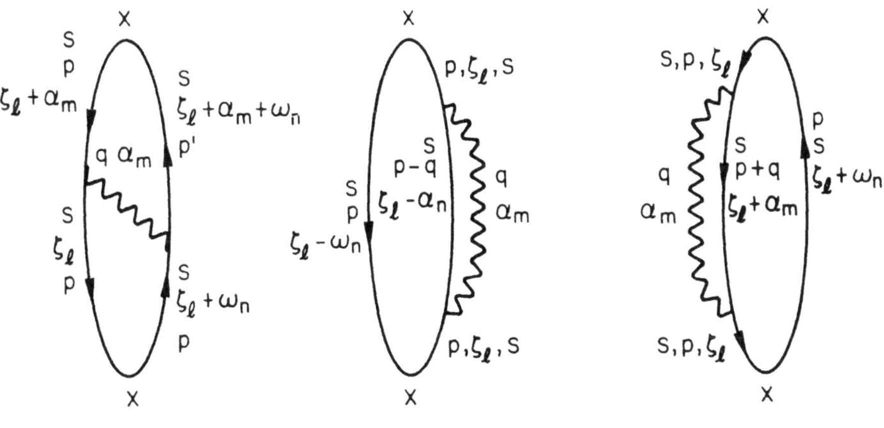

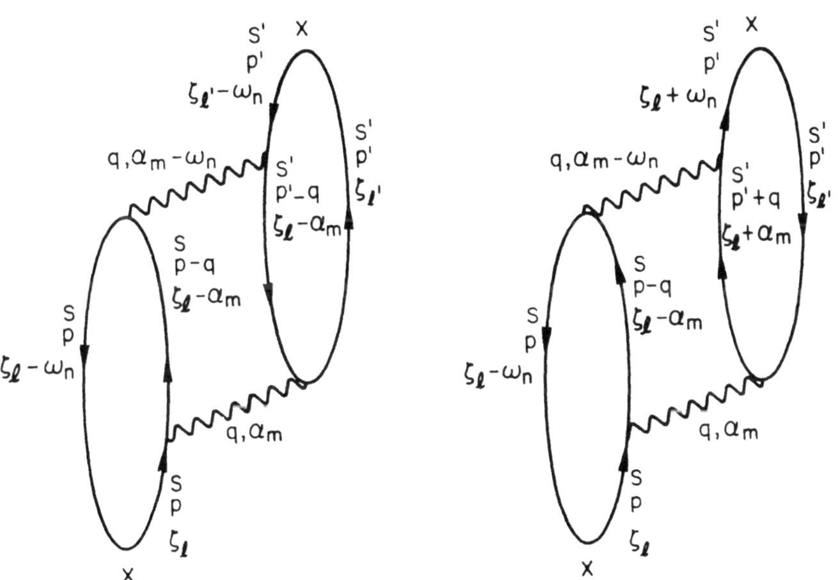

Fig. 1. The class of diagrams which contribute to the high-frequency conductivity.

The basic rules for the perturbation expansion of M(k,n) are given by Luttinger and Ward.[5] We use their diagram technique and indicate by a solid line with labels s, $\vec{p}$, and ζ_ℓ the s-species free particle propagator of wave vector $\vec{p}$ and "energy" ζ_ℓ (we restrict ourself to fermions only)

$$G_p(\zeta_\ell, s) = [\zeta_\ell(s) - \epsilon_p(s)]^{-1}$$

$$\zeta_\ell(s) = i\pi(2\ell+1)\frac{1}{\beta} + \mu_s;\ \ell = 0,\ \pm 1,\ \pm 2,\ \ldots.$$

$$\epsilon_p(s) = \hbar^2 p^2/2m_s \tag{20}$$

and by dotted line the interaction $4\pi/k^2$. To each vertex we assign a charge e_s given by the s label of the particle.

In the high-frequency long-wavelength region we take into account a generalized version of the diagrams considered by Perel and Eliashberg.[7] Our generalization corresponds to considering all species in equivalent manner, see Fig. 1. The wavy line represents the effective potential shown in Fig. 2 and it is given by

$$U_q(\alpha_m) = \frac{4\pi}{q^2} + \frac{4\pi}{q^2}\sum_s e_s^2 \frac{1}{V}\sum_{\vec{p}}\frac{1}{\beta}\sum_\ell G_{\vec{p}+\vec{q}}(\zeta_\ell+\alpha_m, s)G_{\vec{p}}(\zeta_\ell, s)U_q(\alpha_m) \tag{21}$$

where $\alpha_m = i2\pi m(1/\beta)$, $m = 0,\ \pm 1,\ \pm 2, \ldots$. We can now cast Eq. (21) into the form

$$U_q(\alpha_m) = \frac{4\pi}{q^2}\left[1 - \frac{4\pi}{q^2}\sum_s e_s^2\, Q_q(\alpha_m, s)\right]^{-1} \tag{22}$$

where

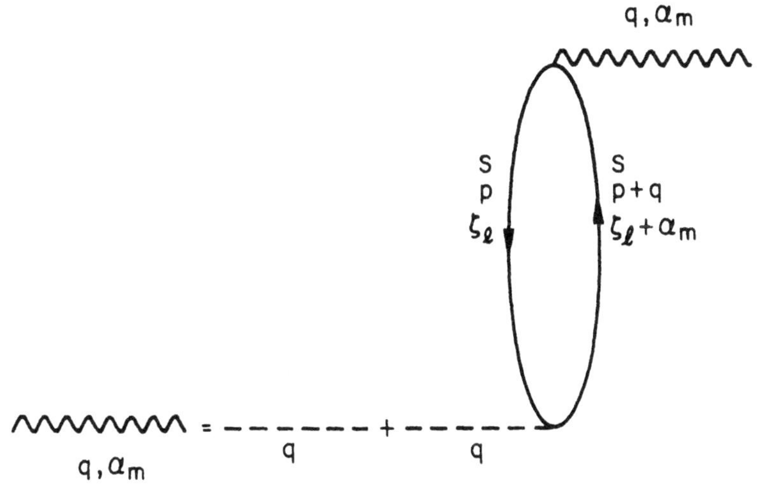

Fig. 2. The integral equation for the effective interaction.

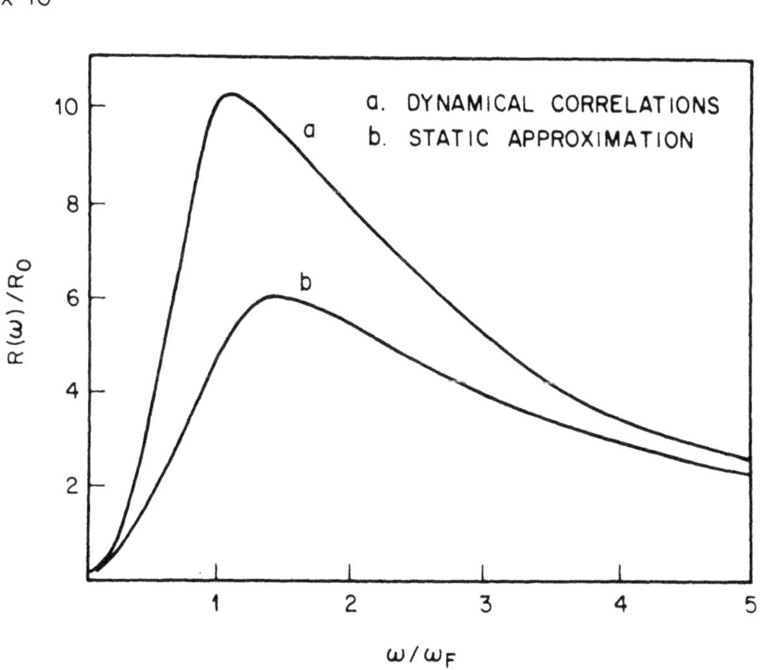

Fig. 3. The resistivity $R(\omega)$ in terms of R_0 for electronic $r_s=0.2$ and mass ratio $\alpha=3$.

$$Q_q(\alpha_m, s) = \frac{1}{V}\sum_{\vec{p}}\frac{1}{\beta}\sum_{\ell} G_{\vec{p}+\vec{q}}(\zeta_\ell+\alpha_m, s)G_{\vec{p}}(\zeta_\ell, s)$$

$$= \frac{1}{(2\pi)^3}\int d\vec{p}\,\frac{n_{\vec{p}+\vec{q}}(s)-n_{\vec{p}}(s)}{\epsilon_{\vec{p}+\vec{q}}(s)-\epsilon_{\vec{p}}(s)-\alpha_m} \tag{23}$$

and

$$n_{\vec{p}}(s) = [\exp(\beta\epsilon_p(s)-\beta\mu_s) + 1]^{-1} \tag{24}$$

is the Fermi distribution of the s species.

We now calculate the contribution of the diagrams, given in Fig. 1. We assert that these are the leading asymptotic terms for long wavelength. Using the prescription of Luttinger and Ward[5] for the many species system under consideration we obtain

$$M_{\mu\nu}(\omega_n) = \frac{\hbar^2}{V}\sum_{s,s'}\frac{e_s e_{s'}}{m_s m_{s'}}\sum_{\vec{p},\vec{p}'} p_\mu p_\nu' \sum_{i=1}^{5} K^{(i)}_{pp'}(\omega_n, s, s'),$$

$$\omega_n = i2\pi n/\hbar\beta, \tag{25}$$

where $K^{(i)}_{pp}(\omega_n, s, s')$ corresponds to the ith diagram of Fig. 1

$$K^{(1)}_{pp'}(\omega_n, s, s') = \delta_{ss'}\frac{e_s^2}{\beta^2 V}\sum_{m,\ell} U_{\vec{p}-\vec{p}'}(\alpha_m)G_p(\zeta_\ell s)G_p(\zeta_\ell+\omega_n s)$$

$$G_{p'}(\zeta_\ell+\omega_n+\alpha_m, s)G_{p'}(\zeta_\ell+\alpha_m, s)$$

$$K^{(2)}_{pp'}(\omega_n, s, s') = \delta_{ss'}\delta_{pp'} \frac{e_s^2}{\beta^2 V} \sum_{m,\ell} \sum_{q} U_q(\alpha_m) \left[G_p(\zeta_\ell, s)\right]^2$$

$$G_{\vec{p}-\vec{q}}(\zeta_\ell - \alpha_m, s) G_p(\zeta_\ell - \omega_n, s)$$

$$K^{(3)}_{pp'}(\omega_n, s, s') = \delta_{ss'}\delta_{pp'} \frac{e_s^2}{\beta^2 V} \sum_{m,\ell} \sum_{q} U_q(\alpha_m) \left[G_p(\zeta_\ell, s)\right]^2$$

$$G_{\vec{p}+\vec{q}}(\zeta_\ell + \alpha_m, s) G_p(\zeta_\ell + \omega_n, s)$$

$$K^{(4)}_{pp'}(\omega_n, s, s') = e_s^2 e_{s'}^2 \frac{1}{V^2\beta^3} \sum_{m,\ell,\ell'} \sum_{q} U_q(\alpha_m) U_q(\alpha_m - \omega_n) G_p(\zeta_\ell, s)$$

$$G_p(\zeta_\ell - \omega_n, s) G_{\vec{p}-\vec{q}}(\zeta_\ell - \alpha_m, s) G_{p'}(\zeta_{\ell'}, s')$$

$$G_{p'}(\zeta_{\ell'} - \omega_n, s') G_{\vec{p}' - \vec{q}}(\zeta_{\ell'} - \alpha_m, s')$$

$$K^{(5)}_{pp'}(\omega_n, s, s') = e_s^2 e_{s'}^2 \frac{1}{V^2\beta^2} \sum_{m,\ell,\ell'} \sum_{q} U_q(\alpha_m) U_q(\alpha_m - \omega_n) G_p(\zeta_\ell, s)$$

$$G_p(\zeta_\ell - \omega_n, s) G_{\vec{p}-\vec{q}}(\zeta_\ell - \alpha_m, s) G_{p'}(\zeta_\ell, s')$$

$$G_{p'}(\zeta_{\ell'} + \omega_n, s') G_{\vec{p}' + \vec{q}}(\zeta_{\ell'} + \alpha_m, s')$$

We now carry out the summation over ℓ and ℓ' by converting the sums into integrals. After considerable manipulations we obtain

$$M_{\mu\nu}(\omega_n) = -\frac{1}{2}\int\frac{d\vec{q}}{(2\pi)^3}\, q_\mu q_\nu \sum_{s,s'} e_s^2 e_{s'}^2$$

$$\frac{e_s}{m_s}\left(\frac{e_s}{m_s}-\frac{e_{s'}}{m_{s'}}\right) F_q(\omega, s, s') \qquad (27)$$

where

$$F_q(\omega_n, s, s') = \frac{\hbar^2}{\omega_n^2}\frac{1}{\beta}\sum_m U_q(\alpha_m)U_q(\alpha_m+\omega_n)$$

$$\left[Q_q(\alpha_m+\omega_n, s)-Q_q(\alpha_m, s)\right]\left[Q_q(\alpha_m+\omega_n, s')-Q_q(\alpha_m, s')\right] \qquad (28)$$

Equations (27) and (28) are not suitable for the analytical continuation to the upper z half-plane. In order to perform the continuation one has first to evaluate the summation over m.[7] After some algebra we obtain

$$F_q(\omega, s, s') = \frac{P}{i\pi\omega^2}\int dx\, H(x)\Big\{U_q^+(x)U_q^+(x+\hbar\omega)\left[Q_q^+(x+\hbar\omega, s)-Q_q^+(x, s)\right]$$

$$\left[Q_q^+(x+\hbar\omega, s')-Q_q^+(x, s')\right] - U_q^-(x)U_q^+(x+\hbar\omega)$$

$$\left[Q_q^+(x+\hbar\omega, s)-Q_q^-(x, s)\right]\left[Q_q^+(x+\hbar\omega, s')-Q_q^-(x, s')\right]\Big\} \qquad (29)$$

In Eq. (28) we denote by P the principal value,

$$H(x) = \frac{1}{2}\,\mathrm{Coth}(\beta x/2) \qquad (30)$$

and

$$Q_q^{\pm}(x, s) = \int \frac{d\vec{p}}{(2\pi)^3} \frac{n_{p+q}(s)-n_p(s)}{\epsilon_{\vec{p}+\vec{q}}(s)-\epsilon_p(s)-x\mp i\epsilon} \tag{31}$$

$U_q^{\pm}(x)$ is determined by Eq. (22), replacing $Q_q(\alpha_m, s)$ by $Q_q^{\pm}(x,s)$.

Now we make use of Eqs. (5), (10), (27), (28), and (29) to obtain our final result for the complex conductivity.

$$\sigma(\omega) = i\sigma_0(\omega) + i\,\frac{4\pi}{3\omega} \sum_{ss'} e_s^2 e_{s'}^2\, \frac{e_s}{m_s}\left(\frac{e_s}{m_s} - \frac{e_{s'}}{m_{s'}}\right) \frac{1}{(2\pi)^3} \int dq q^4$$

$$\frac{1}{2\pi i}\, P \int dx\; H(x) U_q^{+}(x+\hbar\omega) \Big\{ U_q^{+}(x) \left[Q_q^{+}(x+\hbar\omega, s) - Q_q^{+}(x, s)\right]$$

$$\left[Q_q^{+}(x+\hbar\omega, s') - Q_q^{+}(x, s')\right] - U_q^{-}(x) \left[Q_q^{+}(x+\hbar\omega, s) - Q_q^{-}(x, s)\right]$$

$$\left[Q_q^{+}(x+\hbar\omega, s') - Q_q^{-}(x, s')\right]\Big\}. \tag{32}$$

In the last equation use has been made of the fact that Q_q depends only on the absolute value of q, and that $\sigma_{\mu\nu}(\omega) = \delta_{\mu\nu}\sigma(\omega)$. This result is rather complicated but in principle can be evaluated analytically or numerically for specific problems. It is clear that the result is applicable for both classical and quantum plasmas for any mass ratio of the species of the system, and for temperatures, where the average potential energy of interaction per particle is smaller than the average kinetic energy.

IV. CALCULATIONS OF THE RESISTIVITY

We consider a system with equal number of negatively and positively charged particles interacting with a homogeneous time dependent electric field. Our starting point is the expression for the conductivity given by Eq. (32). Using this result, for two species, i.e. electrons and holes, and taking into account the symmetry of their expression with respect to the mutual interchanging of electron-hole, we obtain:

$$\sigma = i\sigma_0(\omega) + \frac{2e_e^2e_h^2}{3\pi\omega^3}\left(\frac{e_e}{m_e} - \frac{e_h}{m_h}\right)^2 \int dq\ P\int dx\ \mathrm{Coth}\ \frac{\beta x}{2}\ F_q(x,\omega) \tag{33}$$

where: e_e, $m_e(e_h, m_h)$ are respectively the electric charge and mass of the electron (hole) and $\hbar$ is taken to be one.

$$\beta = \frac{1}{k_BT}\ ; \quad \sigma_0 = \frac{1}{4\pi\omega}\ (\omega_{pe}^2+\omega_{ph}^2)$$

P represents the principle value integral and $F_q(x,\omega)$ is given by:

$$F_q(x,\omega) = \mathcal{E}_q^{-1}(x+\omega)\{\mathcal{E}_q^{-1}(x)[Q_q(x+\omega)-Q_q(x)][B_q(x+\omega)-B_q(x)]$$

$$- \mathcal{E}_q^{*-1}(x)[Q_q(x+\omega)-Q_q^*(x)][B_q(x+\omega)-B_q^*(x)]\}. \tag{34}$$

Here $Q_q(x)(B_q(x))$ represents the pair density fluctuation of the electrons (holes) given by

$$Q_q(x) = \sum_k \frac{f_{k+q}-f_k}{\epsilon_{k+q}-\epsilon_k-\omega-i\eta} \tag{35}$$

where ϵ_k are electronic kinetic energies and f_k is the electronic Fermi-Dirac distribution function. A similar expression stands for the hole density fluctuations. $\mathcal{E}_q(x)$ is the dielectric function of the electron-hole system and is given by

$$\mathcal{E}_q(x) = 1 - \frac{4\pi e_e^2}{q^2}\ Q_q(x) - \frac{4\pi e_h^2}{q^2}\ B_q(x) \tag{36}$$

The absorption properties of the plasma can be best described by the real part of the resistivity i.e.

$$R = \mathrm{Re}\,\frac{1}{\sigma}\ .$$

The resistivity, as it is well known, has the same frequency dependence as the collision frequency. For the case of $\omega\tau > 1$ we write:

$$R = \sigma_0^{-2}\ \mathrm{Re}\ \frac{2e_e^2e_h^2}{3\pi\omega^3}\left(\frac{e_e}{m_e} - \frac{e_h}{m_h}\right)^2 \int dq\ P\int dx\ \mathrm{Coth}\ \frac{\beta x}{2}\ F_q(x,\omega)$$

$$= \frac{32\ \pi e_e^2e_h^2(e_e/m_e - e_h/m_h)^2}{3(\omega_{pe}^2 + \omega_{ph}^2)^2\omega}\ \mathrm{Re}\int dq\ P\int dx\ \mathrm{Coth}\ \frac{\beta x}{2}\ F_q(x,\omega) \tag{37}$$

We shall next investigate the effect of the small mass ratio (i.e. $m_e/m_h \lesssim 1$) on the conductivity at low temperatures. Let us first take the limiting case of $\beta \to \infty$ and also treat the screening effect due to the self-consistent field statically. This approximation is taken at this time in order to illustrate the basic difference between light and massive holes. The proposed approximation simplifies drastically $F_q(x,\omega)$ in Eq. (34). Here we replace $\mathcal{E}_q(x)$ by $\mathcal{E}_q(0)$, thus in this limit Eq. (34) reads

$$F_q(x,\omega) \approx \mathcal{E}_q^{-2}(0)\{[Q_q(x+\omega)-Q_q(x)][B_q(x+\omega)-B_q(x)]$$

$$- [Q_q(x+\omega)-Q_q^*(x)][B_q(x+\omega)-B_q^*(x)]\} \tag{38}$$

We also may use the analytical properties of $Q_q(x)$, $B_q(x)$ and rewrite our expression for R which now reads

$$R = \frac{C}{\omega}\int \frac{dg}{\mathcal{E}_q^2} P \int dx \left[\text{Coth}\,\frac{\beta x}{2} - \text{Coth}\,\frac{\beta(x+\omega)}{2}\right]$$

$$\text{Im}\, Q_q(x+\omega)\ \text{Im}\, B_q(x) \tag{39a}$$

$$C = 32\,\pi e_e^2 e_h^2 \left(\frac{e_e}{m_e} - \frac{e_h}{m_h}\right)^2 / 3(\omega_{pe}^2 + \omega_{ph}^2)^2 \quad .$$

In the limit of zero temperature our result becomes

$$R = \frac{C}{\omega}\int \frac{dq}{\mathcal{E}_q^2} P \int_0^{\omega} dx\ \text{Im}\, B_q(x)\ \text{Im}\, Q_q(\omega-x) \ . \tag{39b}$$

We would like now to consider qualitatively the case of a massive hole. Here the Im $B_q(x)$ will be given by a function which peaks in the vicinity of $x \sim 0$ and is proportional to n, the average particle density. We thus can approximate in the integrand of Eq. (39b) the term Im $B_q(x)$ by an expression proportional to $n\delta(x)$ and obtain that the resistivity is given by an integral over q of an integrand proportional to Im $Q_q(\omega)/\mathcal{E}_q^2(0)$. This is the expected result. We next study the frequency behavior of the resistivity for mass ratios of order unity. Let us limit ourselves to the case of small external frequency ω. Here, as we shall see shortly the range of integration over q is limited between zero and $2q_F$. Also, Im $B_q(x)$ and Im $Q_q(\omega-x)$ can be taken to be proportional to x and $(\omega-x)$ respectively. Substituting these values in Eq. (39b) will lead to the frequency dependence of the resistivity:

$$R \sim \frac{1}{\omega}\int_0^{\omega} x(\omega-x)dx = \frac{\omega^2}{6} \tag{40}$$

This ω^2 behavior of the resistivity is due to lack of phase-space for the absorption process. We should remember that we consider electron-hole systems in which both electrons and holes have their own Fermi sphere. The photon is absorbed by mutually exciting an electron pair and a hole pair (Note: here the hole is not the vacancy in the electronic Fermi sea left by the excitation of an electron). The energy supplied by the photon is $\hbar\omega$, a negligible quantity, thus the two pairs must be in the vicinity of their Fermi sea where the density of states is negligibly small. The momentum to this process is supplied by the Coulomb collision of the electron and the hole which in our formulation is given by q. Thus the momentum q is limited to be in the range between zero and $2q_F$. As stated above the density of states vanishes with vanishing ω and it can be shown, using density of state arguments, that the joint density of states for the electron-hole excitation divided by ω is proportional to ω^2 as we find in Eq. (8).

We shall next evaluate the resistivity, as given by Eqs. (34) and (37). Here we take into account the effects of the dynamical screening of the plasma, which is a complex function of q and ω. Thus one does not necessarily expect that the integrand in Eq. (37) will depend solely on the imaginary part of the hole and the electron pair distribution function respectively as it appears in Eq. (39). Indeed we find that when the electron and hole can be thought of as scattered by a retarded potential, Eq. (39) is generalized to include terms proportional to the product of two imaginary parts of electron (hole) density fluctuations.

The calculations of the resistivity using our Eqs. (34) and (37) are now straightforward but tedious. We shall present our final result:

$$R = R_0 \frac{1}{\omega} \int_0^\infty dy y^4 \int_0^\omega dx \left| \mathcal{E}_y(x)\right|^{-2} \left|\mathcal{E}_y(\omega-x)\right|^{-2} \left\{(F_1+F_2)\epsilon_y^I(\omega-x)E_y^I(x) + G\epsilon_y^I(\omega-x)\epsilon_y^I(x) + HE_y^I(\omega-x)E_y^I(x)\right\} \tag{41}$$

where

$$F_1 = \mathcal{E}_y^R(\omega-x)\,\mathcal{E}_y^R(x)$$

$$F_2 = -\left[\epsilon_y^R(\omega-x)-\epsilon_y^R(x)\right]\left[E_y^R(\omega-x)-E_y^R(x)\right] + \left[\epsilon_y^{I^2}(x)+E_y^{I^2}(x)\right] + \mathcal{E}_y^R(\omega-x)\left\{\left[\epsilon_y^R(\omega-x)-\epsilon_y^R(x)\right] + \left[E_y^R(\omega-x)-E_y^R(x)\right]\right\} \tag{42}$$

$$G = \frac{1}{2}[E_y^R(\omega-x)-E_y^R(x)];\ H = \frac{1}{2}[\epsilon_y^R(x-\omega)-\epsilon_y^R(x)].$$

In Eqs. (41), (42) ω, x are normalized to ω_F, $y = q/2q_F$, the superscripts R, I represent the Real Part and the Imaginary Part of the dielectric function respectively. Also $\epsilon(E)$ represents the electron (hole) dielectric function and

$$R_0 = 32\,\omega_{pe}^2 \left(1 + \frac{m_e}{m_h}\right)^2 \omega_F/(\omega_{pe}^2 + \omega_{ph}^2)^2. \tag{43}$$

It is of interest to note that for large q the real part of the dielectric function can be approximated by unity and the imaginary part becomes small, both for electrons and holes. In this case F_2, G, H will not contribute much. Here the main contribution will come from the term proportional to F_1. This term resembles our result given in Eq. (39), however it includes dynamic screening. Numerical evaluation should reveal the relative contribution stemming from the various terms and also determine the importance of dynamical correlations on the resistivity.

V. DISCUSSION

In this paper we have calculated the resistivity for plasma systems composed of negatively and positively charged particles. We particularly focus our attention on the dynamical correlation effects for small mass-ratio of the two species. Using our result given by Eqs. (41), (42) and (43), we compute the resistivity, in the zero-temperature limit for various values of r_s and mass ratio, as a function of the incident light frequency. We find, as expected, that for extremely low and high frequencies, the static approximation, as described in the text, is applicable. However, for most regions of interest, dynamical correlation effects cannot be omitted. Our computational results are presented in Figs. 3-7. In Fig. 3 we present the resistivity for the case of electronic $r_s = 0.2$ and mass ratio $\alpha = m_h/m_e = 3$. Here the effect of the dynamical correlations is to increase the resistivity in the intermediate frequency region. In Fig. 4 we plot the resistivity for the electronic $r_s = 1$ and $\alpha = 3$. The effect of larger r_s results in the increase of the resistivity. The curve in Fig. 5 represents the resistivity for $r_s = 0.2$ and $\alpha = 6$. Here the resistivity begins to develop a shoulder at the low frequency part of the curve. In Fig. 6 we plot $R(\omega)$ for

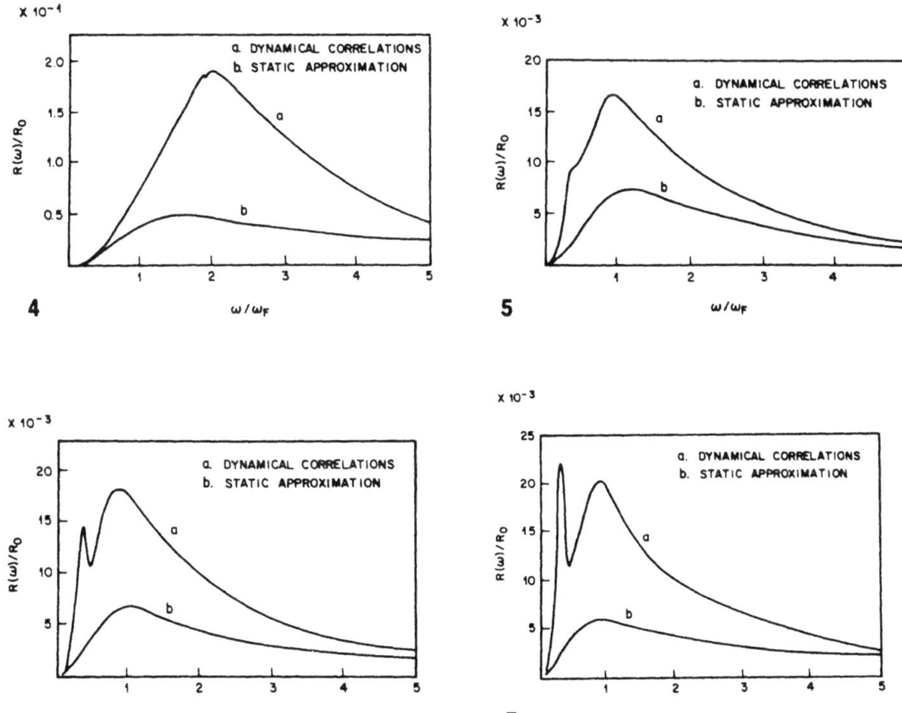

Fig. 4. The resistivity R(ω) in terms of R_0 for electronic r_s=1 and mass ratio α=3.

Fig. 5. The resistivity R(ω) in terms of R_0 for electronic r_s=0.2 and mass ratio α=6.

Fig. 6. The resistivity R(ω) in terms of R_0 for electronic r_s=0.2 and mass ratio α=8.

Fig. 7. The resistivity R(ω) in terms of R_0 for electronic r_s=0.2 and mass ratio α=10.

electronic $r_s = 0.2$ and $\alpha = 8$ where a peak in the resistivity appears at low frequencies. This peak is an indication of the low frequency mode of the heavy species screened by the light species, the well-known ion acoustic mode. Here the absorption process of the photon consists predominantly of coupling to the "light" electrons which in turn scatter from the coherent density fluctuations of the "heavy" holes. In Fig. 7 we plot $R(\omega)$ for electronic $r_s = 0.2$ and $\alpha = 10$ where the peak is more pronounced.

In conclusion we have presented the calculation for the conductivity to the lowest order in the plasma parameter r_s for plasmas in which the mass ratios of the negatively and positively charged particle are small. We calculate the resistivity for various values of the parameters r_s and α and show a pronounced effect due to the dynamic correlations.

REFERENCES

1. C. Oberman, A. Ron and J. Dawson, Phys. Fluids 5, 1514 (1962). V. I. Perel and G. M. Eliashberg, Zh. Eksperim. i Teor. Fiz. 41, 886 (1961)[Soviet Physics - JETP 14, 633, (1962)]. D. F. Dubois, V. Gilinsky and M. G. Kivelson, Phys. Rev. 129, 2376 (1963). A. Ron and N. Tzoar, Phys. Rev. Lett. 10, 45, (1963) and Phys. Rev. 131, 12, (1963).

2. Ya. Pokrovskii, Phys. Stat. Solidi 11 385 (1972). Session 9 in the "Proceedings of the Eleventh International Conference on the Physics of Semiconductors", Warsaw (1972), PWW-Polish Scientific Publisher, Warszawa (1972). Mark N. Gurnee, Maurice Glicksman and Phil Won Yu, Solid State Communications 11, 11 (1972); V. Marello, T. F. Lee, R. N. Silver, T. C. McGill and J. W. Mayer, Phys. Rev. Letters 31, 539 (1973); W. F. Brinkman and T. M. Rice, Phys. Rev. B7, 1508 (1973).

3. R. S. Markiewicz, J. P. Wolfe and C. D. Jeffries, Phys. Rev. Lett. 32, 1357 (1974); J. P. Wolfe, R. S. Markiewicz, C. Kittel and C. D. Jeffries, Phys. Rev. Lett. 34, 275 (1975).

4. R. Kubo, J. Phys. Soc. Japan 12, 570 (1957).

5. J. M. Luttinger and J. C. Ward, Phys. Rev. 118, 1417 (1960), A. A. Abrikosov, L. P. Gorkov and F. E. Dzyaloshinsky, Zh. Eksperim i Teor. Fiz. 36, 900 (1959) [translation: Soviet Phys.- JETP 9, 636 (1959)].

6. R. Balescu, Phys. Fluids 4, 94 (1961).

7. See P. I. Perel and G. M. Eliashberg, Ref. 1.

ASPECTS OF TRANSPORT IN LIQUID METALS

M.P. Tosi

Istituto di Fisica dell'Universita' e GNSM del CNR

Roma, Italy

1. INTRODUCTION

The assumption of weak electron-ion interactions in a 'simple' liquid metal has led, on the one hand, to the calculation of electronic transport properties via the weak scattering formalism (Ziman (1), Baym (2))and, on the other, to the introduction of effective pair interactions between the ions in which the electrons enter through the electron-gas dielectric function. Several properties of simple liquid metals can be understood on this basis. For example, the effective-interaction model appears to account for important features of the neutron inelastic scattering spectrum of a system such as liquid rubidium (Rahman (3)), just as it gives a good account of phonons in the solid alkali metals (Price et al(4)).

It is nevertheless becoming increasingly apparent that this simple picture needs to be transcended in some respects. With regard to the electronic properties, one may recall that:

(i) a careful comparison between X-ray and neutron diffraction data from liquid metals (Egelstaff et al. (5)) indicates the existence of substantial ordering in the conduction electron system. An analogous suggestion has been made for the solid in connection with X-ray inelastic scattering data (Platzman and Eisenberger (6)), which display a fairly well defined plasmon peak at wave vectors well past the cut-off wave number of electron gas theory. Preliminary theoretical discussions of the static electronic pair structure in liquid metals have been presented by Watabe and Hasegawa (7) and by Chihara (8), and extensive calculations have been carried out on the related system of electron-hole drops in semiconductors (Vashishta et al. (9)).

(ii) strong scattering situations in electronic transport are being attacked (Rousseau et al. (10), Jones (11)) by taking advantage of the long-known relation between inverse transport coefficients and time correlations of random conjugate forces (Nyquist (12), Mori (13); see also Kubo (14)). It may be noted in this connection that the theory of atomic dynamics in monatomic liquids is currently in the stage of analyzing this relation on the basis of computer simulations of liquid argon (see, for instance, Copley and Lovesey (15)).

Similarly, a number of ionic properties of a liquid metal cannot be accounted for by simulating the system as a one-component fluid of pseudoatoms, but require explicit consideration of its two--component nature. Focussing specifically on the hydrodynamic regime,

(i) the pseudoatom model does not account for the effect of the conduction electrons on sound wave damping, which is well known in the solid where it provides one of the tools for studying the Fermi surface, nor does it incorporate the Wiedemann-Franz law which is experimentally obeyed to high accuracy in liquid metals;

(ii) observed anomalies in the mass dependence of the shear viscosity of liquid lithium isotopes (Ban et al. (16)) may similarly involve specific electronic effects (March and Tosi (17));

(iii) weak electron-ion scattering theory does not seem to explain the Haeffner effect, that is the electromigration of light isotopes to the anode in liquid-metal isotopic mixtures (Parrinello et al. (18)). In this instance, and more generally for treatments of electro-osmonis and of electromigration and thermal diffusion in liquid alloys, a force-force correlation approach seems again called for.

The aim of these lectures is, first of all, to present the formulation of the theory of a liquid metal as a two-component electron-ion fluid, as recently developed (March and Tosi (17), Tosi et al. (19)). It may be claimed, correctly, that the general theory is already formulated in the work of Mori (13), but our discussion presents the abstract theory in an alternative, and perhaps more transparent, light, and is at any rate the first step towards concrete applications to the systems of present interest. For the sake of (relative) simplicity the discussion omits to consider entropy fluctuations, but it will be apparent from the context how these may be included. The theory provides a unified basis to discuss problems of structure and thermodynamics, and of dynamics and mass and charge transport, in an approximation which should still be reasonable in practice as the specific heats ratio is about 1.2÷1.3 in liquid metals near their freezing point. We shall then show how the theory reduces to weak scattering theory and give a brief discussion of electromigration and the Haeffner effect in this framework.

2. THE ELECTRON GAS

We first illustrate the approach by applying it to a discussion of the dielectric properties of the electron gas. We consider the equation of motion for the density matrix $\langle \psi^+(\underline{x},t)\psi(\underline{x}',t)\rangle$ in a weak external potential $V_e(\underline{x},t)$, which reads

$$\left[i\hbar \frac{\partial}{\partial t} - \frac{\hbar^2}{2m}(\nabla^2_{\underline{x}} - \nabla^2_{\underline{x}'}) + V_e(\underline{x},t) - V_e(\underline{x}',t)\right] \langle \psi^+(\underline{x},t)\psi(\underline{x}',t)\rangle$$

$$+ \int d\underline{x}''\left[v(\underline{x}-\underline{x}'') - v(\underline{x}'-\underline{x}'')\right] \langle \psi^+(\underline{x},t)\psi(\underline{x}',t)\,\varrho(\underline{x}'',t)\rangle = 0 \quad (2.1)$$

where $v(\underline{x}-\underline{x}'')$ is the Coulomb potential and $\varrho(\underline{x},t)$ is the particle density operator. Equation (2.1) is equivalent to an infinite set of equations of motion for observable physical quantities, akin to those entering the theory of atomic dynamics in fluids. Explicitly, we introduce the Wigner distribution function $f(\underline{p},\underline{R},t)$ defined by (see, for instance Baym and Kadanoff (20))

$$f(\underline{p},\underline{R},t) = \int d\underline{r}\ \exp(i\ \underline{p}\cdot\underline{r}/\hbar)\,\langle \psi^+(\underline{R}+\tfrac{1}{2}\underline{r},t)\ \psi(\underline{R}-\tfrac{1}{2}\underline{r},t)\rangle \quad (2.2)$$

from which we can construct physical observables such as the particle density,

$$n(\underline{R},t) = \sum_{\underline{p}} f(\underline{p},\underline{R},t), \quad (2.3)$$

the particle current density,

$$\underline{j}(\underline{R},t) = \frac{1}{m}\sum_{\underline{p}} \underline{p}\ f(\underline{p},\underline{R},t), \quad (2.4)$$

and the kinetic tensor,

$$\pi(\underline{R},t) = \frac{1}{m} \sum_{\underline{p}} \underline{p}\,\underline{p}\, f(\underline{p},\underline{R},t). \tag{2.5}$$

By expanding eqn. (2.1) about its diagonal (namely, in powers of $\underline{r} = \underline{x}-\underline{x}'$) and equating to zero the coefficients of successive powers of $\underline{r}$, we find first the usual continuity equation,

$$\partial n(\underline{R},t)/\partial t + \underline{\nabla}\cdot\underline{j}(\underline{R},t) = 0\ ; \tag{2.6}$$

next, the equation of motion for the current density,

$$m\,\partial \underline{j}(\underline{R},t)/\partial t + \underline{\nabla}\cdot\pi(\underline{R},t) + n(\underline{R},t)\,\underline{\nabla}\, V_e(\underline{R},t) + \int d\underline{x}''\,\underline{\nabla}\, v(\underline{R}-\underline{x}'')\,\langle \rho(\underline{R},t)\,\rho(\underline{x}'',t)\rangle = 0; \tag{2.7}$$

thirdly, the equation of motion for the kinetic tensor,

$$\frac{1}{2}\,\partial \pi(\underline{R},t)/\partial t + (2m^2)^{-1}\,\underline{\nabla}\cdot \sum_{\underline{p}} \underline{p}\,\underline{p}\,\underline{p}\, f(\underline{p},\underline{R},t) + \underline{j}(\underline{R},t)\,\underline{\nabla}\, V_e(\underline{R},t) + \int d\underline{x}''\,\underline{\nabla} v(\underline{R}-\underline{x}'')\,\langle \underline{j}(\underline{R},t)\,\rho(\underline{x}'',t)\rangle = 0\ . \tag{2.8}$$

Equation (2.7) is the generalized Navier-Stokes equation for the fluid, while the trace of eqn. (2.8) yields the energy transport equation (17).

Our aim is to derive from the above set of equations the density response function $\chi(\underline{R},t)$ of the system, defined by

$$n(\underline{R},t) = \iint d\underline{R}'\, dt'\, \chi(\underline{R}-\underline{R}',\, t-t')\, V_e(\underline{R}',t') \tag{2.9}$$

or, equivalently

$$\chi(\underline{R}-\underline{R}',t-t') = \delta n(\underline{R},t)/\,\delta V_e(\underline{R}',t') \tag{2,10}$$

where the symbol δ denotes the functional derivative. In the hydrodynamic limit, the conservation principles ensure that the particle density, the current density and the energy density form a complete basic set of dynamic variables (Kadanoff and Martin (21)). As stated in the introduction, we simplify the discussion by taking account only of the particle density and the current density, which are furthermore related by the continuity equation (2.6). We can thus rewrite eqn. (2.7) as an equation of motion for the particle density, in the form (Singwi et al. (22))

$$m\,\partial^2 n(\underline{R},t)/\partial t^2 = \nabla_\alpha \nabla_\beta\, \pi_{\alpha\beta}(\underline{R},t) + n(\underline{R},t)\,\nabla^2\left[V_e(\underline{R},t) + \int d\underline{x}\, v(\underline{R}-\underline{x})\, n(\underline{x},t)\right]$$
$$+ \int d\underline{x}\; \underline{\nabla}\cdot\left[\underline{\nabla} v(\underline{R}-\underline{x})\, \langle \rho(\underline{R},t)\, \rho(\underline{x},t)\rangle_c\right] \qquad (2.11)$$

where we have also separated explicitly the long range effects of the Coulomb potential by writing the two-body correlation function in eqn. (2.7) as

$$\langle \rho(\underline{R},t)\, \rho(\underline{x},t)\rangle = n(\underline{R},t)\, n(\underline{x},t) + \langle \rho(\underline{R},t)\, \rho(\underline{x},t)\rangle_c. \qquad (2.12)$$

A formal solution of eqn. (2.11), after linearization, can now be achieved by writing for the effect of the external field on the kinetic tensor and on the 'cluster' correlation function

$$\delta\pi_{\alpha\beta}(\underline{R},t) = \iint d\underline{y}\; d\tau \left.\frac{\delta\pi_{\alpha\beta}(\underline{R},t)}{\delta n(\underline{y},\tau)}\right|_{eq} \delta n(\underline{y},\tau)$$
$$\equiv \iint d\underline{y}\; d\tau\; \lambda_{\alpha\beta}(\underline{R}-y,t-\tau)\; \delta n(\underline{y},\tau) \qquad (2.13)$$

and

$$\delta\langle \rho(\underline{R},t)\rho(\underline{x},t)\rangle_c = \iint d\underline{y}\; d\tau \left.\frac{\delta\langle \rho(R,t)\rho(x,t)\rangle_c}{\delta n(\underline{y},\tau)}\right|_{eq} \delta n(\underline{y},\tau)$$
$$= \iint d\underline{y}\; d\tau\; F(\underline{R}-\underline{x},\underline{x}-\underline{y},t-\tau)\, \delta n(\underline{y},\tau) \qquad (2.14)$$

The solution of eqn. (2.11) in Fourier transform then yields

$$\chi(k,\omega) = \frac{nk^2}{m}\left[\omega^2 - \omega_p^2 - \frac{1}{m}\tilde{\Gamma}(k,\omega)\right]^{-1} \qquad (2.15)$$

where $\omega_p = (4\pi ne^2/m)^{\frac{1}{2}}$ is the usual plasma frequency, and

$$\tilde{\Gamma}(k,\omega) = k_\alpha k_\beta \lambda_{\alpha\beta}(\underline{k},\omega) + \sum_{\underline{k}'} \underline{k}\cdot\underline{k}'\; v(k')\; F(\underline{k}-\underline{k}',\underline{k},\omega). \qquad (2.16)$$

Equations (2.15) and (2.16) embody the 'generalized hydrodynamics' of the fluid (in the absence of entropy fluctuations), in which the function $\tilde{\Gamma}(k,\omega)$ plays the role of a generalized thermodynamic and transport coefficient. In particular, in the thermodynamic limit the functional derivatives become local in space and time, and one easily recovers the expression

$$\lim_{k\to 0} \tilde{\Gamma}(k,o) = (nK)^{-1}\, k^2 \qquad (2.17)$$

where K is the compressibility.

To clarify the connection of this formalism with the formulation of transport given by Mori (13) on the basis of the 'generalized Langevin equation', we introduce the conductivity $\sigma(k,\omega)$ as the coefficient relating the current density $\underline{j}(\underline{k},\omega)$ to the local field driving the current. In the long wavelength limit the latter is determined by the Hartree potential,

$$
\begin{aligned}
V_H(\underline{k},\omega) &= V_e(\underline{k},\omega) + v(k)\, n(\underline{k},\omega) \\
&= V_e(\underline{k},\omega)/\varepsilon(k,\omega)
\end{aligned}
\tag{2.18}
$$

where $\varepsilon(k,\omega)$ is the usual dielectric function, obtained from (2.15) as

$$
\varepsilon(k,\omega) = 1 - \omega_p^2 \left[\omega^2 - \tilde{\Gamma}(k,\omega)/m\right]^{-1}. \tag{2.19}
$$

At finite k, however, additional local effects, such as diffusion in the gradient of electrochemical potential caused by charge separation, become relevant. These effects are accounted for if one writes the local potential as

$$
V_{loc}(\underline{k},\omega) = V_e(\underline{k},\omega) - n(\underline{k},\omega)/\chi(k,o) \tag{2.20}
$$

where the second term, which describes the potential set up by charge separation, reduces to the expression (2.18) for $k \to o$, according to eqns. (2.15) and (2.17), but for finite k also includes the local effects associated with the 'thermodynamic' term $\tilde{\Gamma}(k,o)$. The conductivity $\sigma(k,\omega)$, defined by

$$
\underline{j}(\underline{k},\omega) = -\sigma(k,\omega)\; i\underline{k}\; V_{loc}(\underline{k},\omega)/e^2, \tag{2.21}
$$

is then found from eqns. (2.6), (2.15) and (2.20) to have the expression

$$
\sigma(k,\omega) = \frac{ne^2}{m}\left[-i\omega + \gamma(k,\omega)\right]^{-1} \tag{2.22}
$$

with

$$
\gamma(k,\omega) = i\left[\tilde{\Gamma}(k,\omega) - \tilde{\Gamma}(k,o)\right]/(m\omega). \tag{2.23}
$$

This latter function, which plays the same role as Mori's memory function, is here expressed through eqns. (2.13) and (2.14) in terms of the perturbations of the kinetic tensor and of the internal forces in the liquid due to the application of a weak external field. It is actually convenient for its evaluation to rewrite eqn. (2.16), using the continuity equation in operator form, as

$$\tilde{\Gamma}(k,\omega) = \frac{\omega}{k}\,\mathrm{F.T.}\left\{ \nabla_\alpha \nabla_\beta \left.\frac{\delta\pi_{\alpha\beta}(\underline{R},t)}{\delta j(\underline{y},\tau)}\right|_{eq} + \int d\underline{x}\; \nabla_\alpha\left[\nabla_\alpha v(\underline{R}-\underline{x})\,\frac{\delta\langle\rho(\underline{R},T)\rho(\underline{x},t)\rangle_c}{\delta j(\underline{y},\tau)}\right]_{eq}\right\}_{\underline{k},\omega}. \tag{2.24}$$

The functional derivatives in this equation can then be treated as local in time in the evaluation of the d.c. limit of eqn. (2.23).

To indicate how the calculation could be continued in principle, let us rewrite the functional derivative entering eqn. (2.14) in the form

$$\frac{\delta\langle\rho(\underline{R},t)\rho(\underline{x},t)\rangle_c}{\delta n(\underline{y},\tau)} = \iint d\underline{y}'d\tau'\,\frac{\delta\langle\rho(R,t)\rho(x,t)\rangle_c}{\delta V_e(\underline{y}',\tau')}\,\frac{\delta V_e(\underline{y}',\tau')}{\delta n(\underline{y},\tau)}$$

$$= -\frac{i}{\hbar}\iint d\underline{y}'d\tau'\,\langle T[\rho(\underline{y}',\tau')\rho(\underline{R},t)\rho(\underline{x},t)]\rangle_c\,\chi^{-1}(\underline{y}'-\underline{y},\tau'-\tau) \tag{2.25}$$

or

$$F(\underline{k}-\underline{k}',\underline{k},\omega) = -\frac{i}{\hbar}\,T(\underline{k}-\underline{k}',\underline{k},\omega)/\chi(k,\omega), \tag{2.26}$$

where $T(\underline{k}-\underline{k}',\underline{k},\omega)$ is the Fourier transform of the dynamic triplet function introduced in eqn. (2.25). The appearance of the inverse response function in eqn. (2.25) indicates that the functional derivative is to be evaluated as a triplet function in which the vertex at $\rho(\underline{y},\tau)$ is to be taken as a proper vertex; namely, as a triplet function in which the densities $\rho(\underline{R},t)$ and $\rho(\underline{x},t)$ correlate with the <u>screened</u> density at $(\underline{y},\tau)$. Similarly, the evaluation of the functional derivative entering eqn. (2.13) involves correlations between the kinetic tensor operator and the screened particle density. To attempt an evaluation of these quantities we should obviously make recourse to their own equations of motion, obtainable for the former from the equation of motion of the two-body density matrix and given for the latter by eqn. (2.8). It is at any rate evident that these functional derivatives, being screened response functions expressing the response to the local field rather than to the bare external field, will admit a structure analogous to that of eqn. (2.22). One thus obtains a continued-fraction structure for the conductivity (Mori (24)).

The above results can be used to discuss a number of properties of the electron gas, but we shall only briefly note the following points:

(i) in the hydrodynamic limit, momentum conservation ensures that the resistivity vanishes. $\tilde{\Gamma}(k,\omega)$ behaves as k^2 at long wavelength and finite frequency, and its real and imaginary parts give, respectively, the dispersion and the damping of the plasmon excitation. Following Goodman and Sjölander (25), one may adopt the empirical form

$$\lim_{k\to 0} \tilde{\Gamma}(k,\omega) = k^2 \left[\alpha_\infty + (\alpha_0 - \alpha_\infty)\,(1 - i\omega\tau)^{-1}\right] \tag{2.27}$$

where τ is an appropriate relaxation time, of the order of ω_p^{-1}, while α_0 is determined by eqn. (2.17) and α_∞ is determined from the third moment sum rule as

$$\alpha_\infty = 2\,T + \frac{4}{15}\,U\,, \tag{2.28}$$

T and U being the kinetic and the potential energy per electron. For an application to plasmons in a classical one-component plasma, see Abramo and Parrinello (26).

(ii) at arbitrary k and ω, a number of the available approximate expressions for the dielectric function of the electron gas follow straightforwardly from the above formalism. For example, the RPA expression (Bohm and Pines (27)) is obtained by neglecting the term (2.14) and by evaluating the term (2.13) as for an ideal Fermi gas responding to the Hartree potential, when one finds

$$\delta\pi_{\alpha\beta}(\underline{k},\omega) = \sum_{\underline{p}} \frac{p_\alpha p_\beta}{m} \; \frac{f(\underline{p}-\frac{1}{2}\underline{k}) - f(\underline{p}+\frac{1}{2}\underline{k})}{\omega - \underline{p}\cdot\underline{k}/m + i0^+} \, V_H(\underline{k},\omega)\,, \tag{2.29}$$

$f(\underline{p})$ being the Fermi distribution function. More refined theories (see, for instance Hubbard (28), Singwi et al. (29), Vashishta and Singwi (30)) include the short range correlations between the electrons expressed by the term (2.14) through appropriate approximations on the triplet function entering this term.

3. THE ELECTRON-ION FLUID

A. Equations of Motion and Response Functions

The extension of the foregoing treatment to a two-component electron-ion fluid is straightforward. From the equations of motion for the density matrices of ions and electrons, which are coupled by the electron-ion interaction, continuity equations and generalized Navier-Stokes equations for the two components are derived by an expansion about the diagonal. These equations are formally solved by the functional derivative technique illustrated above to find expressions for the matrix of density response functions, defined by writing

$$n_{\ell}(\underline{k},\omega) = \sum_j \chi_{\ell j}(k,\omega) V_j^e(\underline{k},\omega) \tag{3.1}$$

where the indices refer to the two components and $V_j^e(\underline{k},\omega)$ is a weak external potential coupled to the density fluctuations of the j-th component. The result is (19)

$$\chi_{\ell\ell}(k,\omega) = \frac{n_\ell k^2}{m_\ell}\left[\omega^2 - n_{\bar{\ell}} k^2 v_{\bar{\ell}\bar{\ell}}(k)/m_{\bar{\ell}} - \tilde{\Gamma}_{\bar{\ell}\bar{\ell}}(k,\omega)/m_{\bar{\ell}}\right] / \Gamma(k,\omega) \tag{3.2}$$

$$\chi_{\ell\bar{\ell}}(k,\omega) = \frac{n_{\bar{\ell}} k^2}{m_\ell m_{\bar{\ell}}}\left[n_\ell k^2 v_{\ell\bar{\ell}}(k) + \tilde{\Gamma}_{\ell\bar{\ell}}(k,\omega)\right] / \Gamma(k,\omega) \tag{3.3}$$

where $\bar{\ell}$ denotes the component of a type different from ℓ, and

$$\Gamma(k,\omega) = \det\left[\omega^2 \delta_{\ell j} - n_\ell k^2 v_{\ell j}(k)/m_\ell - \tilde{\Gamma}_{\ell j}(k,\omega)/m_\ell\right]. \tag{3.4}$$

Here, in a notation analogous to eqns. (2.13) and (2.14),

$$\tilde{\Gamma}_{\ell j}(k,\omega) = k_\alpha k_\beta \lambda^{\ell j}_{\alpha\beta}(\underline{k},\omega) + \sum_s \sum_{\underline{k}'} \underline{k}\cdot\underline{k}' v_{\ell s}(k') F_{\ell s j}(\underline{k}-\underline{k}',\underline{k},\omega). \tag{3.5}$$

The quantities $v_{\ell j}(k)$ represent the bare potentials between the various pairs of particles, that we have taken as independent of energy.

The above expressions admit a simple physical interpretation in terms of effective single-component response functions and effective potentials. We define an effective single-component response function by an obvious extension of eqn. (2.15),

$$\chi_{\ell}^{eff}(k,\omega) = \frac{n_\ell k^2}{m_\ell}\left[\omega^2 - n_\ell k^2 v_{\ell\ell}(k)/m_\ell - \tilde{\Gamma}_{\ell\ell}(k,\omega)/m_\ell\right]^{-1} \quad (3.6)$$

and the corresponding 'screened' response function

$$\tilde{\chi}_{\ell}^{eff}(k,\omega) = \frac{n_\ell k^2}{m_\ell}\left[\omega^2 - \tilde{\Gamma}_{\ell\ell}(k,\omega)/m_\ell\right]^{-1} \quad (3.7)$$

which relates the response to the Hartree potential seen by the ℓ -th component. Of course, the full effects of the electron-ion interaction still enter implicitly through $\tilde{\Gamma}_{\ell\ell}(k,\omega)$. Similarly, we define effective potentials as

$$v_{\ell\ell}^{eff}(k,\omega) = v_{\ell\ell}(k) + |v_{\ell\bar{\ell}}^{eff}(k,\omega)|^2 \chi_{\bar{\ell}}^{eff}(k,\omega) \quad (3.8)$$

$$v_{\ell\bar{\ell}}^{eff}(k,\omega) = v_{\ell\bar{\ell}}(k) + \tilde{\Gamma}_{\ell\bar{\ell}}(k,\omega)/(n_\ell k^2) \,. \quad (3.9)$$

The expression (3.8) will be recognized as formally identical to the weak-interaction expression for the effective ion-ion potential in a metal, modified by non-linear effects entering through the effective single-component response function and through the effective electron-ion interaction given in eqn. (3.9). In terms of these quantities eqns. (3.2) and (3.3) can be written

$$\chi_{\ell\ell}(k,\omega) = \tilde{\chi}_{\ell}^{eff}(k,\omega)\left[1 - v_{\ell\ell}^{eff}(k,\omega)\,\tilde{\chi}_{\ell}^{eff}(k,\omega)\right]^{-1} \quad (3.10)$$

$$\chi_{\ell\bar{\ell}}(k,\omega) = \chi_{\ell}^{eff}(k,\omega)\, v_{\ell\bar{\ell}}^{eff}(k,\omega)\, \chi_{\bar{\ell}\bar{\ell}}(k,\omega) \,. \quad (3.11)$$

Thus, according to eqn. (3.10) each component responds to a direct perturbation as if it were effectively a single component with an interparticle potential screened by the other component. Similarly, according to eqn. (3.11) each component responds to a perturbation of the other component as if it were effectively a single component responding to an external field produced by the polarization of the other component, to which it is coupled via a modified interaction potential.

B. Thermodynamics

The above form of the response functions is particularly suitable for a discussion of the thermodynamic limit ($\omega = o$ and $k \to o$). The functional derivatives become local in space and time in this limit (Fröhlich (31)), that is

$$\delta\pi^{\ell}_{\alpha\beta}(\underline{R},t) = \sum_{s} \frac{\partial \pi^{\ell}_{\alpha\beta}}{\partial n_s} \delta n_s(\underline{R},t) + \ldots , \tag{3.12}$$

$$\delta\langle \rho_{\ell}(\underline{R},t)\, \rho_j(\underline{x},t)\rangle_c = \frac{1}{2}\Big[\gamma_{\ell j}(r)\,\delta n_{\ell}(\underline{R},t) + \gamma_{j\ell}(r)\,\delta n_j(\underline{R},t)\Big] + \ldots \tag{3.13}$$

with $\underline{r} = \underline{R}-\underline{x}$ and

$$\gamma_{\ell j}(r) = \partial\langle \rho_{\ell}(\underline{R})\, \rho_j(\underline{x})\rangle_c / \partial n_{\ell} \quad . \tag{3.14}$$

The coefficients $\tilde{\tilde{\Gamma}}_{\ell j}(k,o)$ behave as k^2 in this limit, and the bare potentials in eqns. (3.2) and (3.3) are dominated by their Coulomb part. One accordingly finds (19)

$$\lim_{k\to o}\Big[\chi_{ii}(k,o) = Z^{-1}\chi_{ie}(k,o) = Z^{-2}\chi_{ee}(k,o)\Big] = -n_i^2 K_T \tag{3.15}$$

where Z is the ionic valence and K_T the isothermal compressibility. The partial structure factors $S_{\ell j}(k)$ are related in the same limit by the overall neutrality conditions, and by assuming that the ions behave classically one has

$$\lim_{k\to o}\Big[\chi_{\ell j}(k,o) = -(n_{\ell} n_j)^{\frac{1}{2}} \beta S_{\ell j}(k)\Big] = -n_{\ell} n_j K_T \tag{3.16}$$

that is, the electrons also behave classically in these conditions.

C. Generalized Hydrodynamics and Hydrodynamic Limit

The most appropriate dynamic variables for the electron-ion fluid are obviously the mass density fluctuations and the charge density fluctuations. We therefore construct linear combinations of the response functions (3.2) and (3.3) giving the response of the mass density to a gravitational potential,

$$\chi_{MM}(k,\omega) = \sum_{\ell j} m_{\ell} m_j \chi_{\ell j}(k,\omega), \tag{3.17}$$

the response of the charge density to an electric potential,

$$\chi_{QQ}(k,\omega) = \sum_{\ell j} e_{\ell} e_j \chi_{\ell j}(k,\omega), \tag{3.18}$$

and the nondiagonal response describing coupling between mass and charge density fluctuations,

$$\chi_{MQ}(k,\omega) = \chi_{QM}(k,\omega) = \frac{1}{2}\sum_{\ell j}(e_\ell m_j + e_j m_\ell)\,\chi_{\ell j}(k,\omega). \qquad (3.19)$$

These are given by

$$\chi_{MM}(k,\omega) = n_i M k^2\left[\omega^2-\omega_p^2-\alpha_p(k,\omega)\right]/\Gamma(k,\omega), \qquad (3.20)$$

$$\chi_{QQ}(k,\omega) = \frac{\omega_p^2 k^2}{4\pi}\left[\omega^2-\alpha_s(k,\omega)\right]/\Gamma(k,\omega), \qquad (3.21)$$

and

$$\chi_{MQ}(k,\omega) = eMk^2(n_i n_e/m_i m_e)^{\frac{1}{2}}\alpha_r(k,\omega)/\Gamma(k,\omega) \qquad (3.22)$$

where $M = m_i + Zm_e$, $\omega_p^2 = 4\pi e^2(Z^2 n_i/m_i + n_e/m_e)$, and the α's are suitable linear combinations of the Γ's. We have incorporated in the $\tilde{\Gamma}$'s also the non-Coulomb parts of the Hartree potentials. The denominator $\Gamma(k,\omega)$ can also be rewritten as

$$\Gamma(k,\omega) = \left[\omega^2-\alpha_s(k,\omega)\right]\left[\omega^2-\omega_p^2-\alpha_p(k,\omega)\right] - \left[\alpha_r(k,\omega)\right]^2 . \qquad (3.23)$$

The significance of the α's is evident from eqns. (3.19)-(3.23). The function $\alpha_r(k,\omega)$ arises because of coupling between mass and charge density fluctuations, which is rigorously negligible only in the long wavelength limit where $\alpha_r(k,\omega)$ is found to behave as k^2. The function $\alpha_s(k,\omega)$ then gives in this limit the frequency and damping of sound waves, while $\alpha_p(k,\omega)$ gives the frequency shift and damping, as well as the dispersion, of the plasmon waves. The same function gives the longitudinal conductivity $\sigma(o,\omega)$ and thus the electrical resistance of the liquid metal when calculated in the low frequency limit.

A special problem arises in an electron-ion liquid in the hydrodynamic limit (small k and ω) as compared with normal liquids, because of the special character of the electronic component (quantum liquid with a continuum of single-particle excited states without a gap, and Fermi velocity much larger than sound velocity). If we take first the limit $k\to o$, as is usually done for normal liquids, we exit from the continuum of single-particle excited states in the (ω-k) plane, which is bound by Pauli principle requirements to the frequency range $0\leqslant \omega \leqslant kv_F$, for small k. We would then be missing the damping of sound waves by electron-hole creation processes, which are certainly operative since $s \ll v_F$. One must thus retain terms in the electronic contributions at long wavelength for which

$kv_F > \omega$. There are then two possible limiting behaviors for the electrons, depending on their mean free path ℓ_e. If $k\ell_e \gg 1$, then the electrons behave quasi-statically rather than hydrodynamically: this is the situation that one tries to realize for studies of the Fermi surface in solids by sound attenuation, since the electronic damping of sound waves can then be distinguished from normal anharmonic damping. In the opposite limit, $k\,\ell_e \ll 1$, as is the case for liquid metals at the wavelengths of interest in ultrasonic experiments, both the electrons and the ions behave hydrodynamically. The electronic contribution to viscosity processes has the same frequency and wavevector dependence as for a normal liquid, but will contribute to anomalous mass dependences of the transport coefficients in liquid-metal isotopes.

We refer the interested reader to ref. (19) for a detailed discussion of this subject, and focus instead on the calculation of the resistivity.

4. ELECTRIC TRANSPORT

A. Formula for Electrical Resistivity

The calculation of the conductivity is considerably simplified in the long wavelength limit, where the coupling between mass and charge fluctuations vanishes and the local field effects are fully described by the Hartree potential. The dielectric function of the liquid metal, as determined by the charge response function of eqn. (3.21), has the form

$$\varepsilon(o,\omega) = 1 - \omega_p^2 \left[\omega^2 - \alpha_p(o,\omega)\right]^{-1} \tag{4.1}$$

and the conductivity is given by

$$\sigma(o,\omega) = \frac{n_e e^2}{m_e} \frac{M}{m_i} \left[-i\omega + M\gamma(\omega)/(n_e m_i m_e)\right]^{-1} \tag{4.2}$$

with

$$\gamma(\omega) = n_e \int d\underline{r}\; \hat{k}\cdot \underline{\nabla} v_{ie}(r) \int d(t-\tau)\, e^{i\omega(t-\tau)} \frac{\delta\langle \rho_i(\underline{R},t)\,\rho_e(\underline{x},t)\rangle_c}{\delta j_e(\tau)} \tag{4.3}$$

This is the only term in $\alpha_p(k,\omega)$ which survives in the long wavelength limit when the functional derivatives are allowed to become local in space, and clearly describes both the frequency shift and the damping of the plasmon excitation due to the electron-ion interaction. The Drude-Zener theory, which is remarkably successful for liquid metals (see, for example Faber (32)) follows if $\gamma(\omega)$ is assumed to vary only slowly with frequency over the range of ω of interest (often $0.1 < \omega \lesssim 3$).

In the low frequency limit, as appropriate for the calculation of the d.c. conductivity, the functional derivatives can be handled as fully local in the velocities of the particles (Fröhlich (31)), that is

$$\delta\langle \rho_\ell(\underline{R},t)\, \rho_j(\underline{x},t)\rangle_c = \left[v_j(\underline{R},t) - v_\ell(\underline{R},t)\right]\sigma_{j\ell}(\underline{r})+\ldots \quad (4.4)$$

where $v_\ell(\underline{R},t) = j_\ell(\underline{R},t)/n_\ell$ and

$$\sigma_{j\ell}(\underline{r}) = -\sigma_{\ell j}(\underline{r}) = \partial\langle \rho_\ell(\underline{R})\, \rho_j(\underline{x})\rangle_c/\partial v_j. \quad (4.5)$$

This leads to an expression for the relaxation time in the conductivity, $\sigma = n_e e^2 \tau/m_e$, which reads

$$\tau^{-1} = (n_e m_e)^{-1}\int d\underline{r}(\hat{k}\cdot\underline{\nabla})v_{ie}(r)\; \partial\langle \rho_i(\underline{R})\, \rho_e(\underline{x})\rangle_c/\partial v_e\,. \quad (4.6)$$

The d.c. resistivity is thus expressed through the electron-ion force associated with the creation of a uniform electronic drift, a result which is correct but not directly useful unless the calculation is continued. It is nevertheless evident that the liquid-metal specimen must be thought of as immersed in an infinite bath of free electrons, as was found necessary for applications of the force-force correlation formula to resistivity calculations (Jones (11)). The same boundary condition is implicit in weak-scattering calculations of the resistivity based on the formalism of inelastic scattering of a beam of incident particles in the Born approximation (Baym (2)), to which we now turn.

B. Reduction to Ziman-Baym theory

The following approximations must be invoked to reduce the expression (4.6) for the resistivity to the weak-scattering formula:

(i) for weak electron-ion interactions, the imaginary part of the non-equilibrium electron-ion correlation function, which alone is of interest for the resistivity calculation, can be factorized to lowest order in the form (Kadanoff and Baym (20))

$$\mathcal{Im}\left[\mathrm{F.T.}\left\{\langle \rho_i(\underline{R},t)\,\rho_e(\underline{x},t)\rangle_c\right\}_{\underline{q}}\right]=$$

$$=\frac{1}{2}\,v_{ie}(\underline{q})\int_{-\infty}^{\infty}\frac{d\omega}{2\pi}\left[S'_{ii}(\underline{q},\omega)S'_{ee}(-\underline{q},-\omega)-S'_{ii}(-\underline{q},-\omega)S'_{ee}(\underline{q},\omega)\right]\quad(4.7)$$

where $S'_{ii}(\underline{q},\omega)$ and $S'_{ee}(\underline{q},\omega)$ are the dynamic structure factors for ions and electrons in the non-equilibrium state;

(ii) in the Born-Oppenheimer approximation we assume that the equilibrium rates of the ionic system are much faster than the electron-ion scattering rates, when we replace $S'_{ii}(\underline{q},\omega)$ by its equilibrium value;

(iii) for weak electron-electron interactions, we replace $S'_{ee}(\underline{q},\omega)$ by its RPA value,

$$S'_{ee}(\underline{q},\omega)=\left[S'_{ee}(\underline{q},\omega)\right]_{free}/|\varepsilon(\underline{q},\omega)|^2 \quad (4.8)$$

where the non-equilibrium structure factor for free electrons is given by

$$\left[S'_{ee}(\underline{q},\omega)\right]_{free}=2\pi\sum_{\underline{p}} f'(\underline{p})\left[1-f'(\underline{p}+\underline{q})\right]\,\delta(\omega-\varepsilon_{\underline{p}+\underline{q}}+\varepsilon_{\underline{p}}) \quad (4.9)$$

in terms of the distorted Fermi distribution

$$f'(\underline{p}) = f_{eq}(\varepsilon_{\underline{p}} - \underline{p}\cdot\underline{v}_e). \quad (4.10)$$

The result of these approximations is

$$\tau^{-1}=(n_e m_e)^{-1}\sum_{\underline{p}\underline{q}}\hat{k}\cdot\underline{q}\int_{-\infty}^{\infty}d\omega\;\delta(\omega-\varepsilon_{\underline{p}+\underline{q}}+\varepsilon_{\underline{p}})\;|v_{ie}^{sc}(\underline{q},\omega)|^2$$
$$S_{ii}(\underline{q},\omega)\,\frac{\partial}{\partial v_e}\left\{f'(\underline{p}+\underline{q})\left[1-f'(\underline{p})\right]\right\}, \quad (4.11)$$

which agrees with Baym's formula for electrical resistivity. Evidently, the phenomenon is here viewed as inelastic scattering of a beam of free electrons by the equilibrium density fluctuations of the ionic system, the coupling between the two systems being given by the screened electron-ion potential.

5. ELECTROMIGRATION AND HAEFFNER EFFECT

Discussions of electromigration and the Haeffner effect have been given by a number of workers (see, for example Epstein and Dickey (33), Epstein and Paskin (34) and references given there; see also Faber (32)). The Haeffner effect, that is the separation of isotopes which occurs when an electric current is passed through a metal, is particularly significant from a theoretical viewpoint, in that in the numerous liquid metals studied so far the lighter isotopes concentrate at the anode, the heavier at the cathode. The predictions of the weak inelastic scattering theory are hardly compatible with the apparent generality of the effect (Parrinello et al. (18)).

In a binary liquid metal mixture, the contribution of ionic component 1, say, to the mass concentration current is related to the driving force on this component by

$$j_x^{(1)} = \left[\lim_{\omega \to 0} \lim_{k \to 0} \frac{i\omega}{k^2} \chi_{1x}(k,\omega) \right] F_1 \qquad (5.1)$$

where $\chi_{1x}(k,\omega)$ is the response function which couples the density fluctuations of component 1 to the mass concentration fluctuations. Use of the Kubo relations for the dynamic structure factors in the mixture (Bhatia et al. (35)) yields

$$j_x^{(1)} = -m_2 D_{12} F_1 / Z \qquad (5.2)$$

where D_{12} is the mutual diffusion constant and $Z = \partial^2 G/\partial c^2$, G being the Gibbs free energy of the mixture. In an isotopic mixture $Z = k_B T/(c_1 c_2)$. The total concentration current can therefore be written as

$$j_x = m_1 m_2 D_{12} \left(\frac{F_2}{m_2} - \frac{F_1}{m_1}\right)/Z \quad . \qquad (5.3)$$

The total force on ionic component 1, on the other hand, is the sum of the direct term z_1 e E and of the 'wind' force exerted by the flowing electrons, both via direct scattering processes against system 1 and via scattering processes against system 2 in which the momentum and energy lost by the electrons end up in system 1. According to eqn. (4.6), which can be written in this case as

$$\rho = (n_e e^2)^{-1} \frac{\partial}{\partial j_e} \Big[\int d\underline{r}\, \hat{k}\cdot\underline{\nabla} v_{1e}(r) \langle \rho_1(\underline{R})\, \rho_e(\underline{x}) \rangle_c + \int d\underline{r}\, \hat{k}\cdot\underline{\nabla} v_{2e}(r) \langle \rho_2(\underline{R})\, \rho_e(\underline{x}) \rangle_c \Big], \tag{5.4}$$

we can define, for any given concentration of the mixture, 'partial resistivities' ρ_1 and ρ_2 such that the total resistivity is $c_1\rho_1 + c_2\rho_2$ and the wind forces are expressed as

$$F_1^w = -(c_1 z_1 + c_2 z_2) e\, E\, \rho_1/(c_1\rho_1 + c_2\rho_2), \tag{5.5}$$

with a similar expression for F_2^w. We then find for the mass concentration current

$$j_x = \frac{m_1 m_2 D_{12}}{Z} \left(\frac{c_2}{m_2} + \frac{c_1}{m_1}\right) \frac{z_2\rho_1 - z_1\rho_2}{c_1\rho_1 + c_2\rho_2}\, e\, E \tag{5.6}$$

The observed sign of the Haeffner effect, with $z_1 = z_2$, requires $\rho_1 > \rho_2$ if $m_1 < m_2$.

The 'partial resistivities' are easily calculated by the weak scattering formalism of section 4.B, when one finds aside from a trivial factor

$$\rho_1 = \langle S_{11}(\underline{k},\omega) \left| v_{e1}^{sc}(\underline{k},\omega) \right|^2 + \left(\frac{c_2}{c_1}\right)^{\frac{1}{2}} S_{21}(\underline{k},\omega) v_{e1}^{sc}(\underline{k},\omega) v_{e2}^{sc}(-\underline{k},-\omega) \rangle, \tag{5.7}$$

with ρ_2 being obtained by interchanging 1 with 2. Here $S_{11}(\underline{k},\omega)$ and $S_{21}(\underline{k},\omega)$ are the partial dynamic structure factor for component 1 and the off-diagonal structure factor, and the average in (5.7) is to be taken as in eqn. (4.11). For an isotopic mixture, taking $v_{e1}^{sc}(\underline{k},\omega) = v_{e2}^{sc}(\underline{k},\omega)$ one finds

$$\rho_1 - \rho_2 = \langle S_{Nc}(\underline{k},\omega) \left| v_{e1}^{sc}(\underline{k},\omega) \right|^2 \rangle / (c_1 c_2) \tag{5.8}$$

where $S_{Nc}(\underline{k},\omega)$ describes dynamic correlations between density and concentration fluctuations. If one now goes to the Ziman quasi-elastic scattering limit of the theory, the result involves the static structure factor $S_{Nc}(k)$, which vanishes identically for a classical isotopic mixture. The Haeffner effect is thus seen to be crucially dependent in weak scattering theory on the energy dependence of the screened electron-ion potential and/or on quantum corrections, leading to an inelastic scattering contribution. The generality of the Haeffner effect makes it, however, unlikely that it can depend on the details of energy-dependent electron-ion interactions or on quantum corrections which also involve the detailed electron-ion interaction. In effect, quantum corrections in a one-phonon model for the dynamic structure factor lead to a decrease in resistivity upon addition of a light isotope, a result which is consistent with the experiments of Dugdale et al. (36) on solid lithium isotopes.

The validity of eqn. (5.6) appears not to be limited to the weak scattering regime, and though it is clearly not possible, from a measurement of the resistivity as a function of concentration in an isotopic liquid metal mixture, to separate ρ_1 and ρ_2 uniquely, it would nevertheless be of interest to examine experimentally the departure from the explicit linear dependence of the resistivity on concentration. If such departures are small, namely the implicit dependence of ρ_1 and ρ_2 on concentration is weak, it should be feasible to test eqn. (5.6) experimentally. On the other hand, this result has only considered the 'systematic' effect of the electronic current in driving the ionic interdiffusion, while the detailed mechanism of the Haeffner effect should involve correlations between 'random forces' for interdiffusion and 'random forces' for electron scattering. Enhanced fluctuations of the wind force at fluctuations of local environment which favour interdiffusion would explain the effect if the lighter ions, as seems reasonable, play a slightly more active role in interdiffusion.

ACKNOWLEDGEMENTS

I wish to thank Professor N.H.March and Dr. M.Parrinello for helpful conversations. It is also a pleasure to acknowledge the hospitality of the Physics Department of the University of Messina during the preparation of this manuscript, and to thank Mr. G.Salvati for typing it.

REFERENCES

1. Ziman, J.M. (1961). Phil. Mag. 6, 1013.
2. Baym, G. (1964). Phys. Rev. 135, A1691.
3. Rahman, A. (1974). Phys. Rev. Letters 32, 52 and Phys. Rev. A9, 1667.
4. Price, D.L., Singwi, K.S. and Tosi, M.P. (1970). Phys.Rev. B2,2983.

5. Egelstaff, P.A., March, N.H. and McGill, N.C. (1974). Canad. J. Phys. 52, 1651.
6. Platzman, P.M. and Eisenberger, P. (1974). Phys. Rev. Letters 33, 152.
7. Watabe, M. and Hasegawa, H. (1973). Second Int. Conf. on Liquid Metals (Taylor and Francis, London).
8. Chihara, J. (1973). Second Int. Conf. on Liquid Metals (Taylor and Francis, London).
9. Vashishta, P., Bhattacharyya, P. and Singwi, K.S. (1974). Phys. Rev. B10, 5108.
10. Rousseau, J.S., Stoddart, J.C. and March, N.H. (1972). J. Phys. C5, L175.
11. Jones, W. (1974). J. Phys. C7, 3357.
12. Nyquist, H. (1928). Phys. Rev. 32, 110.
13. Mori, H. (1965). Progr. Theor. Phys. 33, 423.
14. Kubo, R. (1966). Rep. Progr. Phys. 29, 255.
15. Copley, J.R.D. and Lovesey, S.L. (1975). Rep. Progr. Phys., in the press.
16. Ban, N.T., Randall, C.M. and Montgomery, D.J. (1962). Phys. Rev. 128, 6.
17. March, N.H. and Tosi, M.P. (1973). Ann. Phys. (NY) 81, 414.
18. Parrinello, M., Tosi, M.P. and March, N.H. (1975). Lett. N. Cimento, in the press.
19. Tosi, M.P., Parrinello, M. and March, N.H. (1974). N. Cimento 23B, 135.
20. Kadanoff, L.P. and Baym, G. (1962). Quantum Statistical Mechanics (Benjamin, New York).
21. Kadanoff, L.P. and Martin, P.C. (1963). Ann. Phys.(NY) 24, 419.
22. Singwi, K.S., Sjölander, A., Tosi, M.P. and Land, R.H. (1970). Phys. Rev. B1, 1044.
23. Pines, D. and Nozières, P. (1966). The theory of Quantum Liquids (Benjamin, New York).
24. Mori, H. (1965). Progr. Theor. Phys. 34, 499.
25. Goodman, B. and Sjölander, A. (1973). Phys. Rev. B8, 200.
26. Abramo, M.C. and Parrinello, M. (1975). Lett. N. Cimento, in the press.
27. Bohm, D. and Pines, D. (1953). Phys. Rev. 92, 609.
28. Hubbard, J. (1957). Proc. Roy. Soc. A243, 336.
29. Singwi, K.S., Tosi, M.P., Land, R.H. and Sjölander, A. (1968). Phys. Rev. 176, 589.
30. Vashishta, P. and Singwi, K.S. (1972). Phys. Rev. B6, 875.
31. Fröhlich, H. (1967). Physica 37, 215.

32. Faber, T.C. (1972). An Introduction to the Theory of Liquid Metals (Cambridge).
33. Epstein, S.G. and Dickey, J.M. (1970). Phys. Rev. B1, 2442.
34. Epstein, S.G.and Paskin, A. (1967). Phys. Letters 24A, 309.
35. Bhatia, A.B., Thornton, D.E. and March, N.H. (1975). J. Phys. Chem. Liquids, in the press.
36. Dugdale, J.S., Gugan, D. and Okumura, K. (1961). Proc. Roy. Soc. A263, 407.

STRUCTURAL EFFECTS ON SUPERCONDUCTIVITY

K. L. Ngai

Naval Research Laboratory

Washington, D. C. 20375

1. INTRODUCTION

The aim of the article is to discuss some aspects of the effects of structure and structural transformations of solids on their superconductive properties.

By structural effects we mean here only the enhancement or suppression of superconductivity in materials for which surfaces are important. These include (1) thin films of metal; (2) multiple superposed thin films of metal, semiconductor, or dielectric; (3) granular metal films made by evaporating metal in an oxygen atmosphere with grain size of say 30 Å; (4) quenched alloy thin films like AℓGe; (5) metal-semiconductor eutectic alloys such as Aℓ-Si; and (6) layered two-dimensional electron gas as closely approximated by the layered transition metal dichalcogenides. These systems have the common characteristic that interfaces appear in abundance. In the case of (6) even the bulk of the material can be considered as a superposition of nearly two-dimensional configurations. Interfaces effects can be expected to be important in determining the electronic transport and superconductive properties of these materials. In particular, surface excitations and their interactions with the layered electron gas can have a significant effect on the electronic properties. Surface excitations include the surface plasmons, excitons and phonons, often collectively called surface polaritons. Although the phonon mechanism of superconductivity has been firmly established, there has been a continued interest in the possibilities of other mechanisms. One such mechanism, the exciton mechanism of Ginzburg[1,2] has recently been reconsidered[3] for a particular model of a thin metal layer on a semiconductor surface. In this model, the metal electrons tunnel into the semiconductor where exchange of virtual excitons takes place and gives rise to attractive pairing interactions in analogy to the phonon mechanism. Substantial enhancement of the superconductive transition temperature T_c due to the exciton mechanism was predicted by calculations.[3] The validity of these results have been questioned[4,5] and the exciton-mechanism is still a subject of controversy. In this work we consider yet another mechanism for superconductivity: the surface plasmon mechanism. Section 2 will be devoted to a detailed discussion of this mechanism.

In Section 3 an entirely different, but not unrelated in spirit, subject of the effect of lattice instability on the electronic properties of the tungsten bronzes. The tungsten bronzes form a large class of interesting nonstoichiometric compounds which have the formula M_xWO_3 where M is usually a metal and $0 < x < 1$. They are known to exist in cubic, hexagonal, and two different tetragonal structures.[6] Some of these materials have been found to be superconducting[7] in the hexagonal and one of the tetragonal structures, and recently an increase in the superconducting transition temperature of the higher composition tetragonal sodium tungsten bronze has been observed as x is decreased toward the value at which the transition to the lower composition tetragonal structure occurs.[8] It was suggested in Reference 8 that this increase is due to softening of the lattice mode which corresponds to a certain geometrical operation[9] that transforms one tetragonal structure into the other. This structural transformation is different from the behavior of displacive ferroelectrics such as $BaTiO_3$ where a phonon mode condenses. Instead this transformation requires a concomitant change in bonding configurations. The picture we have is that as the lattice mode softens to condense, electronic tunneling occurs between the two bonding configurations. Electron tunneling adds a new dimension to the problem and novel effects on electronic properties near the structural transformation can be expected. In Section 3 we will introduce a simple model of the composition dependence of the free energy and the soft phonon frequency, from which the observed dependence of T_c on x can be obtained. The structural transition is treated as a change in bonding configurations, so that the model is similar to the two-well configuration space model utilized recently[10] to treat amorphous materials. In the present case the tunneling between configurations is assisted by the soft phonon, which is itself treated in a generalization of the anharmonic model[11] of lattice instabilities. We find that as x is decreased toward x_c, the critical composition for transition between structures, the phonon softening at first strongly enhances T_c. Near x_c however, the coupling between the soft phonon and the configuration change suppresses the softening, and T_c increases more slowly. In fact, for some values of the parameters, a maximum T_c is observed. In this simple model, then, the effect of the soft phonon on T_c is explicitly calculated for a system where the superconducting and structural transitions can occur simultaneously.

2. SURFACE PLASMON MECHANISM OF SUPERCONDUCTIVITY

2.1 Surface Plasmon Mechanism

We examine the surface plasmon mechanism of superconductivity for a model of periodic layered electron gas, in which the effective attractive interaction between electrons comes from virtual excitation of surface plasmons (SP). Since the electron-SP interaction is completely determined and the SP excitations in the layered structures can be calculated from first principles, the model will enable us to evaluate the importance of the SP mechanism. The choice of the multiple thin-films system for consideration of the SP mechanism is motivated by the presence of many SP modes. In particular, a group of low lying SP modes have properties that suggest that they may induce superconductivity. It should be pointed out that the SP-mechanism for superconductivity has been suggested in the past by Cohen and Douglas[12] and more recently by Anderson.[13] The calculations we shall describe in this work is, to our knowledge, the first quantitative estimate of the magnitude of the SP mechanism. In this connection we should mention that in metals with overlapping narrow d or f-bands and very broad s or p-band, Fröhlich[14] has found that the screening of the d-band plasma oscillation by the s electrons leads to the existence of an acoustic plasmon branch. The ω versus q dispersion relation is linear at small q. Gutfreund and Unna[15] have calculated the dispersion relation for the acoustic plasmon in the long wavelength limit. Rothwarf[16] has used the acoustic plasmon to explain various normal state properties of transition metals and alloys such as Nb_3Sn. Fröhlich[14] has suggested that the acoustic plasmon modes could lead to superconductivity, but unfortunately the transition temperature cannot be estimated due to lack of information of the parameters in transition metals and alloys. Experimentally, there is no strong evidence of this Fröhlich's mechanism of superconductivity. We mention these works here because of the apparent analogy with our SP mechanism. The important difference between our case and Fröhlich's consideration is that we are interested in layered electron gas where interfaces dominate the bulk properties of the material. As we shall see, the results indicate that the SP mechanism should be observable in experiments under optimum conditions.

First let us discuss qualitatively some of the properties of SP as well as their interaction with electrons. SP have been observed and studied in various geometries such as single planar interfaces, thin films, tunneling junction configurations, spherical and cylindrical geometries, etc. SP play an important role in many phenomena such as reflection and transition of em waves, high energy electron scattering, transition radiation, low energy electron diffraction, photoemission, tunneling (both normal and superconducting), field ion emission, etc.[17] Thus the existence of SP's and their interactions with electrons has been established.

For the geometry of many superposed planar films, we consider here, each SP is characterized by a wave vector $\vec{Q}$ parallel to the interfaces and by another number specifying the relative charge distribution on each surface.[18] For a periodic arrangement of films this number is a continuous variable, i.e. a quasi-momentum along the perpendicular direction; it will be denoted by α. The eigenfrequency of each eigenmode $\omega_\alpha^{(\pm)}(\vec{Q})$ depends on $\vec{Q}$ and α. For a given set of values of $\vec{Q}$ and α there are two eigenmodes denoted by ($\pm$) corresponding to symmetric or antisymmetric charge distribution within each plasma film. Thus for each $\vec{Q}$ two separate bands are formed, the lower one corresponding to the symmetric (+) distribution and the higher one corresponding to the antisymmetric charge distribution. For low $\vec{Q}$ the modes of the lower band exhibit a linear dispersion relation $\omega_\alpha^{(+)} = \bar{c}_\alpha Q$ with a phase velocity $\bar{c}_\alpha$ which satisfies the inequalities $\bar{c}_{min} < \bar{c}_\alpha < \bar{c}_{max}$. For d_i, $d_m \ll \lambda_p$, $\bar{c}_{min} \approx \frac{1}{2}\,\omega_p\,(d_i d_m)^{1/2}$ and $\bar{c}_{max} \approx c[d_i/(d_i + d_m)]^{1/2}$ where d_i and d_m are the thicknesses of the insulating and metallic films respectively, ω_p is the plasma frequency, c the velocity of light and $\lambda_p \equiv k_p^{-1} = c/\omega_p$. These SP (to be referred thereafter as acoustic SP) are good candidates for mediating superconductivity. Such acoustic SP have been observed in tunneling geometries: the so-called Swihart's mode[19] appearing in superimposed superconducting tunnel junctions is an acoustic SP.[18,20] The low eigenfrequency of these acoustic SP is due to a near cancellation within the plasma films of the fields created by the surface charges; as a result, the restoring force is very weak and the eigenfrequency very low. Note that, for extremely thin films, (d_i, d_m of the order of Å) $\bar{c}_{min}$ can be as low as 10^7 cm/sec.

There is so far no direct experimental observation of these acoustic SP in periodic film configurations we consider here. It would be rather difficult to detect the acoustic SP by the standard method of high energy electron loss[17] since the acoustic SP band would appear as a short tail (of the order of 1 eV or less) in the distribution of the primary beam. On the other hand experiments measuring directly the radiation due to the coupling of these SP to photons through surface roughness[17] are expected to be successful. Also experiments employing the idea of attenuated total reflection[17] should show the existence of these acoustic SP and in particular should determine $\bar{c}_{max}$ and possibly $\bar{c}_{min}$ for not so thin films.

SP, (or plasmons in general) are associated with electromagnetic fields and thus they interact with electrons. One way[17] to find this interaction in the electro-static approximation is to express the field ϕ associated with the plasmon in terms of plasmon creation and annihilation operators and then multiply by the charge of the electron, i.e. $H_{e-SP} = e\phi$.

SP mediated electron-electron attraction is created as follows: An electron interacts and excites SP through H_{e-SP}; another electron interacts with the excited SP again through H_{e-SP} and thus an effective electron-electron interaction results. One possible approach to the problem, attempted earlier,[21] is to find the electron-surface plasmon interaction and express it in a second quantized form:[17]

$$H_{e-SP} = \sum_{qk} [W_{qk} \, c^+_{q+K} \, c_k \, b_q + c \cdot c \cdot] \ ,$$

where c^+_{q+k} c_k are electron creation and annihilation operators, b_q is an SP annihilation operator, and W_{qk} is the strength of the interaction. Having H_{e-SP}, which is of identical form as the electron-phonon Hamiltonian, one can proceed as in the phonon exchange case[21] and calculate λ_{SP}, which is the average (over the Fermi surface) of the surface plasmon mediated electron-electron attraction. The effect on the transition temperature T_c can be calculated as in Ref. 3 (where λ_{ex} will be replaced by λ_{SP}) provided that the quantities μ, the effective Coulomb repulsion, and λ, the average phonon mediated attraction, are known.

The procedure just outlined is incomplete. The appearance of surface plasmons modifies the properties of the bulk plasmons which contribute significantly to the value of μ, the effective Coulomb repulsion. Thus, the appearance of λ_{SP} is always associated with a change in the value of μ. To become more quantitative, consider the definition of μ.

$$\mu \equiv N(E_f) \, \frac{1}{S_f} \int \frac{V_q}{\varepsilon(q,o)} \, dS_f \ , \tag{1}$$

where $N(E_f)$ is the density of states per unit volume per spin, S_f is the area of the Fermi surface, V_q is the matrix element of the Coulomb interaction $e^2/|\vec{r}_1-\vec{r}_2|$ between electronic states lying at the Fermi surface and $\varepsilon(q,o)$ is the dielectric function at $\omega = o$. Using the general relation

$$\frac{1}{\varepsilon(q,o)} = 1 + \frac{2}{\pi} \int \frac{d\omega}{\omega} \, \mathrm{Im} \, \frac{1}{\varepsilon(q,\omega)} \ , \tag{2}$$

and the fact that, for the bulk case,

$$\mathrm{Im} \, \frac{1}{\varepsilon(q,\omega)} = \left[\mathrm{Im} \, \frac{1}{\varepsilon(q,\omega)} \right]_{BP} + \left[\mathrm{Im} \, \frac{1}{\varepsilon(q,\omega)} \right]_{PH} \ , \tag{3}$$

where the subscripts BP and PH indicate bulk plasmon and particle-hole contributions respectively, we obtain

$$\mu = \mu_{bare} + \mu_{BP} + \mu_{PH} \ , \tag{4}$$

for the bulk case. The quantities μ_{BP}, μ_{BH} are negative while μ_{bare} is positive. Actually, both μ_{bare} and μ_{BP} are infinite while their sum is finite. For the case where surfaces are present Eq. (3) should be written as

$$\mathrm{Im}\,\frac{V_q}{\varepsilon(q,\omega)} = \left[\mathrm{Im}\,\frac{V_q}{\varepsilon(q,\omega)}\right]'_{BP} + \left[\mathrm{Im}\,\frac{V_q}{\varepsilon(q,\omega)}\right]_{SP} + \left[\mathrm{Im}\,\frac{V_q}{\varepsilon(q,\omega)}\right]'_{PH}, \qquad (3')$$

which combined with Eqs. (2) and (1) leads to

$$\mu' = \mu'_{bare} + \mu'_{BP} + \mu_{SP} + \mu'_{PH} \quad , \qquad (4')$$

where the subscripts SP indicate surface plasmon contribution. Actually, the quantity μ_{SP} is negative and $\mu_{SP} = -\lambda_{SP}$. The quantities μ'_{bare}, μ'_{BP}, and μ'_{PH} are different from μ_{bare}, μ_{BP}, and μ_{PH}. One expects that $\mu'_{PH} \approx \mu_{PH}$. The conclusion is that the calculation of λ_{SP} (or μ_{SP}) does not provide a complete answer for the surface modification of the electron-electron interaction; the quantity $\mu'_{bare} + \mu'_{BP} + \mu'_{PH}$ should be calculated as well. We have examined this proposal and found the procedure, though possible, is rather difficult. Results of attempts to calculate the surface modification of the electron-electron interaction (or the surface plasmon mechanism) is too involved and will not be reported in this work. Instead we give a succinct account of calculations on the surface plasmon mechanism of a exactly soluble model in the next section.

2.2 Exactly Soluble Model

We have found a model where the quantity $\frac{V_q}{\varepsilon(q,o)}$ can be calculated without further approximations than those used in computing μ in the bulk case[22] (namely, the omission of band structure effects and the use of RPA). For this model μ' can then be calculated analytically and compared with the bulk value μ. This model involves a periodic arrangement of parallel two dimensional conducting sheets at a distance d apart.[23] It is a limiting case of the periodic alternating metal-dielectric film geometry.[21,24]

The electronic eigenfunctions for the geometry under consideration are of the form

$$\psi_{\vec{Q}\alpha}(\vec{r}) = \frac{1}{\sqrt{A(2n+1)}} \sum_{n=-N}^{N} e^{i\alpha nd}\, e^{i\vec{Q}\vec{R}}\, [\delta(z-nd)]^{1/2}, \qquad (5)$$

where $\vec{Q}$ is the parallel wave number, $\hbar\alpha$ is a quasi momentum along the perpendicular (i·e, z) direction, $\vec{r} = \vec{R} + \vec{z}_o z$, A is the area of each metallic sheet, and 2N + 1 (N → ∞) is the total number of sheets. $V_q \equiv V_{q\alpha}$ is by definition

$$V_{\vec{Q}\alpha} = \iint \psi_{\vec{Q}_1\alpha_1}(\vec{r}_1)\, \psi^*_{\vec{Q}_1-\vec{Q},\alpha_1-\alpha}(\vec{r}_1)\, \frac{e^2}{|\vec{r}_1-\vec{r}_2|} \cdot$$

$$\psi_{\vec{Q}_2\alpha_2}(\vec{r}_2)\, \psi^*_{\vec{Q}_2+\vec{Q},\alpha_2+\alpha}(\vec{r}_2)\, d^3r_1 d^3r_2 \tag{6}$$

The integrations and summations can be done analytically and the final result is

$$V_{\vec{Q}\alpha} = \frac{2\pi e^2 d}{Q} \frac{\sinh Qd}{\cosh Qd - \cos\alpha d} \tag{7}$$

Note that for Qd, $\alpha d \ll 1$ $V_{\vec{Q}\alpha} \to \frac{4\pi e^2}{q^2}$ where $q^2 = Q^2 + \alpha^2$, as expected.

The dielectric function $\varepsilon(Q,\alpha,\omega)$ for the present geometry can be calculated by considering an applied electric field $\vec{E}_{o,\vec{Q}\alpha\omega}$ which together with the induced field $\vec{E}_{\vec{Q}\alpha\omega}$ produces the total field $\vec{E}_{\vec{Q}\alpha\omega}$. The polarization $\vec{P}_{\vec{Q}\alpha\omega}$ is given by $\vec{P}_{\vec{Q}\alpha\omega} = \chi\,(Q,\omega)\,\vec{E}_{\vec{Q}\alpha\omega}$, where $\chi(Q,\omega)$ is the polarizability of a two dimensional electron gas.[25] $\chi(Q,\omega)$ has been calculated by Stern.[26] One starts with $\vec{E}_{\vec{Q}\alpha\omega}$; then $\vec{P}_{Q\alpha\omega}$ is obtained from which $\vec{E}_{ind,\vec{Q}\alpha\omega}$ can be calculated. Then, the dielectric function is given by

$$\varepsilon(Q,\alpha,\omega) = \varepsilon_i \frac{\vec{E}_{\vec{Q},\alpha\omega} - \vec{E}_{ind,\vec{Q}\alpha\omega}}{\vec{E}_{\vec{Q}\alpha,\omega}} \tag{8}$$

The final result for ε is

$$\varepsilon(Q,\alpha,\omega) = \varepsilon_i + 2\pi Q\chi(Q,\omega)\, \frac{\sinh Qd}{\cosh Qd - \cos\alpha d}\,, \tag{9}$$

where ε_i is the dielectric constant of the substance between the conducting planes. Eqs. (7) and (9) generalize to the multiple-sheet geometry the results of Stern[26] obtained for a single metal sheet. Eqs. (9) and (7) can be used to study collective elementary excitations, screening of point charges, etc. in the present model structure.

Combining Eqs. (1), (7), and (9) and taking into account that the Fermi surface in the present case is a circular cylinder of radius $k_F = (2\pi N_2)^{1/2}$ and height $2\pi/d$, where N_2 is the two dimensional electron density in each metallic sheet, we obtain for μ'

$$\mu' = \frac{\rho_F d}{8\pi^2} \int_{-\pi/d}^{\pi/d} d\alpha \int_0^{2\pi} d\theta\, V_{Q\alpha}/\varepsilon(Q,\alpha,\omega = 0) \tag{10}$$

where

$$Q = 2k_F \sin\frac{\theta}{2} \tag{11}$$

and

$$\rho_F = \frac{2}{(2\pi)^3} \int \frac{dS_k}{|\nabla\varepsilon|} = m^*/\pi d\hbar^2 \tag{12}$$

is the density of states at the Fermi surface. Here m* is the electron effective mass. Further computations lead us to

$$\mu' = \frac{xk_F}{4\pi^2} \int_{-\pi/y}^{\pi/y} d\theta \int_0^{\pi} d\phi\, \sinh(2y\sin\phi) \cdot$$

$$\{2\sin\phi[\cosh(2y\sin\phi) - \cos y\theta] + x\sinh(2y\sin\phi)\} \tag{13}$$

or, after one integration

$$\mu' = \frac{x}{2\pi} \int_0^{\pi} d\phi\, \{4\sin^2\phi + x^2 + 4x\sin\phi\coth(2y\sin\phi)\}^{-1/2} \tag{14}$$

where $y \equiv k_F d$, $x = \frac{s}{k_F}$, $s = \frac{2e^2 m^*}{\hbar^2 \varepsilon_i}$.

The quantity μ' has been calculated by performing the integral in Eq. (14) numerically as a function of N_2 or equivalently N_3, where N_3 is the average three dimensional density ($N_2 = N_3 d$) with d and $\varepsilon_i \frac{m}{m^*}$ as parameters; m is the free electron mass.

In Fig. 1, μ' is plotted as a function of N_3, where $N_3 = N_2/d$ with d = 9·5 Å. This value of d is a typical one for intercalated layered structures. Results are presented for three different values of the parameter $\varepsilon_i \frac{m}{m^*}$, i.e. $\varepsilon_i\, m/m^* = 1, 10, 100$. To show the dependence on d when $\varepsilon_i\, m/m^*$ and N_2 remain constant, we present results for the case $d = \infty$ as a function of $N_3 = N_2/d$ with d = 9·5 Å. For comparison, the 3-D μ as calculated by Morel and Anderson[11] is also given in Fig. 1.

Smaller values of d correspond to smaller values of μ'; this can also be seen from Eq. (14). However, the difference is appreciable only for $k_F d < 1$. This is expected to be so, since the effects of the nearby sheets on electrons in a given sheet

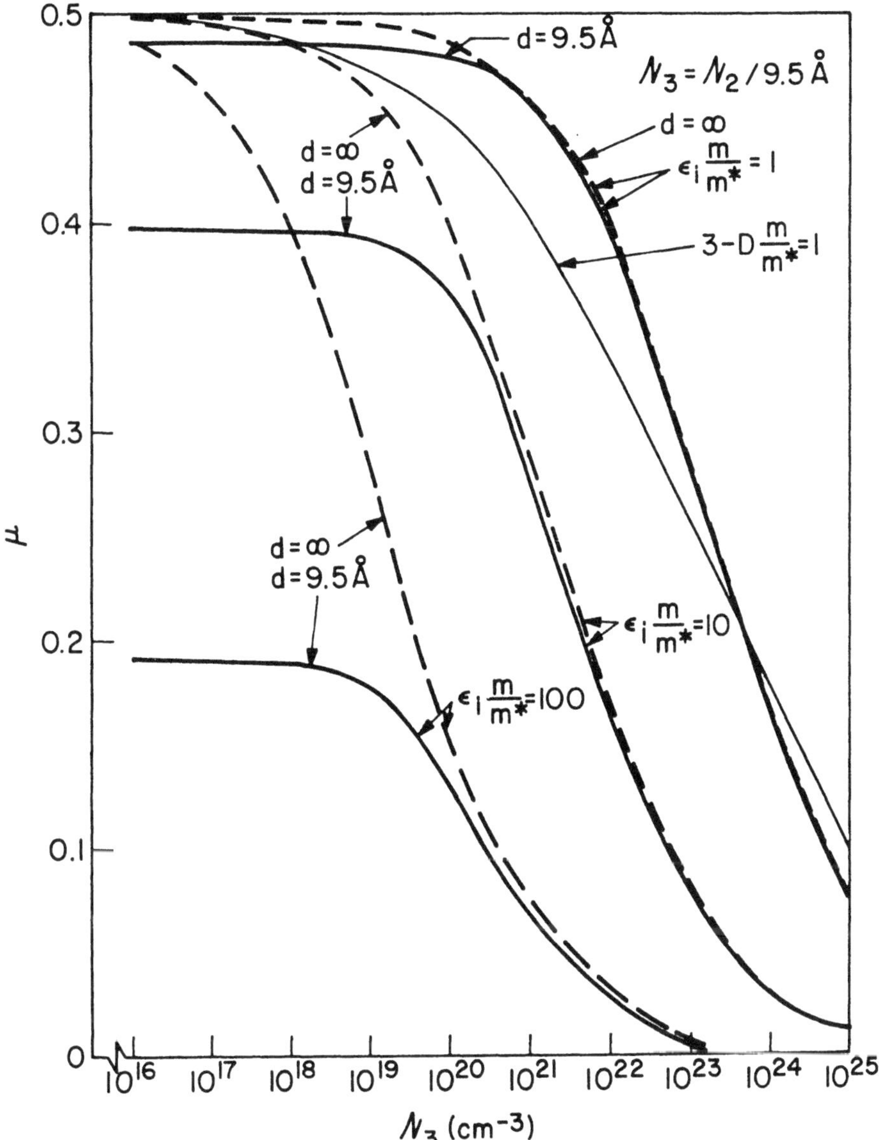

Fig. 1. The Coulomb repulsion μ plotted as a function of the average 3-D electron density N_3, for the periodic geometry of metallic parallel sheets at a distance d apart (continuous and broken heavy lines) and for the uniform 3-D case (light continuous line). The 2-D electron density N_2 is related to N_3 by $N_2 = N_3d$ with d = 9.5Å. ε_i is the dielectric constant of the material between the metallic sheets and m/m* is the ratio of the free to the effective electron mass.

are important only as long as the distance d is less than the screening length which is of the order of $1/k_F$.

For ε_i $m/m^* = 1$, μ' is very close to the 3-D values of μ. Actually $\mu' < \mu$ for low and high N_3 and $\mu' > \mu$ for intermediate values of N_3. As ε_i m/m^* increases, there is a quite significant drop in the values of μ' as can be seen in Fig. 1.

To find the effect on T_c [the superconducting transition temperature] due to changes in μ, we employ McMillan's formula[27] for T_c

$$T_c = \frac{\Theta}{1.45} \exp\left[\frac{1.04\ (1+\lambda)}{\lambda-\mu^*(1+.62\lambda)}\right] \tag{15}$$

where Θ is the Debye temperature, λ is the phonon attraction, $\mu^* = \mu/(1+\mu \ln\frac{E_F}{\omega_o})$, E_F is the Fermi energy and ω_o a typical phonon energy. Typically $\ln\frac{E_F}{\omega_o}$ is about 6 although values as low as 3 are conceivable. For the extreme case of $\ln\frac{E_F}{\omega_o} = 3$ and $\varepsilon_i \frac{m}{m^*} = 100$, μ^* can change from .16 in the 3-D case to .02 in the geometry under consideration (we assume that $N_3 = 10^{22}$ cm^{-3}). For the more typical case of $\ln\frac{E_F}{\omega_o} = 6$ and $\varepsilon_i \frac{m}{m^*} = 10$, μ^* can change from .11 in the 3-D case to .085 in the geometry under consideration (again assuming $N_3 = 10^{22}$ cm^{-3}). In Fig. 2 we plot T_c according to Eq. (15) as a function[28] of λ for the above four values of μ^* and for $\Theta = 290$K. The increase of T_c corresponding to the reduction of μ^* from .16 to .02 is an extreme case. More typical is the increase of T_c corresponding to the reduction from $\mu^* = .11$ to $\mu^* = .085$.

It is interesting to compare these predictions of the dependence on dielectric constant with some of the features of superconductivity in metal-semiconductor eutectic alloys[29] (microscopic array of domains of metal and semiconductor). In metal-semiconductor eutectic alloys such as Al-Si with Si having a sizable dielectric constant, a significant increase in the superconducting transition temperature was observed. On the other hand, for a eutectic system between two metallic phases Al - Al_2Cu, no enhancement in T_c was observed. Similar behavior seems to hold also for quenched ℓGe alloy thin films[30] where T_c is enhanced by a factor of four as compared with bulk Aℓ.

2.3 Acoustic Surface Plasmons

It is useful to elucidate the possible physical origin of the significant enhancement in T_c of the layered system by a study of the nature of the surface plasmon excitations in a periodic layered electron gas.

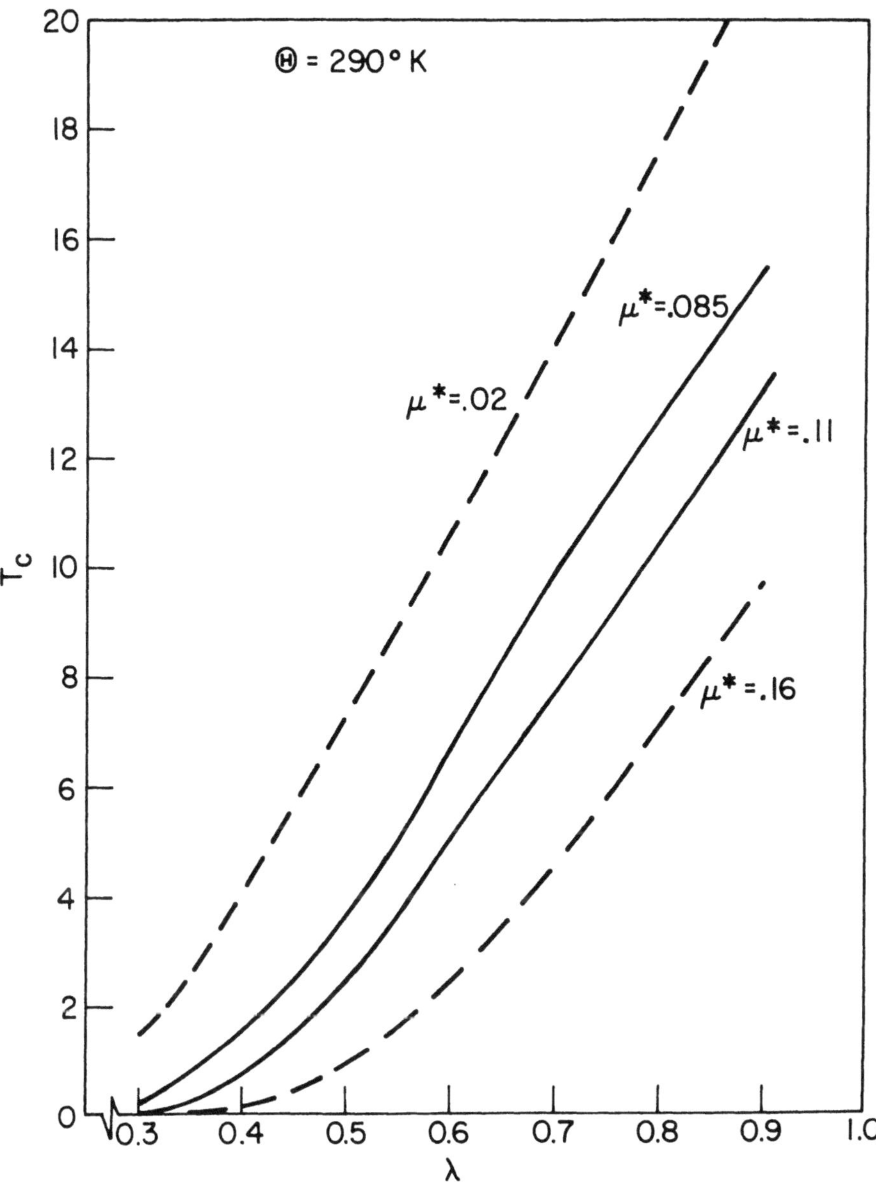

Fig. 2. The superconducting transition temperature plotted as a function of the phonon attraction λ, for various values of the effective Coulomb repulsion μ*. Θ is the Debye temperature.

Consider the infinitely periodic thin films system depicted in Fig. 3. The periodic structure consists of alternating metal and insulating films of thicknesses d_m and d_i respectively. The period is $d = d_i + d_m$. Periodicity implies that the eigensolutions of SP oscillations must obey the Floquet-Bloch theorem; namely any SP field component ϕ satisfies the condition

$$\phi_\alpha (z + d) = e^{i\alpha d} \phi_\alpha(z)$$

α is the quasimomentum in the z-direction. The complete dispersion relations of both the higher and the lower acoustic SP bands including retardation effects have been given earlier.[17] In the acoustic band a SP mode has linear dispersion

$$\omega_\alpha(\vec{q}) = \bar{c}_\alpha \mathbf{Q}$$

with

$$\bar{c}_\alpha = c[d_i/(d_i + 2\lambda_p \cdot \frac{\exp (2k_p d_m) + 1-2 \cos (\alpha d) \exp (k_p d_m)}{\exp (2k_p d_m) - 1}]^{1/2} \tag{16}$$

valid for $\mathbf{Q} < 1/d_m$, $1/d_i$. Here c is the velocity of light in the dielectric and ω_p is the bulk plasma frequency of the metal and $\lambda_p = 1/k_p$. If $k_p d_m << 1$, and $\alpha d \sim \pi$ Eq. (16) reduces to a form that can be obtained with electrostatic theory. The reason for this has been discussed elsewhere.[17] We have in mind thin films system such that both $k_p d_m$ and $k_p d_i$ are small; then Eq. (16) for $\alpha d \sim \pi$ becomes

$$\bar{c}_\alpha = \omega_p (d_i d_m)^{1/2}/2|\sin \frac{\alpha d}{2}| \tag{17}$$

For $\alpha = \pi/d$, d_m, $d_i \approx 10$ Å and $10^{15} < \omega_p < 10^{16}$ sec^{-1}, typical of real metals, we see that

$$5 \times 10^7 \lesssim \bar{c}_{min} \lesssim 5 \times 10^8 \text{ cm/sec.}$$

is appreciably smaller than the velocity of light.

It should be emphasized that the surface plasmons described by Eq. (16) are obtained[17] by assuming the metal and dielectric films can be described by their bulk dielectric functions. The only restriction on the thickness d_i of the insulating film is that $d_i >> \delta$ where δ is the length which determines how well defined the interfaces are (see Fig. 3); δ is usually less than 1 Å while d_i is larger or equal to 5 Å. For the metallic film $\varepsilon_m(\omega)$ is taken to be of the form

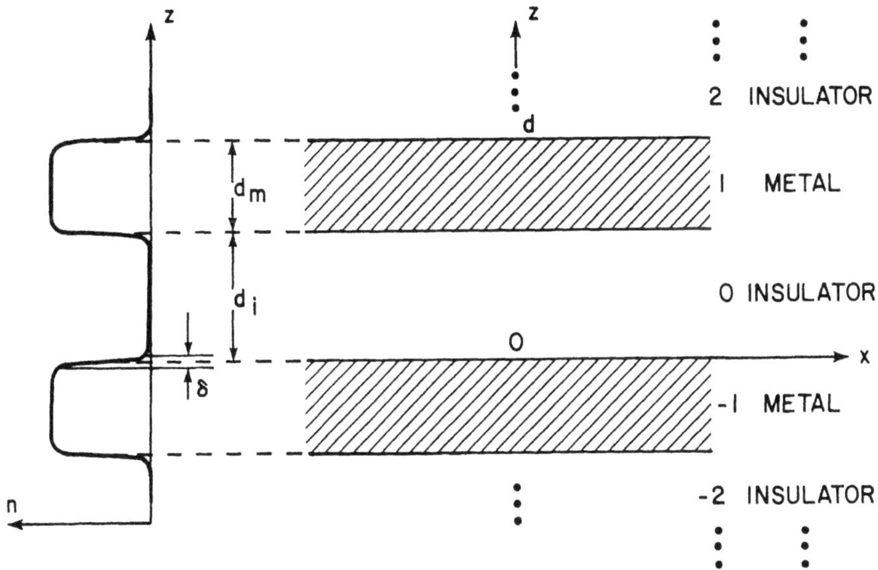

Fig. 3. Geometry and electron distribution of the periodic metal, insulating film configuration. δ is a measure of how well defined the interfaces are.

$$\varepsilon_m(\omega) = 1 - \frac{\omega_p^2}{\omega^2} \frac{1}{1-i\omega\tau} \tag{18}$$

where τ is the collision time. Since we consider thin films of near atomic dimension, it seems, on first observation that the validity of this description is questionable. A more careful investigation of this question reveals that the answer depends on the type of the eigenmode. Eigenmodes where the charge motion is parallel to the interfaces can be described by the macroscopic dielectric function[18] even in the case of $d_m \to 0$, as long as q is less than a cutoff value of $q_c \sim k_F$ where k_F is the Fermi momentum. On the other hand the eigenmodes involving charge motion perpendicular to the films disappear for extremely thin films and as a consequence cannot be described by considering the metallic films as macroscopic. In the present case the modes of the lower band belong to the first group and can be described by the macroscopic approach we have adopted here. The modes of the higher band belong to the second group and should disappear for extremely thin metallic films. For this reason we do not consider at all the higher band in what follows.

An explicit way to demonstrate the correctness of the above statements is to consider the case $d_m \to 0$. This case can be studied from first principles as a two dimensional electron gas.[26,23] The same limit can be obtained from the present macroscopic approach by allowing $d_m \to 0$ while keeping $d_m n$ finite where n is the electron density. By comparing the two approaches one can check the correctness of the macroscopic approach. This check has been performed explicitly for the cases of a metal film in vacuum,[26,31] and a metal film on a metallic substrate;[31] for the present case of a periodic arrangement one can easily demonstrate that the modes of the lower band reduce to the modes found by Fetter[23] in periodically arranged planes of two dimensional electron gas.

The SP potential field $\phi(r)$ can be expressed as the sum of products $u_\alpha(\vec{\rho},z)e^{i\alpha z}$ where $u_\alpha(\vec{\rho},z)$ is the periodic part

$$u_\alpha(\vec{\rho},z) = \begin{cases} \sum_{\vec{Q}} (\phi_{1Q\alpha}e^{Qz} + \phi'_{1Qa}c^{-Qz})e^{i\vec{Q}\cdot\vec{\rho}} & 0 < z < d_i \\ \sum_{\vec{Q}} (\phi_{2Q\alpha}e^{Qz} + \phi_{2Q\alpha}e^{-Qz})e^{i\vec{Q}\cdot\vec{\rho}} & d_i < z < d \end{cases} \tag{19}$$

The SP potential field $\phi(\vec{r})$ is a sum

$$\phi(\vec{r}) = \sum_\alpha u_\alpha(\vec{r})e^{i\alpha z} \tag{20}$$

A method of obtaining both the SP dispersion relation and eigensolutions for the associated fields has been given earlier.[31,32] For a review see Ref. 17.

Consistent with our dielectric formulation, we consider a continuum model of an electron gas with a rigid jellium background of positive charge. Electron charge density fluctuation due to electron motion gives rise to the SP electric fields. Let n_i be the electron number density and m_i the effective mass in the i-th metallic region, $\delta\rho_i(\vec{r}_i)$ the charge density fluctuation from the equilibrium value of $\rho_i = n_i e$, $\vec{R}_i(\vec{r}_i)$ the vector displacement of the electron gas at $\vec{r}_i$, and $\vec{\pi}_i(\vec{r}_i)$ its time derivative,

$$\vec{\pi}_i(\vec{r}_i) = \partial\vec{R}_i(\vec{r}_i)/\partial t \tag{21}$$

The energy of the SP partly resides in the electric fields and partly in the kinetic energy of the electron motion. The electric field energy density is given by $\frac{1}{2}\,\delta\rho(r_i)\;\phi(\vec{r}_i)$, where $\phi(\vec{r}_i)$ is the scalar potential. The Lagrangian density in the i-th region is

$$\Lambda(\vec{r}_i) = \frac{1}{2}\, n_i m_i \vec{\pi}_i(\vec{r}_i)\cdot \vec{\pi}_i(\vec{r}_i) - \frac{1}{2}\,\delta\rho_i(\vec{r}_i)\;\phi(\vec{r}_i) \tag{22}$$

and the total Lagrangian L for the system is

$$L = \sum_i \int \Lambda(\vec{r}_i)\,d\vec{r}_i \tag{23}$$

Charge fluctuations for SP mode can only reside on the interfaces. These surface charges must also satisfy the Bloch periodic condition. Referring to Fig. 3, two interfaces labeled as (10) and (21) occur in a unit cell. Then the surface charge fluctuation $\sigma(\vec{r})$ has the form

$$\sigma(\vec{r}) = \sum_\alpha \left[u_\alpha^{(10)}(\vec{\rho},z) + u_\alpha^{(21)}(\vec{\rho},z)\right]e^{i\alpha z}$$
$$-\infty < z < \infty \tag{24}$$

The cell periodic functions $u_\alpha^{(10)}$ and $u_\alpha^{(21)}$ when Fourier analyzed in the $\vec{\rho}$ plane have the form

$$u_\alpha^{(ij)}(\vec{\rho},z) = \sum_{\vec{Q}} \sigma_{\vec{Q}\alpha}^{(ij)}(z)\,e^{i\vec{Q}\cdot\vec{\rho}} \tag{25}$$

where the pair superscript (ij) is either (01) or (21). The surface charge $u_\alpha^{(ij)}(\vec{\rho},z)$ is given by the discontinuity of the normal component of the electric fields. This enables us to express $\sigma_{\vec{Q}\alpha}^{(ij)}$ in terms of $\phi_{\vec{Q}\alpha}$ as follows:

$$\sigma_{\vec{Q}\alpha}^{(10)} = -\frac{1}{4\pi}\,Q\,[\phi_{1\vec{Q}\alpha} - \phi'_{1\vec{Q}\alpha} - e^{-i\alpha d}(\phi_{2\vec{Q}\alpha}\,e^{Qd}$$
$$- \phi'_{2\vec{Q}\alpha}\,e^{-Qd})]\;\delta(z) \tag{26}$$

and

$$\sigma_{\vec{Q}\alpha}^{(21)} = \frac{1}{4\pi}\,Q\,[\phi_{1Q\alpha}\,e^{Qd_i} - \phi'_{1\vec{Q}\alpha}\,e^{-Qd_i}\,.$$
$$- \phi_{2Q\alpha}\,e^{Qd_i} + \phi'_{2Q\alpha}\,e^{-Qd_i}]\;\delta(z-d_i) \tag{27}$$

The SP field energy is obtained by integrating over all space of the energy density as $\frac{1}{2}\int \sigma(\vec{r})\;\phi(\vec{r})\;d\vec{r}$ substituting from Eqs. 19-20, 24-27 into this expression. Letting A be the area of the film surface and N the number of unit cells in the z-direction, and use is made of the continuity equations at the two interfaces within each unit cell, the field energy can be expressed in terms of one field variable $\phi_{1\vec{Q}\alpha}$ only,

$$\frac{1}{2}\int \sigma(\vec{r})\ \phi(\vec{r})\ d\vec{r} = NA \sum_{\vec{Q}\alpha} \left(\frac{\pi n^2 e^2}{Q}\right) \Phi_{Q\alpha}\ \phi_{1\vec{Q}\alpha}\phi_{1-\vec{Q}-\alpha} \tag{28}$$

$$\Phi_{Q\alpha} = \frac{Q^2}{8\pi^2 n^2 e^2}\ \{(\varepsilon_- e^{Qd_i} - \varepsilon_+\xi_+)(e^{Qd_i} + \varepsilon_+\xi_-/\varepsilon_-) - (\varepsilon_- - \varepsilon_+\xi_+\ e^{Qd_i})\} \tag{29}$$

where

$$\varepsilon_\pm = 1 \pm \varepsilon_i/\varepsilon_m \tag{30}$$

$$\xi_\pm = (e^{\pm i\alpha d} - e^{Qd})/(e^{Qd_m} - e^{\pm i\alpha d + Qd_i}) \tag{31}$$

The kinetic energy part of the SP is

$$\frac{1}{2}\int n\ m\ \vec{\pi}(\vec{r})\cdot\vec{\pi}(\vec{r})\ d\vec{r} \tag{32}$$

The electron displacement field $\vec{R}(\vec{r})$ similarly has the form

$$R(\vec{r}) = \sum_{\vec{Q}\alpha} (R_{\vec{Q}\alpha}\ e^{Qz} + R'_{Q\alpha}\ e^{-Qz})\ e^{i\vec{Q}\cdot\vec{\rho}}\ e^{i\alpha z} \tag{33}$$

where the cell-periodic part

$$\sum_{Q} (R_{\vec{Q}\alpha}\ e^{Qz} + R'_{\vec{Q}\alpha}\ e^{-Qz})\ e^{i\vec{Q}\cdot\vec{\rho}} \tag{34}$$

describes the electron displacement in the metallic subcell $d_i < z < d$ of the unit cell $0 < z < d$. Boundary conditions at the interfaces give the relationship

$$-\ ne(R_{\vec{Q}\alpha z}\ e^{Qd_i} + R'_{\vec{Q}\alpha z}\ e^{-Qd_i}) = \sigma^{(21)}_{Q\alpha} \tag{35}$$

and

$$ne(R_{\vec{Q}\alpha z}\ e^{Qd} + R'_{\vec{Q}\alpha z}\ e^{-Qd}) = \sigma^{(10)}_{Q\alpha}\ . \tag{36}$$

$R_{\vec{Q}\alpha z}$ is the component of $R_{\vec{Q}\alpha}$ in the z-direction. The absence of bulk charge fluctuations for SP modes inside the metal implies the condition

$$\mathrm{div}\ R(\vec{r}) = 0 \tag{37}$$

Eqs. 33-37 permit us to write the kinetic energy term in the form,

$$\frac{1}{2}\int nm\ \vec{\pi}(\vec{r})\cdot\vec{\pi}(\vec{r})\ d\vec{r} = NA\sum_{\vec{Q}\alpha}\left(\frac{nm}{2Q}\right)\Psi_{Q\alpha}\ \dot{\phi}_{1\vec{Q}\alpha}\dot{\phi}_{1-\vec{Q}-\alpha} \tag{38}$$

where

$$\Psi_{Q\alpha} = \left(\frac{Q}{4\pi ne}\right)^2 (e^{-Qd_m} - e^{Qd_m})^{-2}$$

$$(e^{2Qd} - e^{2Qd_i})\,|\,(\varepsilon_- - \varepsilon_+\ \xi_+\ e^{Qd_i})\ e^{i\alpha d-Qd_i}$$

$$-(\varepsilon_-\ e^{Qd_i} - \varepsilon_+\ \xi_+)\ e^{-Qd}|^2 + (e^{-2Qd_i} - e^{-2Qd})\,|$$

$$(\varepsilon_-\ e^{Qd_i} - \varepsilon_+\ \xi_+)e^{Qd} - (\varepsilon_- - \varepsilon_+\ \xi_+\ e^{Qd_i})$$

$$e^{i\alpha d + Qd_i}|^2 \tag{39}$$

Collecting the terms in Eqs. (28) and (38), the total Lagrangian of the system is

$$L = NA\sum_{\vec{Q}\alpha}\left[\left(\frac{nm\Psi_{\vec{Q}\alpha}}{2Q}\right)\dot{\phi}_{1\vec{Q}}\ \dot{\phi}_{1-\vec{Q}-\alpha}\ \cdot\right.$$

$$\left.-\ \frac{\pi n^2e^2\Phi_{Q\alpha}}{Q}\ \phi_{1\vec{Q}\alpha}\ \phi_{1-\vec{Q}-\alpha}\right] \tag{40}$$

The momentum $p_{\vec{Q}\alpha}$ conjugate to the field coordinate $\phi_{1\vec{Q}\alpha}$ is given by

$$p_{\vec{Q}\alpha} = \frac{\partial L}{\partial\dot{\phi}_{1Q\alpha}} = (NA\ nm\ \Psi_{Q\alpha}/Q)\ \dot{\phi}_{1-\vec{Q}-\alpha} \tag{41}$$

The Hamiltonian for the SP excitation is

$$H = \sum_{\vec{Q}\alpha}\left[\frac{Q}{2NAnm\Psi_{Q\alpha}}\ p_{Q\alpha}\ p_{-\vec{Q}-\alpha} + \frac{NA\pi n^2e^2\Phi_{Q\alpha}}{Q}\ \phi_{1\vec{Q}\alpha}\phi_{1-\vec{Q}-\alpha}\right] \tag{42}$$

It follows from Eq. 42 that the SP frequency is given by

$$\omega_{Q\alpha} = (\Phi_{Q\alpha}/2\Psi_{Q\alpha})^{\frac{1}{2}}\ \omega_p \tag{43}$$

The SP field is now quantized by imposing the Bose-Einstein commutation relations

$$\left[\phi_{1\vec{Q}\alpha},\ p_{\vec{Q}'\alpha'}\right] = i\hbar\ \delta_{\vec{Q},\vec{Q}'}\cdot\delta_{\alpha,\alpha'} \tag{44}$$

The SP creation and annihilation operators are

$$c^{+}_{\vec{Q}\alpha} = -i\left(\frac{2NA\ nm\,\Psi_{Q\alpha}\hbar\omega_{Q\alpha}}{Q}\right)^{-\frac{1}{2}} p_{\vec{Q}\alpha} + \left(\frac{NA\pi n^2 e^2 \Phi_{Q\alpha}}{Q\hbar\omega_{Q\alpha}}\right)^{\frac{1}{2}}\cdot$$

$$\phi_{1-\vec{Q}-\alpha} \tag{45a}$$

and

$$c_{\vec{Q}\alpha} = i\left(\frac{2NAnm\,\Psi_{Q\alpha}\hbar\omega_{Q\alpha}}{Q}\right)^{-\frac{1}{2}} p_{-\vec{Q}-\alpha} + \left(\frac{NA\pi n^2 e^2 \Phi_{Q\alpha}}{Q\hbar\omega_{Q\alpha}}\right)^{\frac{1}{2}}\cdot$$

$$\phi_{1\vec{Q}\alpha} \tag{45b}$$

The Hamiltonian (Eq. 42) can then be written as

$$H = \sum_{\vec{Q},\alpha} \hbar\omega_{\vec{Q}\alpha}\ (c^{+}_{\vec{Q}\alpha}\ c_{\vec{Q}\alpha} + \tfrac{1}{2}) \tag{46}$$

The electron-SP interaction is

$$H_{e-SP} = e\phi(\vec{r}) \tag{47}$$

The SP wave field $\phi(\vec{r})$ can be written in terms of $c^{+}_{\vec{Q}\alpha}$ and $c_{\vec{Q}\alpha}$, and we have the formula explicitly as

$$H_{e-SP} \equiv \sum_{\vec{Q}\alpha} W_{\vec{Q}\alpha}\ (z)e^{-i(\vec{Q}\cdot\vec{\rho}+\alpha z)}\ (c^{+}_{-\vec{Q}-\alpha} + c_{\vec{Q}\alpha})\cdot$$

$$W_{\vec{Q}\alpha}\ (z) = \left(\frac{2\pi e^2 d\hbar\omega_{\vec{Q}\alpha}}{Q\zeta(Q)V}\right)^{\frac{1}{2}} f_{\vec{Q}\alpha}(z) \tag{48a}$$

where $f_{\vec{Q}\alpha}(z)$ is defined by

$$f_{\vec{Q}\alpha}(z) = \begin{cases} e^{Qz} + (\varepsilon_+ \xi_+/\varepsilon_-) e^{Q(d_i - z)} & 0 \leqslant z \leqslant d_i \\ (e^{i\alpha d}/2)\left[\varepsilon_+(1 + \xi_+ e^{Qd_i}) e^{-Q(d-z)} + \right. \\ \left. + \varepsilon_- (1 + \varepsilon_+^2 \xi_+ e^{Qd_i}/\varepsilon_-^2) e^{Q(d-z)}\right] \\ d_i < z < d \end{cases} \tag{48b}$$

where $\varepsilon_\pm$ are as given by Eq. (30) but with its frequency variable replaced by $\omega_{\vec{Q}\alpha}$ and

$$\zeta(Q) = 8(\pi ne/Q)^2 \Phi_{Q\alpha} \tag{49}$$

is the same as the expression inside the curly bracket in Eq. (29). V, the volume of the periodic structure, is equal to the product NAd.

Before we proceed further to examine the electron-SP interaction, we return to Eq. (43) which is the expression of the SP frequency. At long wave lengths, such that $Qd \ll 1$, the quantities $_{Q\alpha}$ and $_{Q\alpha}$ can be expanded in powers of Qd_i, Qd_m and Qd. The leading terms of these expansions are

$$\Phi_{Q\alpha} = Q^3 d_i/2(\pi ne)^2 + \ldots\ldots \tag{50}$$

$$\Psi_{Q\alpha} = Q\sin^2(\frac{\alpha d}{2})/d_m(\pi ne)^2 + \ldots\ldots \tag{51}$$

Substitution of these into Eq. (43) yields $\omega_{Q\alpha} = \bar{c}_\alpha Q$ with $\bar{c}_\alpha$ as given by Eq. (17) as it should.

The basic result of this section is Eq. (48) expressing the el-acoustic SP interaction in terms of acoustic SP creation and annihilation operators. The latter are characterized by $\vec{Q}$ and α. For the SP corresponding to $\alpha \approx \pi/d$ the charge distribution at the interfaces is as follows: The two interfaces of the same metallic film carry the same charge while neighboring films carry almost opposite charges. This configuration creates very weak fields inside the metal films and rather strong fields in the dielectric regions. This charge distribution is consistent with the low frequency of these modes. It should be mentioned that for these modes and for rather thin films (d_i, $d_m \ll \lambda_p \equiv 1/k_p$) electrostatics is a very good approximation. On the other hand, for $\alpha \approx 0$ the surface charges on the different metallic films are in phase and these modes resemble an ordinary photon with a phase velocity reduced by a factor of $d_i/(d_i + d_m)^{\frac{1}{2}}$. These modes require, of course, the complete set of Maxwell's equations for their descrip-

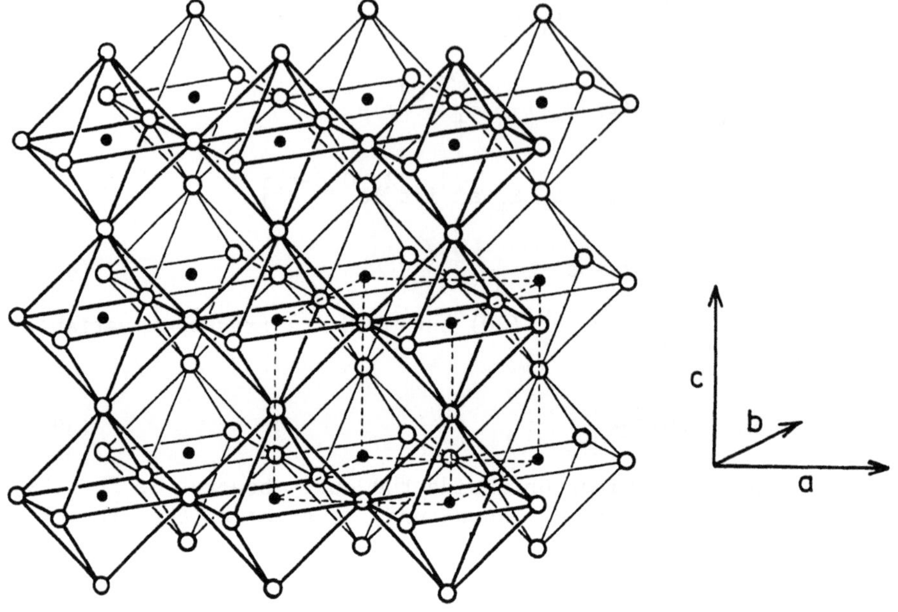

Fig. 4. Matrix of WO_6 octahedra for the cubic perovskite-related structure of M_xWO_3. Open circles represent O atoms, filled circles W atoms. M atoms (not shown) would occupy the body centered position in the cubic unit cell.

tion. We note that, in addition to the SP, the present system sustains bulk plasmons as well, involving charge motion parallel to the planes and exhibiting an eigenfrequency of $\omega = \omega_p$.

The appearance of the acoustic SP modes and their interactions H_{e-SP} with electrons thus leads one to expect enhancement in superconductivity for the layered electron gas system. We can proceed to evaluate the electron-SP interaction parameter λ_{SP} from the expression given explicitly for H_{e-SP}. However, for reasons already given earlier, the accompanying surface modification of the quantities μ'_{bare} and μ'_{BP} need also be known to complete the enhancement in T_c due to the surface plasmon mechanism. We shall not present results of these calculations in these lectures. Nevertheless, for the layered electron gas system, the dramatic effect of interfaces

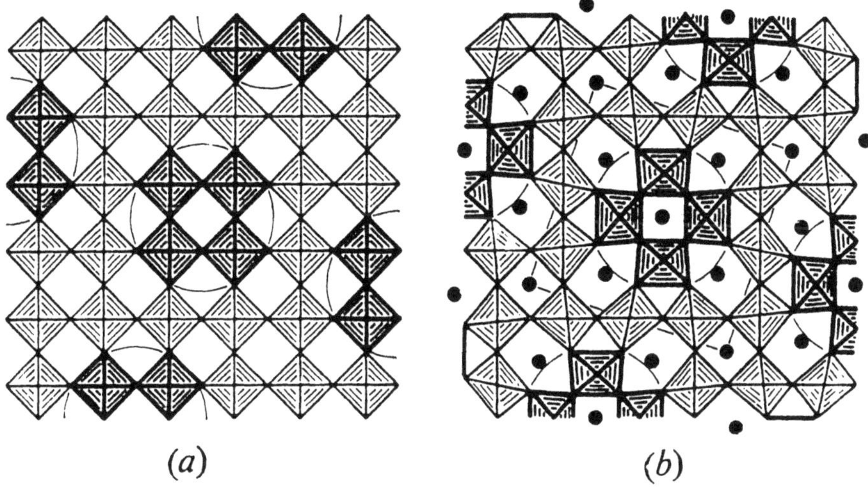

Fig. 5(a). Matrix of WO_6 octahedra in T2 structure, projected on a a - b plane.

Fig. 5(b). Matrix of WO_6 octahedra in T1 structure. Circles identify groups of octahedra which have been rotated by 45^0 to change 5(a) to 5(b). Filled circles represent M atom sites, not necessarily all occupied.

on the electron-electron interaction can be seen by the appearance of entirely new collective excitations, i.e. the acoustic SP modes.

3. SUPERCONDUCTIVITY IN SODIUM TUNGSTEN BRONZE

3.1 Free Energy Model

For values of x in the upper portion of the range $0 < x < 1$, many of the tungsten bronzes take on a cubic perovskite-related structure. The matrix of WO_6 octahedra for this structure is shown in Fig. 4. The M atoms (not shown) would occupy the body centered position in the cubic unit cell. For very small values of x, a tetragonal structure with very few M atoms present is found. The matrix of WO_6 octahedra for this structure is illustrated in

Fig. 5a, where a view in the a-b plane is shown. The other tetragonal structure is shown in Fig. 5b, along with a simple geometrical operation[9] which takes one between the two tetragonal structures. This operation consists of the rigid rotation of each of the groups of four circled WO_6 octahedra by 45° about the center of the unit. Hereafter we refer to the rotated structure as T1 (Fig. 5b) and the simple tetragonal structure as T2 (Fig. 5a). The M atoms occupy the "tunnels" between the arrays of WO_6 octahedra, so that the essential difference between T1 and T2 is the replacement of eight square tunnels by four pentagonal ones and four triangular ones per unit cell. Of course by rotating differently neighboring groups of octahedra one can form hexagonal tunnels as well, so that many other structures in addition to T1 can be obtained from T2, as pointed out in Ref. 9. The stability of a particular structure will depend upon the free energy of the M atoms in the tunnels found in that structure. For example, if M is sodium, one might expect that the sodium atoms would prefer the larger pentagonal tunnels, since the Na-O distance in the square tunnels is only 1.95A°,[33] whereas the Na-O bond length in $NaWO_3$ is 2.73A°.[34] One may take this effect into account by writing the free energy for a given configuration $\{\vec{r}_i\}$ of the octahedra as

$$F\{\vec{r}_i\} = W\{\vec{r}_i\} + N\Sigma_\alpha n_\alpha\{r_i\}M_\alpha, \tag{52}$$

where α is summed over square, triangular, pentagonal, and hexagonal tunnels, N is the total number of (square) tunnels in the T2 structure, $n_\alpha\{\vec{r}_i\}$ is the fraction of occupied α type tunnels in the $\{\vec{r}_i\}$ configuration, M_α is the free energy of the M atom in an α type tunnel, and $W\{\vec{r}_i\}$ is the free energy of the WO_6 octahedra, connected as in configuration $\{\vec{r}_i\}$, and in the absence of M. Now $F\{\vec{r}_i\}$ is to be minimized with respect to the constraint:

$$\Sigma_\alpha n_\alpha\{\vec{r}_i\} = x \tag{53}$$

We note that the M_α will depend strongly on the size of M. A small atom such as lithium will be easily accomodated in the square tunnels. However, for an atom as large as rubidium one would expect the free energy in the hexagonal configuration to be the lowest. Comparing the M-O bond lengths[34] found in similar compounds with the size of the various tunnels,[33] one can understand qualitatively the occurrence (under ordinary conditions)[35] of the lithium tungsten bronze in only the cubic and T2 structures. This fact is due to the small reduction in energy, $N\Sigma_\alpha n_\alpha\{T1\}M_\alpha - N\Sigma_\alpha n_\alpha\{T2\}M_\alpha$, for Li, which cannot compensate for the larger increase in the WO_6 octahedra configurational energy, $W\{T1\} - W\{T2\}$. Similar reasoning explains why sodium tungsten bronze occurs only in cubic, T1 and T2 structures, the potassium in cubic, T1, T2, and hexagonal, and the rubidium in only hexagonal. Moreover, from the constraint Eq. (52) one sees that T1 cannot be stable for $x > \cdot 6$[36] and that the hexagonal phase[37] cannot be stable for $x > \cdot 33$.

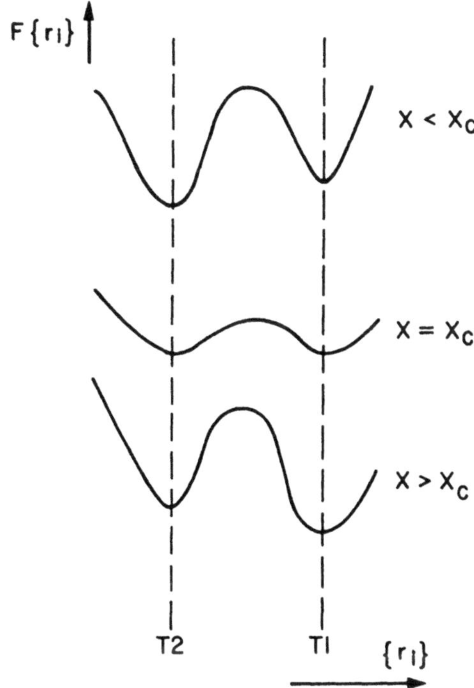

Fig. 6. Compositional dependence of the configurational free energy, $F\{\vec{r}_i\}$. The ordered phases T1 and T2 shown in Fig. 4 are local minima. Intermediate configurations are represented schematically as a quantum barrier. The rotational modes depicted in Fig. 5 correspond to specific excursions about the local minima in $F\{\vec{r}_i\}$.

For the remainder of this paper we will focus our attention on the sodium tungsten bronze. In this case, $M_p < M_s << M_t$. The triangular tunnels are so small that n_t can be neglected. Thus the free energy takes the form

$$F\{T1\} = W\{T1\} + N\left[n_s\{T1\}M_s + n_p\{T1\}M_p\right] \tag{54a}$$

$$(n_s + n_p = x)$$

$$F\{T2\} = W\{T2\} + NxM_s \tag{54b}$$

where $W\{T2\} < W\{T1\}$. For small values of x, T2 is clearly the stable structure, since the increase in W energy will not be compensated by the M terms in Eq. (54). However, as x increases, the energy lowering obtained by increasing n_p and decreasing n_s will at some point (x_c) compensate the difference $W\{T1\} - W\{T2\}$, as illustrated in Fig. 6. For values of $x > x_c$, but less than x 6, T1 should be the stable structure. Experimentally, x_c is found[35] to be about .2, and x_{max} about .5. To complete the model, we note that the rotation shown in Fig. 5 takes one from T2 to T1 and vice versa. However, the change in configurations requires the breaking of both W-O and M-O bonds, so that the transition does not result from a simple condensation of the phonon mode corresponding to the rotation. It follows, however, from the fact that $F\{\vec{r}_i\}$ is linear in x that this phonon frequency will soften in the manner

$$\omega_o^2 = c^2 (x - x_o) \tag{55}$$

as x approaches x_c.[38] Physically this corresponds to the weakening of the force constants for the rotational mode as the M atoms are depleted from the pentagonal tunnels. x_o is expected to be greater than zero, since the rotation should be unstable at $x = 0$, when T2 is by far the low free energy configuration. Eq. (55) is the analog of the usual anharmonic soft mode,[11] with two notable differences. In contrast to the anharmonic case, in this system one may obtain a softening even in the harmonic approximation.[38] Also, the parameter which determines the phonon frequency is the M concentration rather than the temperature.

3.2 Phonon - Configuration Coupling

As one approaches x_c, the two wells in $F\{\vec{r}_i\}$ will be close in energy, and tunneling between configurations may occur, assisted by the ω_o phonon. The effect of this process on the phonon frequency must be considered. For this purpose we treat the two configurations as a pseudospin ½ system and use the following model Hamiltonian:

$$H = \omega_o a^+ a + \Delta E S^z + V(x)(S^+ a^+ + S^- a), \tag{56}$$

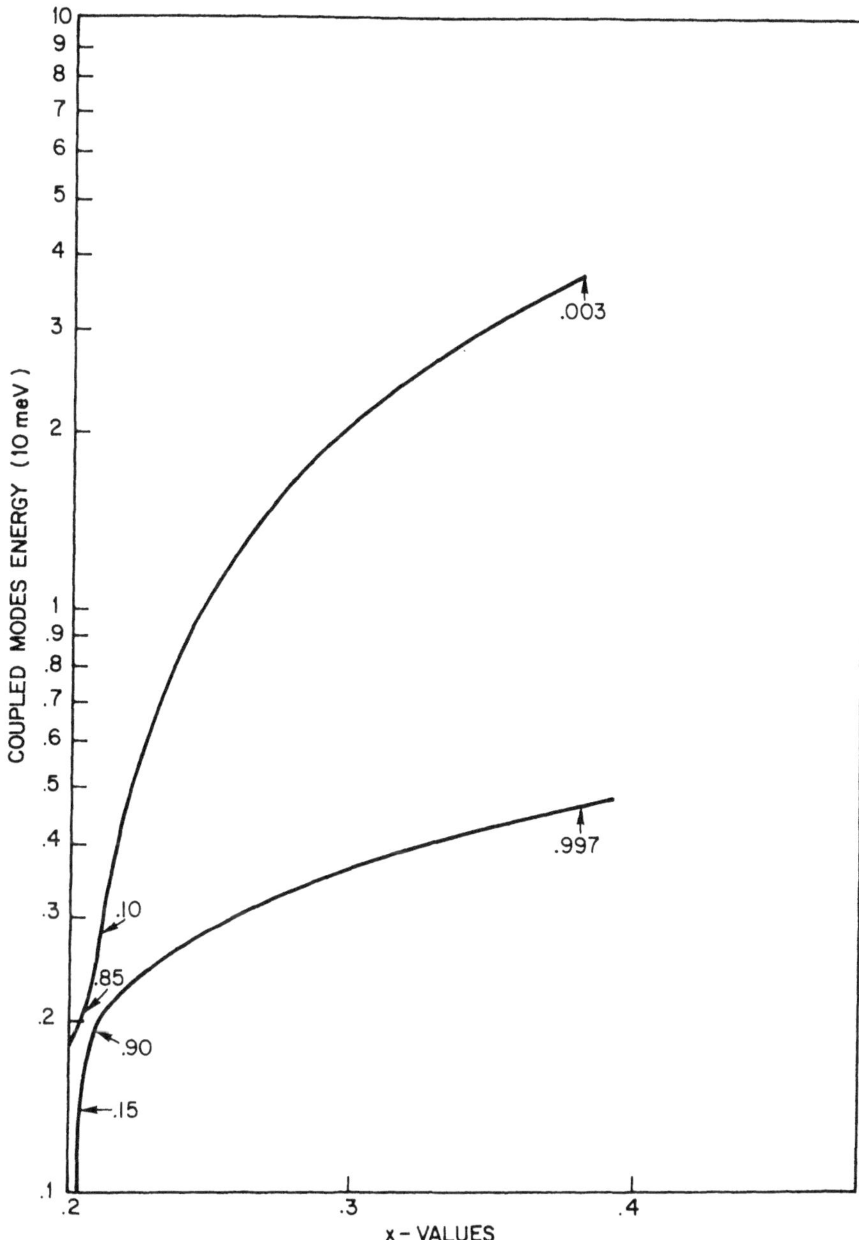

Fig. 7(a). The coupled modes described by Eq. (60) with $x_o = .17$, $x_c = .1975$, $V_o = .19$ meV, $\xi = 10$, c = 10 meV, $\varepsilon = 200$ meV, the weak coupling limit.

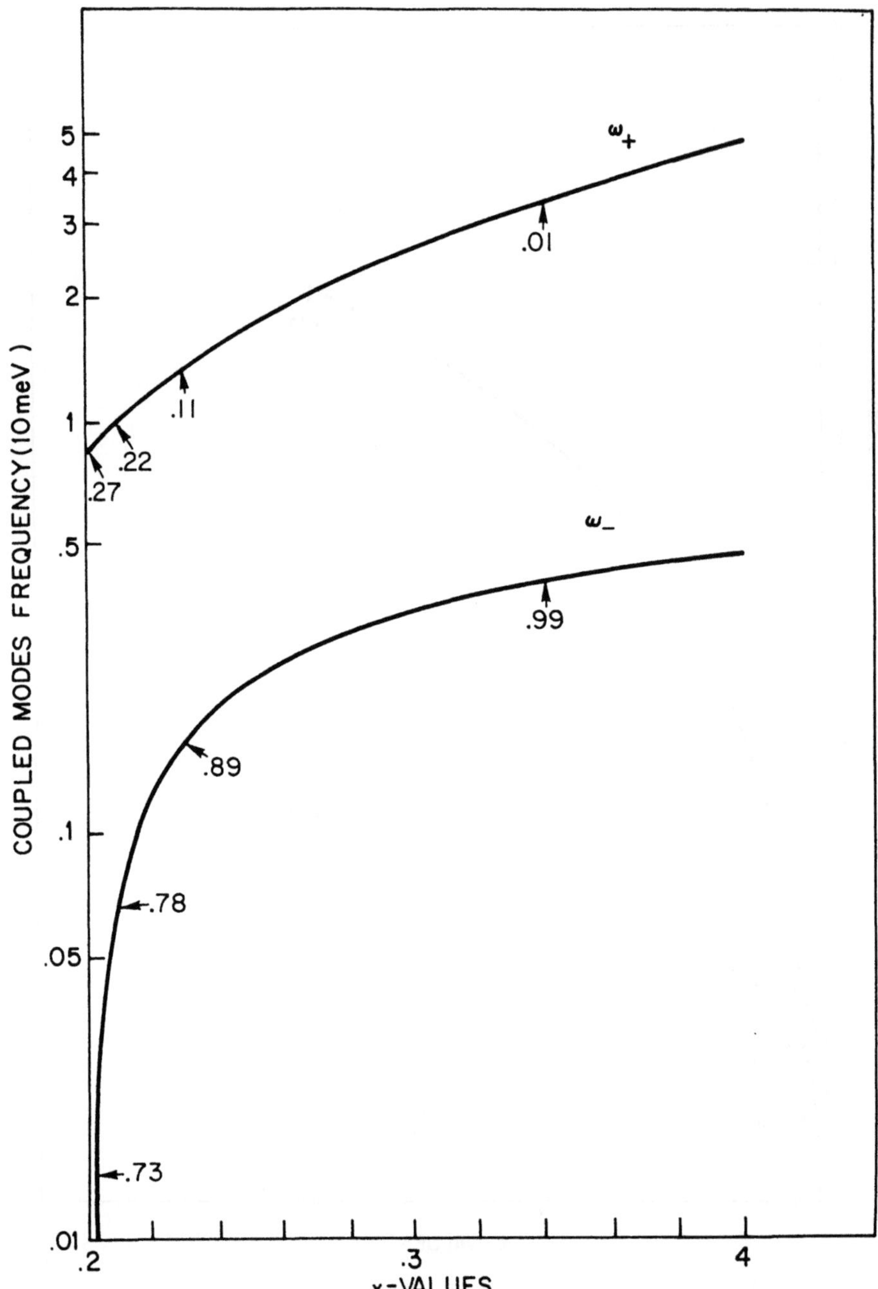

Fig. 7(b). Corresponds to x_o = .14, x_c = .17, V_o = 3.9 meV, ξ = 1, c = 10 meV, ε = 200 meV, the strong coupling limit. The numbers above and below the curves indicate the amount of phonon character to the mode. The coupled modes depend little on the choice of ξ from ξ = 1 to ξ = 10.

where ω_o is given by the Eq. (4), $\Delta E = \varepsilon(x - x_c)$ is the difference between the free energies of the T1 and T2 configurations (assumed to be linear in x), and V(x) is the matrix element for the configuration tunneling.[39] This matrix element, which increases as x approaches x_c^+ due to the flattening of the T1 well as ω_o softens, is assumed to have the form $V_o \exp(-\xi x)$. The effect of the configuration coupling term on the phonon frequency may be obtained from the phonon self energy,[40] according to

$$\Sigma_{ph}(\omega) = \frac{V^2(x)}{\omega - \varepsilon(x - x_c) + i\delta}, \tag{57}$$

which gives for the phonon propagator[40] ($\omega > 0$)

$$D(\omega) = \left[\omega - c\,(x - x_o)^{1/2} - \frac{V^2(x)}{\omega - \varepsilon(x - x_c)} + i\delta\right]^{-1}. \tag{58}$$

Thus

$$D(\omega) = \frac{\omega - \varepsilon(x - x_c)}{(\omega - \omega_+)\,(\omega - \omega_-) + i\delta} \tag{59}$$

where

$$\omega_\pm = \left\{c(x - x_o)^{\frac{1}{2}} + \varepsilon(x - x_c)\right\}/2 \pm \tfrac{1}{2}\left\{\left[c(x - x_o)^{\frac{1}{2}}\right.\right. \cdot$$

$$\left.\left. - \varepsilon(x - x_c)\right]^2 + 4V^2(x)\right\}^{\frac{1}{2}} \tag{60}$$

Equation (60) represents a simple linear coupling of two modes, corresponding to the phonon and the configurational change. The lower branch, ω_-, will be required to vanish at x = .2, where the T1 to T2 transition is observed to occur. Since the transition is driven by the configuration change, rather than phonon condensation, we have the condition $x_o < x_c \lesssim .2$. It is convenient to define two limits, that of strong coupling (V(x) > c) and that of weak coupling (V(x) << c). The two branches given by Eq. (59) are plotted in Fig. 7 vs. x for both cases. The residues of $D(\omega)$ at $\omega_\pm$ are given by:

$$A_\pm \equiv \operatorname{Res} D(\omega_\pm) = \tfrac{1}{2}\left\{1 \pm \frac{c(x - x_o)^{\frac{1}{2}} - \varepsilon(x - x_c)}{\left[(c(x - x_o)^{\frac{1}{2}} - \varepsilon(x - x_c)^2 + 4V^2(x)\right]^{\frac{1}{2}}}\right\} \tag{61}$$

$A_\pm$ determines the amount of phonon character at $\omega_\pm$ and is also shown in Fig. 7. The results of the coupling are that the phonon softening is suppressed near x = .2, and that the amount of phonon character in the ω_- mode near x = .2 in the weak coupling case is very small. This is to be expected, since it is the configuration change, and not the soft phonon, which drives the transition.

3.3 Superconducting Transition Temperature

Now that we have calculated the x dependence of the soft phonon frequency, taking into account the important effects of the configuration coupling near the T1 - T2 transition, we may investigate the effect of this dependence on the superconducting transition temperature in the T1 phase, assuming that this phonon makes the major contribution to the variation of T_c with x. For this purpose, we will utilize McMillan's formulation[41] of T_c in terms of the electron-phonon spectral function $\alpha^2F(\omega)$. According to Ref. 15:

$$T_c = \frac{\langle\omega\rangle}{1.2} \exp\left[\frac{-1.04\ (1+\lambda)}{\lambda - \mu^* - 0.62\ \lambda\mu^*}\right], \tag{62}$$

where μ^* is the Coulomb pseudopotential,

$$\lambda = 2 \int_0^\infty \frac{\alpha^2 F(\omega)d\omega}{\omega} \tag{63a}$$

$$\langle\omega\rangle = \frac{2}{\lambda} \int_0^\infty \alpha^2 F(\omega)d\omega \tag{63b}$$

and $\alpha^2F(\omega)$ is defined by:[42]

$$\alpha^2 F(\omega) \equiv \frac{-\frac{1}{\pi} \sum_{\vec{k},\vec{Q}} |M(\vec{k} \to \vec{k}+\vec{Q}|^2 \mathrm{Im}D(\vec{Q},\omega)\ \delta(\varepsilon_{\vec{k}+\vec{Q}})\ \delta(\varepsilon_{\vec{k}})}{\sum_{\vec{k}} \delta(\varepsilon_{\vec{k}})} \tag{64}$$

In Eq. (64) M is the matrix element for an electron-phonon scattering to occur between electron states $\vec{k}$ and $\vec{k}+\vec{Q}$ on the Fermi surface, and D $(\vec{Q},\omega)$ is the propagator for a phonon of momentum $\vec{Q}$. Assuming that the important variation of T_c with x is due to the soft phonon treated in the previous section, and that the density of electronic states is proportional to x,[43] we obtain for the x dependence of λ and $\langle\omega\rangle$:

$$\lambda = \frac{\alpha x}{(x-x_o)^{\frac{1}{2}}} \left(\frac{A_+}{\omega_+} + \frac{A_-}{\omega_-}\right) \tag{65a}$$

$$\langle\omega\rangle = \beta\left\{\frac{A_+}{\omega_+} + \frac{A_-}{\omega_-}\right\}^{-1} \tag{65b}$$

where α and β are constants. Assuming that the density of states dominates the x-dependence of the Coulomb pseudopotential, we take $\mu^* = .15$ x. For $x \sim 1$ this gives a value close to that predicted by the nearly free electron model and observed in many systems.[41] For x not too close to .2, $A_- \sim 1$, $A_+ \sim 0$, $\omega_- \sim \omega_o = c(x-x_o)^{\frac{1}{2}}$, so that (neglecting for the moment μ^* and noting that $\lambda \ll 1$)

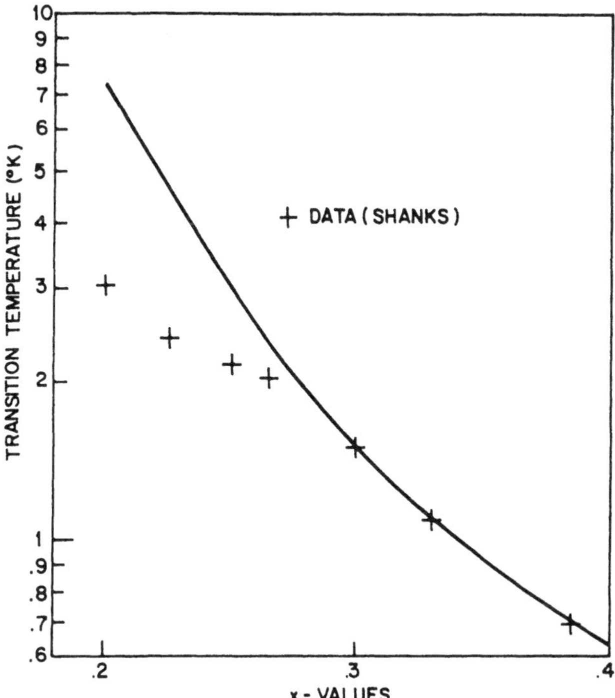

Fig. 8. Numerical evaluation (solid line) of Eq. (66) and the data (circles) of Ref. 8 vs. x. Parameters A and B in Eq. (66) are determined by fitting the high x end of the data points.

T_c will be of the form:

$$T_c(x) = A(x - x_o)^{\frac{1}{2}} e^{B/x} \quad (x >> .2). \tag{66}$$

In Fig. 8, we have plotted Shanks' data[8] on T_c (x) for sodium tungsten bronze in the T1 phase, along with Eq. (66), where A and B are obtained by fitting to the two highest x points. The fit is rather good for $.225 < x < .4$, but for x near .2, Eq. (66) is seen to seriously overestimate the enhancement of T_c due to the soft phonon. We note, however, that for values of x not too close to the T1 - T2 transition, a simple soft phonon enhancement of the electron-phonon coupling can account for the observed enhancement of T_c.

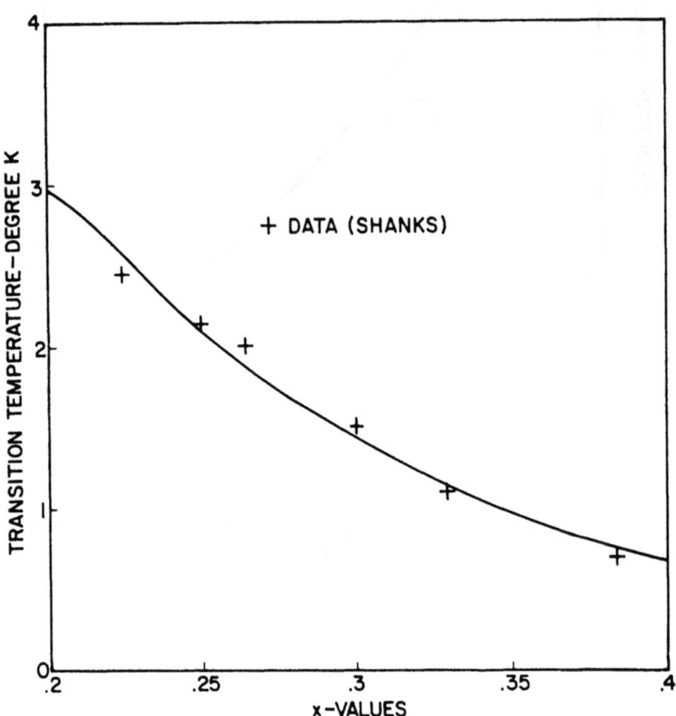

Fig. 9(a). Numerical evaluations of Eq. (62) for the same sets of parameters as used in Fig. 7(a): weak coupling. Data points are again taken from Ref. 8. The result depends little on the choice of ξ from ξ = 1 to ξ = 10.

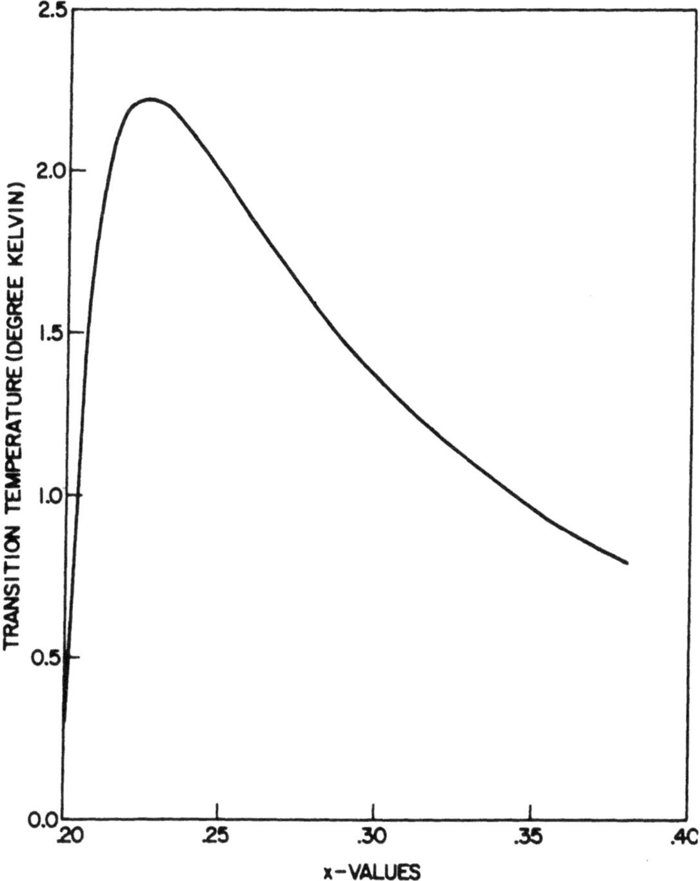

Fig. 9(b). Strong coupling case with the same set of parameters as in Fig. 7(b).

For x near .2, we must go back to Eq. (65), and include both ω_+ and ω_-, for in this region the phonon-configuration coupling will be essential. We can see from Fig. 7 that its effect will be to suppress somewhat the strong enhancement of T_c obtained from the bare soft phonon near x = .2. In Fig. 9 we show numerical results for $T_c(x)$ vs. x with the same sets of values of the parameters as in Fig. 7. The configuration-phonon coupling indeed suppresses the enhancement of T_c. In addition a maximum value of T_c is obtained. In the weak coupling limit, Fig. 9a, this maximum occurs very close to the transitional composition, is not shown, and may not be observable. The choice of parameters of Fig. 9(a) gives good agreement with the data of Ref. 8 for the T1 - T2 transition in Na_xWO_3. In the strong coupling case, shown in Fig. 9(b), $T_c(x)$ vs. x develops a pronounced maximum for x significantly larger than the critical value (assumed here to be x = .2). One might speculate that the reason for the weak coupling behavior in the sodium case is that since T2 is the most favorable configuration for the WO_6 octahedra alone, the barrier illustrated in Fig. 6 should be large. For transitions between two distorted phases, such as T1 and hexagonal for example, the barrier should be smaller due to the fact that the WO_6 free energies of the two phases are almost the same: Thus the strong coupling behavior illustrated in Fig. 9(b) might be expected for transitions not involving the T2 or cubic phases.[44]

To summarize, we have treated the T1 - T2 phase transition in sodium tungsten bronze with a two-well configuration space model for the free energy. Tunneling between the configurations is allowed via a particular phonon, which softens as x approaches the critical composition. Since the change in configurations involves a change in the bonding, the transition is not accomplished simply by the condensation of the soft phonon as, for example, in the case of the Peierls transition[45] or the usual anharmonic soft phonon.[11] Moreover, the phonon softening is obtained here even in the harmonic approximation, since the force constants depend upon x.[38] An interesting feature of the tungsten bronze system is that since the phonon softening occurs as a function of composition rather than temperature, it is possible to have the superconducting transition and lattice instability occur simultaneously. Also, since the phonon involved is essentially a localized excitation, it has a greater effect on T_c than a soft mode with a particular wavevector. In the latter case only a small fraction of the phonon density of states is involved in the enhancement of the electron-phonon interaction.

We have utilized a simple model of both the soft phonon itself and of the phonon-configuration coupling, and conclude for Na_xWO_3 that this coupling is essential very near the transition. Within the simple model we have explicitly calculated the dependence of T_c on x, and have obtained good agreement with the data of Ref. 8. As x is decreased toward the critical composition, the phonon softening enhances T_c dramatically, but this softening is suppressed by the phonon-configuration coupling very near the transition, and T_c then increases more slowly, exhibiting a visible maximum in the strong coupling case. It will be of great interest to directly observe the soft mode that assists the T1 - T2 transition. Earlier Raman scattering studies[46] of metallic tungsten bronzes have been restricted to the higher frequency range at room temperature. More detailed measurements, either by light scattering or neutron scattering, may be necessary to experimentally confirm the x-dependent softening of the rotational mode.

ACKNOWLEDGMENT

It is a pleasure for me to acknowledge the invaluable contributions and collaboration of Professor E. N. Economou to Section 2, and Dr. R. Silberglitt to Section 3 of these notes.

REFERENCES

1. V. L. Ginzburg, Usp. Fiz. Nauk 101, 185 (1970) [Sov. Phys. - Usp. 13, 335 (1970)].

2. V. L. Ginzburg and D. A. Kirzhnits, in Soviet-American Symposium on Electron Theory of Solids, Leningrad, 1971 (unpublished).

3. D. Allender, J. Bray and J. Bardeen, Phys. Rev. B7, 1020 (1973).

4. J. C. Inkson and P. W. Anderson, Phys. Rev. Comments and Addenda B8, 4429 (1973).

5. D. Allender, J. Bray and J. Bardeen, ibid, p. 4434.

6. A. D. Wadsley in "Non-Stoichiometric Compounds", L. Mandelcorn, ed., Academic Press, New York (1964).

7. C. J. Raub, A. R. Sweedler, M. A. Jensen, S. Broadsten and B. T. Matthias, Phys. Rev. Letters 13, 746 (1964); A. R. Sweedler, C. J. Raub and B. T. Matthias, Phys. Letters 15, 108 (1965); J. P. Remeika, T. H. Geballe, B. T. Matthias, A. S. Cooper, G. W. Hall and E. M. Kelly, Phys. Letters 24A, 565 (1967).

8. H. R. Shanks, Solid State Commun. 15, 753 (1974).

9. B. C. Hyde and M. O'Keefe, Acta Cryst. A29, 243 (1973).

10. P. W. Anderson, B. I. Halperin and C. Varma, Phil. Mag. 25, 1 (1972).

11. W. Cochran, Adv. Phys. 9, 387 (1960); P. W. Anderson, Fiz. Dielect., G. I. Skanawi, ed., Academy of Sciences, Moscow (1960).

12. M. H. Cohen and D. H. Douglass, Jr., Phys. Rev. Letters 19, 118 (1967).

13. P. W. Anderson in Proceedings of the NATO Advanced Study Institute on Elementary Excitations in Atoms, Molecules and Solids, Antwerp, Belgium (1973).

14. H. Frohlich, J. Phys. C1, 544 (1968).

15. H. Gutfreund and Y. Unna, J. Phys. Chem. Solids 34, 1523 (1973).

16. A. Rothwarf, Phys. Rev. B2, 3560 (1970).

17. For a review see E. N. Economou and K. L. Ngai, to appear as a chapter in "Advances in Chemical Physics", edited by S. A. Rice and I. Prigogine, John Wiley and Sons, New York (1974).

18. E. N. Economou, Phys. Rev. 182, 539 (1969).

19. J. C. Swihart, J. Appl. Phys, 32, 461 (1961).

20. K. L. Ngai, Phys. Rev. 182, 555 (1969).

21. K. L. Ngai and E. N. Economou, unpublished.

22. P. Morel and P. W. Anderson, Phys. Rev. 125, 1263 (1962).

23. A. L. Fetter, to be published.

24. E. N. Economou, Phys. Rev. 182, 539 (1969).

25. $\chi(Q,\omega)$ does not depend on α because in the RPA χ depends only on the electronic eigenenergy $E_{Q\alpha} = \frac{h^2Q^2}{2m}$ which is independent on α.

26. F. Stern, Phys. Rev. Letters 18, 546 (1967).

27. W. L. McMillan, Phys. Rev. 167, 331 (1968).

28. The quantity λ is expected to change as we go from the bulk to a very thin film structure. Consequently, a calculation of T_c in a given layered material requires the calculation of λ for this layered structure as well as the calculation of μ.

29. C. C. Tsuei and W. L. Johnson, Phys. Rev. B9, 4742 (1974).

30. G. Deutscher, J. P. Farges, F. Meunier and P. Nedellec, Phys. Letters 35A, 265 (1971).

31. K. L. Ngai and E. N. Economou, Phys. Rev. B4, 2132 (1971); K. L. Ngai, E. N. Economou and Morrel H. Cohen, Phys. Rev. Letters 24, 61 (1970).

32. K. L. Ngai, E. N. Economou and Morrel H. Cohen, Phys. Rev. Letters 22, 1375 (1969).

33. A. J. Bevolo, H. R. Shanks, P. H. Sidles and G. C. Danielson, Phys. Rev. B9, 3220 (1974).

34. J. C. Slater, "Quantum Theory of Molecules and Solids", Vol. 2, McGraw Hill, New York (1965).

35. H. R. Shanks, Jour. Cryst. Growth 13/14, 433 (1972).

36. Since the triangular tunnels are too small to contain M atoms, the T1 configuration has available per unit cell one square tunnel and four pentagonal ones, whereas T2 has nine squares. The outer tunnels are shared with other unit cells, so the ratio becomes 3:5, hence x_{max}(T1) = .6.

37. The hexagonal phase consists of only triangular and hexagonal tunnels (see Ref. 6, p. 139), and thus cannot be derived geometrically from T2 in the same manner as T1. The free energy arguments made in the text are nonetheless valid for this phase.

38. Eq. (4) assumes uncorrelated behavior of the M atoms. We note that a linear dependence of ω^2 on x has been observed for a TO phonon in the nonstoichiometric ferroelectric PZT by G. Burns and B. A. Scott, Phys. Rev. Letters 25, 1191 (1970), as described in A. Pinczuk, Solid State Comm. 12, 1035 (1973).

39. Contributions from all possible configurational paths between T1 and T2, or from all combinations of phases of rotating groups of WO_6 octahedra, are included in the tunneling matrix element V(x).

40. A. A. Abrikosov, L. P. Gorkov and I. E. Dzyaloshinskii, "Quantum Field Theoretical Methods in Statistical Physics", second edition, Pergamon, New York (1965).

41. W. L. McMillan, Phys. Rev. 167, 331 (1968). The general form of this equation can be derived analytically from the Eliashberg gap equations, although to obtain the constants numerical calculations based on the phonon spectrum of Nb were used. The results obtained in this paper do not depend on the exact values of these constants.

42. D. J. Scalapino in "Superconductivity", R. D. Parks, ed., Dekker, New York (1969).

43. In Ref. 8 it is pointed out that specific heat measurements on cubic Na_xWO_3 yield N(0) proportional to x, and that preliminary susceptibility measurements in the T1 phase of Na_xWO_3 give the same result. In a rigid band model this would require an extremely rapid variation of N(E) vs. E.

44. Preliminary data on $T_c(x)$ in the hexagonal phase for K_xWO_3 and Rb_xWO_3 suggests behavior of the type shown in Fig. 6(b), H. R. Shanks and M. J. Sienko, private communications.

45. M. J. Rice and S. Strassler, Solid State Commun. 13, 125 (1973); P. A. Lee, T. M. Rice and P. W. Anderson, Solid State Commun. 14, 703 (1974).

46. J. F. Scott, R. F. Leheny, J. P. Remeika and A. R. Sweedler, Phys. Rev. B2, 3883 (1970).

CLASSICAL TRANSPORT IN SMALL-GAP SEMICONDUCTORS

Wlodek Zawadzki

Institute of Physics, Polish Academy of Sciences

Warsaw, Poland

1. INTRODUCTION

One can probably claim that small-gap semiconductors are "better" semiconductors than others; they have smallest effective masses and highest mobilities, and being most sensitive to doping, temperature, magnetic field and pressure, they exhibit typically semiconducting features but to a higher degree. The purpose of these lectures is to review the theory and experiment concerning electron scattering and transport phenomena in these materials. We shall mostly consider III - V and II - VI intermetallic compounds, although the number of small-gap semiconductors includes many more materials.

Our analysis, restricted to the nonquantum region of magnetic fields and omitting hot-carrier effects, is based on the classical Boltzmann equation, which turned out to be exceptionally useful in taking into account the complexities of the band structure, and it is this aspect of transport phenomena that will be emphasized throughout.

We define small-gap semiconductors as materials in which the value of electron energy ε, as counted from the bottom of the conduction band, can become comparable to the energy gap ε_0. The electron effective mass m_e is approximately proportional to the gap, so that the small gap means also the small mass. Since, in turn, the density of states $\rho(\varepsilon) \sim (m_e)^{3/2}$, we deal in small-gap materials with small density of states, and even a relatively small number of the electrons populates the conduction band into quite high energies above the band edge. In this situation the proximity of the valence band is strongly felt and, as a result, the conduction electrons have rather special properties. The most striking feature is that the $\varepsilon(k)$ relation in the band becomes strongly nonparabolic and the Bloch wavefunctions are strongly $\vec{k}$ - dependent. In general the carriers in small-gap materials combine the properties of atomic and free electrons, which opens a variety of new possibilities.

The formalism is developed for a spherical energy band with arbitrary $\varepsilon(k)$dependence and it can be reduced to the description of standard bands by simplifying the final formulas. We shall be concerned with conduction band theory and experiments, since the transport phenomena in degenerate or nearly degenerate valence bands are not really understood as yet. The limited scope of the article does not allow to present all numerous details of the problem. Thus, we work out the formalism using experimental results mainly to illustrate the methods. For the same reason we do not discuss all the variations of thermomagnetic effects as well as the phenomena due to gradients of carrier concentration.

Our approach is based on the relaxation time approximation. Considerable amount of work has been devoted to the limitations of this concept. There are many attempts to go beyond it, some of them successful. Still, it seems that the relaxation time approximation gives the only working general theory. In the section devoted to scattering we mention the restrictions on its validity. However, it has been the general experience that the Boltzmann equation with scattering term in the form of the relaxation time is valid much further than it is supposed to.

A few words should be said about the highlights in the history of the subject. Wilson [1] in his well known book on metals derived many formulas in terms of the general density of states specializing them subsequently for a band of the standard form. Probably the first experimental indication of a strong band's non--parabolicity in a semiconductor was given in 1954 by Hrostowski et al. [2] from optical data in InSb. The first serious attempt to take into account the general shape of energy bands,both in the transport integrals and the relaxation time, was made by Barrie [3], who was also the first to recognize that the usual effective mass, described for a parabolic band by the second derivative of energy with respect to momentum, is not valid in the general case. Similar problems were independently studied by Radcliffe [4]. In 1957 Kane [5] published his famous paper, deriving explicitly non-parabolic $\varepsilon(k)$ dependence for the conduction band in InSb, which greatly stimulated experimental and theoretical investigations on the effects of non-parabolicity. Almost at the same time there appeared farreaching and flawless analyses of Ehrenreich [6,7] who considered scattering of electrons by heavy holes and polar optical modes in intrinsic InSb and correlated the theory with existing experimental data. Kane's band model for InSb was directly confirmed by Spitzer and Fan [8], and Smith et al.[9] from measurements of the free-carrier reflectivity and the Faraday rotation. Shortly after, Kolodziejczak showed that the transport theory for non-parabolic but spherical bands can be developed in quite a general way, both for dc [10] and optical [11] phenomena, describing the parabolic band as a special case. This was followed by an important paper with Sosnowski [12], in which it was demonstrated that band's non-parabolicity can have an essential influence on thermoelectric effects, reversing even the sign of some of them. Copious theoretical work in Warsaw extended the formalism to other phenomena and more complex energy bands. The theory was in turn followed by experimental investigations with III-V, II-VI and IV-VI intermetallic compounds, mainly in Leningrad and Warsaw. An observation of Rodot [13] that the thermoelectric power at high magnetic fields does not depend on relaxation time turned out to be particularly useful in tracing shapes of energy bands in many materials. In the last few years there has been a big increase of interest in the subject, due to the appearence of zero-gap materials of α - Sn -type (Groves and Paul [14]) and the theoretical progress in the field. The relaxation times for various scattering modes in the non-parabolic range of electron energies were calculated explicitly for InSb-type compounds by Korenblit and Sherstobitov [15] and by Zawadzki and Szymańska [16], who also used the theory to describe all existing transport data in InSb [17]. A similar work was carried out by Ravich and Morgovskii [18] for lead chalcogenides. Recently Liu and coworkers [19,20] predicted singularities in the dielectric constant of zero--gap materials, which allowed Broerman [21] to explain very high electron mobilities observed in these crystals at low temperatures. The latest progress in the field has been summarized at the Warsaw Semiconductor Conference [22]. An overall review of transport problems in small-gap materials is given in ref.[23], while questions particular to lead chalcogenides have been discussed by Ravich and coworkers [24].

2. GENERAL FORMALISM

We consider a spherical energy band with arbitrary non-parabolicity $\varepsilon(\vec{k}) = \varepsilon(k)$, where ε is the electron energy and $\vec{k}$ its wave vector. Let us calculate a statistical integral of an arbitrary function of energy $\chi(\varepsilon)$ over the band [25]. Observing that in the spherical coordinate system $d^3k = k^2(dk/d\varepsilon)d\varepsilon d\Omega$, we obtain after two-fold integration by parts,

$$\frac{2}{(2\pi)^3}\int f_o\chi d^3k = \frac{1}{\pi^2}\int f_o\chi\frac{dk^3}{d\varepsilon}\frac{d\varepsilon}{dk}dk = \frac{1}{3\pi^2}\int\left(-\frac{\partial f_o}{\partial\varepsilon}\right)\chi k^3 d\varepsilon -$$

$$-\frac{1}{3\pi^2}\int f_o\frac{d\chi}{d\varepsilon}k^3 d\varepsilon = \frac{1}{3\pi^2}\langle\chi\rangle - \frac{1}{3\pi^2}\left\langle\int_0^\varepsilon \chi' d\xi\right\rangle \quad (2.1)$$

The bracket notation, which shall be used in the following, is defined as

$$\langle A\rangle = \int_0^\infty \left(-\frac{\partial f_o}{\partial\varepsilon}\right)\hat{A}k^3(\varepsilon)d\varepsilon \quad (2.2)$$

$$\text{where}\quad f_o = [\exp(z-\eta)+1]^{-1} \quad (2.3)$$

is the Fermi-Dirac distribution function. $z = \varepsilon/k_0T$ and $\eta = \zeta/k_0T$ are the reduced energy and the reduced Fermi energy, respectively. The order of factors in the integrand (2.2) should be observed, since $\hat{A}$ is in general an operator. For a system obeying the Maxwell-Boltzmann (M-B) statistics, the integral (2.1) can be calculated simply, using the relation

$f_o = -\partial f_o/\partial z$, to give

$$\frac{2}{(2\pi)^3}\int f_o\chi d^3k = \frac{1}{3\pi^2}\left\langle\chi\frac{d}{dz}\right\rangle \quad (2.4)$$

According to the definition (2.2), this means

$$\left\langle\chi\frac{d}{dz}\right\rangle = \int_0^\infty\left(-\frac{\partial f_o}{\partial z}\right)\chi(\varepsilon)\frac{d}{dz}k^3(\varepsilon)d\varepsilon \quad (2.5)$$

We shall be often using the average values, defined as

$$\overline{A} = \langle A \rangle / \langle 1 \rangle \tag{2.6}$$

Region of high degeneracy of the electron gas (big η values) is of special interest. In this case one can avoid the numerical integrations by using a generalized Bethe-Sommerfeld expansion

$$\langle A \rangle = 2\sum_{l=o}^{\infty} \left. \frac{d^{2l}C}{dz^{2l}} \right|_{z=\eta} \frac{\pi^{2l} \left| (2^{2l-1}-1)B_{2l} \right|}{(2l)!} \tag{2.7}$$

where $C = Ak^3(\varepsilon)$, and B_{2l} are the Bernoulli numbers. In general this expansion omits a contribution of the order of exp $(-\eta)$. For large η values the series is quickly converging and one can retain only the first two terms, i.e.

$$\langle A \rangle \approx \left(C + \frac{\pi^2}{6} \frac{d^2C}{dz^2}\right)_{z=\eta} \tag{2.8}$$

The nonparabolicity of the band is due to the fact that $k^2(\varepsilon)$ is not in general proportional to ε. For not too large values of energy, a formal expansion gives

$$k^2(\varepsilon) = \sum_{l=1} \lambda_l \varepsilon^l = \lambda_1 \left(\varepsilon + \sum_{l=2} \frac{\lambda_l}{\lambda_1} \varepsilon^l\right) \tag{2.9}$$

where $\lambda_l = \frac{1}{l!} \left(\frac{d^l k^2}{d\varepsilon^l}\right)_{\varepsilon=o}$ (2.10)

It can be seen that for a parabolic band all λ_l but first vanish. If the expansion (2.9) is quickly converging, one may use the first nonparabolic approximation

$$k^2(\varepsilon) = \lambda_1 \varepsilon \, (1+\lambda\varepsilon) \tag{2.11}$$

where $\lambda = \lambda_2/\lambda_1$. In the following we shall bo mostly concerned with energy bands described by Eq. (2.11).

Let us now calculate the velocity af a carrier in the band,

$$v_i = \frac{1}{\hbar}\frac{\partial \varepsilon}{\partial k_i} = \frac{1}{\hbar}\frac{d\varepsilon}{dk}\frac{\partial k}{\partial k_i} = \frac{1}{\hbar}\frac{d\varepsilon}{dk}\frac{k_i}{k} = \frac{1}{\hbar k}\frac{d\varepsilon}{dk}\delta_{ij}k_j \tag{2.12}$$

where we use the sum convention over repeated indices. Defining the inverse effective mass tensor, relating velocity to pseudomomentum

$$v_i = m_{ij}^{-1} \hbar k_j \tag{2.13}$$

we obtain

$$m^{-1}_{ij} = \frac{1}{\hbar^2 k} \frac{d\varepsilon}{dk} \delta_{ij} \tag{2.14}$$

Thus, the effective mass $1/m = (1/\hbar^2 k)(d\varepsilon/dk)$ has scalar properties in a spherical energy band. It can be seen that, in general, the effective mass depends on energy, unless the band is parabolic. One can also define an effective mass tensor relating force to acceleration

$$\frac{1}{M_{ij}} = \frac{1}{\hbar^2} \frac{\partial^2 \varepsilon}{\partial k_i \partial k_j} \tag{2.15}$$

It can be verified, that for the parabolic band the two masses coincide. However, in general they are not identical and it is the momentum effective mass that should be regarded as basic quantity. This is due to the fact that, as we shall demonstrate, it is the momentum mass that appears in the description of transport phenomena and all other properties of the electron gas in a band with arbitrary non-parabolicity. In this sense the common use of the M^{-1}_{ij} effective mass in the theory of semiconductors and metals with parabolic band structures should be regarded as accidental.

3. REAL CONDUCTION BANDS

Now we review briefly the real conduction bands of important small-gap materials in the vicinity of the lowest minimum in the Brillouin zone, in order to verify to what extent the assumed model is adequate for their description.

3.1 InSb-type III-V Compounds

The band structure near the center of the Brillouin zone has been described by Kane [5], who took exactly into account $\vec{k}.\vec{p}$ interaction of the s - like conduction Γ_6 level with the p - like Γ_8 valence level, and the p - like Γ_7 level split-off by the spin-orbit interaction, treating all other bands as a perturbation according to the effective mass approximation procedure. For not very high energies the effect of higher bands is negligible, and the conduction and valence bands are described by the following equation (the zero of energy is at the bottom of the conduction band),

$$\varepsilon'(\varepsilon'+\varepsilon_o)(\varepsilon'+\varepsilon_o+\Delta)-k^2P^2(\varepsilon'+\varepsilon_o+\tfrac{2}{3}\Delta)=0 \tag{3.1}$$

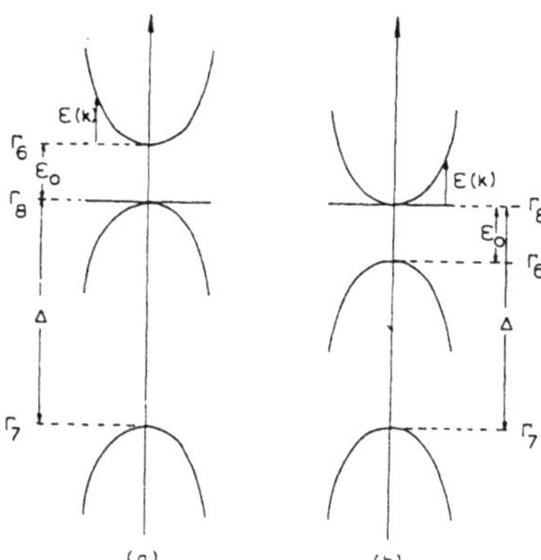

Fig.1 The energy band structure at the Γ point of the Brillouin zone: a) InSb-type semiconductors, b) HgTe-type semiconductors. In the three-level model of Γ_6, Γ_8, and Γ_7 levels the shape of the conduction band in both cases is almost the same.

where $\varepsilon' = \varepsilon - \hbar^2 k^2/2m_0$. Here m_0 - the free electron mass, ε_0 - the energy gap, Δ - the spin-orbit splitting, and $P = -i(\hbar/m_0)\langle S|p_z|Z\rangle$ is the matrix element of momentum, accounting for the coupling between the conduction and valence bands. For $\varepsilon' \ll \varepsilon_0 + (2/3)\Delta$ the resulting equation becomes

$$\varepsilon(1+\varepsilon/\varepsilon_0) = \hbar^2 k^2/2m_e \qquad (3.3)$$

and $m = m_e\,(1+2\varepsilon/\varepsilon_0)$ (3.5)

with $1/m_e = (4P^2/3\hbar^2\varepsilon_0)\,(\frac{2}{3}\Delta+\varepsilon_0)\,/\,(\Delta+\varepsilon_0)$ (3.4)

The small free-electron term has been neglected. This dispersion relation is of the form (2.11) with $\lambda = 1/\varepsilon_0$. Solving Eq.(3.2) we get for electrons

$$\varepsilon = -\frac{\varepsilon_0}{2} + \left[\left(\frac{\varepsilon_0}{2}\right)^2 + \varepsilon_0\frac{\hbar^2 k^2}{2m_e}\right]^{1/2} \qquad (3.5)$$

Keeping the free-electron term, it can be seen that the effective mass at ε=o is $1/M = 1/m_0 + 1/m_e$, but since $m_e \ll m_0$ this term is really small and the above approximation is justified. In InSb at low temperature $m_e = 0.0145 m_0$, $\varepsilon_0 = 0.236$ eV, and $\Delta \approx 0.9$ eV [26], so that to a good approximation one can assume $\Delta \ll \varepsilon_0$, and simplify the above relations. Since the effect of the split-off valence level on the $\varepsilon(k)$ dependence in the conduction band is opposite to the effect of higher bands, it has been the general experience with InSb and other small-gap materials that a good approximation is obtained by simultaneously assuming an infinite Δ and neglecting the effect of higher bands.

Kane's theoretical description is essentially valid for the conduction bands of other III-V compounds, although the relative influence of other bands may be stronger than in InSb. Eq. (3.4) may be used to estimate P^2, if m_e, ε_0 and Δ are known. In various III-V compounds the value of P^2 turned out to be almost the same: $E_p = (2m_0/\hbar^2)P^2 \approx 20$ eV. Sometimes a quantity $Q = (2/3)E_p$ is used.

Fig. 2 presents a compilation of experimental data on the concentration dependence of the electron effective mass in InSb at room temperature. The solid line

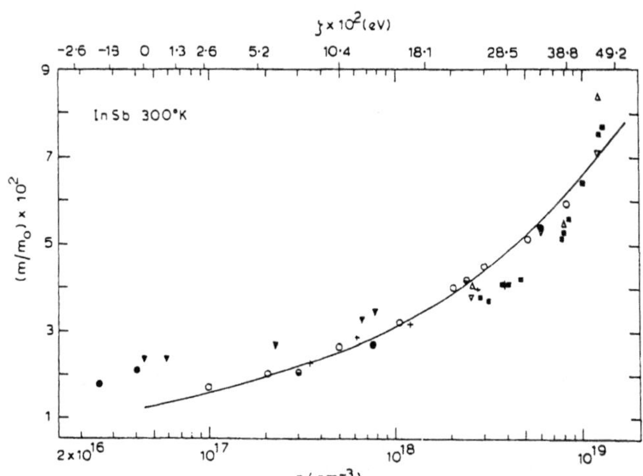

Fig.2 Electron effective mass in InSb at room temperature, as measured by various authors, versus free electron concentration. The solid line, calculated for the Kane two-level model, represents the mass values at the Fermi energy, as indicated on the upper abscissa. At lower electron concentrations (non-degenerate limit) the mass averages obtained from optical measurements are higher than those determined from thermomagnetic experiments.

is calculated using the two-band model, that is Eq.(3.3) with ε_o = 0.20 eV and m_e = $0.0126m_o$, and it indicates the effective mass at the Fermi level for a given electron concentration. The Fermi energies shown on the upper abscissa correspond to the electron concentration given below. It can be seen that the two-level model adequately describes the experimental data until highest achievable values of the Fermi energy. At lower concentrations the electron gas is not strongly degenerate and, due to the nonparabolicity of the band, various effects yield different mass averages.

3.2. HgTe - type Materials

Until present HgTe, HgSe, β -HgS, and some $Hg_{1-x}Cd_xTe$ alloys are known to have the band ordering of α -Sn proposed by Groves and Paul [14]. In this ordering Γ_6 -like level is below Γ_8 -like level and forms the light-hole band. The light-carrier subband of Γ_8 which is the light-hole band in InSb, here forms the conduction band, whereas the heavy-carrier subband of Γ_8 remains the valence band of heavy holes. Thus, the conduction and heavy-hole bands are degenerate at k=o and the thermal gap ε_g is zero at all temperatures. Still, the interacting levels Γ_6 and Γ_8 are separated by a finite energy ε_o. This situation is demonstrated in Fig. 1 b. A procedure analogous to that for InSb yields,

$$\varepsilon'(\varepsilon'+\varepsilon_o)(\varepsilon'+\Delta) - k^2P^2(\varepsilon'+\frac{2}{3}\Delta) = 0 \qquad (3.7)$$

and for $\varepsilon' \ll 2\Delta/3$ the dispersion relation (3.5) is valid, with $1/m_e = 4P^2/3\hbar^2\varepsilon_o$.

Fig. 3 presents the effective mass of conduction electrons at the Fermi energy as a function of the electron concentration in the band. The solid line is calculated with the use of the parameters: ε_o=0.22 eV, P=7.2 eVcm , and Δ=0.45 eV [27], including higher bands to the order of k^2 terms. Inverted Kane's band model describes the experimental data up to the highest achievable concentration of $n = 3\times10^{19}$ cm^{-3}, for wich the Fermi energy rises about 0.7 ev above the bottom of the band.

Solid solutions of mercury and cadmium telluride offer very interesting possibilities, CdTe having the InSb-type sequence of levels. By changing the chemical composition of mixed crystals $Hg_{1-x}Cd_xTe$ one can go continuously from HgTe to CdTe level ordering and at

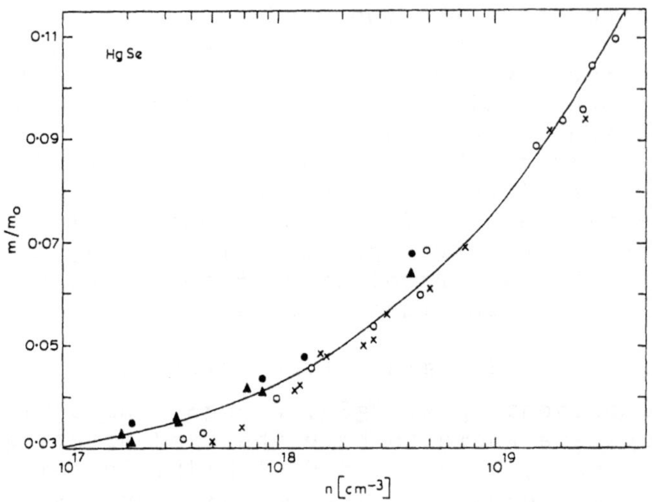

Fig.3 Electron effective mass, as measured by various authors, in HgSe *versus* free electron concentration. The solid line is calculated for the Groves and Paul three-level model, including higher bands to k^2 terms. [After W.Zawadzki, Proc. 11 Intern.Conf. Phys. Semicond.,Pol.Scient.Publ. Warsaw 1972, p. 87].

$x \approx 17$ % molar (at 4° K) the energy separation between Γ_8 and Γ_6 disappears, with triple degeneracy at $k = 0$. At this point m_e tends to zero together with ε_0. As follows from Eqs. (3.4) and (3.5), in this case

$$\varepsilon = (2/3)^{1/2} P.k \tag{3.8}$$

i.e. the electron energy is proportional to its momentum. The effective mass becomes

$$m = (3/2)^{1/2} (\hbar^2/P).k \tag{3.9}$$

Such a dispersion relation has been actually observed experimentally by Gałazka and Sosnowski [28]. A band of this kind can be described with Eq. (2.11) by putting a large value of λ .

3.3 Lead Chalcogenides

The conduction band minima in PbS, PbSe, PbTe, PbSnTe, and PbSnSe occur at the L point of the Brillouin zone, the complete band having cubic symmetry. It appears that for not too high values of electron energy the conduction bands can be described by nonparabolic spheroids,

$$\varepsilon(1+\frac{\varepsilon}{\varepsilon_o}) = \frac{\hbar^2 k_\perp^2}{2m_\perp} + \frac{\hbar^2 k_\parallel^2}{2m_\parallel} \qquad (3.10)$$

the subscripts referring to the main spheroid axis. The transport theory can be formulated also in this case, using general description of the band in the form [29]: $\gamma(\varepsilon) = a_{\alpha\beta}k_\alpha k_\beta$, where arbitrary function $\gamma(\varepsilon)$ takes into account band's nonparabolicity. However, the procedure in this case is somewhat tedious mathematically and we do not report it here [30].

4. THERMODYNAMIC PROPERTIES OF ELECTRON GAS [31]

Concentration of free electrons in the band per unit volume of the crystal is obtained from Eq. (2.1) by putting $\chi(\varepsilon) = 1$,

$$n = (1/3\pi^2)\ \langle 1\rangle \qquad (4.1)$$

For a nonparabolic band described by Eq. (3.2), this becomes

$$n = \frac{1}{3\pi^2}\left(\frac{2m_e k_o T}{\hbar^2}\right)^{3/2}\ {}^{o}L_{o}^{3/2}(\eta,\beta) \qquad (4.2)$$

where $\beta = k_oT/\varepsilon_o$ and the generalized Fermi integrals ${}^{n}L_{k}^{m}$ are defined (Kołodziejczak [32])

$${}^{n}L_{k}^{m}(\eta,\beta) = \int_0^\infty \left(-\frac{\partial f_o}{\partial z}\right) z^n (z+\beta z^2)^m (1+2\beta z)^k dz \qquad (4.3)$$

Their properties have been investigated in ref. [33]. A transition to the parabolic band is made by putting $\beta = o$. In the region of M-B statistics one transforms Eq. (4.2) into

$$n_{M-B} = \frac{1}{4\pi^2}\left(\frac{2m_e k_o T}{\hbar^2}\right)^{3/2} \beta^{-1/2} \exp(\eta+\frac{1}{2\beta}) K_2(\frac{1}{2\beta}) \qquad (4.4)$$

where K_2 is the Mcdonald function [34]. Using the asymptotic relation $K_n(x) \approx (\pi/2x)\exp(-x)$ for large x, one can obtain for $\beta \to o$ the classical exponential formula for nondegenerate carrier density in a parabolic band.

Internal energy of the free carriers in the band is obtained by putting in the formula $\chi(\varepsilon) = \varepsilon$. Hence,

$$U = 4\pi k_o T \left(<z> - <\int_0^z d\xi>\right) \quad (4.5)$$

In case of M-B statistics one finds from Eq. (2.4),

$$U_{M-B} = nk_o T \left(\overline{z\frac{d}{dz}}\right) \quad (4.6)$$

This can also be expressed by the Mcdonald functions in the form,

$$\frac{U_{M-B}}{n} = \frac{3}{2} k_o T + \frac{k_o T}{2\beta} \left[K_1\left(\frac{1}{2\beta}\right)/K_2\left(\frac{1}{2\beta}\right) + 3\beta - 1\right] \quad (4.7)$$

It can be seen from Eq.(4.7) that for a nonparabolic band we do not have the classical value of $U_{M-B} = (3/2)nk_o T$. The latter is obtained in the limit of $\beta \to o$.

The free energy of the system at a constant volume can be found similarly to the internal energy [35], to give

$$F = nk_o T \left[\eta - \frac{2}{3}\frac{<z>}{<1>} + \frac{2}{3} < \int_0^z d\xi > / <1> \right] \quad (4.8)$$

and

$$F_{M-B} = nk_o T \left[\eta - \frac{2}{3}\left(\overline{z\frac{d}{dz}}\right)\right] \quad (4.9)$$

The entropy is

$$S = nk_o \left[\frac{5}{3}\frac{<z>}{<1>} - \frac{5}{3} < \int_0^z d\xi > / <1> - \eta \right] \quad (4.10)$$

$$S_{M-B} = nk_o \left[\frac{5}{3} \left(\overline{z\frac{d}{dz}}\right) - \eta\right] \quad (4.11)$$

The specific heat at a constant volume and carrier density independent of temperature can be calculated using the relation

$$(\partial f_o/\partial k_o T) = (-\partial f_o/\partial \varepsilon)(\varepsilon - \zeta) \quad (4.12)$$

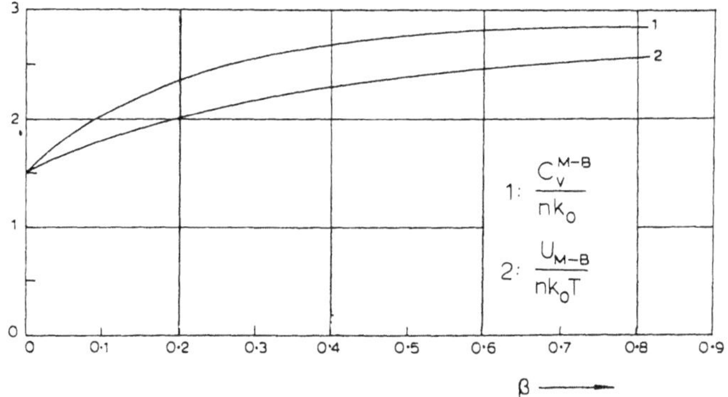

Fig. 4 Mean internal energy U and specific heat C_V of the nondegenerate electron gas (Maxwell-Boltzmann statistics) in a first-order nonparabolic band versus $\beta = k_o T/\varepsilon_o$. For $\beta = o$ the classical values of (3/2) are obtained.

We obtain

$$C_v = \frac{2}{(2\pi)^3}\frac{d}{dT}\int f_o \varepsilon d^3k = k_o n(\overline{\frac{d}{dz}})\;(\overline{z^2}-\bar{z}^2) \qquad (4.13)$$

where

$$\overline{z^2} = \langle z^2 \frac{d}{dz}\rangle / \langle\frac{d}{dz}\rangle \qquad \bar{z} = \langle z\frac{d}{dz}\rangle / \langle\frac{d}{dz}\rangle \qquad (4.14)$$

For M-B statistics

$$C_V^{M-B} = k_o n\;(\overline{z^2}-\bar{z}^2) \qquad (4.15)$$

In terms of the L integrals

$$\frac{C_v}{nk_o} = \frac{3}{2}\,\frac{{}^oL_1^{1/2}}{{}^oL_o^{3/2}}\left[\frac{{}^2L_1^{1/2}}{{}^oL_1^{1/2}} - \left(\frac{{}^1L_1^{1/2}}{{}^oL_1^{1/2}}\right)^2\right] \qquad (4.16)$$

Again, in the general case we do not obtain the classical value $C_V^{M-B} = (3/2)nk_o$; this is true only for $\beta = o$, i.e.

for the parabolic band. The mean internal energy and the specific heat *versus* $\beta = k_0T/\varepsilon_0$ for the electron gas obeying M-B statistics are plotted in Fig.4.

Expansion (2.8) can be used to evaluate the specific heat in strongly degenerate conditions. The result is,

$$C_V = \left(\frac{\pi}{3}\right)^{2/3} \frac{k_o^2 T}{\hbar^2} \quad m\, n^{1/3} \tag{4.17}$$

where m is defined in Eq.(2.14). This is a direct generalization of the formula for a parabolic band, with a constant effective mass replaced in the above formula by the variable mass taken at the Fermi energy.

5. THE BOLTZMANN TRANSPORT EQUATION

In this section we solve the linearized Boltzmann transport equation for electrons in a spherical band with arbitrary non-parabolicity, in the presence of external gradients and a magnetic field. We do not elaborate on details of the procedure, since similar methods are presented in many texts [36]. Limitations and assumptions introduced by representing collision term in a form of the relaxation time are discussed later.

The general equation for the electron distribution function in the stationary state is,

$$\frac{df}{dt} = \frac{\partial f}{\partial t} + \left(\frac{\partial f}{\partial t}\right)_{\text{fields}} + \left(\frac{\partial f}{\partial t}\right)_{\text{coll.}} = 0 \tag{5.1}$$

describing the total variation of the distribution $f(\vec{r},\vec{k})$ due to external forces (fields) and internal forces (collisions). The term accounting for the influence of external fields is,

$$(\partial f/\partial t)_{\text{fields}} = \dot{\vec{k}}.\nabla_k f + \vec{v}.\nabla_r f \tag{5.2}$$

We write the collision term using an isotropic, energy-dependent relaxation time

$$(\partial f/\partial t)_{\text{coll.}} = (f - f_o)/\tau \tag{5.3}$$

f_0 is the distribution function in absence of external fields, given by Eq. (2.3). For the non-equilibrium concentrations present in the material, the Fermi level is to be replaced by a quasi-Fermi level, separate for each band. In the stationary state, if the external fields do not depend on time, there is $\partial f/\partial t=0$, and the equation (5.1) may be written as follows,

$$(1/\hbar)\vec{F}.\nabla_k f+\vec{v}.\nabla_r f+\Delta f/\tau=0 \qquad (5.4)$$

where $\vec{F}$ is the external force. In the presence of applied electric and magnetic fields $\vec{F}$ is the Lorentz force,

$$\vec{F} = q\,\vec{E} + (q/c)\,(\vec{v}\times\vec{H}) \qquad (5.5)$$

where q = - e for electrons, e being the absolute value of the electron charge. A solution of Eq. (5.4) is assumed to be of the form,

$$f = f_o + \Phi\,(\vec{k},\vec{L})\,(\partial f_o/\partial\varepsilon) \qquad (5.6)$$

$\vec{L}$ denotes the external fields (magnetic field and the gradients). A restriction to not very strong fields allows us to assume,

$$\nabla_r f\approx\nabla_r f_o \qquad (5.7)$$

Let us also consider the gradients of temperature and carrier concentration in the crystal. Then,

$$\frac{df_o}{dx_1} = -\frac{\partial f_o}{\partial\varepsilon}\left[\frac{\varepsilon-\zeta}{T}\frac{dT}{dx_1}+\left(\frac{\partial\zeta}{\partial T}\right)_n\frac{dT}{dx_1}+\left(\frac{\partial\zeta}{\partial n}\right)_T\frac{dn}{dx_1}\right] \qquad (5.8)$$

As it has been mentioned, we want to consider the linear Boltzmann equation. To this end, the electric field is assumed to be not very strong, so that one is in the range of Ohm's law. In this connection the non-linear terms $E.\Phi$ and $E.\nabla\Phi$ are neglected, and finally an equation for the unknown function Φ is obtained as,

$$(q/\hbar c)\,(\vec{v}\times\vec{H}).\nabla_k\Phi+\vec{v}.\vec{W}+\Phi/\tau=0 \qquad (5.9)$$

where

$$\vec{W}=q\vec{E}'-C\nabla_r T-\left(\frac{\partial\zeta}{\partial n}\right)_T\nabla_r n \qquad (5.10)$$

with

$$C=(\varepsilon-\zeta)/T \quad \text{and} \quad \vec{E}'=\vec{E}-(1/q)(\nabla_r\zeta)_n \qquad (5.11)$$

Let us look for the solution of Eq.(5.9) in the form,

$$\Phi(\varepsilon) = \vec{X}(\varepsilon).\vec{k}(\varepsilon) \qquad (5.12)$$

Remembering, that $\vec{k} = (m/\hbar)\vec{v}$, one obtains

$$\left(\frac{q}{\hbar c}\right)(\vec{v}\times\vec{H}).\vec{\chi}+\vec{W}.\vec{v}-\frac{m}{\hbar\tau}\vec{\chi}.\vec{v}=0 \qquad (5.13)$$

The velocity $\vec{v}$ can now be eliminated and we finally get,

$$\vec{\chi}=\vec{A}+\vec{\phi}\times\vec{\chi} \qquad (5.14)$$

with $\vec{\phi}=u\vec{H}/c$ and $\vec{A}=(\hbar\tau/m)\vec{W}$. Clearly for $H=0$ there is simply $\vec{\chi}=\vec{A}$. However, for $H\neq 0$ the problem remains to determine $\vec{\chi}$ from the above equation. Upon observing that $\vec{\phi}.\vec{\chi}=\vec{\phi}.\vec{A}$, one has

$$\vec{\chi} = \vec{A}+\vec{\phi}\times\vec{\chi} = \vec{A}+\vec{\phi}\times(\vec{A}+\vec{\phi}\times\vec{\chi}) = \vec{A}+\vec{\phi}\times\vec{A}+\vec{\phi}\times(\vec{\phi}\times\vec{\chi}) =$$

$$= \vec{A}+(\vec{\phi}\times\vec{A})+(\vec{\phi}.\vec{\chi})\vec{\phi}-\phi^2\vec{\chi} = \vec{A}+(\vec{\phi}\times\vec{A})+(\vec{\phi}.\vec{A})\vec{\phi}-\phi^2\vec{\chi}$$

So that, $$\vec{\chi} = \frac{\vec{A}+\vec{\phi}\times\vec{A}+(\vec{\phi}.\vec{A})\vec{\phi}}{1+\phi^2} \qquad (5.15)$$

In the following we will use a quantity $\vec{x} = (m/\hbar\tau)\vec{\chi}$, which is given by Eq. (5.15) with $\vec{A}$ replaced by $\vec{W}$,

$$x_\nu=g_{\nu\mu}W_\mu \qquad (5.16)$$

where $$g_{\nu\mu}=(\delta_{\nu\mu}-\varepsilon_{\nu\mu\zeta}\phi_\zeta+\phi_\nu\phi_\mu)\frac{1}{1+\phi^2} \qquad (5.17)$$

$\delta_{\nu\mu}$ is the Kronecker delta function and $\varepsilon_{\nu\mu\zeta}$ the completely antisymmetric Levi-Civita tensor.

As follows from the above reasoning, the correction to the distribution function, due to external and internal forces, is

$$f_1 = (-\partial f_o/\partial\varepsilon)\vec{\chi}.\vec{k} \qquad (5.18)$$

One can now calculate the vector of electric current,

$$I_\nu = \frac{2}{(2\pi)^3}\int f_1 q v_\nu d^3k = \frac{2}{(2\pi)^3}\int \left(-\frac{\partial f_o}{\partial \varepsilon}\right)\frac{\hbar\tau}{m}$$

$$\times (\vec{X}.\vec{k})k_\nu \frac{\hbar q}{m} d^3k = \frac{2}{(2\pi)^3}\int \left(-\frac{\partial f_o}{\partial \varepsilon}\right)u x_i \frac{k_i k_\nu}{k^2} k^3 d\varepsilon d\Omega =$$

$$= \frac{1}{3\pi^2}\int \left(-\frac{\partial f_o}{\partial \varepsilon}\right)u x_\nu k^3 d\varepsilon \qquad (5.19)$$

and the vector of heat current (electronic part),

$$Q_\nu^e = \frac{1}{(2\pi)^3}\int f_1(\varepsilon-\zeta)v_\nu d^3k = \frac{1}{3\pi^2 q}\int \left(-\frac{\partial f_o}{\partial \varepsilon}\right)u(\varepsilon-\zeta)x_\nu k^3 d\varepsilon \qquad (5.20)$$

where $u=q\tau/m$ is carrier's mobility. Using Eq. (5.10) we can now express the currents by external fields,

$$I_\nu = \sigma_{\nu\mu}E'_\mu - \theta_{\nu\mu}\frac{\partial T}{\partial x_\mu} - qD_{\nu\mu}\frac{\partial n}{\partial x_\mu} \qquad (5.21)$$

$$Q_\nu = \chi_{\nu\mu}E'_\mu - (\xi_{\nu\mu}+\kappa^1_{\nu\mu})\frac{\partial T}{\partial x_\mu} - qG_{\nu\mu}\frac{\partial n}{\partial x_\mu} \qquad (5.22)$$

where $\chi_{\nu\mu}(\vec{H}) = T\theta_{\mu\nu}(-\vec{H})$ (5.23)

$$D_{\nu\mu} = \frac{1}{q^2}\left(\frac{\partial\zeta}{\partial n}\right)_T \sigma_{\nu\mu} \qquad G_{\nu\mu} = \frac{T}{q^2}\left(\frac{\partial\zeta}{\partial n}\right)_T \theta_{\nu\mu} \qquad (5.24)$$

and the three basic transport tensors are,

$$\sigma_{\nu\mu}=\frac{q}{3\pi^2}<ug_{\nu\mu}>=qn\frac{<ug_{\nu\mu}>}{<1>} \tag{5.25}$$

$$\theta_{\nu\mu}=k_o n\frac{<(z-\eta)ug_{\nu\mu}>}{<1>} \qquad \xi_{\nu\mu}=\frac{k_o^2T}{q}n\frac{<(z-\eta)^2ug_{\nu\mu}>}{<1>} \tag{5.26}$$

In Eq. (5.22) we have included also the lattice part of the heat conductivity $\kappa^{l}_{\nu\mu}$. In the following we shall omit the effects due to carrier concentration gradients, described by the diffusion tensors $D_{\nu\mu}$ and $G_{\nu\mu}$ (cf. ref. [37]).

Omitting the diffusion terms, Eqs. (5.21) and (5.22) can be rewritten in a more traditional, phenomenological form. Expressing the field intensities by the currents, we get

$$E'_{\nu}=(\rho_{\nu\mu}+\alpha_{\nu\alpha}\kappa^{-1}_{\alpha\beta}\pi_{\beta\mu})I_{\mu}-\alpha_{\nu\alpha}\kappa^{-1}_{\alpha\mu}Q_{\mu} \tag{5.27}$$

$$\frac{\partial T}{\partial x_{\mu}}=\kappa^{-1}_{\nu\mu}\pi_{\alpha\mu}I_{\mu}-\kappa^{-1}_{\nu\mu}Q_{\mu} \tag{5.28}$$

with

$$\rho_{\nu\mu}=\sigma^{-1}_{\nu\mu} \qquad \text{resistivity tensor,}$$

$$\alpha_{\nu\mu}=\sigma^{-1}_{\nu\beta}\theta_{\beta\mu} \qquad \text{thermoelectric tensor}$$

$$\pi_{\nu\mu}=\chi_{\nu\beta}\sigma^{-1}_{\beta\mu} \qquad \text{Peltier tensor} \qquad (5.29)$$

$$\kappa_{\nu\mu}=\xi_{\nu\mu}-\chi_{\nu\beta}\alpha_{\beta\mu}+\kappa^{1}_{\nu\mu} \qquad \text{thermal conductivity tensor}$$

where the relation $\pi_{\nu\mu}(H) = T\alpha_{\mu\nu}(-H)$ holds. It can be seen that all the phenomenological transport tensors in Eqs. (5.27) and (5.28) can be obtained by multiplication and addition of basic transport tensors $\sigma_{\nu\mu}$, $\theta_{\nu\mu}$ and $\xi_{\nu\mu}$ and their inverse tensors. Eqs.(5.27) and (5.28) with the help of Eqs. (5.23), (5.25) and (5.26) form the final set which will be used for discussion of the transport phenomena.

Assuming isothermal conditions: $\partial T/\partial x_{\nu}=0$, one obtains from Eqs. (5.27) and (5.28): $E = \rho_{\nu\mu}I_{\mu}$. Thus, the tensor $\rho_{\nu\mu}$ describes the galvanomagnetic effects (the electrical resistance, the transverse and planar Hall effects, and the magnetoresistance) in isothermal conditions.

Tensor $\rho_{\nu\mu}+\alpha_{\nu\mu}\kappa^{-1}_{\alpha\beta}\pi_{\beta\mu}$ describes the galvanomagnetic effects under adiabatic conditions.

Tensor $\alpha_{\nu\mu}$ describes thermoelectric power, and the longitudinal, planar and transverse Nernst-Ettingshausen effects.

Tensor $\kappa^{-1}_{\nu\alpha}$ describes the thermal resistivity, and the longitudinal, planar and transverse Maggi-Righi-Leduc effects.

Tensor $\kappa^{-1}_{\nu\alpha}\pi_{\alpha\mu}$ describes the Nernst effect, and the longitudinal, planar and transverse Ettingshausen-Nernst effects under adiabatic conditions.

For a spherical energy band one can set the magnetic field along the z direction without loss of generality. This gives the $g_{\nu\mu}$ tensor in the form,

$$g_{\nu\mu} = \begin{pmatrix} \frac{1}{1+\phi^2} & -\frac{\phi}{1+\phi^2} & 0 \\ \frac{\phi}{1+\phi^2} & \frac{1}{1+\phi^2} & 0 \\ 0 & 0 & 1 \end{pmatrix} \qquad (5.30)$$

The electric conductivity is now easily found from Eq. (5.25), and its inverse tensor, i.e. the electrical resistivity becomes,

$$\rho_{\nu\mu} = \frac{1}{qnB} \begin{pmatrix} \left(\overline{\frac{u}{1+\phi^2}}\right) & \left(\overline{\frac{u\phi}{1+\phi^2}}\right) & 0 \\ -\left(\overline{\frac{u\phi}{1+\phi^2}}\right) & \left(\overline{\frac{u}{1+\phi^2}}\right) & 0 \\ 0 & 0 & \frac{1}{\overline{u}}B \end{pmatrix} \qquad (5.31)$$

where $$B = \left(\overline{\frac{u}{1+\phi^2}}\right)^2 + \left(\overline{\frac{u\phi}{1+\phi^2}}\right)^2 \qquad (5.32)$$

and the average values are defined in Eq. (26). The other phenomenological transport tensors can be easily calculated for arbitrary magnetic field strength.

Regions of weak and strong magnetic fields are traditionally defined by the conditions $\phi \ll 1$ and $\phi \gg 1$ respectively. The physical sense of these conditions is revealed by **observing**, that $\phi = uH/c = \omega_c \tau$, where $\omega_c = eH/mc$ is the cyclotron frequency. Thus, the condition of strong fields means that during the time between two consecutive collisions (given approximately by τ) an electron manages to make more than one full circle in his cyclotron orbit. In the quantum mechanical picture one can write this condition as $\hbar\omega_c \gg k/\tau$, meaning that the separation between Landau levels $\hbar\omega_c$ is larger than the level broadening due to collision relaxation $\hbar/\tau$. In this situation one should expect quantum effects to appear. The only way to assure the validity of the classical (non quantum) description is to have the temperature sufficiently high, so that the condition $\hbar\omega_c << k_0 T$ is satisfied. In this case the electrons participating in transport are distributed over many Landau levels and the average transport behavior is still quasi-classical.

6. VALIDITY OF THE RELAXATION TIME APPROXIMATION

In this section we consider scattering of electrons in small-gap semiconductors. We do not intend to give derivations (those can be found in original publications), but we would like to mention problems arising in the relaxation time approximation and its application. We also give final expressions describing relaxation times for main scattering mechanisms, which are necessary to proceed with the discussion of various transport effects.

Free electrons in a crystal can be scattered by localized imperfections of the lattice: charged impurities, neutral impurities and dislocations; by other free carriers: electrons and holes; and by lattice vibrations: polar optical, non-polar optical, piezoacoustic and acoustic interactions. By polar scattering we mean the scattering of electrons due to their interaction with a longitudinal optical branch of the lattice vibrations which polarize the crystal when the two atoms in a unit cell are not alike.

One has to distinguish between elastic scattering processes in which the energy of an electron is, approximately at least, conserved,and non-elastic collisions in which not only the momentum but also the energy is markedly changed. If, which is usually the case, the effective mass of heavy holes is much larger than that of electrons, so that an electron-hole collision may be regarded as elastic, the only important non-elastic processes are due to optical phonon mode and the electron--electron scattering. The main reason for distinguishing the non-elastic processes is that the relaxation time approximation cannot be applied here. Scattering by optical phonons can be non-elastic since the phonon energy $\hbar\omega_{op}$ is not always negligible compared with that of an electron. Since the optical polar mode appears to be present in all polar semiconductors,the problem of treating the transport phenomena in this case has received a fair amount of attention. For higher temperatures, which satisfy relation $k_0 T > \hbar\omega_{op} = k_0\Theta$, the energy of the majority of electrons is much higher than the phonon energy, so that the collisions are approximately elastic and the relaxation time can be introduced in the usual way. It turns out that also for very low temperatures $T<<\theta$ the relaxation time concept is valid due to the fact that at those temperatures the time during which an electron remains in the phonon-excited state is very short. The intermediate region of temperatures, most difficult to deal with, is usually treated by variational techniques.

Ehrenreich [6,7], who first emphasized the importance of polar scattering in InSb and similar materials, generalized variational procedure of Howarth and Sondheimer [38,39] for Kane's shape of the conduction band and calculated the elctron mobility and the thermoelectric power in intrinsic InSb below room temperature. Korenblit and Sherstobitov [40] and Tamarchenko et al. [41] considered quasi-elastic optical scattering in small-gap materials for strongly degenerate electron gas characterized by $\zeta>>\hbar\omega_{op}$.

Electron - electron (e - e) non-elastic collisions do not change the momentum and the energy of the electron gas as a whole, but they redistribute momenta and energies of separate carriers, thus affecting other scattering modes. Actually, e-e collisions lower the mobilities connected with charged impurities and optical phonons in the non-degenerate electron gas. In the degenerate gas the redistribution of momenta does not affect the mobility since all electrons participating in the transport have approximately the same energy. However, the thermal properties are markedly affected even in degenerate conditions since they essentially depend on the difference between "warm" and "cool" carriers (with respect to the Fermi level).

7. RELAXATION TIMES FOR ELASTIC SCATTERING MODES

We turn now to the main topic,that is to the relaxation time for electrons in small-gap materials. We shall be mostly interested in main collision processes, namely: ionized impurity and heavy hole, polar optical (for $T>\theta$), and acoustic scattering mechanisms.

As is well known, the relaxation time for electrons in a spherical band can be expressed in the general form,

$$1/\tau(\varepsilon)\sim\rho(\varepsilon)W(\varepsilon)$$

where $\rho(\varepsilon)$ is the density of states per unit energy and $W(\varepsilon)$ the transport scattering probability. The density of states is in general,

$$\rho(\varepsilon)=\frac{1}{\pi^2}k^2\frac{dk}{d\varepsilon} \tag{7.1}$$

For a parabolic band this gives $\rho(\varepsilon)=(1/\pi^2)(m_e/\hbar^2)k$, whereas for an arbitrary spherical band using the definition (2.14) one has $\rho(\varepsilon)=(1/\pi^2)(m/\hbar^2)k$. Thus both results look the same, with m_e replaced by m in the general case. However, a complete and consistent theory of electron scattering in small-gap materials should take into account not only the actual $\varepsilon(k)$ relation in the band but also appropriate electron wave functions for the band in question, their symmetry and $\vec{k}$ - dependence. In other words the non-standard structure of the conduction band affects the relaxation time not only via $\rho(\varepsilon)$ but also via $W(\varepsilon)$. The older scattering theory, taking into account only the proper density of states has been summarized by Kolodziejczak and Sosnowski [12].

7.1 InSb - type III - V compounds

As it has been mentioned in Section 3, when the electron energy becomes comparable to the energy gap one can not decouple the conduction and valence levels in the k.p theory. Instead, one deals with the case of nearly degenerate levels, which have to be taken into account exactly. Treating the interaction between the Γ_6, Γ_8, and Γ_7 levels, and neglecting contributions from all other bands, one obtains the following Bloch functions for the conduction electrons,

$$\Psi_{\vec{k},j_z}(\vec{r}) = u_{\vec{k},j_z}(\vec{r})\exp(i\vec{k}\vec{r}) \tag{7.2}$$

where the $\vec{k}$ - dependent periodic parts are [16]

$$u_{\vec{k},1/2}=(iaS+\frac{b-c\sqrt{2}}{2}\frac{k_+}{k}R_- -\frac{b+c\sqrt{2}}{2}\frac{k_-}{k}R_+ -c\frac{k_z}{k}Z)\uparrow +$$
$$+b(\frac{k_z}{k}R_+ -\frac{k_+}{k}\frac{Z}{\sqrt{2}})\downarrow$$

$$u_{\vec{k},-1/2}=(iaS-\frac{b+c\sqrt{2}}{2}\frac{k_+}{k}R_-+\frac{b-c\sqrt{2}}{2}\frac{k_-}{k}R_+-c\frac{k_z}{k}Z)\downarrow-$$

$$-b(\frac{k_z}{k}R_--\frac{k_-}{k}\frac{Z}{\sqrt{2}})\uparrow \tag{7.3}$$

with $k_\pm = k_x \pm ik_y$ and $R_\pm=(X\pm iY)\sqrt{2}$. The arrows $\uparrow$ and $\downarrow$ are the spin-up and spin-down functions, and S,X,Y,Z denote s - like, p - like periodic amplitudes at the Γ point. The total angular momentum $j = 1/2$ is quantized along the z direction. For InSb - like crystals the coefficients are given by simple formulas in the energy range $\varepsilon < \varepsilon_o+(2/3)\Delta$

$$a^2=1-L \qquad b^2=L/3 \qquad c^2=2L/3 \tag{7.4}$$

with $L=\varepsilon/(\varepsilon_o+2\varepsilon)$

It can be seen from the above expressions that for $\varepsilon \simeq \varepsilon_0$ the periodic Bloch amplitudes are strongly $\vec{k}$ dependent and that they become mixtures of s and p - - like components. Also due to the strong spin-orbit interaction, which splits the Γ_8 and Γ_7 levels, the Bloch amplitudes are not pure spin functions $(b \neq o)$. The above properties of the wave functions introduce new features into the scattering processes. For example, due to the fact that the spin and the coordinate variables are intermingled, the spin-flip (strictly speaking $\Delta j_z = \pm 1$) scattering processes become possible for all modes. In the following we quote the results of ref. [16,43] based on Kane's three-level model.

Charged Center Scattering (CC). Scattering by ionized impurities and heavy holes is calculated for the screened Coulomb potential. The electron relaxation time is obtained in the form

$$\tau_{cc}=\frac{1}{2\pi}\frac{\kappa_o^2\hbar}{e^3N_i}\frac{1}{F_{cc}}\frac{d\varepsilon}{dk}k \tag{7.5}$$

where κ_0 is the static dielectric constant N_i the concentration of scattering centers. F_{cc} is to a good approximation given by

$$F_{cc}(\xi,L)=\ln(\xi+1)-\frac{\xi}{\xi+1}-L(4-L)[1-\frac{2}{\xi}\ln(\xi+1)]+$$

$$+\frac{3}{2}L^2(1-\frac{4}{\xi}) \tag{7.6}$$

where $\xi=(2k\lambda)^2$. Quantity λ is the screening length describing screening of impurity ions by the free electrons. For the energy band described by Eq.(3.2) λ is (Kolodziejczak [32]),

$$\frac{1}{\lambda^2}=\frac{2}{\pi}\frac{e^2}{\kappa_o k_o T}\left(\frac{2m_e k_o T}{\hbar^2}\right)^{3/2} {}^{o}L_1^{1/2}(\eta,\beta) \qquad (7.7)$$

An expression for the screening length in a spherical band with arbitrary nonparabolicity can be found in ref. [16].

Appearance of terms involving L in Eq. (7.6) is due to the above mentioned mixing of p - like components into the conduction band wave function. For large ε_0 the mixing is negligible: $L\approx 0$ and the well known formula for the parabolic band is obtained. The above expressions can also be applied to the scattering of electrons by heavy holes, with κ_0 replaced by κ_∞. In III-V compounds the two dielectric constants differ only slightly and it is possible to describe both scattering modes by Eq. (7.5) putting for the number of scattering centers: $N_i+p=n$, the electron concentration in the conduction band.

<u>Polar Optical Scattering (OP)</u>. It is calculated using Fröhlich's Hamiltonian for the polar interaction between electrons and optical phonons. According to Ehrenreich [7] the screening of initial interaction by other free electrons introduces two effects. First, there appears the following dispersion in the longitudinal optical branch

$$\omega_1^2(q)=\omega^2\left(1+\frac{\kappa_\infty}{\kappa_o}\frac{1}{q^2\lambda_\infty^2}\right) \Big/ \left(1+\frac{1}{q^2\lambda_\infty^2}\right) \qquad (7.8)$$

and, secondly, the initial interaction U' is weakened, according to

$$U = U'/\left(1+\frac{1}{q^2\lambda_\infty^2}\right) \qquad (7.9)$$

where λ_∞ is the screening length given by Eq. (7.7) with κ_0 replaced by κ_∞. In III-V compounds $0.8<\kappa_\infty/\kappa_0<1$ and to a good approximation $\omega(q)\approx\omega$ and $\lambda_\infty\approx\lambda_0$. Still, one deals with the modified interaction (7.9). The final formula for the relaxation time due to the polar optical scattering takes the form

$$\tau_{op} = \frac{1}{8\pi} \frac{M\Omega (k_0 T)^2}{\hbar (ee^\star)^2 k_0 T} \frac{1}{F_{op}} \frac{d\varepsilon}{dk} \tag{7.10}$$

where M is the reduced mass of ions $1/M = 1/m_1 + 1/m_2$, Ω the volume of the unit cell, θ the Debye temperature for longitudinal optical branch $k_0\theta = \hbar\omega_{op}$, $e^\star$ the effective ionic charge, defined as

$$(e^\star)^2 = \frac{\Omega M \omega_{op}^2}{4\pi} \left(\frac{1}{\kappa_\infty} - \frac{1}{\kappa_0}\right) \tag{7.11}$$

F_{op} is to a good approximation given by

$$F_{op} = 1 - \frac{2}{\xi}\ln(\xi+1) - \frac{1}{2}(4L - L^2)(1 - \frac{4}{\xi}) + L^2(1 - \frac{3}{\xi}) \tag{7.12}$$

where ξ and L are defined as before. Again $L \approx 0$ describes a transition to the parabolic band, and $\xi \to \infty$ corresponds to the no-screening limit.

Acoustic Scattering (AC). It is calculated on the basis of the Bloch deformable-ion model. It should be emphasized that the calculation of accoustic scattering in this case has to go beyond the simple deformation potential approximation, since the electron wave functions contain the periodic amplitudes belonging to Γ_6, Γ_8 and Γ_7 levels, which move independently under deformation. (In the following we put all deformation potentials equal). Due to mixing of s and p components, the transverse modes also couple to electrons. The account of spin-flip processes gives all in all six different transitions contributing to acoustic scattering. The final formula for the relaxation time is

$$\tau_{ac} = \frac{\pi\hbar\rho}{k_0 T E^2} \frac{v^2}{F_{ac}} \frac{d\varepsilon}{dk} k^{-2} \tag{7.13}$$

$$\text{with } \frac{v^2}{F_{ac}^2} = \frac{v_\parallel^2}{F_{ac}^\parallel} + \frac{v_\perp^2}{F_{ac}^\perp} \tag{7.14}$$

where ρ is the crystal density, E is the deformation potential constant, and $v_\parallel$, $v_\perp$ denote the sound velocities for longitudinal and transverse modes, respectively .

$$F_{ac}^{\parallel} = (1-\frac{5}{3}L)^2+\frac{73}{180}L^2 \qquad (7.15)$$

and $$F_{ac}^{\perp} = \frac{11}{60}L^2 \qquad (7.16)$$

7.2. HgTe - type Materials

The electron wave functions have the same general form as in InSb type semiconductors, i.e. they are given by Eqs. (7.2) and (7.3). The coefficients, however, are different

$$a^2=L \qquad b^2=(1/3)(1-L) \qquad c^2=(2/3)(1-L) \qquad (7.17)$$

with $L=\varepsilon/(\varepsilon_0+2\varepsilon)$, expressing the fact that now for $\varepsilon=o$ the wave functions have pure p-type symmetry. In addition to the usual problems arising in small-gap semiconductors one faces in symmetry-induced zero-gap materials special features connected with the degenerate character of the conduction band at $k=o$.

Ionized Impurity Scattering. Liu and Brust [19] calculating the static dielectric function in the random phase approximation found that, due to the Γ_8 symmetry of the conduction band near the edge and the symmetry-induced zero gap between the conduction and the heavy-hole bands, the dielectric function $\kappa(q)$ in a pure sample at $T = 0°K$ exhibits a singularity of the type $1/q$, as q approaches zero. Next Liu and Tosatti [20] showed that impurity electrons remove this singularity leaving a finite dielectric constant which is strongly dependent on impurity concentration and greatly enhanced over the background dielectric constant. According to their treatment, the dielectric function describing the Coulomb interaction between an ionized impurity and an electron,

$$V(\vec{r})=-\frac{e^2}{2\pi^2}\int e^{i\vec{q}\vec{r}}\frac{1}{\kappa(q)q^2}d^3q \qquad (7.18)$$

can be conveniently divided into

$$\kappa(q) = \kappa_o+4\pi\alpha^{inter}(q)+4\pi\alpha^{intra}(q) \qquad (7.19)$$

the interband polarization arising from excitations between the degenerate bands Γ_8 near $k=o$, the intraband polarization of the Thomas-Fermi type, and the background dielectric constant κ_0 due to all other bands. A calculation for a parabolic band gives to a good approximation for small q,

$$4\pi\alpha^{inter} = \frac{8e^2\mu m_o}{\pi\hbar^2 k_F}\left(1-a\frac{q^2}{k_F^2}\right) \tag{7.20}$$

where k_F is the Fermi momentum, μ the effective mass ratio m_e/m_{hh}, and $a \approx 1/12$. Broerman [44] has shown that this is a good approximation also for Kane's model. The above results were obtained for $T = 0°K$. Broerman [45] examined numerically the static dielectric function at non-zero temperatures and showed that thermally excited carriers also remove the singularity, leaving a dielectric constant which is strongly temperature and impurity concentration dependent.

When calculating the ionized impurity scattering, the dielectric function given by the above equations must be used for the effective Coulomb interaction defined in Eq. (7.18). If the q - dependent term in Eq. (7.20) is neglected (it has been shown in ref. [43] that this does not influence strongly the final results)the calculation of scattering in HgTe-type materials with the concentration dependent dielectric constant can be carried out similarly as for InSb,with due regard to the inverted band structure. As follows from the comparison of the electron wave functions in both cases, the relaxation time due to ionized impurity scattering in this approximation can be obtained from formula (7.5) with L in Eq. (7.6) replaced by $1-L$. This gives for large ξ values

$$F_{imp} = \ln(\xi+1) - \frac{\xi}{\xi+1} - (1-L)(3+L)\left[1-\frac{2}{\xi}\ln(\xi+1)\right] + \frac{3}{2}(1-L)^2\left(1-\frac{4}{\xi}\right) \tag{7.21}$$

In the low energy limit of $L \approx 0$ the above expression gives a considerably lower value of F_{imp} than that for InSb-type semiconductors, which corresponds to higher mobilities even for the same values of material constants. The strong enhancement of the dielectric constant with decreasing electron concentration, as described by Eq. (7.20), weakens the Coulomb interaction (7.18) thus increasing the relaxation time, but on the other hand it increases the screening radius λ of Eq. (7.7), which has the opposite effect. The net enhancement of mobility due to concentration dependence of the dielectric constant is approximately threefold, as shown in the following section.

Polar Optical Scattering. Since at higher temperatures the static and the high frequency dielectric constants have no singularities, it seems that the polar optical scattering can be calculated as in the case of InSb-type semiconductors. As follows from comparison of the electron wave functions in both cases, the final result of Eq. (7.10) may be taken over directly for HgTe-type materials with L in Eq. (7.12) replaced by $1-L$. This gives, in approximation for $\xi > 10$

$$F_{op} = 1 - \frac{2}{\xi}\ln(\xi+1) - \frac{1}{2}(1-L)(3+L)(1-\frac{4}{\xi}) + (1-L)^2(1-\frac{3}{\xi}) \tag{7.22}$$

The formula is valid for electron energies $\varepsilon << 2\Delta/3$. In the low energy limit of $L \approx 0$ the above expression gives the value of mobility approximately twice as high as that obtained using Eq. (7.12) for III-V compounds.

Acoustic Scattering. Acoustic scattering in HgTe-type materials can be calculated by the same method which has been used for InSb. If, as before, one assumes all four deformation potentials equal (but only in this case!), the resulting formulas are the same as for III-V compounds, cf. Eq. (7.13), with L replaced by $1-L$ [43]. Hence one obtains

$$F_{ac}^{\parallel} = \frac{51}{60} - \frac{91}{30}L + \frac{191}{60}L^2 \qquad F_{ac}^{\perp} = \frac{11}{60}(1-L)^2 \tag{7.23}$$

In most materials $v_{\parallel}^2/v_{\perp}^2 \approx 2$, and at the band edge ($L \approx 0$) the transverse modes contribute about 30% to the total acoustic scattering.

8. ELECTRON MOBILITY

We shall not discuss here the galvanomagnetic phenomena in various configurations, as they are well known. Instead, we consider directly the electron mobility, a fundamental quantity in all transport phenomena. General expression for the electron mobility in case of a single scattering mode can be obtained using Eq. (2.14) for the effective mass and expressions for the relaxation time. For InSb-type materials one obtains in general

$$u_r = \frac{e\tau_r}{m} = u_{ro}\frac{1}{F_r}\left(\frac{d\varepsilon}{dk}\right)^2 k^r \tag{8.1}$$

where $r = +1$ for ionized impurity scattering, $r = -1$ for polar optical scattering, and $r = -3$ for acoustic scattering. The values of u_{ro} can be easily deduced from above equation. As follows from the considerations in the preceding section, the mobilities in HgTe-type materials are also given by the same general formula. Consequently, the general expression (8.1) for the electron mobilities will be used in further considerations. It should be emphasized that, due to the energy

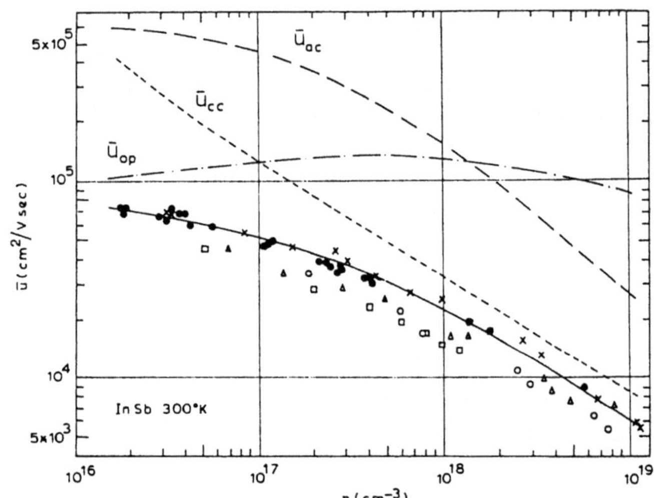

Fig. 5 Electron mobility in n - InSb at room temperature, as measured by various authors, versus free electron concentration. Dashed lines denote the theoretical mobilities for charged center, polar optical, and acoustic scattering modes, solid line shows the theoretical mobility for the mixed mode. [After W. Zawadzki, Proc. 11 Intern. Conf. Phys. Semicond., Pol. Sci. Publ., Warsaw 1972, p. 87].

dependence of the effective mass, it is the mobility that plays the role of basic quantity in the transport theory for small-gap semiconductors. If, which is usually the case, one deals simultaneously with different scattering mechanisms, the total mobility is given by the well known relation

$$1/u = 1/u_{cc} + 1/u_{op} + 1/u_{ac} \tag{8.2}$$

According to Eq. (5.25) the ohmic mobility is

$$\bar{u} = \langle u \rangle / \langle 1 \rangle \tag{8.3}$$

In general one has to compute it numerically using the definition (2.2) of the brackets and the expressions quoted in the previous sections. It is instructive, however, to consider a high-degeneracy region of high electron concentrations or low temperatures, where one can omit the numerical integrations. A good approximat-

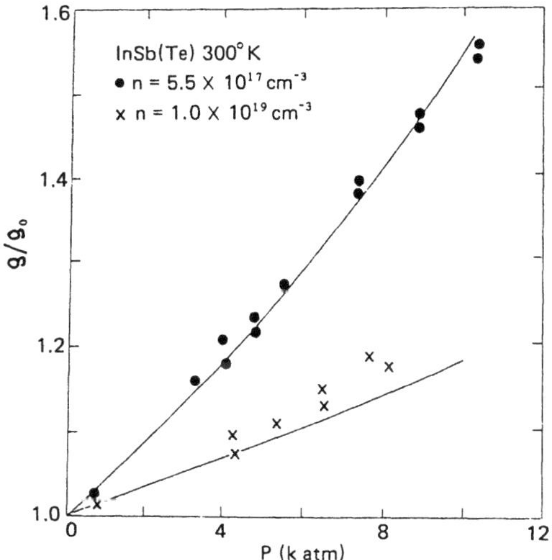

Fig. 6 Change of relative resistance in Te-doped InSb at room temperature versus hydrostatic pressure. The solid lines are theoretical. [After E. Litwin-Staszewska, S. Porowski, and A.S. Filipchenko, Phys. Stat. Solidi (b), 48, 519 (1971)].

ion for mobility in this case is obtained by putting in Eq. (2.2): $(-\partial f_0/\partial\varepsilon)=\delta(\varepsilon-\zeta)$ the Dirac delta function. Then,

$$k_F = (3\pi^2 n)^{1/3} \tag{8.4}$$

and using $d\varepsilon/dk = \hbar^2 km$ one can express the mobilities in the following form [17]

$$\bar{u}_{cc} = \frac{3\pi}{2}\ \frac{\kappa^2\hbar^3}{e^3}\ \frac{1}{F_{cc}}\ \frac{1}{m^2} \tag{8.5}$$

$$\bar{u}_{op} = \frac{1}{4}\left(\frac{3}{\pi}\right)^{1/3} \frac{\Omega M\hbar(k_0\theta)^2}{e(e^\star)^2 k_0 T}\ \frac{1}{F_{op}}\ \frac{n^{1/3}}{m^2} \tag{8.6}$$

$$\bar{u}_{ac} = \left(\frac{\pi}{3}\right)^{1/3} \frac{e\rho\hbar^3}{E^2 k_0 T}\ \frac{v^2}{F_{ac}}\ \frac{1}{m^2 n^{1/3}} \tag{8.7}$$

where now $\xi = \pi^2\left(\frac{3}{\pi}\right)^{1/3} \frac{\kappa\hbar^2}{e^2}\ \frac{n^{1/3}}{m}$ and $L = 1 - \frac{m_e}{m}$ (8.8)

The Fermi energy and the effective mass are :

$$\zeta = (1/2\varepsilon_0)(\sqrt{D}-1) \qquad m = m_e\sqrt{D} \tag{8.9}$$

where D as a function of electron density is obtained from Eq. (4.2) in the form [32]

$$D = 1 + 2(3\pi^2)^{2/3}(\hbar^2/m_e\varepsilon_0)n^{2/3} \tag{8.10}$$

It can be seen that the concentration dependence of $\bar{u}_{op}$ is the weakest and that of $\bar{u}_{ac}$ the strongest of the three. Consequently, one would expect qualitatively the optical mode to be of importance at lower electron concentrations and the acoustic mode at higher ones.

Fig. 5 presents a compilation of experimental data, as obtained by various authors, for the Hall mobility in InSb at room temperature as a function of the electron concentration in the conduction band. The solid line represents the theoretical total mobility computed on the basis of the above formalism with the following values of the parameters: $m_e = 0.0126\ m_0$, $\kappa_0 = 17.5$, $e^\star = 0.14\ e$,

$\theta=283°K$, $E=14.6$ eV , and the dashed lines denote theoretical single-mode mobilities $\bar{u}_{cc}$, $\bar{u}_{op}$ and $\bar{u}_{ac}$ calculated for the same values of material constants. The value of the deformation potential was the main adjustable parameter in the theory, determined by the best over-all fit for the mobility and other transport effects [17] . In the intrinsic material the polar OP mode, giving the lowest mobility, constitutes the dominant scattering mechanism with the electron-hole collisions also contributing to the total mobility. With growing electron concentration the role of CC scattering grows rapidly. However, the mobility due to AC mode decreases even more strongly with n and for the concentrations above 2×10^{18} cm-3 this mode becomes more important than OP scattering, so that at the highest achievable concentrations it is the CC and AC modes which determine the scattering processes.

Hydrostatic pressure experiments can serve as a particularly sensitive test of the scattering theory in small-gap materials since, as follows explicitly from Eqs. (7.3) and (7.4), by changing the energy gap one

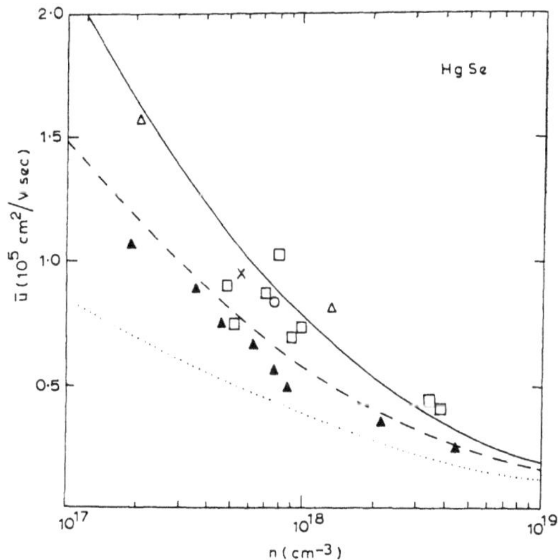

Fig. 7 Low-temperature electron mobility in n - HgSe, as measured by various authors, versus free electron concentration. The dotted line is theoretical for ionized impurity scattering in a conduction band with Γ_6 band-edge symmetry, the dashed one is the same for Γ_8 band-edge symmetry. The solid line takes into account the anomalous static dielectric function. [After J.G. Broerman, Phys. Rev. B 2, 1818 (1970)].

changes not only the $\varepsilon(k)$ relation but also the wave functions which, in turn, affects the scattering probability. Fig. 6 presents results of Litwin-Staszewska et al.[46] on the pressure dependence of resistivity in Te-doped n - InSb at room temperature. Since the electron concentration does not depend on pressure in the investigated range, the change of resistivity is all due to the change of electron mobility. The solid lines in Fig. 6 , calculated according to above theory with the same values of parameters as those used in Fig. 5 for mobility, are seen to be in good agreement with experiment for both weakly and highly Te - doped samples. Since it is the acoustic mode which is mainly responsible for the change of mobility with pressure, the agreement between the theory and experiment seems to confirm the above value of the deformation potential for the Γ_6 conduction band in InSb.

Fig. 7 presents the electron mobility in HgSe versus carrier concentration at liquid nitrogen temperature. The solid line shows the theoretical mobility due to ionized impurity scattering calculated by Broerman [21] , taking into account the concentration dependent dielectric constant and the Γ_8 symmetry of the conduction band, with the values of material constants established in ref. [27] (cf. Fig. 3). The dashed line shows the theory for a concentration-independent dielectric constant. A comparison between the two calculations and the experimental data definitely confirms the dielectric singularity in this symmetry-induced zero-gap semiconductor.

9. THERMOELECTRIC POWER

The thermomagnetic power α in absence of magnetic field can be obtained from Eqs.(5.29),(5.30) and (5.31),

$$\alpha(o)=\alpha_{11}(o) = \rho_{1\nu}\theta_{\nu 1} = \rho_{11}\theta_{11} = \frac{k_o}{T}\left(\frac{\langle zu\rangle}{\langle u\rangle}-\eta\right) \qquad (9.1)$$

Using the definition of the brackets and the expressions for mobilities the above formula can be computed numerically for a single or combined scattering mode.

The influence of band's nonparabolicity is stronger at higher electron concentrations, where the Fermi level probes higher energies in the band. For such conditions the integrals represented by brackets can be calculated using expansion (2.8) applicable at high degeneracy of the electron gas. Since in such conditions the expansion is quickly converging, the first two terms will be used for the <zu> integral and only the first term for the <u> integral. This gives the first nonvanishing approximation. Since, in general

$$\frac{d^2(zC)}{dz^2} = z\frac{d^2C}{dz^2} + 2\frac{dC}{dz} \tag{9.2}$$

one obtains from Eq. (9.1) for electrons,

$$\alpha(o) = -\frac{k_o}{e}\frac{\pi^2}{3}k_oT\left(\frac{1}{c}\frac{dC}{d\varepsilon}\right)_{\varepsilon=\zeta} \tag{9.3}$$

where, in the considered case, $C=uk^3$. This yields,

$$\alpha(o) = -\frac{k_o}{e}\frac{\pi^2}{3}k_oT\left(\frac{1}{u}\frac{du}{d\varepsilon} + \frac{3}{k}\frac{dk}{d\varepsilon}\right)_{\varepsilon=\zeta} \tag{9.4}$$

In order to proceed further one has to specify the mobility as a function of energy. First, we consider a single scattering mode. Using the general expression (8.1) one has for any single mode,

$$\frac{du_r}{d\varepsilon} = u_{ro}\frac{1}{F_r}\frac{d\varepsilon}{dk}k^{r-1}\left[r - \frac{k}{F_r}\frac{dF_r}{dk} + 2k\left(\frac{d\varepsilon}{dk}\right)^{-1}\frac{d^2\varepsilon}{dk^2}\right] \tag{9.5}$$

Defining a quantity (Kolodziejczak and Sosnowski [12]),

$$1 = \frac{k^3}{m}\frac{dm}{d(k^3)} = \frac{1}{3}\left[1 - k\left(\frac{d\varepsilon}{dk}\right)^{-1}\frac{d^2\varepsilon}{dk^2}\right] \tag{9.6}$$

we have

$$\frac{1}{u_r}\frac{du_r}{d\varepsilon} = \frac{dk}{d\varepsilon}\frac{1}{k}\left[r - \frac{k}{F_r}\frac{dF_r}{dk} + 2 - 6 1\right] \tag{9.7}$$

Using Eqs. (2.14) and (8.4) one obtains finally the thermoelectric power for a single scattering mode in highly degenerate conditions

$$\alpha_r(o)=-\frac{1}{3}\left(\frac{\pi}{3}\right)^{2/3}\frac{k_o^2T}{e\hbar^2}\frac{m}{n^{2/3}}(r'+5-6l) \qquad (9.8)$$

where

$$r'=r-\frac{k}{F_r}\frac{dF_r}{dk} \qquad (9.9)$$

denotes an effective scattering index for the scattering mode in question.

Now we consider a combined mode of scattering. Suppose, one deals with two different modes represented by u_1 and u_2. In this case the total mobility is given by Eq. (8.2). Hence

$$u=(1/u_1+1/u_2)^{-1} \qquad (9.10)$$

We are to substitute this total mobility in the general formula (9.4). Since,

$$\frac{1}{u}\frac{du}{d\varepsilon} = u\frac{dk}{d\varepsilon}\frac{1}{k}\left[\frac{1}{u_1}(r_1'+2-6l)+\frac{1}{u_2}(r_2'+2-6l)\right] \qquad (9.11)$$

the general formula for the thermoelectric power in highly degenerate conditions is obtained in the following form

$$\alpha(o)=-\frac{1}{3}\left(\frac{\pi}{3}\right)^{2/3}\frac{k_o^2T}{\hbar^2e}\frac{m}{n^{2/3}}(R+5-6l) \qquad (9.12)$$

where

$$R=\frac{r_1'/u_1+r_2'/u_2}{1/u_1+1/u_2} \qquad (9.13)$$

is the effective scattering index for the combined scattering mode. r_1' and r_2' are given by Eq. (9.9). The above expression for R can be further generalized for more than two scattering modes, provided they can be described by the general formula (8.1). If $u_1 \ll u_2$, then $R \approx r_1'$, and the above expression reduces to the single-mode formula (9.8).

As follows from Eqs. (9.8) or (9.12), in highly degenerate conditions the nonparabolicity of the band affects the thermoelectric power also through the appearance of the parameter l , which can reach the value of 1/3 for the first-order nonparabolicity. It can in principle change the sign of the effect. In general the thermoelectric power is sensitive to the $\varepsilon(k)$ shape of the band via the $n(\zeta)$, $m(\zeta)$ and $l(\zeta)$ dependences, but not very sensitive to the nature of scattering. This can be seen directly from Eq. (9.12) by observing, that the large factor 5 next to the scattering index R dominates small possible changes of the latter resulting from different relative importance of various scattering modes.

Fig. 8 shows the thermoelectric power in n - InSb at room temperature, as measured by various authors, together with the theoretical curve computed using formula (9.1) for the mixed mode of charged center, polar and acoustic scattering. The same set of parameters was used in the calculation of mobility in the previous section. To get some insight into the problem we consider high-degeneracy range. According to Eqs.(9.12) and (9.13) the scattering enters via the index R . We shall

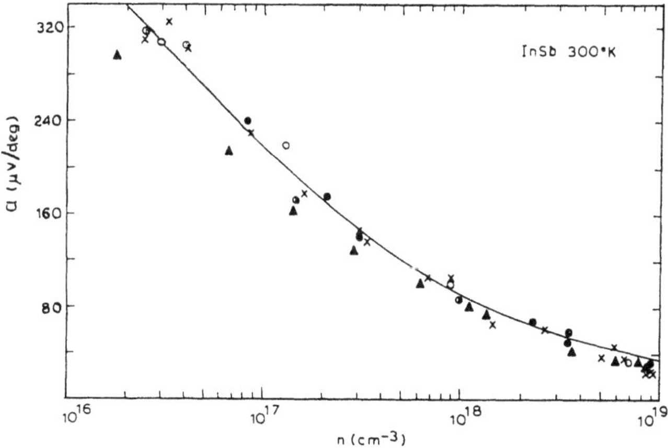

Fig. 8 Thermoelectric power in n - InSb at room temperature, as measured by various authors, versus free electron concentration. The solid line is calculated for the mixed mode of charged center, polar optical and acoustic scattering. [After W. Zawadzki and W.Szymańska, J. Phys. Chem. Solids 32, 1151 (1971)].

calculate R for the electron concentration $n = 1.1 \times 10^{19}$ cm^{-3} and quote the arithmetics, as it is instructive. For the above concentration one has : $\bar{u}_{cc} = 7.90 \times 10^3$, $\bar{u}_{op} = 8.65 \times 10^4$, $\bar{u}_{ac} = 2.43 \times 10^4$ cm^2/V sec, and $r'_{cc} = 0.27$, $r'_{op} = -1.39$ and $r'_{ac} = -2.66$. Hence

$$R = \frac{\dfrac{r'_{cc}}{\bar{u}_{cc}} + \dfrac{r'_{op}}{\bar{u}_{op}} + \dfrac{r'_{ac}}{\bar{u}_{ac}}}{\dfrac{1}{\bar{u}_{cc}} + \dfrac{1}{\bar{u}_{op}} + \dfrac{1}{\bar{u}_{ac}}} = \frac{3.42 - 1.61 - 10.25}{12.65 + 1.16 + 4.11} = -0.47 \qquad (9.14)$$

This value of R inserted into formula (9.12) gives the result which is in agreement with the experimental data in this concentration range. As it can be seen, in spite of the fact that the CC scattering gives the largest contribution to the mobility, it is the AC mode which mainly determines the sign of R . This is due to the large absolute value of r'_{ac} or, in more physical terms, to the strong energy dependence of AC scattering. Quite similar results were obtained by Rode [47] ,

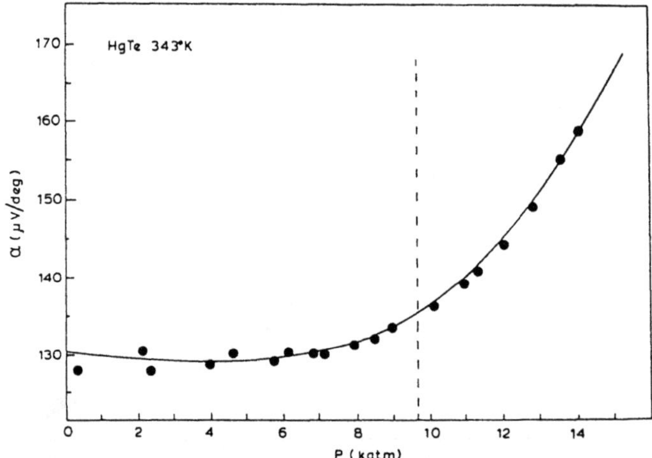

Fig. 9 Thermoelectric power in intrinsic HgTe at 343° K versus hydrostatic pressure. The solid line is theoretical. A steep increase of α above $P = 9.6$ katm is due to transition from HgTe-type level ordering $\varepsilon(\Gamma_6) < \varepsilon(\Gamma_8)$ to InSb-type ordering $\varepsilon(\Gamma_6) > \varepsilon(\Gamma_8)$.[After R.Piotrzkowski and S.Porowski, in "II-VI Semiconducting Compounds" (D.G.Thomas, Ed.), Benjamin, New York 1967, p. 1090].

who also described successfully the temperature variation of α in the range of 200° - 750° K.

Measurements of the thermoelectric power in HgTe as a function of hydrostatic pressure by Piotrzkowski et al. [48] have played an important role in establishing the Groves-Paul model of band structure for this material. The energy separation $\varepsilon(\Gamma_6)-\varepsilon(\Gamma_8)$ increases with pressure and in the vicinity of room temperature one can go over from a zero-gap semimetal $(\varepsilon(\Gamma_6)-\varepsilon(\Gamma_8))$ to a real-gap semiconductor $(\varepsilon(\Gamma_6)-\varepsilon(\Gamma_8))$ by increasing pressure. Fig. 9 shows $\alpha(p)$ in intrinsic HgTe at 343° K. The weak change of α at low pressures is explained by the decrease of electron mass with decreasing $|\varepsilon_0|$, compensated by the decrease of intrinsic electron concentration. The rapid growth of α above the pressure of 9.6 katm is a result of opening the thermal gap. In this range of pressures both the decrease of intrinsic concentration with increasing energy gap, and the increase of the effective mass, enhance the thermoelectric power. The solid, theoretical curve, is not very sensitive to the description of scattering but reflects mainly the transition between the two band structures. A similar behavior of α has been observed as a function of temperature.

Now we consider a magnetic field dependence of the thermoelectric power. It is easily seen that $\alpha_{33}(H) = \rho_{33}\theta_{33} = \alpha(o)$, so that the longitudinal field has no influence on the Seebeck coefficient. Considering the effect of a transverse field, we calculate directly the high field limit $\phi >> 1$. This gives

$$\alpha(\infty) = \alpha_{11}(H\to\infty) = \rho_{1\nu}\theta_{\nu 1} \approx \rho_{12}\theta_{21} = \frac{k_o}{q}\left(\frac{\langle z\rangle}{\langle 1\rangle} - \eta\right) \qquad (9.15)$$

Thus, the thermoelectric power at high transverse fields reaches a saturation value, which does not depend on scattering.

For the first-order non-parabolic band $\alpha(\infty)$ can be written for any degeneracy in terms of L integrals

$$\alpha(\infty) = (k_o/q)({}^1L_o^{3/2}/{}^oL_o^{3/2} - \eta) \qquad (9.16)$$

In case of strong degeneracy one can again use expansion (2.8) for the <z> integral. This way we obtain from Eq. (9.15) for electrons [12,49]

$$\alpha(\infty)=\pi^2 \frac{k_o^2T}{q}\left(\frac{1}{k}\frac{dk}{d\varepsilon}\right)_\zeta = -\left(\frac{\pi}{3}\right)^{2/3}\frac{k_o^2T}{e\hbar^2}\frac{m}{n^{2/3}} \qquad (9.17)$$

This is a very simple result depending neither on the scattering index R nor 1 . The nonparabolicity enters here only through $m(n)$ .This provides a very direct method of determining the effective mass value, if one finds simultaneously the free electron concentration from the Hall effect at high fields (Rodot [13]). Having $1/m=(1/\hbar^2 k)(d\varepsilon/dk)$ and $k=(3\pi^2 n)^{1/3}$ one can then find the $\varepsilon(k)$ relation by a simple integration.

The effective mass determined from Eqs.(9.16) and (9.17) is practically identical with the mass value at the Fermi level, even in the nondegenerate electron gas. This in addition makes the above procedure very useful, and many effective mass values quoted in section 3 were obtained by this method.

10. THE NERNST - ETTINGSHAUSEN EFFECTS

The transverse Nernst-Ettingshausen effect means traditionally a creation of an electric field perpendicular to the thermal gradient in the presence of a transverse magnetic field, similarly to the Hall effect configuration. Thus, we define the N-E coefficient at low fields as $\vec{E}=P_{N-E}\cdot\vec{H}\times\nabla T$. In our configuration this is equivalent to $E_2=\alpha_{21}(\partial T/\partial x_1)$. Since

$$\alpha_{21}=\rho_{21}\theta_{11}+\rho_{22}\theta_{21} \qquad (10.1)$$

we obtain for low magnetic fields

$$P_{N-E}=-\frac{k_o}{q}\,a_r\,\frac{\bar{u}}{c}\left(\frac{\langle zu^2\rangle}{\langle u^2\rangle}-\frac{\langle zu\rangle}{\langle u\rangle}\right) \qquad (10.2)$$

where $a_r=\langle u^2\rangle\langle 1\rangle/\langle u\rangle^2$ is the Hall scattering factor. The sign in Eq. (10.2) is chosen in such a way that for electrons ($q=-|e|$ and $\bar{u}=-|\bar{u}|$) the sign of the effect agrees with that conventionally measured.

In order to compute P_{N-E} for arbitrary degeneracy

one has to insert the expression for mobility into formula (10.2) and integrate numerically. At high degeneracy of the electron gas one can use again the expansion (2.8). This gives

$$\frac{k_o}{q}\left(\frac{\langle zu^2\rangle}{\langle u^2\rangle}-\frac{\langle zu\rangle}{\langle u\rangle}\right)\approx\frac{\pi^2}{3}\frac{k_o^2T}{q}\left(\frac{1}{u}\frac{du}{d\varepsilon}\right)_\zeta \tag{10.3}$$

Now we can use the procedure presented in Eqs. (9.5) to (9.13), to obtain,

$$P_{N-E}=-\frac{1}{3}\left(\frac{\pi}{3}\right)^{2/3}\frac{k_o^2T}{q\hbar^2}\frac{\bar{u}}{c}\frac{m}{n^{2/3}}(R+2-6l) \tag{10.4}$$

In terms of $\alpha(\infty)$,

$$P_{N-e}=(|\bar{u}|/3c)\alpha(\infty)(R+2-6l) \tag{10.5}$$

One can now relate P_{N-E}, $\alpha(o)$, $\alpha(\infty)$ and the Hall mobility $R_H\sigma=\bar{u}.a_r$ in degenerate electron gas

$$\frac{P_{N-E}}{R_H\sigma}+\alpha(o)=\alpha(\infty)=\left(\frac{\pi}{3}\right)^{2/3}\frac{k_o^2T}{q\hbar^2}\frac{m}{n^{2/3}} \tag{10.6}$$

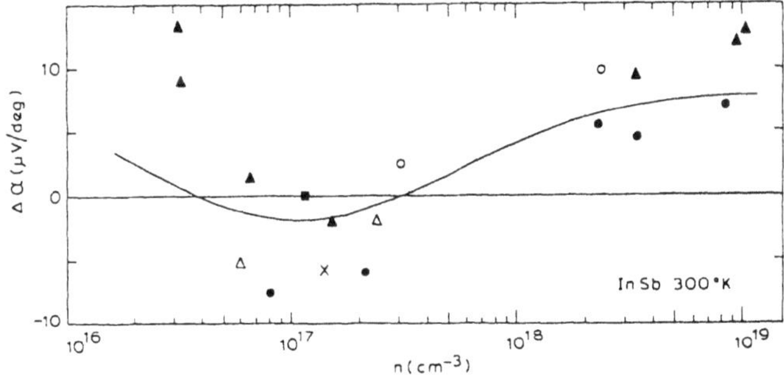

Fig. 10 The longitudinal Nernst-Ettingshausen effect in Te-doped n - InSb at room temperature, as measured by various authors, *versus* free electron concentration. The solid line is calculated for the mixed mode of charged center, polar optical, and acoustic scattering. [After W. Zawadzki, Proc.XI Intern.Conf.Phys.Semicond., Pol.Sci.Publ., Warsaw 1972, p.87] .

Thus, the above combination of quantities measured experimentally at low magnetic fields provides another way to determine the effective mass value avoiding considerations of carrier's scattering.

The longitudinal Nernst-Ettingshausen effect describes a change of the thermoelectric power in a transverse magnetic field: $\Delta\alpha(H)=\alpha(H)-\alpha(o)$. We shall be interested only in the high-magnetic-field limit, at which $\alpha(H)$ reaches the saturation value in the classical description. Thus, according to the definition and Eqs.(9.1) and (9.4), one gets for electrons,

$$\Delta\alpha\equiv\alpha(\infty)-\alpha(o)=\frac{k_o}{e}\left(\frac{\langle zu\rangle}{\langle u\rangle}-\frac{\langle z\rangle}{\langle 1\rangle}\right) \qquad (10.7)$$

At high degeneracy, one can either apply expansion (2.8) to the integrals in the above expression, or use directly Eqs. (9.12) and (9.17) for $\alpha(o)$ and $\alpha(\infty)$. This gives for electrons

$$-\Delta\alpha=-\frac{1}{3}\left(\frac{\pi}{3}\right)^{2/3}\frac{k_o^2T}{e\hbar^2}\frac{m}{n^{2/3}}(R+2-61) \qquad (10.8)$$

It can be seen that the transverse N-E effect at low magnetic fields and longitudinal N-E effect at high fields give very similar information. At high degeneracy they can be easily related to each other,

$$P_{N-E}=-\Delta\alpha|u|/c \qquad (10.9)$$

This relation is approximately fulfilled for arbitrary degeneracy of the electron gas [17].

As follows from Eqs. (10.2) and (10.7) the N-E effects are particularly sensitive to the shape of $u(\varepsilon)$ i.e. to the nature of scattering, since only due to the energy dependence of mobility they do not vanish. For a parabolic band, both in the nondegenerate and strongly degenerate limits, both effects calculated for a single scattering mode are proportional to the scattering index p where $u_p=u_{po}\varepsilon^p$. For high degeneracy this can be seen from Eqs. (10.4) and (10.8) putting $1\equiv0$ and $R+2=2p$. Hence, the sign of the effect can in principle distinguish between different scattering modes : for acoustic phonons $(p=-1/2)$ the effects should be positive, for polar scattering $(p=+1/2)$ they should be negative, and for ionized impurity scattering $(p=+3/2)$ the effects should have large negative values. As follows from the previous discussion, for high electron energies in a nonparabolic band: $2-\delta 1 \approx 0$, and the effects are almost directly proportional to the scattering index R. In case of arbitrary degeneracy things are more complicated, but the N-E effects remain the most sensitive classical tools in the investigations of electron scattering.

The history of long, hot controversy on the interpretation of the N-E effects in InSb is described in ref [17]. Here we interprete the data consistently with all other transport effects, as described above. First we consider the region of high electron concentrations, where formulas (10.4) and (10.8) may be employed. As it has been shown before, Eq. (9.14) for $n = 1.1\times10^{19}$ cm^{-3} the effective scattering index $R \approx -0.47$. Since for this electron concentration $\delta 1 \approx 1.93$, it follows that the sign of the effect is positive. As mentioned

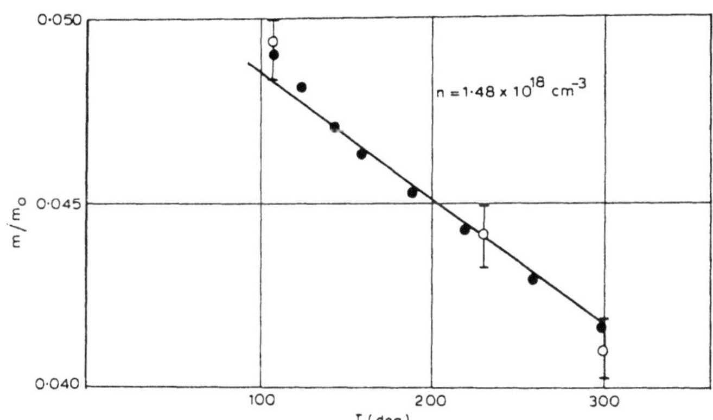

Fig. 11 Electron effective mass in HgTe at the Fermi energy corresponding to free electron concentration $n = 1.5\times10^{18}$ cm^{-3} versus temperature: • - data obtained from the thermoelectric power in a strong transverse magnetic field, ϙ - data obtained from low-field measurements of the transverse Nernst-Ettingshausen effect, the thermoelectric power, and the Hall mobility.[After A.Jedrzejczak, Doctoral Thesis, Institute of Physics, Polish Acad. of Sciences (1972), unpublished].

before, this value of R is determined essentially by CC and AC scattering. However it comes out to be rather close to that for polar OP mode ($r_{op} = -1$) Fig. 10 presents the longitudinal N-E effect in n-InSb at room temperature, as measured by various authors, together with the theoretical curve calculated numerically on the basis of the above theory for the combined mode of CC + OP + AC scattering. At high electron concentrations the positive sign of the effect is mainly due to the influence of AC mode. Going to lower impurity concentrations the effect changes sign to negative due to the combined effect of CC + OP modes. At lowest concentrations the calculated curve goes again to positive values, which seems rather unexpected. However, with the above choice of the deformation potential constant it can be seen from Fig. 5 that at $n = 2 \times 10^{16}$ cm $^{-3}$ the mobilities for CC and AC scattering become again comparable, and the influence of acoustic mode is felt once again. The second change of sign into positive values at low electron concentration was first predicted theoretically [17] and then confirmed experimentally. The discrepancy between the theory and experiment at lowest concentrations is probably due to the contribution of holes into the measured values. (The latter can not be completely eliminated in this quasi-intrinsic region of conductivity). The overall agreement between the theory and experiment for the longitudinal N-E effect should be regarded as a crucial test for the validity of the theoretical description of electron scattering in Te-doped InSb.

The transverse N-E effect was employed in HgTe by Jedrzejczak [50] together with other measurements, according to Eq. (10.6), to determine the temperature dependence of the electron effective mass. Fig. 11 presents the data for one sample, together with the mass values determined from $\alpha(\infty)$. Due to high precision of all measurements the results obtained by both methods coincide quite well. The change of the mass with temperature is due to both the lattice dilatation and phonon contribution.

11. THERMAL CONDUCTIVITY

We calculate now the electronic part of thermal conductivity in absence of a magnetic field. All transport tensors have in this case scalar properties, so that $\kappa^e(o) = \xi - \chi\alpha$, and one obtains

$$\kappa^{e}(0)=\frac{k_0^2}{q^2}\,T\sigma\left[\frac{\langle z^2u\rangle}{\langle u\rangle}-\frac{\langle zu\rangle^2}{\langle u\rangle^2}\right] \tag{11.1}$$

where the electric conductivity $\sigma=nq\bar{u}$. The Lorenz number is $L=\kappa^{e}/\sigma$.

In order to investigate behavior of L in the high-degeneracy range we use again expansion (2.8) To the accuracy of the second derivative with respect to η this gives, in general

$$\langle A\rangle=C+\frac{\pi^2}{6}\,\frac{d^2C}{d\eta^2} \tag{11.2}$$

$$\langle zA\rangle=\eta C+\frac{\pi^2}{3}\,\frac{dC}{d\eta}+\frac{\pi^2}{6}\,\eta\,\frac{d^2C}{d\eta^2} \tag{11.3}$$

$$\langle z^2A\rangle=\eta^2C=\frac{\pi^2}{3}C+\frac{\pi^2}{3}\,2\eta\frac{dC}{d\eta}+\frac{\pi^2}{6}\,\eta^2\,\frac{d^2C}{d\eta^2}+\frac{7\pi^4}{30}\,\frac{d^2C}{d\eta^2} \tag{11.4}$$

where $C=A(\eta)k^3(\eta)$. One can use the above expansions to calculate L . This gives

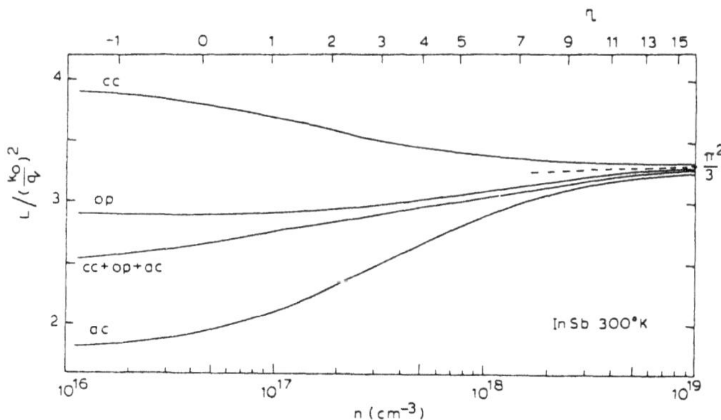

Fig. 12 Lorenz number in n - InSb at room temperature versus free electron concentration, calculated for single and combined scattering modes. At high degeneracy of the electron gas the classical value of $\pi^2/3$ is obtained regardless of scattering mechanism and shape of energy band, as long as the scattering is elastic.

$$L=\left(\frac{k_o}{q}\right)^2 T\ \left\{\frac{\pi^2}{3} - \frac{\pi^4}{18}\left[\ \frac{2}{C^2}\left(\frac{dC}{d\eta}\right)^2 - \frac{26}{5}\frac{1}{C}\frac{d^2C}{d\eta^2}\ \right]\right\} \qquad (11.5)$$

It is easy to show that the second term in the above equation is of the order of $1/\eta^2$. Hence in the limit of strong degeneracy (large η values) the classical value of $(k_0/q)^2T(\pi^2/3)$ is obtained regardless of band shape and scattering mechanism.

Fig. 12 presents theoretical values of the Lorenz number for n - InSb as a function of electron concentration at room temperature for single modes of CC, OP and AC scattering, as well as for the combined mode, calculated on the basis of the above theory and Eq. (11.1). In agreement with the above considerations, in the high-degeneracy limit one obtains the classical Sommerfeld value of $\pi^2/3$. Moreover, the results in the nondegenerate limit of small η are also quite close to the classical values for the parabolic band $L_{cc}^{par} = 4(k_o/q)^2$, $L_{op}^{par} = 3(k_o/q)^2$, and

$L_{ac}^{par} = 2(k_o/q)^2$. This is understandable, since

for electron energies near the band edge the nonparabolicity does not play a significant role. One can expect the influence of nonparabolicity to be important only for doped samples at high temperatures, when the carriers are distributed high in the band and the electron gas is not strongly degenerate.

Since in the strongly degenerate region one should always obtain $L.(q/k_0)^2 = \pi^2/3$ for elastic collisions, any deviations from this value may be used as a measure of non-elasticity of electron scattering. In order to separate experimentally the electron thermal conductivity κ^e from that of the lattice κ^l two methods have been used. First is the change of κ^e in transverse magnetic field (the Maggi-Righi-Leduc effect). As follows from the theory, for low magnetic fields the thermal conductivity is: $\kappa^e(H)=\kappa^e(o)-aH^2$, and for high magnetic fields it tends to zero. By proper extrapolation one can determine the electronic part of thermal conductivity as the one which can be suppressed by magnetic field (Shalyt et al. [51].) The second method uses the fact that κ^e depends much more strongly on doping than K^l so that one can separate them by investigating the thermal conductivity as a function of donor concentration and temperature [52].

12. FINAL REMARKS

More than twenty years of research in the field of transport phenomena in small-gap materials have been filled with controversies and misinterpretations. It is only recently that the hard-won realisation has come, as to what extent the nonstandard band structure affects the electron scattering and transport. It seems, however, that we are still at the beginning. First of all, a serious attempt to understand the electron scattering has been made for only two materials: InSb and PbTe, and even this work is far from completion. Thus, one will certainly see more and more materials coming under serious consideration, with all the problems and possibilities that each of them brings. Secondly, one would like to see all the new effects particular to small-gap materials not only in transport phenomena, where they are partly lost in the overall behavior. We have particularly in mind the spin-flip transitions, dielectric function anomalies and interaction of transverse acoustic waves with conduction electrons. Optical and acoustic phenomena as well as effective use of quantizing magnetic field seem particularly promising as tools of further research in this areas.

ACKNOWLEDGMENTS AND APOLOGIES

I would like to acknowledge and apologize to all authors, whose experimental results have been used in the figures of this article, and whom I could not quote for the lack of space. I am obliged to Dr. A. Jedrzejczak for making available his unpublished data.

REFERENCES

1. A.H. Wilson, "The Theory of Metals", Second Ed., Cambridge Univer.Press, 1953.
2. H.J. Hrostowski, G.H. Wheatley, and W.F. Flood, Phys. Rev. 95, 1683 (1954).
3. R. Barrie and J.T. Edmond, J. Electronics, 1,161 (1955). R. Barrie, Proc.Phys.Soc. London, B 69 553 (1956). R.Barrie, Phys.Rev. 103,1581 (1956).
4. J.M. Radcliffe, Proc.Phys.Soc. A 68, 675 (1955).
5. E.O. Kane, J.Phys.Chem.Solids,1, 249 (1957).
6. H. Ehrenreich, J.Phys.Chem. Solids, 2, 131 (1957).
7. H. Ehrenreich, J.Phys.Chem. Solids, 9, 129 (1959).
8. W.G. Spitzer and H.Y. Fan, Phys.Rev. 106, 882 (1957).
9. S.D. Smith, T.S. Moss, and K.W. Taylor, J.Phys.Chem. Solids 11, 131 (1959).
10. J. Kołodziejczak, Acta Phys.Polon. 20, 379 (1961).
11. J. Kołodziejczak, Acta Phys.Polon. 21, 637 (1962).
12. J. Kołodziejczak and L.Sosnowski, Acta Phys.Polon. 21, 399 (1962). J. Kołodziejczak, L.Sosnowski, and W.Zawadzki, Proc.Intern.Conf.Phys.Semicond. Exeter 1962, p.94.
13. M. Rodot, J.Phys.Chem. Solids 8, 358 (1959).
14. S. Groves and W. Paul., Phys.Rev.Letters 11, 194 (1963); Proc. VII Intern.Conf.Phys.Semicond. Dunod, Paris, 1964, p. 41.
15. L.L. Korenblit and V.E. Sherstobitov, Fiz. Tekh. Poluprov. 2, 675 (1968). This paper contains numerous errors and misprints.
16. W. Zawadzki and W. Szymańska, Phys.Stat.Solidi (b), 45, 415 (1971).
17. W. Zawadzki and W. Szymańska, J.Phys.Chem.Solids 32, 1151 (1971).
18. Yu.I. Ravich and L.Ya. Morgovskii, Fiz. Tekh. Poluprov. 3, 1528 (1969).
19. L. Liu and D. Brust, Phys. Rev. 173, 777 (1968).
20. L. Liu and E. Tossatti, Phys.Rev. B,2, 1926(1970).
21. J.G. Broerman, Phys.Rev. B,2, 1818 (1970).
22. Proc. 11 Intern. Conf. Phys. Semicond., Polish Scient. Publishers, Warsaw, 1972.
23. W. Zawadzki, Advances in Physics 23, 435 (1974).
24. Yu.I. Ravich, B.A. Efimova, and V.I. Tamarchenko, Phys. Stat.Sol (b) 43, 11 (1971); 43, 453 (1971).
25. W. Zawadzki and J. Kołodziejczak, Phys.Stat.Sol. 6, 409 (1964).
26. C.R. Pidgeon and R.N. Brown, Phys.Rev. 146, 575 (1966).

27. R.R. Gałazka, D.G. Seiler, and W.M. Becker, in "Physics of Semimetals and Narrow-Gap Semiconductors", Ed. D.L. Carter and R.T. Bate, Pergamon Press 1971, p. 481.
28. R.R.Gałazka and L. Sosnowski, Phys.Stat.Solidi 20, 113 (1967).
29. S. Zukotyński and J. Kołodziejczak, Phys.Stat.Sol. 3, 990 (1963); J. Kołodziejczak and S. Zukotyński, Phys.Stat.Sol. 5, 145 (1964).
30. W.Zawadzki and J. Kołodziejczak, Phys.Stat. Sol. 6,419 (1964); W. Zawadzki, Phys.Stat.Sol. 8, 739 (1965).
31. Due to a formal analogy between the first-order non-parabolic ε(k) dependence of Eq. (3.5) and the energy momentum relation for the free relativistic electrons (cf. W. Zawadzki, in "Optical Properties of Solids" E.D.Haidemenakis, Ed., Gordon and Breach 1970, p. 179), the thermodynamic properties described in this chapter are formally identical with those of Jüttner's relativistic gas of fermions [F. Jüttner' Z. Physik 47, 542 (1928); H. Arzelièes, "Thermodynamique Relativiste et Quantique", Gauthier--Villars, Paris 1968].
32. J. Kołodziejczak, Proc.Intern.Conf.Phys.Semicond., Prague 1960, p. 950.
33. W. Zawadzki, R. Kowalczyk , and J. Kołodziejczak, Phys.Stat.Sol., 10, 513 (1965).
34. H.H. McDonald, Proc. London Math. Soc. 30, 167 (1899). E.T. Whittaker and G.N. Watson, " A Course of Modern Analysis" Cambridge 1927.
35. A.H. Wilson, "Thermodynamics and Statistical Mechanics", Cambridge Univ.Press (1957).
36. A.I. Anselm, " Introduction to the Theory of Semiconductors", GIFML, Moscow 1962 (in Russian).
37. W. Zawadzki, Phys.Stat.Sol., 3, 692(1963); 3, 1006 (1963).
38. D.J. Howarth and E.H. Sondheimer, Proc. Roy. Soc. A 219, 53 (1953).
39. B.F. Lewis and E.H. Sondheimer, Proc.Roy.Soc. A 227, 241 (1954).
40. L.L. Korenblit and V.E. Sherstobitov, Fiz.Tekh.Poluprov. 2, 688 (1968).
41. V.I. Tamarchenko, Yu.I.Ravich, L.Ya.Morgovskii, and I.V. Dubrovskaya, Fiz.Tverd.Tela 11, 3206 (1969).
42. P. Kacman and W. Zawadzki , Phys.Stat.Sol.(b), 47 629 (1971).

43. W. Szymańska, P.Bogusławski, and W. Zawadzki , Phys. Stat.Sol. (b), 65, 641 (1974).
44. J.G. Broerman, Phys.Rev. B 1, 4568 (1970).
45. J.G. Broerman, Phys.Rev.Lett. 25, 1658 (1970).
46. E. Litwin-Staszewska, S. Porowski, and A.A. Filipchenko, Phys.Stat.Sol.(b), 48,519 (1971).
47. D.L. Rode, Phys.Rev. B 3, 3287 (1971).
48. R. Piotrzkowski, S.Porowski, Z.Dziuba, J.Ginter, W.Giriat, and L.Sosnowski, Phys.Stat.Sol. 8, K.135 (1965).
49. I.V. Motchan, Yu.N. Obraztsov, and T.V. Smirnova, Fiz.Tverd.Tela 4, 1021 (1962).
50. A. Jedrzejczak, PhD Thesis, Institute of Physics, Polish Academy of Sciences, Warsaw 1972. Unpublished.
51. V.M. Muzhdaba and S.S. Shalyt, Fiz.Tverd.Tela 8, 3727 (1966).
52. Yu.I.Ravich, I.A. Smirnov, and V.V.Tikhonov, Fiz. Tekh. Poluprov., 1, 206 (1967).

Part III
Amorphous Semiconductors

SOLVED AND UNSOLVED PROBLEMS FOR NON-CRYSTALLINE MATERIALS

N.F. MOTT

Department of Physics, University of Cambridge

Cambridge, England

These lectures will, I hope, interrelate with those given by others, particularly by Butcher, Nagels and Emin. My particular aim is to show you how open-ended is the subject of conduction in non-crystalline materials and what enormous disagreement exists in the interpretation of experiment.

I would like to speak first about metals, and particularly about simple ones for which the Fermi surface is nearly spherical. For pure metals the resistivity is due to interaction with phonons, but this is not important in non-crystalline materials and I shall discuss the resistance due to impurities. We introduce the scattering cross-section $I(\theta)$, due to each impurity, for electrons at the Fermi energy. Then the mean free path L is, as shown in any text book,

$$\frac{1}{L} = N \int I(\theta)\ (1-\cos\theta) 2\pi \sin\theta d\theta \tag{1}$$

and the conductivity

$$ne^2L/mv = S_F\ e^2L\ /\ 12\ \pi^3\hbar \qquad (2)$$

where n is the number of electrons per unit volume, v the velocity at the Fermi surface and S_F the Fermi surface area. If the scattering potential V is small compared with the Fermi energy, then the Born approximation can be used and $I(\theta)$ is proportional to V^2. V is of course the difference between the potential of the scatterer and that of the host matrix; thus the additional resistivity due to 1% of Zn, Ga and Sn in Cu is proportional to 1, 4, 9(Mott, Jones, Theory of the properties of metals and alloys, p.293).

The oldest treatment of the resistivity of the disordered system is that of Nordheim in 1931 (Annalen d. Physik 9, 641) in a random alloy. Here one takes Nx atoms with potential V_A and $N(1-x)$ with potential V_B; the average potential is

$$x\,V_A + (1-x)V_B$$

and the deviation at an A atom is $(1-x)(V_B - V_A)$ and at a B atom $x(V_B - V_A)$. These potentials each scatter the electron wave, so that the resistivity is proportional to

$$[x(1-x)^2+(1-x)x^2](\Delta V)^2=x(1-x)(\Delta V)^2$$

where $\Delta V = V_B - V_A$.

This is a weak scattering theory, valid only in the Born approximation limit, so that L is great compared with the distance between atoms. A major theme of these lectures is what happens when this is not so.

Another weak scattering theory is that given by Ziman for liquid metals. Here each atom is considered a weak scatterer, not the difference between its potential and anything else. The scattering cross-section $I(\theta)$ may be calculated using the Born approximation from the pseudopotential. The mean free path is given by

$$\frac{1}{L} = N \int S(q)(1-\cos\theta)I(\theta)2\pi\sin\theta d\theta \quad ,$$

where $S(q)$ is the structure factor given by

$$S(q) = N^{-1} \int \{1+\exp(iqR)\}^2 \; P(R)d^3x \; ,$$

$P(R)d^3X$ being the probability that another atom is at a distance R from a given atom. The main success of the Ziman theory is in explaining the temperature dependence of the conductivity in terms of the experimental dependence of $S(q)$ on T, determined by X-ray or neutron diffraction.

We may note that L for most metals is considerably greater than $\underline{a}$ (the distance between atoms). For mercury, a poor conductor, $\sigma \approx 10{,}000$ ohm^{-1} cm^{-1}, and $L \approx 7$Å. When σ drops below 3,000 , as for liquid Te, the condition $L \gg a$ is not valid and the problem is no longer a weak scattering one.

A system much studied is that introduced by Anderson (Phys. Rev. <u>109</u>, 1492 (1958), namely a cubic array of potential wells giving a tight-binding band of width B , with a random potential between the limits $\pm\frac{1}{2} V_o$ in each well. Suppose the band is half full, with one electron per well, and we work with non-interacting electrons (no Mott transitions!). Then if $V_o \ll B$, the Born approximation gives

$$\frac{a}{L} = \left(\frac{V_o}{B}\right)^2 \times \text{const.}$$

When V_o and B become comparable, $a/L \sim 1$ and again we no longer have a weak scattering problem.

Multiple scattering theory should then be able to give the resistivity, but I do not know of a usable solution; we need one. The present author, Friedman and Hindley use a short cut, and suppose that, when $L \sim a$ the familiar Bloch tight-binding wave function

$$\psi = \Sigma \exp(ika_n)\phi_n(x,y,z) ,$$

where k is the wave-vector, a_n a lattice point and ϕ_n an atomic wave function at the site n becomes

$$\psi = \Sigma \exp(i\eta_n)\phi_n , \qquad (3)$$

where the η_n are random phases. Our next problem is to calculate the conductivity in this case. This is done by means of the so-called Kubo-Greenwood formula. One calculates the optical absorption by radiation of small frequency ω due to electrons within an energy $\hbar\omega$ of the Fermi energy, and from this one deduces

$\sigma(\omega)$, the conductivity at frequency ω . The calculation is entirely elementary; the trick is to add the contribution from each atom to the matrix element of d/dx using the random phases. The Kubo-Greenwood formula in general gives, for $\omega \to 0$

$$\sigma = (2\pi e^2\hbar^3/m^2) \, |D_E|^2_{av} \, \{N(E)\}^2 \qquad (4)$$

where D_E is the matrix element

$$D_E = \int \psi_E \left(\frac{\partial}{\partial x}\right) \psi_{E'} \, d^3x \qquad (5)$$

and the "av" means an average of E,E' over all states at the Fermi energy. For the case $L \sim a$ when we use the random-phase wave function (3), we set

$$D_E = N^{1/2} \delta_E$$

where δ_E is the matrix element (5) for a single well.

If $N(E)$ is the free electron density of states, it is not difficult to show that the conductivity reduces approximately to the form (2) with $L \sim a$ or putting in the value for S_F ,

$$\sigma = \frac{1}{3}\frac{e^2}{\hbar a} \qquad \sim 3000 \ \Omega^{-1} cm^{-1}$$

But if, with the Anderson model, $V_o > B$, the density of states approximates to $1/a^3V_o$ instead of $1/a^3B$ in the middle of the band, so that

$$\sigma = \frac{1}{3}\frac{e^2}{\hbar a}\left(\frac{B}{V_o}\right)^2 \qquad (6)$$

This important result comes from the presence of $\{N(E)\}^2$ in (4), the density $N(E)$ (for frequency ω) appearing in the initial and final state of the electron. It follows that metallic behaviour below 3000 $\Omega^{-1}cm^{-1}$ can occur; in fact one finds, in many systems in which an electron moves in a non-crystalline field, conductivities constant down to low T as small as 700 $\Omega^{-1}cm^{-1}$, but not lower. It looks as if something went wrong when (B/V_o) dropped below 1/2.

What goes wrong is "Anderson localization". Anderson in 1958 showed that when B/V_o drops below a certain critical value, now believed to be about 1/2 (or 1 in two dimensions), the wave functions become <u>localized</u>. This means that instead of the form (3), the wave functions take the form

$$\sum_n \exp(i\eta_n)\phi_n \exp\{-\alpha|r-r_s|\} \qquad (7)$$

r_s being some point in space and α a constant which is zero when B/V_o has the critical value and increases as

$$\left[\left(\frac{B}{V_o}\right)_{crit} - \left(\frac{B}{V_o}\right)\right]^{0.6} \qquad (8)$$

The concept of Anderson localization was slow in gaining acceptance. People felt that a continuous range of localized states was impossible, because an electron with a quantized energy in one such state would always find another with exactly the same energy at a great enough distance. This turns out to be wrong; two states cannot have the same energy if there is any overlap between them; the degeneracy will always be removed.

We call a degenerate electron gas for which the states are localized at the Fermi energy a "Fermi glass". Its properties are the following:

1) $\sigma(\omega)$, the conductivity at frequency ω , is proportional to ω^2 at $T = 0$.

2) At finite temperatures electrons will move from one state to another by absorbing energy from phonons. In this case

$$\sigma = C\, e^{-B/T^{1/4}} \qquad (9)$$

($T^{1/3}$ in two dimensions). The simplest argument is that the probability of a hop over a distance R is

$$C \exp(-2\alpha R - W/kT) \,, \qquad (10)$$

and, if the electron moves a distance R , it has R^3/a^3 sites to choose from; it will choose that which has the lowest energy, and thus

$$W \propto W_0 a^3/R^3$$

The most frequent hops are thus these for which $2\alpha R - W_0 a^3/R^3kT$ is a minimum, and this gives the form (9).

3) The density of states, linear specific heat and Pauli paramagnetism are unaltered by Anderson localization.

In a system where (V_0/B) is less than the critical value, it will nonetheless follow that states in the tail of the band are localized. An energy, denoted by E_c and called the "mobility edge" separates localized from non-localized states. For energies E below E_c, states are localized and wave functions have the form (7), α tending to zero as $E_c - E$ tends to zero. For a degenerate electron gas of which the Fermi energy lies below E_c, the d.c. conductivity tends to zero with T according to equation (9). But equation (9) will only express the conductivity at low T; at higher temperatures current may be caused by electrons excited to E_c, giving a conductivity of the form

$$\sigma = \sigma_0 \exp\{-(E_c-E_f)/kT\} , \qquad (11)$$

A number of systems exist in which it is possible, by varying the composition or in other ways, to change the number of free electrons and hence the Fermi energy in such a way that $E_f - E_c$. Examples are a system such as $La_xSr_{1-x}VO_3$, where the number of electrons in the vanadium d^1 band varies with x, and the disorder is provided by the random positions of the two ions La^{3+} and Sr^{2+}. Another is the impurity band of Si:P, where the number of electrons can be changed by compensation (i.e. the addition of acceptors). Another is the inversion layer of a MOSFET (metal-oxide-surface-field-effect transistor), where E_f for the two-dimensional electron gas varies with the gate voltage. These experiments will be described in the lecture. In all cases a metal-insulator transition is observed, which has been called an Anderson transition. The resistivity-temperature behaviour is as in fig. 1.

A controversial point in the theory is the value

of the conductivity just before localization occurs; the author has called it the "minimum metallic conductivity", denoted it by σ_{min} ; also it is easily shown that in region A of fig. 1

$$\sigma = \sigma_{min} \exp(-E/kT) \quad , \quad E = E_c - E_f \ .$$

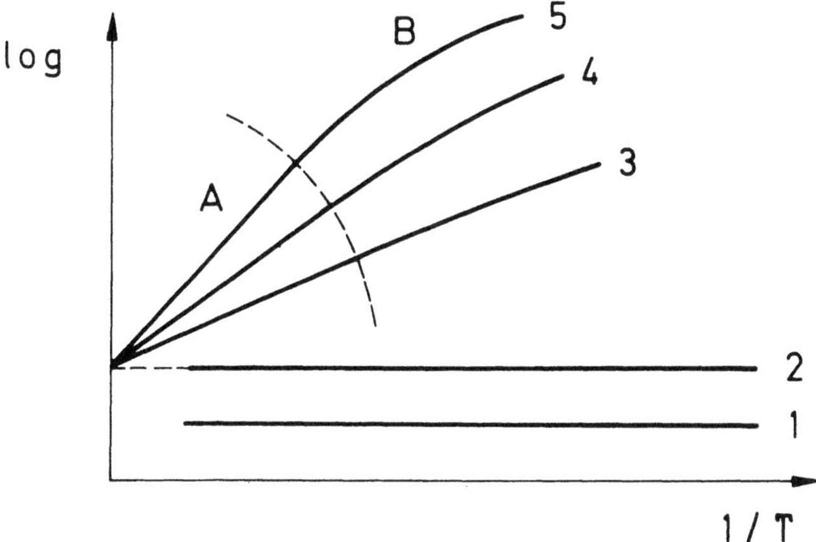

Fig. 1. Log (resistivity) against 1/T for an Anderson transition. The curves 1,2, ... 5 are for increasing disorder, or decreasing Fermi energy. 1 and 2 are metallic, ($E_F > E_c$) and 3, 4, 5 are for $E_c < E_F$. For these the region A is when the conductivity follows the equation

$$\sigma = \sigma_{min} \exp\{-(E_c - E_f)/kT\}$$

and the region B

$$\sigma = A \exp(-B/T^{1/4}) \ .$$

In two dimensions $T^{1/4}$ is to be replaced by $T^{1/3}$.

The author has taken it to be given by (6) when (B/V_o) has the critical value for Anderson localization; if this is 1/2 , it is

$$\sim 0.1\ e^2/\hbar a$$

(though elaboration of the theory can give values of the constant somewhat lower). Cohen and co-workers in a series of papers have maintained that long-range fluctuations in the potential must give a zero value of σ_{min}, because metallic conduction must take place along classical percolation channels. I do not think these arguments are correct unless the potential energy has large long-range fluctuations.

To test the theory, one ought to choose a situation where the hydrogen radius and the range of the potential fluctuations are large, so that α^{-1} , the radius of the localized states, is large. There are two reasons for this:

1) If α^{-1} is comparable with the lattice parameter, an electron in a localized state will always distort its surroundings. Thus the hopping energy in equation (9) will contain a term of polaron type, except at low T .

2) One has to ask for a degenerate electron gas whether the Anderson localized states are singly or doubly occupied.
   If the interaction $\langle e^2/r_{12}\rangle$ (the Hubbard U) between electrons is neglected, these states are of course doubly occupied, just like the Bloch functions in a metal; introduction of the Hubbard U must mean that states at the Fermi energy are singly occupied, but if α^{-1} is large, this will be so only very near E_F .

The two systems most useful for verifying the theory are

(a) Impurity bands in doped and compensated Ge and Si.

(b) Two dimensional conduction at the Si/SiO_2 interface.

In the former, $\underline{a}$ in eq. (12) is the distance between donors, and $\sigma_{min} \sim 10\ \Omega^{-1}\ cm^{-1}$; this seems in accord with experiment. Also the $T^{-1/4}$ behaviour has been verified. In the latter, since $\sigma_{min} \sim 0.1\ e^2/\hbar$, σ_{min} appears to be a universal constant independent of the form of disorder which in this case is due to random charges in the oxide. $1/T^{1/3}$ behaviour is observed.

Behaviour of the type shown in fig. 1 is also observed in many systems where the condition $\alpha^{-1} > \underline{a}$ is not satisfied. These include $La_xSn_{1-x}VO_3$ and related compounds, pyrolitic carbons and many others. It is possible that the electron gas here should be described as a "degenerate gas of small polarons". At low temperatures the polaron formation will simply decrease the band-width, and lead to Anderson localization for smaller random fields V_o than would otherwise be the case; it would not - and apparently does not - affect the minimum metallic conductivity. On the other hand at higher temperatures a polaron hopping energy ought to be observed.

There are many unsolved problems in this area. I would like to list a few of them.

1) Is the author's treatment of the minimum metallic conductivity correct ? The evidence seems to favour it. If so, how should one treat the transition to long-range fluctuations, when classical percolation theory is applicable ?

2) A Hall coefficient is observed in the region A of fig. 1; it is always negative, and the Hall and conductivity mobilities are unequal, for reasons explained by Friedman. But in the hopping $T^{-1/4}$ region, the Hall effect is observed to disappear. We are not quite sure why. This is in marked contrast with polaron hopping.

3) What does interaction between electrons do to variable-range hopping, and should $1/T^{1/4}$ still be expected ? The problem arises because the potential in which an electron moves depends on where the other electrons are. An electron in a localized state raises the energy of neighbouring states, which will therefore be unoccupied. The electron therefore builds an "electronic polaron" round it, which it must carry

with it when it hops.

Amorphous Semiconductors

In the reports of most conferences on conduction in non-crystalline materials, the largest number of papers deal with evaporated or sputtered films of amorphous silicon, germanium and the chalcogenide glasses (e.g. As_2Te_3). Thin film technology plays an important part here; sometimes the films contain voids (holes of macroscopic size, probably charged, which may give rise to large-range fluctuations of potential). The questions one can ask here are

1) How can there be an energy gap at all ? The conventional theory of energy gaps depend on Bragg reflection from the crystal lattice. I would say that no fully satisfactory theory exists, but the most striking fact about glasses is that they are transparent (As_2Te_3 is in the infra-red). So a gap, and conduction and valence bands, exist.

2) Can one envisage an "ideal glass" e.g. amorphous Si or As_2Te_3, in which all bonds are satisfied ? We assume that this is so, but that real amorphous materials and glasses, like crystals, contain point objects (dangling bonds). These lead to <u>deep</u> donor and acceptor states near mid-gap.

3) It is difficult in non-crystalline semiconductors (but not impossible in amorphous Si) to produce shallow donors; this was first shown by Kolomiets for the chalcogenides. It is supposed that the structure of the glass is such that all electrons are taken up in bonds.

4) Is it correct to generalize the Anderson model and suppose that conduction and valence bands, due to absence of long-range order, have a range of localized states leading up to a mobility edge ? This was first proposed by Cohen, Fritzsche and Ovshinsky (1969). There is a limited amount of evidence that this is so, but no calculations have been made of what the range of localized states should be. Experiments suggest 0.1 to 0.2 eV.

5) Do carriers in band-edge localized states form polarons ? Emin and co-workers suggest that this is so for holes in chalcogenide glasses. Part of the evidence is that, if $\sigma = \sigma_0 \exp(-E/kT)$ describes

the conduction, the thermopower behaves like

$$S = (k/e)\,[E'/kT + \text{const}]$$

and $E \neq E'$, the difference $E - E'$ being ~ 0.15 eV. It is suggested that this represents a polaron hopping energy. Other explanations are however possible, as suggested by Nagels*, and I think the question is open.

For glow-discharge-deposited Si, formed from Silane (SiH_4), there seem to be no voids, and the density of gap states has been investigated by many techniques, particularly by Spear*. These seem to be due to divacencies, and to be similar to these in crystalline Si. There is little evidence for any large Stokes shift or polaron formation. In evaporated Silicon containing voids there is a high concentration of divacencies (leading to $1/T^{1/4}$ hopping at low T) and (charged) voids. If as I believe these lead to long-range fluctuations of potential, percolation theory rather than that of a mobility edge E_c should be applicable. Some evidence will be presented.

Chalcogenides have been a puzzle for a long time. There is very strong evidence for "gap states", which pin the Fermi energy in mid-gap. On the other hand no Curie paramagnetism or esr signal is observed. A recent paper by P.W. Anderson (Phys. Rev. Letters 34, 953 (1975) may give the clue. I believe dangling bonds do exist in these glasses, but that a lone pair Se or Te orbital from which an electron is missing at a defect site forms a very strong bond with a neighbouring Se atom, and that an orbital with two holes forms an even stronger bond. It is then possible that all dangling bonds contain either two or no electrons. These speculations are explored by N.F. Mott, E.A. Davis and R.A. Street, Phil. Mag. (1975), in press.

* Articles by Nagels and by Spear will be found in "Amorphous and Liquid Semiconductors", ed. J. Stuke and W. Brenig.

REFERENCES

The author's two books are:

- N.F. Mott and E.A. Davis, Electronic Processes in Non-crystalline materials, Oxford (1971).
- N.F. Mott, Metal-Insulator Transitions, Taylor & Francis (1974).

See also

Electronic and Structure Properties of Amorphous Semiconductors, ed. P.G. Le Comber and J. Mort, Academic Press (1973).

Much of the experimental data are found in reports of recent conferences, e.g.

Amorphous and Liquid Semiconductors, ed. J. Stuke and W. Brenig, Taylor & Francis (1974) and

Tetrahedrally-bonded Amorphous Semiconductors, American Institute of Physics, ed. M.H. Brodsky, S. Kirkpatrick and D. Weaire (1974).

The experimental work on the Si/SiO_2 interface is described by N.F. Mott, M. Pepper, S. Pollitt, R.H. Wallis and C.J. Adkins, Proc. Roy. Soc. A. (London), <u>345</u>, 169.

THEORY OF HOPPING CONDUCTIVITY IN DISORDERED SEMICONDUCTORS

P.N. Butcher

University of Warwick

Coventry, England

ABSTRACT

The semiclassical theory of hopping conductivity in disordered semiconductors is reviewed. The relationship between hopping conductivity and random walks in random systems is described. The randomness of the system is shown to play a major role in determining the frequency dependence of the ac conductivity. The dc conductivity is controlled by the electron diffusivity which involves the product of the mean hopping frequency and the mean square distance moved per hop. The mean hopping frequency is calculated for simple models and shown to contain no exponential factors. The mean square distance moved per hop is exponentially smaller than the mean square nearest neighbour intersite spacing. It is calculated for simple models by using an approach derived from percolation theory.

1. INTRODUCTION

We are concerned in these lectures with the conduction of electricity by electrons occupying localized states in disordered semiconductors. The emphasis is on effects which are directly related to the randomness of the system. Small polaron hopping in crystals will be discussed by Prof. Emin. An electron in a localized state carries no current, i.e. the expectation value of the electron velocity is zero. Conduction processes involving localized states are therefore entirely determined by transitions of electrons from full states to neighbouring empty states. It is important to emphasise that this basic process changes the position of the electrons in $\underline{r}$-space. It is called "hopping" and conduction involving localized states is usually referred to as "hopping" conductivity.

The hopping mechanism was originally proposed by Conwell[1] and Mott[2,3] to explain the behaviour of the dc conductivity observed in compensated crystalline semiconductors cooled to liquid helium temperatures so as to suppress band conduction. In an n-type crystalline semiconductor the hopping conductivity is due to electrons in localized donor states which are randomly located throughout the crystal. Compensation is necessary to make some of these states empty. The charged acceptors have a second effect: they produce a Coulomb field which randomizes the donor levels. This means that the hops are necessarily inelastic and must be phonon-assisted. Miller and Abrahams[4] calculate the transition rates for this case in a pioneering paper setting out the basis for a semi-classical treatment of hopping conductivity.

Interest in hopping conductivity has grown up again in recent years with the expansion of research on the physical properties of amorphous semiconductors[3]. Some of the electron states in these materials are localized and a hopping mechanism is to be expected for conduction involving these states. Moreover, the observed behaviour of both the ac and dc conductivity in many of these materials is qualitatively similar to that found in compensated crystalline semiconductors[3].

In developing the theory of hopping conductivity we adopt a one-electron semiclassical viewpoint. The localized states are supposed to be centred on fixed sites which may be occupied by one and only one electron. The theory concentrates on transitions between the sites. The localized wave functions serve only to determine where the sites are and, of course, they enter into the formulae for the transition rates. We employ simple approximations to the transition rates which are believed to exhibit their most important features. This approach is naive but fruitful. It embraces much of the current understanding of experimental data. Moreover, by keeping the physics simple we throw into prominence the pronounced effect of the randomness in determining the salient features of both ac and dc hopping conductivity in disordered semi-conductors.

In Section 2 we use the rate equation approach[4-6] to develop a formal expression for the ac hopping conductivity. The formalism has a direct interpretation in terms of random walks in random systems[7,8] which we discuss in Section 3. At high frequencies the pair approximation introduced by Pollak and Geballe[9] reproduces the general features of the experimental data. We describe this approximation in Section 4. It concentrates attention on electrons executing reciprocating hops between anomalously close pairs of sites and breaks down when the frequency tends to zero. We discuss the dc limit in Section 5. It is here that the random walk interpretation of the formalism has most value[8]. The conductivity is related to the diffusivity by the Einstein relation. The diffusivity involves the product of the mean hopping frequency and the mean square displacement per hop. We show in Section 6 that the mean hopping frequency is easily evaluated for simple models. It contains no exponential factors of the type which show up in experimental data on dc hopping conductivity[3]. The exponential factors arise from the mean square displacement per hop which is exponentially smaller than the mean square nearest neighbour intersite spacing. This is a characteristic feature of random walks in random systems[8]. We examine it in detail in Section 7 in the light of the percolation theoretic ideas introduced into hopping conductivity theory by Ambegaokar, Halperin and Langer and others[10-12].

Scher and Lax[13,14] have given an illuminating stochastic treatment of both ac and dc hopping conductivity in an attempt to go beyond the confines of the pair approximation. Their results are identical to those for a random walk in which the site distribution is re-randomized immediately after every hop[8]. We discuss this new type of random walk briefly in Section 8. It is of considerable interest in its own right. However, its application to hopping conductivity is limited because site re-randomization has a drastic effect on both the mean hopping frequency and the mean square distance moved per hop.

2. THE RATE EQUATION FORMALISM

2.1 The Rate Equations in an Applied Potential Field

We consider a finite array of N_s sites in a macroscopic volume Ω and suppose that each site may be occupied by one and only one electron. The sites are to be imagined as located at the mean position of an electron occupying the associated localized state. We label the sites by an integer m and write f_m for the probability that site m is occupied by an electron of either spin orientation. To determine f_m we use the rate equations

$$\frac{df_m}{dt} = \sum_n \left[f_n(1-f_m)R_{nm}-f_m(1-f_n)R_{mn}\right] \tag{1}$$

where R_{mn} is the transition rate from a full site m to an empty site n which we suppose to be independent of spin.

To determine f_m when the system is in uniform thermal equilibrium at temperature T we use the conventional procedures of statistical mechanics. For simplicity we suppose that an electron has only one energy level E_m available to it at site m. Because we allow only single occupancy of the sites, the problem is completely analogous to that of determining the probability of occupation of a donor with a single level in an n-type semiconductor. The result is that f_m reduces to the Fermi-Dirac form

$$f_m^o = [\exp\{\beta(E_m-\zeta)\}+1]^{-1} \tag{2}$$

where $\beta = (k_BT)^{-1}$. We refer to the quantity ζ in eqn. (2) as the chemical potential although, strictly speaking, it is the chemical potential plus $k_BT\ln 2$ because of the spin degeneracy.

Detailed balance prevails in thermal equilibrium, i.e. each term in the sum on the right-hand side of eqn. (1) vanishes separately. It follows from eqn. (2) that the transition rates R_{mn}^o in thermal equilibrium satisfy the detailed balance relation

$$R_{mn}^o/R_{nm}^o = \exp[\beta(E_m-E_n)] \tag{3}$$

Now let us suppose that a weak external potential field $U(\underline{r},t)$ is applied to the system so that the potential at site m with position vector $\underline{r}_m$ is $U_m = U(\underline{r}_m,t)$. We approximate the effect of the applied potential on the transition rates by supposing that R_{mn}/R_{nm} is given by (3) with E_m-E_n replaced by $(E_m+U_m) - (E_n+U_n)$. Then to first order in U_m we have

$$R_{mn}/R_{nm} = [R_{mn}^o/R_{nm}^o][1+\beta(U_m-U_n)] \tag{4}$$

The effect of the applied potential on the ratio of the rates is all that is required to solve the rate equations in first order. Eqn. (4) is certainly valid for a static applied potential and will remain approximately valid at the low frequencies which are of interest in the conductivity problem.

In the presence of the applied potential f_m suffers a perturbation f_m^1 at site m. By substituting $f_m = f_m^o+f_m^1$ in eqn. (1) and using eqns. (2) and (4) we find that f_m^1 is given to first order by the linearized rate equations

$$\frac{df_m^1}{dt} = \sum_n[f_n^1R_{nm}^e - f_m^1R_{mn}^e] + \beta\sum_n[F_nU_nR_{nm}^e - F_mU_mR_{mn}^e] \tag{5}$$

where

$$R_{mn}^e = \Gamma_{mn}/F_m \tag{6}$$

with

$$\Gamma_{mn} = f_m^o(1-f_n^o)R_{mn}^o \tag{7}$$

and

$$F_m = f_m^o(1-f_m^o)$$

$$= -k_B T \, df_m^o/dE_m \tag{8}$$

We see by inspection of eqn. (7) that Γ_{mn} is the electron flux from m to n in thermal equilibrium which is symmetrical in m and n because of detailed balance. We also notice that F_m reduces to f_m^o and R_{mn}^e reduces R_{mn}^o in the non-degenerate limit when $f_m^o \to o$. Consequently the general linearized rate eqns. (5) are formally identical to the corresponding equations for a non-degenerate system with f_m^o replaced by F_m and R_{mn}^o replaced by R_{mn}^e. The notation has been chosen to emphasize this analogy of the general case and the non-degenerate case. The analogy arises because we have linearized the equations. It turns out to be useful to think in these terms and we refer to F_m and R_{mn}^e respectively as the "effective" occupation probability of site m and the "effective" transition rate from m to n.

The formal solution of eqns. (5) is facilitated by the introduction of matrix notation. We write $\underline{f}^1$ and $\underline{U}$ for row matrices whose mth columns are f_m^1 and U_m respectively and define a diagonal square matrix F whose (mm)th element is F_m. Finally, we define a relaxation matrix R whose (mn)th element is given by

$$R_{mn} = R_m^e \, \delta_{mn} - R_{mn}^e \tag{9}$$

where

$$R_m^e = \sum_n R_{mn}^e \tag{10}$$

is the total effective transition rate out of site m. In equations (9) and (10) R_{mn}^e vanishes when m=n and δ_{mn} is the Kronecker δ-symbol. With this notation eqn. (5) becomes simply

$$\frac{d\underline{f}^1}{dt} = - \underline{f}^1 R - \beta \underline{U} F R \tag{11}$$

2.2 The ac Conductivity Formula

In order to derive an expression for the ac conductivity at frequency ω we suppose that $U(\underline{r},t)$ is the potential due to a uniform electric field E applied in the x direction and having a time factor $\exp(-i\omega t)$. Then $U(\underline{r},t) = eEx\,\exp(-i\omega t)$ and $\underline{U} = eE\underline{x}\,\exp(-i\omega t)$ where $\underline{x}$ is a row matrix whose mth column is x_m. Moreover $\underline{f}^1(t) = \underline{f}^1(o)\exp(-i\omega t)$ and eqn. (11) reduces to

$$\underline{f}^1(o)(R-i\omega) = -\beta eE\underline{x}FR \tag{12}$$

The formal solution of eqn. (12) is

$$\underline{f}^1(o) = -\beta eE\underline{x}FRG \tag{13}$$

where the Green matrix G is defined by

$$(R-i\omega)G = 1 \tag{14}$$

Now that eqn. (11) has been solved the formal calculation of the ac conductivity is trivial. The induced dipole moment density in the x-direction is

$$\begin{aligned} P_x(t) &= -e\Omega^{-1} \sum_m f_m^{\,1}(t)x_m \\ &= -e\Omega^{-1}\, \underline{f}^1(t)\tilde{\underline{x}} \end{aligned} \tag{15}$$

where Ω is the volume of the system and we use a tilde to indicate the transpose of a matrix. The x component of the current density is $-i\omega P_x(t)$ and the conductivity at frequency ω is therefore given by

$$\begin{aligned} \sigma(\omega) &= \frac{-i\omega P_x(t)}{E\exp(-i\omega t)} \\ &= -i\omega e^2\beta\Omega^{-1}\, \underline{x}FRG\tilde{\underline{x}} \end{aligned} \tag{16}$$

The disordered systems which we consider are all isotropic and the direction of the x-axis in the above calculation is arbitrary. Most of our subsequent discussion is concerned with the interpretation of the formal result (16).

3. RANDOM WALK INTERPRETATION

3.1 Stochastic Interpretation of G

The matrix G defined by eqn. (14) has a simple stochastic interpretation. Let us consider a single electron performing a random walk over the system of sites. We suppose that no other electrons are present and that the motion of the single electron which is present is controlled by the effective transition rates. If the electron is known to be on site m at time 0 then the probability $P_{mn}(t)$ that it will be on site n at time t is the solution of the homogeneous rate equations

$$\frac{d}{dt} P_{mn}(t) = \sum_p \left[P_{mp}(t)R^e_{pn} - P_{mn}(t)R^e_{np}\right] \tag{17}$$

subject to the initial condition $P_{mn}(o) = \delta_{mn}$.

We may write eqn. (17) in matrix notation as $dP(t)/dt = -P(t)R$ where $P(t)$ has elements $P_{mn}(t)$ and the relaxation matrix R is defined by eqn. (9). In order to incorporate the initial condition $P(o) = 1$ easily into the solution of this equation it is convenient to work with $P(t)\theta(t)$ where $\theta(t)$ is the unit step function. Then we have

$$\frac{d}{dt}\left[P(t)\theta(t)\right] = -P(t)\theta(t)R+\delta(t) \tag{18}$$

where $\delta(t)$ is the Dirac δ-function. The Fourier transform of eqn. (18) is

$$P^+(\omega)\left[R-i\omega\right] = 1 \tag{19}$$

where

$$P^+(\omega) = \int_{-\infty}^{\infty} P(t)\theta(t)e^{i\omega t}dt \tag{20}$$

is the Fourier transform of $P(t)\theta(t)$.

Equation (19) implies that $P^+(\omega)$ is the left-hand inverse of $R-i\omega$. On the other hand, eqn. (14) implies that G is the right-hand inverse of $R-i\omega$. It follows that

$$G = P^{+}(\omega) \tag{21}$$

since the left-hand and right-hand inverses of a finite matrix are identical. Thus G is the Fourier transform of $P(t)\theta(t)$. This is the stochastic interpretation of G which we require.

In the conductivity problem we are concerned with the motion of many electrons. When the electron statistics are non-degenerate each electron moves independently of the others and $R^{e}_{mn} = R^{o}_{mn}$. In that case the motion of the single electron considered above provides a correct description of the motion of any one electron in the many electron system in thermal equilibrium. This is obviously no longer true when the electron statistics are degenerate because the hopping trajectories of different electrons in the many electron system are correlated and $R^{e}_{mn} \neq R^{o}_{mn}$. The motion of the single electron considered above is then artificial. Nevertheless it continues to provide a useful interpretation of G. Equation 21 shows that, whether the electron statistics are degenerate or non-degenerate, G_{mn} is the Fourier transform of $P_{mn}(t)\theta(t)$ calculated by considering a random walk performed by a single electron which is controlled by the effective transition rates. Many of the relations to be developed here become particularly transparent when interpreted in these terms. We shall not bore the reader with continual reminders that the random walk problem is an artificial one in the general case.

3.2 Stochastic Interpretation of $\sigma(\omega)$

Now that we have a stochastic interpretation of G it is only a matter of algebra to rewrite eqn. (16) in a form which also has a stochastic interpretation. In carrying out the algebraic manipulations we use the following relations

$$RG = 1+i\omega G \tag{22}$$

$$1+i\omega G_{mm} = -i\omega \sum_{n \neq m} G_{mn} \tag{23}$$

$$F_m G_{mn} = F_n G_{nm} \tag{24}$$

Equation (22) is simply eqn. (14) rewritten in a more convenient way. To prove the identity (23) we remember that $P_{mn}(t)$ is a probability distribution over n. Consequently

$$\sum_n P_{mn}(t)\theta(t) = \theta(t) \tag{25}$$

Taking the Fourier transform of this equation and using eqn. (21) we obtain eqn. (23). Equation (24) states that the matrix FG is symmetrical. This fact follows easily from eqn. (14) when we recognise that eqn. (6) and the symmetry of Γ_{mn} imply that FR is symmetrical.

To carry out the algebra we substitute for RF in eqn. (16) from eqn. (22) and write out the matrix products in full. The diagonal elements of G may then be eliminated with the aid of eqn. (23) to yield

$$\sigma(\omega) = -i\omega e^2\beta\Omega^{-1}\left[\sum_m x_m F_m \sum_{n=m} (1+i\omega G)_{mn} x_n - \sum_m x_m F_m \sum_{n=m} i\omega G_{mn} x_m\right]$$

$$= \omega^2 e^2 \beta\Omega^{-1} \sum_{mn} F_m G_{mn} x_m (x_n - x_m) \qquad (26)$$

The algebra is completed by symmetrizing the summand in eqn. (26) and using eqn. (24). Thus we obtain the final result

$$\sigma(\omega) = -\frac{1}{2}\omega^2 e^2 \beta\Omega^{-1} \sum_{mn} F_m G_{mn} (x_m - x_n)^2 \qquad (27)$$

The stochastic interpretation of eqn. (29) becomes obvious when we remember that the electron density is

$$n = \Omega^{-1} \sum_m f_m^o \qquad (28)$$

so that, using eqn. (2) and (8), we have

$$\frac{dn}{d\zeta} = \beta\Omega^{-1} \sum_m F_m \qquad (29)$$

We may therefore rewrite eqn. (29) as

$$\sigma(\omega) = e^2 \frac{dn}{d\zeta} D(\omega) \qquad (30)$$

where

$$D(\omega) = -\frac{1}{2}\omega^2 \frac{\sum_m F_m \Delta_m^2(\omega)}{\sum_m F_m} \qquad (31)$$

with

$$\Delta_m^2(\omega) = \sum_n G_{mn}(x_m - x_n)^2 \tag{32}$$

Equation (30) is just the Einstein relation at frequency ω for a degenerate electron system. The quantity $D(\omega)$ is easily recognised as the diffusivity at frequency ω. We see from eqns. (21) and (32) that $\Delta_m^2(\omega)$ is the Fourier transform of $\theta(t)\delta_m^2(t)$ where

$$\delta_m^2(t) = \sum_n P_{mn}(t)(x_m - x_n)^2 \tag{33}$$

is the mean square x-displacement at time t for a single electron setting out from site m at time 0. The effective occupation probability F_m appears in eqn. (31) to take proper account of the electron statistics in averaging $\theta(t)\delta_m^2(t)$ over all possible initial sites in the calculation of the diffusivity. Scher and Lax[13] derive eqns. (30) to (32) from a Kubo formula in the non-degenerate case. It is also possible to calculate $D(\omega)$ directly from the linearized rate equations for the case when there is time-dependent gradient in the electron concentration[7]. The result is eqn. (31).

3.3 Stochastic Interpretation of the Dyson Expansion for G

We may derive a Dyson expansion for G from eqn. (14) by splitting R into a diagonal part R^d, and an off-diagonal part $-R^1$. Taking the term $-R^1G$ over to the right-hand side and iterating we have

$$G = G^o + G^oR^1G^o + G^oR^1G^oR^1G^o + \dots \tag{34}$$

where

$$G^o = (R^d - i\omega)^{-1} \tag{35}$$

is diagonal. We see from eqn. (9) that $R_{mm}^d = R_m^e$ and $R_{mn}^1 = R_{mn}^e$. Hence the matrix elements of G in eqn. (34) are given by

$$G_{mn} = G_{mm}^o\delta_{mn} + G_{mm}^o\left[R_{mn}^e + \sum_p R_{mp}^e G_{pp}^o R_{pn}^e + \dots\right]G_{nn}^o \tag{36}$$

where

$$G_{mn}^o = \left[R_m^e - i\omega\right]^{-1} \tag{37}$$

The individual terms in the series (36) have a stochastic interpretation. The probability that a single electron placed on site m at time 0 will wait there without hopping anywhere else until time t is $\exp(-tR^e_m)$. The Fourier transform of $\theta(t)\exp(-tR^e_m)$ is G^o_{mm}, i.e. the leading term in the series (36) with n=m. Moreover, the probability that the electron will wait on site m until time τ, then make a single hop to site n=m in the time interval $(\tau, d\tau)$ and finally wait on site n until time t is

$$\exp(-\tau R^e_m) R^e_{mn} d\tau \exp\left[-(t-\tau)R^e_n\right] \tag{38}$$

Integrating the expression (38) with respect to τ from o to t, multiplying by $\theta(t)$ and taking the Fourier transform we obtain the leading term in the series (36) when n=m. The higher order terms in the series may be generated in a similar way. Thus we see that the Nth term in the series is the Fourier transform of $\theta(t)$ times the probability that an electron placed on site m at time 0 will be found on site n at time t having made N-1 hops to get there.

4. THE PAIR APPROXIMATION

4.1 General Formulae

Let us write

$$\sigma(\omega) = \sigma_1(\omega) + i\sigma_2(\omega) \tag{39}$$

to separate the real and imaginary parts of $\sigma(\omega)$ and use angular brackets to indicate system averages. We may immediately obtain exact formulae for $\langle\sigma_1(\infty)\rangle$ and $\langle\sigma_2(\infty)\rangle$. When $\omega \to \infty$ we see from eqns. (36) and (37) that the off-diagonal elements of G are given by $G_{mn} = -R_{mn}/\omega^2$. Substituting this result into eqn. (27) and using eqn. (6) we have $\langle\sigma_2(\infty)\rangle = 0$ and

$$\langle\sigma_1(\infty)\rangle = \frac{1}{2}e^2\beta\Omega^{-1}\sum_{mn}\langle\Gamma_{mn}(x_m - x_n)^2\rangle \tag{40}$$

The evaluation of this exact formulae will be completed in some special cases later on. For the moment we merely observe that, because a double sum is invôlved, the result must be proportional to the square of the site density $n_s = N_s/\Omega$. At any finite, non-zero frequency there will be additional terms in an expansion of $\langle\sigma(\omega)\rangle$ in powers of n_s but the dominant term when $n_s \to o$ is also proportional to n_s^2. It is easy to calculate and is known as the pair approximation. Zero frequency is excluded because expansion of $<\sigma(o)>$ in powers of n_s is impossible.

The pair approximation to $<\sigma(\omega)>$ is obtained by selecting the contributions to the Dyson expansion (36) with n=m in which the intermediate sites are all identical to either m or n. The sum of all the terms of this type in the series is

$$G^p_{mn} = G^o_{mm}R^e_{mn}(1+q_{mn}+q^2_{mn}+ \ldots.)G^o_{nn}$$

$$= G^o_{mm}R^e_{mn}G^o_{nn}(1-q_{mn})^{-1} \qquad (41)$$

where $q_{mn} = G^o_{nn}R^e_{nm}G^o_{mm}R^e_{mn}$. Moreover, to order n_s^2 in the final result, we may identify R^e_m and $R_n{}^e$ with R^e_{mn} and R^e_{nm} respectively in evaluating the expression (41) with the aid of eqn. (37). Thus we obtain

$$G^p_{mn} = \frac{iR^e_{mn}\tau_{mn}}{\omega(1-i\omega\tau_{mn})} \qquad (42)$$

where the relaxation time τ_{mn} is given by

$$\tau^{-1}_{mn} = R^e_{mn} + R^e_{nm} \qquad (43)$$

Equation (42) may also be derived by ignoring the presence of all other sites except m and n in eqn. (14) so that we have only a 2x2 matrix to invert. Pollak and Geballe[9] originally introduced the pair approximation in this way for a particular case and Pollak[15-17] has made several extensions of the original analysis. Butcher and Morys[18] have verified that the approximation (42) yields $<\sigma(\omega)>$ to order $n_s{}^2$ when $\omega \neq 0$.

When G^p_{mn} is substituted into eqn. (27) we obtain the system-averaged result

$$<\sigma^p(\omega)> = \frac{1}{2}\, e^2\beta\Omega^{-1}(-i\omega)\sum_{mn} <\frac{F_m R^e_{mn}\tau_{mn}}{1-i\omega\tau_{mn}}(x_m-x_n)^2> \qquad (44)$$

when $\omega \to \infty$ we see that eqn. (44) reproduces the exact result (40). This comes about because we retain the leading term in the off-diagonal series (36) in the pair approximation. When $\omega \to o$, on the other hand, eqn. (44) yields $<\sigma^p(\omega)> = 0$, i.e. the pair approximation fails to account for dc conductivity.

4.2 Energy-Independent Model

The frequency dependence of $\langle\sigma^P(\omega)\rangle$ is well exhibited by a simple model with no energy dependence. The model is defined by

$$F_m = F \tag{45}$$

$$\Gamma_{mn} = \gamma_o \exp(-2\alpha r_{mn}) \tag{46}$$

and the assumption that each site location is independently and uniformly distributed over the volume Ω of the system. In eqn. (45) the quantity F is a constant effective site occupation probability. In eqn. (46) the quantity γ_o is a constant equal to the equilibrium electron flux between two immediately adjacent sites. The reader is cautioned that γ_o is <u>not</u> the maximum effective transition rate which is equal to $\gamma_o F^{-1}$ from eqn. (6). We have assumed that Γ_{mn} falls off exponentially with increasing distance r_{mn} between the sites with a decay constant 2α. In compensated crystalline semiconductors the spatial dependence of Γ_{mn} is indeed dominated by an exponential factor because the thermal equilibrium transition rate involves an overlap integral between exponentially decaying wave functions[4]. In amorphous semiconductors the spatial dependence of Γ_{mn} may actually be much more involved.

With the above assumptions the evaluation of $\langle\sigma_1(\infty)\rangle$ from eqn. (40) is trivial and we have

$$\langle\sigma_1(\infty)\rangle = n_s^2 e^2 \beta \gamma_o 16\pi (2\alpha)^{-5} \tag{47}$$

Using this result to normalize $\langle\sigma^P(\omega)\rangle$ we obtain from eqn. (44)

$$\frac{\langle\sigma^P(\omega)\rangle}{\langle\sigma_1(\infty)\rangle} = \frac{1}{24}(2\alpha)^5(-i\omega\tau_o)\int_o^\infty r^4\left[1-i\omega\tau_o e^{2\alpha r}\right]^{-1} dr \tag{48}$$

where $\tau_o = F/2\gamma_o$ is the reciprocal of twice the maximum effective transition rate. The real and imaginary parts of the integral in eqn. (48) have the character of Fermi integrals. When $\omega\tau_o \ll 1$ they may be evaluated by using the asymptotic method developed by Sommerfeld[19]. The leading terms are obtained by noting that[9,18]

$$-i\left[1-i\omega\tau_o e^{2\alpha r}\right]^{-1} \simeq \frac{\pi}{4\alpha}\,\delta(r-r_\omega) - i\theta(r_\omega - r) \tag{49}$$

where $r_\omega = (2\alpha)^{-1} \ln (\omega\tau_o)^{-1}$ is the intersite separation for which $\omega\tau_{mn} = 1$. Thus we have

$$\frac{<\sigma^p(\omega)>}{<\sigma_1(\infty)>} \simeq \frac{\pi}{48}\omega\tau_o \left[\ell_\omega^{\ 4} - i\frac{2}{5\pi}\ell_\omega^{\ 5}\right] \tag{50}$$

where $\ell_\omega = \ln (\omega\tau_o)^{-1}$. The complete asymptotic series for $<\sigma_1(\omega)>$ and $<\sigma_2(\omega)>$ are given by[20]

$$\frac{<\sigma_1^p(\omega)>}{<\sigma_1(\infty)>} = \frac{\pi}{48}\omega\tau_o \left[\ell_\omega^{\ 4} + \frac{3\pi^2}{2}\ell_\omega^{\ 2} + \frac{5}{16}\pi^4\right] \tag{51}$$

$$\frac{<\sigma_2^p(\omega)>}{<\sigma_1(\infty)>} = -\frac{1}{120}\omega\tau_o \left[\ell_\omega^{\ 5} + \frac{5}{6}\pi^2\ell_\omega^{\ 3} + \frac{7}{48}\pi^4\ell_\omega\right] \tag{52}$$

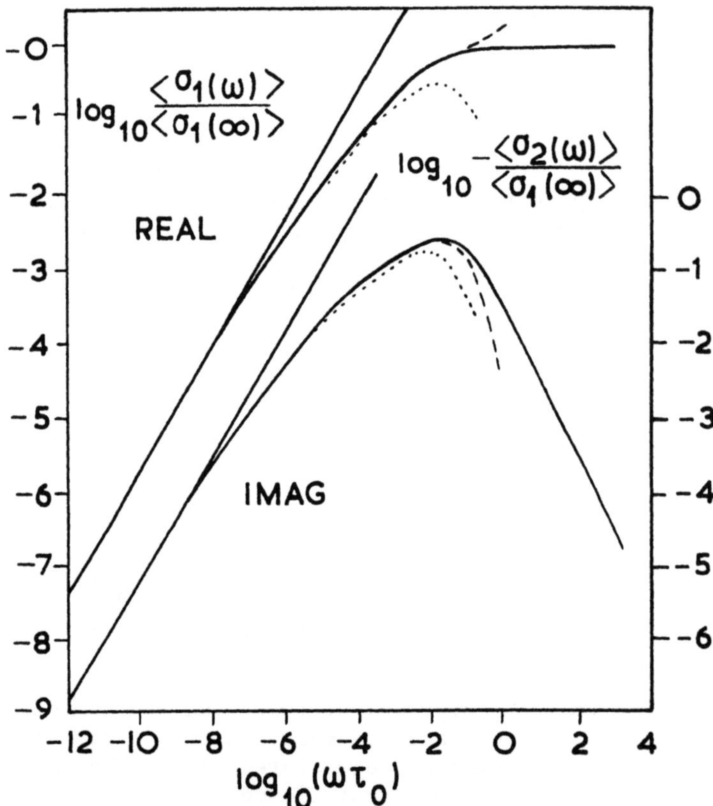

Fig.1. Frequency dependence of the real and imaginary parts of the ac conductivity in the pair approximation[20].

In Fig. 1 we compare the asymptotic formulae (51) and (52) (dashed curves) with exact results (full curves) obtained by numerically integrating the integral in eqn. (48). The dotted curves show the real and imaginary parts of the expression (50) and the straight lines both have a slope of 0.82. We see that the leading terms in the asymptotic expansions are adequate when $\omega\tau_o < 10^{-5}$. Since τ_o is estimated to be in the order of 10^{-12} to 10^{-13} in many materials[3] this implies that eqn. (50) is a good approximation for frequencies ω below about 10 MHz. For higher frequencies the additional terms included in eqns. (51) and (52) are necessary but these again fail for $\omega\tau_o > 10^{-1}$ and the numerical results are required for good accuracy.

At low frequencies both $\langle\sigma_1^p(\omega)\rangle$ and $\langle\sigma_2^p(\omega)\rangle$ are approximately proportional to ω^s with $s \simeq 0.8$. A great deal of experimental data has been successfully interpreted on this simple power law basis[9,21-25]. We may easily use the Kramers-Kronig relations to show that $\langle\sigma_1^p(\omega)\rangle \propto \omega^s$ implies that $\langle\sigma_2^p(\omega)\rangle/\langle\sigma_1^p(\omega)\rangle = -\tan(\frac{1}{2}\pi s)$. Most of the experimental data quoted above is apparently consistent with this relation with a constant value of s. However, we see from eqn. (50) that

$$\frac{-\langle\sigma_2^p(\omega)\rangle}{\langle\sigma_1^p(\omega)\rangle} = -\frac{2}{5\pi}\,\ell_\omega = +0.293\ \lg\,(\omega\tau_o) \qquad (53)$$

We would therefore expect $-\langle\sigma_2^p(\omega)\rangle/\langle\sigma_1^p(\omega)\rangle$ to fall off at the rate of 0.293 per decade increase in frequency. There have been no experimental observations of such a slow decrease of the ratio. Observation of it would provide a check on the model and would also yield unambiguous values of τ_o. In this connection, we also notice that when the right-hand of eqn. (53) is much greater than one it may be written in the form $\tan\left|\frac{1}{2}\,s(\omega)\right|$ where $s(\omega) = 1 - 5\ell_\omega$ is the logarithmic slope of $\langle\sigma_2^p(\omega)\rangle$ derived from eqn. (50), which should also decrease slowly as the frequency increases.

4.3 Energy Dependent Model

We have discussed at length the frequency dependence predicted for the energy-independent model because essentially the same results are obtained when energy dependence is taken into account. Let us confine our attention to the most interesting case when the electron statistics are strongly degenerate. Then, in place of eqn. (45) we must use eqn. (8). To generalize eqn. (46) we follow Ambegaokar, Halperin and Langer[10] and suppose that the hops are single phonon transitions. Then R^o_{mn} contains a factor n_{ph} when

$E_m<E_n$ and a factor $n_{ph}+1$ when $E_m>E_n$. Here

$$n_{ph} = \left[\exp\{\beta|E_m-E_n|\}-1\right]^{-1} \tag{54}$$

is the number of phonons per mode with energy $|E_m-E_n|$ in thermal equilibrium. It is easy to show from eqn. (7) that, provided E_m and E_n are separated from each other and from the chemical potential by a few k_BT, the energy-dependence of Γ_{mn} is dominated by a factor

$$Q(E_m,E_n) = \exp\left[-\tfrac{1}{2}\beta\{|E_m-\zeta|+|E_n-\zeta|+|E_m-E_n|\}\right] \tag{55}$$

For simplicity, we ignore any deviations from the asymptotic form (55) and write

$$\Gamma_{mn} = Q(E_m,E_n)\gamma_o\exp(-2\alpha r_{mn}) \tag{56}$$

as the energy-dependent generalization of eqn. (46). Then γ_o keeps its previous interpretation as the maximum equilibrium electron flux. In carrying out system averages we suppose, as before, that the site locations are independently and uniformly distributed. Moreover, we suppose that the energy of site m is independently distributed with a probability distribution $\rho(E_m)n_s^{-1}$ where $\rho(E_m)$ is the density of states function for the electron system.

With the above assumptions we find from eqn. (40) that

$$\langle\sigma_1(\infty)\rangle = n_s^2e^2\beta\gamma_o 16\pi(2\alpha)^{-5}\langle Q\rangle \tag{57}$$

which differs from our previous result (47) only by the factor

$$\langle Q\rangle = n_s^{-2}\int dE_1\int dE_2 Q(E_1,E_2)\rho(E_1)\rho(E_2) \tag{58}$$

Since $Q(E_1,E_2)$ is large only when both E_1 and E_2 are close to ζ we may extract a factor $\rho(\zeta)^2$ from the integral to obtain $\langle Q\rangle = 6\rho(\zeta)^2/(\beta n_s)^2$. Consequently $\langle\sigma_1(\infty)\rangle$ is now proportional to T instead of T^{-1} (as in eqn. 47) because transitions are only possible between pairs of states within about k_BT of the chemical potential.

When eqn. (56) is substituted into the basic pair approximation formula (44) we have

$$<\sigma^P(\omega) = \tfrac{1}{2}e^2\beta\gamma_o\int\rho(E_1)dE_1\int\rho(E_2)dE_2Q(E_1,E_2)\int\tfrac{4}{3}\pi r^4(-i\omega\tau_{12})(1-i\omega\tau_{12})^{-1}dr \tag{59}$$

where the relaxation time τ_{12} obtained from eqns. (43) and (6) is given by

$$\tau_{12}^{-1} = \left[F_1^{-1}+F_2^{-1}\right]Q(E_1,E_2)\gamma_o\exp(-2\alpha r) \tag{60}$$

Since $Q(E_1,E_2)$ is large only when both E_1 and E_2 are close to ζ we may approximate τ_{12} in eqn. (59) by the value which it has when $E_1 = E_2 = \zeta$. Thus, $\tau_{12} = \tau_o\exp(2\alpha r)$ with $\tau_o = (8\gamma_o)^{-1}$ which, as in the energy-independent case, is just the reciprocal of twice the maximum effective transition rate. The triple integral in eqn. (59) then factors into the space integral in eqn. (48) and the energy integral in eqn. (58). Consequently eqn. (48) remains valid but $<\sigma_1(\infty)>$ is now given by eqn. (57).

Our approximate evaluation of the triple integral in eqn. (59) could be improved but there is no reason to expect any significant changes in the predicted frequency dependence. At high frequencies the value of τ_{12} is immaterial. At low frequencies we may begin by evaluating the space integral using the asymptotic method already employed in the energy-independent case. The energy-dependent relaxation time enters into the result only logarithmically and may be replaced by a suitable mean value before the energy integrals are carried out. This alternative procedure yields the same low frequency form for $<\sigma^P(\omega)>$ apart from a trivial numerical factor[6,18].

4.4 Random Debye Model

The basic structure of the pair approximation (44) is

$$<\sigma^P(\omega)> = -i\omega\left\langle\frac{(\varepsilon_o-\varepsilon_\infty)}{1-i\omega\tau}\right\rangle \tag{61}$$

where the particular values of $\varepsilon_o-\varepsilon_\infty$ and τ appropriate to the rate equation formulation of the problem are obvious by inspection. The notation in eqn. (61) has been chosen to emphasise that this is just what one expects to find in a Debye model of a random dielectric medium[26,27]. One may therefore start ab initio from eqn. (61) and make physically reasonable assumptions about $\varepsilon_0-\varepsilon_\infty$ and τ and their dependence on the random parameters of the system. In the energy-independent case discussed in Section 4.2 we take $\tau = \tau_0\exp(2\alpha r)$ and $\varepsilon_0-\varepsilon_\infty$ proportional to r^2 while the probability distribution for r is proportional to $4\pi r^2$. In the energy-dependent case discussed in Section 4.3 we again take $\tau = \tau_0\exp(2\alpha r)$ but assume that $\varepsilon_0-\varepsilon_\infty$ is proportional to $r^2\ Q(E_1,E_2)$ and the probability distribution for r, E_1 and E_2 is proportional to $4\pi r^2\rho(E_1)\rho(E_2)$. A number of other sets of assumptions of this type exist in the literature which we have cited. All of them lead to frequency-dependences which are similar to that described by eqn. (48). Jonscher[28] has emphasised the universality of ω^s type behaviour and suggests that it requires a correspondingly universal explanation outside the framework of the random Debye model.

5. THE DC LIMIT

When ω=o the ac conductivity and diffusivity are given by eqns. (30) to (33). The quantity of major significance is $\Delta_m^2(\omega)$ which is the Fourier transform of $\delta_m^2(t)\theta(t)$ where $\delta_m^2(t)$ is the mean square x-displacement at time t of an electron setting out from site m at time o. Hence

$$\Delta_m^2(\omega) = \int_{-\infty}^{\infty} \delta_m^2(t)\theta(t)e^{i\omega t}dt \qquad (62)$$

Integrating by parts twice and writing $\delta_m^2(o) = o$ we obtain

$$-\omega^2\Delta_m^2(\omega) = \int_{-\infty}^{\infty}\frac{d}{dt}\left[\theta(t)\frac{d}{dt}\delta_m^2(t)\right]e^{i\omega t}dt \qquad (63)$$

when $\omega\to o$ we therefore have

$$\lim_{\omega\to o}\left[-\omega^2\Delta_m^2(\omega)\right] = d\delta_m^2(\infty)/dt \qquad (64)$$

Now, $\delta_m^2(t)$ obviously becomes very large and independent of the initial site m when $t\to\infty$. The subscript m is therefore superfluous on the right-hand side of eqn. (64) and is dropped in the following analysis. Consequently, when eqn. (64) is substituted into eqn. (31) we obtain the simple result

$$D(o) = \frac{1}{2} \, d\delta^2(\infty)/dt$$

$$= \frac{1}{2} \lim_{t\to\infty} \delta^2(t)/t \qquad (65)$$

It is convenient to multiply and divide the right-hand side of eqn. (65) by $\overline{N}(t)$, the mean number of hops in time t. Then we may write

$$D(o) = \frac{1}{2} \, fd^2 \qquad (66)$$

where

$$f = \lim_{t\to\infty} \overline{N}(t)/t \qquad (67)$$

and

$$d^2 = \lim_{t\to\infty} \delta^2(t)/\overline{N}(t) \qquad (68)$$

Moreover, when t is large, the deviations of the actual number N of hops executed in time t from $\overline{N}(t)$ are negligible. So are the deviations of the mean time $\overline{t}(N)$ of the Nth hop from $\overline{N}(t)/f$. We may therefore write eqns. (67) and (68) in the more convenient form

$$f^{-1} = \lim_{N\to\infty} \overline{t}(N)/N \qquad (69$$

$$d^2 = \lim_{N\to\infty} \delta^2(N)/N \qquad (70)$$

where $\delta^2(N)$ is the mean square x-displacement in a chain of N hops.

We see that f and d^2 are respectively the mean frequency and mean square x-displacement per hop in an infinite chain of hops. The results (66), (69) and (70) are familiar in conventional random walk theory. We may specialize the present theory to that case by supposing that the sites occupy a Bravais lattice and that the array of transition rates has the lattice periodicity. Then the electron is presented with the same spectrum of possible displacements with the same weights at every site. It is not difficult to show that f^{-1} and d^2 reduce to $\overline{t}(1)$ and $\delta^2(1)$ respectively in that case which may be readily evaluated for any particular model. It is a characteristic feature of random walks on random arrays of fixed sites that f and d^2 cannot be expressed in terms of quantities relating to a single hop. Special methods are required for their evaluation which we discuss in the next two Sections.

6. THE MEAN HOPPING FREQUENCY

6.1 General Formula

In our discussion of the dc limit in Section 5 it was tacitly assumed that the system of sites is infinite in extent. This is necessary because $d^2 = o$ for any finite system. However, to calculate f we may consider a finite (but large) system contained in a volume Ω. As $\Omega\to\infty$ the deviation of f from the value appropriate to an infinite system become negligible. We see from eqn. (69) that our primary task is to calculate the asymptotic behaviour of $\bar{t}(N)$ which is the mean time of the Nth hop in a chain of N hops. This quantity becomes independent of the initial site m when $N\to\infty$. When N is finite, however, we write $\bar{t}_m(N)$ for the mean time of the Nth hop in a chain of N hops starting from site m. Then

$$\bar{t}_m(N) = \int_{-\infty}^{\infty} \theta(t) p_m(N,t) t \, dt \tag{71}$$

where $p_m(N,t)$ is the probability that the last hop comes between t and t+dt. We may readily verify that

$$\bar{t}_m(N) = -i \lim_{\omega\to o} \frac{d}{d\omega} \overset{+}{p}_m(N,\omega) \tag{72}$$

where $\overset{+}{p}_m(N,\omega)$ is the Fourier transform of $p_m(N,t)\theta(t)$.

In Section 3.3 we gave a stochastic interpretation of each term in the Dyson expansion (36) of the Green matrix G. In particular: the (N+1)th term in the series is the Fourier transform of $\theta(t)$ times the probability that an electron initially on site m will be found on site n at time t having made N hops to get there. This term includes a factor G^o_{nn} because we allow the electron to arrive at site n before time t and then wait there without hopping anywhere else. When this factor is removed we have the Fourier transform of $\theta(t)$ times the probability distribution of the time of the last hop in a sequence of N hops leading from site m to site n. Summing over n yields $\overset{+}{p}_m(N,\omega)$. Thus we have

$$p_m^+(N,\omega) = \Sigma G^o_{mm} R^e_{mp_1} G^o_{p_1p_1} R^e_{p_1p_2} \ldots\ldots G^o_{p_{N-1}p_{N-1}} R_{p_{N-1}n} \tag{73}$$

where the summation is over all site labels except m. Now, we see from eqn. (37) that

$$\frac{d}{d\omega} G^o_{mm} = i(G^o_{mm})^2 \tag{74}$$

Hence, by differentiating eqn. (73) and taking the limit $\omega \to o$ with the aid of eqn. (74), we see that eqn. (72) gives

$$\bar{t}_m(N) = \Sigma Q_{mp_1} Q_{p_1p_2} \ldots\ldots Q_{p_{N-1}n} \left[\frac{1}{R^e_m} + \frac{1}{R^e_{p_1}} + \ldots\ldots \frac{1}{R^e_{p_{N-1}}} \right] \tag{75}$$

where

$$Q_{pq} = R^e_{pq}/R^e_p \tag{76}$$

Equation (75) may be written in a more convenient form by noting from eqn. (10) that

$$\sum_q Q_{pq} = 1 \tag{77}$$

for all p. Thus we have

$$\bar{t}_m(N) = \frac{1}{R^e_m} + \sum_{p_1} Q_{mp_1} \frac{1}{R_{p_1}} + \sum_{p_1p_2} Q_{mp_1} Q_{p_1p_2} \frac{1}{R^e_{p_2}}$$

$$+ \ldots\ldots \Sigma Q_{mp_1} Q_{p_1p_2} \ldots Q_{p_{N-2}p_{N-1}} \frac{1}{R^e_{p_{N-1}}} \tag{78}$$

where, in the last term the summation is over all p_1, p_2, ... p_{N-1}.

In order to determine the asymptotic behaviour of $\bar{t}_m(N)$ it is useful to consider a sequence of functions $\tau_m(M)$ defined on the sites of the system and constructed in the following way. The zeroth number of the sequence $\tau_m(0)$ is arbitrarily defined on site m. Subsequent members of the sequence are defined by the recurrence relation

$$\tau_m(M+1) = \sum_n Q_{mn}\tau_n(M) \tag{79}$$

We see from eqns. (77) and (79) that $\tau_m(M+1)$ is obtained by averaging $\tau_n(M)$ with the probability distribution Q_{mn}. Hence $\tau_m(M)$, which is obtained by M repetitions of this averaging procedure, becomes independent of both m and M when $M\to\infty$. We write $\tau(\infty)$ for the large M limit of $\tau_m(M)$. To obtain its value we note from eqns. (6), (10) and (76) that

$$Q_{mn} = \Gamma_{mn}/\Gamma_m \tag{80}$$

where

$$\Gamma_m = \sum_n \Gamma_{mn} \tag{81}$$

is the total equilibrium electron flux out of site m. Since Γ_{mn} is symmetrical we may readily verify that $\sum_m \Gamma_m\tau_m(M)$ is independent of M. It follows that

$$\tau(\infty) = \sum_m \Gamma_m\tau_m(0)/\sum_m \Gamma_m \tag{82}$$

The successive terms in the sum (78) are members of just such a sequence with $\tau_m(0) = 1/R_m^e$. Hence, leaving this identification understood, we see that

$$\begin{aligned} \bar{\tau}_m(N) &= \tau_m(0) + \tau_m(1) + \ldots. + \tau_m(N-1) \\ &\to N\tau(\infty) \end{aligned} \tag{83}$$

since the contribution from the early terms in the series, which have not yet reached the asymptotic value $\tau(\infty)$, may be ignored when $N\to\infty$. By substituting this result into eqn. (69) and calculating $\tau(\infty)$ from eqn. (82) with $\tau_m(0) = 1/R_m^e$ we finally obtain for the mean hopping frequency the simple formula

$$f = \sum_m \Gamma_m / \sum_m F_m$$

$$= \sum_m F_m R_m^e / \sum_m F_m \tag{84}$$

To obtain the second form of f we have used the identity $\Gamma_m = F_m R_m^e$ which follows immediately from eqns. (6) and (81).

We see that f is the average over all sites of the total effective transition rate R_m^e. The average is calculated with a weighting factor equal to the effective site occupation probability F_m. For a conventional random walk on a Bravais lattice R_m^e is independent of m and $f = R_m^e$. This is just the reciprocal of the mean time before the first hop which is given by the first term in eqn. (78). For a random walk in a random system we have to evaluate the formula (84). We discuss several simple cases below.

6.2 One Dimensional Chain

We consider a one-dimensional chain of sites strung out on a straight line and suppose that hopping is possible only between nearest neighbours[8]. Then

$$R_m^e = R_{m,m-1}^e + R_{m,m+1}^e \tag{85}$$

We also suppose for simplicity that F_m is independent of m. Then, from eqn. (6), R_{mn}^e is symmetrical because Γ_{mn} is symmetrical. Hence eqn. (84) reduces to

$$f = 2N_s^{-1} \sum_m R_{m,m+1}^e \tag{86}$$

where N_s is the number of sites in the chain.

To evaluate $\langle f \rangle$ in a particular case we write $R_{m,m+1}^e = R_o \exp(-s_m)$ where s_m is uniformly distributed between 0 and $s_o \gg 1$. Then $\langle f \rangle \simeq 2R_o/s_o$.

6.3 Energy Independent Model

We consider now the model defined by eqns. (45) and (46) and the assumption that we are concerned with N_s sites whose locations are independently and uniformly distributed over a large volume Ω[8].

We see from eqns. (6), (45) and (46) that $R^e_{mn} = R_o\exp(-2\alpha r_{mn})$ where $R_o = \gamma_o/F$ and r_{mn} is the distance between sites m and n. Since F_m has a constant value F, eqn. (84) reduces to

$$f = N_s^{-1}\sum_m R^e_m = N_s^{-1}\sum_{mn} R^e_{mn} \tag{87}$$

Taking the system average we obtain

$$\begin{aligned} \langle f\rangle &= N_s^{-1}N_s^2\int_o^\infty R_o\exp(-2\alpha r)4\pi r^2\,\Omega^{-1}dr \\ &= 2R_o\eta \end{aligned} \tag{88}$$

where $\eta = 4\pi n_s(2\alpha)^{-3}$ is a non-dimensional measure of the site density $n_s = N_s/\Omega$.

6.4 Energy-Dependent Model

The energy-dependent model introduced by Ambegaokar, Halperin and Langer[10] is defined by eqns. (55) and (56) and the assumption that the density of states is slowly-varying in the neighbourhood of the chemical potential. The system average of the first form of f in eqn. (84) is easily evaluated. The denominator is obtained by using eqn. (29) and the expression

$$n = \int\rho(E)f^o(E)dE \tag{89}$$

for the electron concentration n. In eqn. (89) $\rho(E)$ is the density of states and $f^o(E)$ is the Fermi-Dirac function (2) with E_m=E. By differentiating eqn. (89) we find immediately that $dn/d\zeta = \rho(\zeta)$ when the electron statistics are highly degenerate. Hence eqn. (29) gives

$$\sum_m F_m = \Omega\beta^{-1}\rho(\zeta) \tag{90}$$

To evaluate the system average of the numerator in the top line of eqn. (84) we suppose that each site location is uniformly and independently distributed over Ω and that the electron energy at each site is independently distributed with a probability density $n_s^{-1}\rho(E)$. Then eqns. (55) and (56) give

$$\langle \sum_m \Gamma_m \rangle = \langle \sum_{mn} \Gamma_{mn} \rangle$$

$$= N_s^2 \int_0^\infty 4\pi r^2 \Omega^{-1} dr \int dE_m \rho(E_m) n_s^{-1}$$

$$\times \int dE_n \rho(E_n) n_s^{-1} \gamma_o \exp(-2\alpha r) Q(E_m, E_n)$$

$$= \Omega(\gamma_o \eta / n_s) 6 [kT\rho(\zeta)]^2 \qquad (91)$$

When eqns. (90) and (91) are substituted in the system average of eqn. (84) we obtain the final result

$$\langle f \rangle = 6\gamma_o \eta \rho(\zeta)/\beta n_s \qquad (92)$$

We may rewrite this equation in a more transparent form by supposing that $\rho(\zeta)$ is constant inside a bandwidth W and is zero everywhere else. Then $\rho(\zeta) = n_s/W$ and $\langle f \rangle = 1.5\ R_o \eta/\beta W$ where $R_o = 4\gamma_o$ is the maximum value of R^e_{mn} in this model. Comparing this result with eqn. (88) we see that the only effect of introducing the energy dependence has been to multiply $\langle f \rangle$ by a factor $0.75/\beta W$.

The principal feature of the results of these elementary calculations requiring emphasis is that in no case does $\langle f \rangle$ contain an exponential factor. The effective transition rates contain exponential factors but these are suppressed by the system averaging.

7. THE MEAN SQUARE DISPLACEMENT PER HOP

7.1 General Formulae

The mean square x-displacement per hop, d^2, is given by eqn. (70) where $\delta^2(N)$ is the mean square x-displacement in a chain of N hops. To calculate $\delta^2(N)$ we consider an electron which starts on site m and hops to a final site n via intermediate sites $p_1, p_2 \cdots p_{N-1}$. The probability that an electron on site p will make its first hop to site q (without regard to the time at which the hop occurs) is Q_{pq} as given by eqn. (76). It follows that

$$d^2 = \lim_{N\to\infty} N^{-1} \Sigma Q_{mp_1} Q_{p_1p_2} \cdots Q_{p_{N-1}m} (x_m - x_n)^2 \qquad (93)$$

where the summation is over all values of p_1, p_2, ... p_{N-1} and m. Formula (93) may be readily evaluated for a conventional isotropic random walk on a Bravais lattice for which Q_{pq} depends only on $|\underline{r}_p-\underline{r}_q|$. In this case the mean square x-displacement in any one hop

$$\delta^2(1) = \sum_q Q_{pq}(x_p-x_q)^2 \tag{94}$$

is independent of p and the mean x-displacement in any one hop vanishes. Hence, by writing

$$x_m-x_n = (x_m-x_{p_1}) + (x_{p_1}-x_{p_2}) + \ldots + (x_{p_{N-1}}-x_m) \tag{95}$$

in eqn. (93) and using the normalization condition (77) we find the familiar result $d^2 = \delta^2(1)$. This argument obviously breaks down completely for a random walk in a random system and we must calculate d^2 in some other way. An alternative expression for d^2 may be obtained by exploiting its relationship to the dc conductivity. We let $\omega \to o$ in eqn. (30), substitute for D(o) from eqn. (66) and solve for d^2 to obtain

$$d^2 = \frac{2\sigma(o)}{e^2 f dn/d\zeta} \tag{96}$$

Since $dn/d\zeta$ and f are given by eqns. (29) and (84) respectively, it only remains to write $\sigma(o)$ in a convenient form.

Miller and Abrahams[4] were the first to emphasise that $\sigma(o)$ may be obtained directly by exploiting the analogy between the steady state rate equations and Kirchoff's equations for a conductance network. When ω=o the perturbed rate equations (12) reduce to

$$-\left[\underline{f}^{(1)}(0) + \beta eE\underline{x}F\right]R = o \tag{97}$$

where $\underline{f}^{(1)}(0)$ is the row matrix of perturbations of the site occupation probabilities produced by a dc electric field E in the x-direction. Equations (97) take the required form when we write

$$\underline{f}^{(1)}(0) = -\beta e\left[\underline{V}+E\underline{x}\right]F \tag{98}$$

where the row matrix $\underline{V}$ remains to be determined. Thus we have $\beta e\underline{V}FR = 0$, i.e., using eqns. (6) and (9),

$$\sum_n g_{mn}(V_m - V_n) = 0 \tag{99}$$

where

$$g_{mn} = e^2 \beta \Gamma_{mn} \tag{100}$$

Equations (99) are Kirchoff's equations for the voltages V_n on site n in a conductance network with conductance g_{mn} connecting sites m and n.

Thus, the dc hopping network problem has a dc conductance network analogue. The analogy is complete. We may readily verify that the net electric current flowing from m to n is indeed $g_{mn}(V_m - V_n)$. Moreover, since $f_m^{(1)}(0)$ must remain finite at infinity, we see from eqn. (98) that eqns. (99) must be solved subject to the asymptotic boundary condition $V_m \sim -Ex_m$. We conclude that $\sigma(0)$ may be identified with the macroscopic conductivity of a system of microscopic conductances g_{mn} subjected to an average applied electric field E in the x-direction. To write down an explicit formula for $\sigma(0)$ on the basis of this analogy we equate the power dissipated in the individual conductances within a finite, but large, volume Ω to the Joule heating rate $\sigma(0)E^2\Omega$. Then we have

$$\sigma(0) = \frac{1}{2} E^{-2} \Omega^{-1} \sum_{mn} g_{mn}(V_m - V_n)^2 \tag{101}$$

where sites m and n must both be inside Ω and the factor of 0.5 avoids double counting.

When we substitute for $dn/d\zeta$, f and $\sigma(0)$ in eqn. (96) from eqns. (29), (84) and (101) respectively we obtain

$$d^2 = E^{-2} \sum_{mn} g_{mn}(V_m - V_n)^2 / \sum_{mn} g_{mn} \tag{102}$$

Thus, d^2 is the mean square voltage drop per unit applied electric field calculated with the conductance as a weighting factor. This is the most transparent general formula for d^2. However, in our subsequent analysis of particular models we use eqn. (96) as it stands because the evaluation of f and <f> has already been discussed in Section 6 while the evaluation of $dn/d\zeta$ is trivial and the evaluation of $\sigma(0)$ and $\langle\sigma(0)\rangle$ from eqn. (101) is of interest in its own right and has been discussed by other authors.

7.2 One-Dimensional Chain

We consider first of all the one-dimensional chain model with nearest neighbour hopping which was introduced in Section 6.2. There is no need to use the general formula (101) to determine $\sigma(0)$ because the equivalent conductance network is an infinite chain of conductors connected in series which may be treated by elementary means. The conductance of a large segment of the chain containing N_s sites is equal to the reciprocal of the sum of the reciprocals of the individual conductances in the segment. Moreover, the voltage drop across the segment is equal to $N_s aE$ where a is the mean nearest neighbour separation. Hence the conductivity of the chain is

$$\sigma(0) = N_s a\left[\sum_m g^{-1}_{m,m+1}\right]^{-1}$$

$$= N_s ae^2\beta F\left[\sum_m (R^e_{m,m+1})^{-1}\right]^{-1} \quad (103)$$

where we have used eqns. (6) and (100) to eliminate $g_{m,m+1}$ in favour of $R^e_{m,m+1}$, remembering that E_m takes a constant value F for this model. Moreover, we see from eqn. (29) that $dn/d\zeta = \beta F/a$. By substituting these results together with eqn. (86) for f into eqn. (96) we obtain

$$d^2 = a^2\left[N_s^{-1}\sum_m R^e_{m,m+1} N_s^{-1}\sum_m (R^e_{m,m+1})^{-1}\right]^{-1} \quad (104)$$

An elementary application of Schwartz's inequality[29] shows that the denominator in eqn. (104) is greater than or equal to one. Equality holds good only when $R^e_{m,m+1}$ is independent of m. In that case we are concerned with a conventional random walk and $d^2 = a^2$. Any spread of the values of $R^e_{m,m+1}$ makes $d^2<a^2$ and a large spread makes $d^2<<a^2$. To illustrate this point we write $R^e_{m,m+1} = R_o\exp(-s_m)$, as in Section 6.2, where s_m is independently and uniformly distributed between o and $s_o>>1$. The site averages in the denominator of eqn. (104) may then be replaced by system averages. A trivial calculation gives $d^2 = (a/s_o)^2\exp(-s_o)$ where the exponential factor comes from the reciprocal effective transition rates in the denominator of eqn. (104).

7.3 Introduction of the Critical Percolation Exponent

We begin to consider now the evaluation of d^2 for more complicated models in which g_{mn} is a function of the distance r_{mn} between sites m and n and the energies E_m and E_n of the localized electron states associated with the sites. It is supposed for simplicity that each site location is independently and uniformly distributed over a volume Ω and that each site energy is independently distributed with a probability density $\rho(E)n_s^{-1}$ where $\rho(E)$ is the density of states and n_s is the density of sites. Taking the system average of the numerator and the denominator separately in eqn. (96) we have

$$\langle d^2\rangle = \frac{2\langle\sigma(0)\rangle}{e^2\langle f\rangle dn/d\zeta} \tag{105}$$

where, from eqn. (101),

$$\langle\sigma(0)\rangle = 2\pi E^{-2}\int\rho(E_m)dE_m\int\rho(E_n)dE_n\int g_{mn}\langle(V_m-V_n)^2\rangle' r^2 dr \tag{106}$$

The prime on the angular bracket in eqn. (106) indicates that the quantity enclosed is to be averaged over all stochastic variables except $r\equiv r_{mn}$, E_m and E_n. Our subsequent analysis is based on eqns. (105) and (106) to facilitate comparison with the results for $\langle\sigma(0)\rangle$ obtained by other authors. We note, however, that $\langle d^2\rangle$ may be expressed in a simpler form by system averaging eqn. (102). Thus we find that $\langle d^2\rangle$ is the average of $E^{-2}\langle(V_m-V_n)^2\rangle'$ calculated with the weighting factor $\rho(E_m)\rho(E_n)g_{mn}r^2$.

Our main task is to investigate the behaviour of $\langle\sigma(0)\rangle$ in eqn. (106). It is convenient to write g_{mn} in the form $g_o\exp(-s)$ where g_o is the maximum value of g_{mn} and s is a positive function of r, E_m and E_n. Analytical progress is possible only when the characteristic length and energy involved in s tend to zero. To study this limit without loss of generality we consider a scaled system in which s is replaced by μs and $\mu\to\infty$. For an energy-independent model with $s = 2\alpha r$, the scaling is equivalent to decreasing the characteristic length α^{-1} by a factor μ^{-1}. For an energy-dependent model with $s = 2\alpha r + 0.5\beta(|E_m-\zeta|+|E_n-\zeta|+|E_m-E_n|)$, it is equivalent to reducing both α^{-1} and the characteristic energy $k_B T$ by a common factor μ^{-1}. When s has a more general form, consideration of the scaled system provides a convenient way of introducing a parameter μ such that the scaled exponent μs becomes increasingly rapidly varying as $\mu\to\infty$. The formulae which we derive apply to the scaled system in this limit. They are applicable to the original system, for which $\mu=1$, provided that the characteristic length and energy involved in s are sufficiently small.

Ambegaokar, Halperin and Langer[10] conjectured that the system averaged dc conductivity of the scaled system takes the form

$$<\sigma(0)>_{\mu} = \sigma_{o}(\mu)\exp(-\mu s_{p}) \tag{107}$$

when $\mu\to\infty$ where the pre-factor $\sigma_o(\mu)$ is slowly-varying compared to the exponential. The quantity s_p is the "critical percolation exponent" of the original system which is defined in the following way. Let us remove from the original network all conductances with s greater than some specified value s_o. The s_p is the least value of s_o for which the remaining conductance network contains an infinite cluster of connected sites. To verify this conjecture we rewrite eqn. (106) in the form (107) for the scaled network. Thus we obtain for the pre-factor the formula

$$\sigma_o(\mu) = 2\pi E^{-2}\int \rho(E_m)dE_m \int \rho(E_n)dE_n \int D_{mn} r^2 dr \tag{108}$$

where

$$D_{mn} = g_{mn}(\mu)<(V_m-V_n)^2> \tag{109}$$

with

$$g_{mn}(\mu) = g_o\exp\left[\mu(s_p-s)\right] \tag{110}$$

It only remains to show that $\sigma_o(\mu)$ is slowly-varying when $\mu\to\infty$.

We see from Kirchoff's equations (99) and the associated boundary condition that multiplying all the conductances in the scaled network by a common factor has no effect on the voltages V_m. Consequently D_{mn} in eqn. (109) is the power dissipated in $g_{mn}(\mu)$ when the scaled network is modified by replacing $g_o\exp(-\mu s)$ by $g_{mn}(\mu)$ for all m and n. Let us pick a particular $g_{mn}(\mu)$ in the modified, scaled network and consider the sub-network consisting of conductances with magnitudes greater than or equal to $g_{mn}(\mu)$ which are connected to $g_{mn}(\mu)$ by continuous chains of conductances in the same category. When $s>s_p$ this sub-network has a non-zero probability of being infinite in extent. The behaviour of $<(V_m-V_n)^2>$ is then dominated by the boundary condition associated with Kirchoff's equations (99) and involves no exponential factors. Consequently, D_{mn} behaves like $g_{mn}(\mu)$ when $s>s_p$ and $\mu\to\infty$. On the other hand, when $s<s_p$ the sub-network is finite with probability one and the influence of the boundary condition on V_m-V_n is removed. Moreover,

the conductances in the sub-network become infinitely large when $\mu\to\infty$ while the conductances outside it include those with $s=s_p$ which are independent of μ. It follows that $V_m-V_n\to 0$ as $\mu\to\infty$ so as to keep the mean square current $I^2_{mn} = g^2_{mn}(\mu)\langle(V_m-V_n)^2\rangle$ free from exponential factors. Consequently, $D_{mn} = I^2_{mn}/g_{mn}(\mu)$ behaves like the reciprocal of $g_{mn}(\mu)$ when $s<s_p$ and $\mu\to\infty$. We conclude that D_{mn} falls off exponentially on either side of the critical percolation surface $s=s_p$ in (r,E_m,E_n)-space from a maximum value which contains no exponential factors. No vestige of the exponential character of D_{mn} will survive the integration in eqn. (108) when $\mu\to\infty$ and therefore $\sigma_0(\mu)$ is slowly-varying in this limit.

7.4 Approximations to the Critical Percolation Exponent

In classical bond percolation problems in r dimensions the average number of open bonds associated with each site at the percolation threshold is close to $r/(r-1)$[30]. Since each bond is associated with two sites we therefore have the percolation criterion $2B/S = r/(r-1)$ where B and S are respectively the number of open bonds and the number of sites per unit volume. In random conductance networks we would expect a similar criterion to determine the percolation threshold. We must, however, allow for the fact that some sites may be automatically isolated by the condition $s<s_p$. Automatic isolation does not occur in classical percolation problems but often arises in hopping conductivity models. When $s = 2\alpha r+0.5\beta(|E_m-\zeta|+|E_n-\zeta|+|E_m-E_n|)$, for example, any site m with $|E_m-\zeta|> s_p/\beta$ is automatically isolated. Since automatically isolated sites cannot possibly belong to an infinite cluster we exclude them in formulating the percolation criterion. Then we have

$$N(s_p) = 2B/S' = N_p \tag{111}$$

where $N(s_p)$ is the average number of conductances with $s<s_p$ associated with the sites which are not automatically isolated and N_p is a number in the order of unity.

In eqn. (111), B is the number of conductances per unit volume with $s<s_p$ and S' is the number of sites per unit volume which are not automatically isolated. Hence

$$S' = \int \rho(E_m)dE_m \tag{112}$$

and

$$2B = \int \rho(E_m)N(E_m, s_p)dE_m \tag{113}$$

where

$$N(E_m, s_p) = \int \rho(E_n)dE_n \int 4\pi r^2 dr \tag{114}$$

is the mean number of conductances with $s<s_p$ emerging from a site with energy E_m. Values of E_m leading to automatic isolation are excluded from the integration in eqns. (112) and (113) while in eqn. (114) the integration is over values of E_n and r for which $s<s_p$. We may rewrite eqn. (113) in a form which is more convenient for subsequent calculations by substituting from eqn. (114). Thus we have

$$2B = \int \rho(E_m)dE_m \int \rho(E_n)dE_n \int 4\pi r^2 dr \tag{115}$$

where the integration is over values of E_m, E_n and r for which $s<s_p$. When eqns. (112) and (113) are substituted into eqn. (111) we see that $N(s_p)$ is the average of $N(E_m, s_p)$ calculated with the weighting factor $\rho(E_m)$. Maschke et al[31] and Pollak[11] have used instead the weighting factor $\rho(E_m)N(E_m, s_p)$ which was suggested by an alternative approximate percolation criterion derived by Pollak. The use of this more complicated weighting factor does not have a significant effect on s_p.

The simplest case to consider is the energy-independent model described in detail in Section 4.2. We have $s = 2\alpha r$ so that automatic isolation does not arise and, of course, $N(E_m, s_p)$ is independent of E_m. Since the sites are uniformly distributed in r-space with a number density n_s we have $N(s_p) = n_s 4\pi r_p^3/3$ where $r_p = s_p/2\alpha$. Hence eqn. (111) yields $s_p = (6N_p\alpha^3/\pi n_s)^{1/3}$. The numerical calculations of Ambegaokar et al[32], Kirkijärvi[33] and Seager and Pike[34] all confirm this formula with $N_p = 2.4$, 2.78 and 2.68 respectively. We see from the discussion given in Section 7.3 that the behaviour of $\langle\sigma(0)\rangle$ is dominated by a factor $\exp(-s_p)$ when $\alpha\to\infty$. We also see from eqn. (105) that the behaviour of $\langle d^2\rangle$ is dominated by the same factor because (from eqns. (29) and 88)) neither $\langle f\rangle$ nor $dn/d\zeta$ involve any exponentials for the energy-independent model.

In discussing energy-dependent models it is convenient to simplify the notation by measuring the site energies from the chemical potential. Then, as in Section 4.3, we write $s = 2\alpha r+0.5\beta(|E_m|+|E_n|+|E_m-E_n|)$. There are a number of different cases to be discussed depending on the energy-dependence of the density of states. They are all derivable by trivial modifications of the analysis for the density of states function introduced by Grant and Davis[35] to discuss hopping in a non-degenerate conduction band tail. Thus we have $\rho(E) = 0$ when $E<E_A$ and $\rho(E) = \rho_o(E-E_A)^q$ when $E>E_A$ where E_A (the conduction band edge) and q are positive constants. Sites with $E_m>s_p/\beta$ are automatically isolated so that eqn. (112) gives

$$S = \frac{\rho_o}{q+1}\left[\frac{s_p^*}{\beta}\right]^{q+1} \tag{116}$$

where $s_p^* = s_p-\beta E_A$. The general form of 2B may be seen immediately by making the change of variables $R = 2\alpha r/s_p^*$, $W_m = \beta(E_m-E_A)/s_p^*$ and $W_n = \beta(E_n-E_A)/s_p^*$ in eqn. (115). Thus, 2B is proportional to

$$\rho_o^2(2\alpha)^{-3}(s_p^*)^{2q+5}\beta^{-2(q+1)} \tag{117}$$

with a constant of proportionality which depends only on q. It follows from eqns. (111) and (116) and (117) that s_p^* is proportional to the (q+4)th root of $N_p\alpha^3\beta^{q+1}/\rho_o$, where $N_p \sim 1$. The integral in eqn. (115) may be most easily evaluated exactly by performing the energy integrations first. We find that

$$2B = \frac{4\pi}{(2\alpha)^3}\left(\frac{\rho_o}{q+1}\right)^2\frac{(s_p^*)^{2q+5}}{\beta^{2(q+1)}}\left[(q+2)(2q+3)2q+5)\right]^{-1} \tag{118}$$

and $s_p = \beta E_A+s_p^*$ is therefore given by

$$s_p = \beta E_A+\left[4(q+1)(q+2)(2q+3)(2q+5)N_p\alpha^3\beta^{q+1}/\pi\rho_o\right]^{\frac{1}{q+4}} \tag{119}$$

Equation (119) gives the exponent in $\langle\sigma(0)\rangle$ for the band tail model when α and β both tend to zero. Since the electron statistics are non-degenerate, we see from eqn. (89) that $dn/d\zeta = \beta n$ which involves an exponential factor $\exp(-\beta E_A)$. It is easy to verify that the same exponential appears in the system average of both the numerator and the denominator of eqn. (84). Hence, once again, $\langle f\rangle$ contains no exponential factors. It therefore follows from eqn. (105) that the behaviour of $\langle d^2\rangle$ is dominated by an exponential of the second term in eqn. (119). Grant and Davis[35] estimate the diffusivity exponent to be proportional to the 4th root of $N_p\alpha^3\beta^{q+1}/\rho_o$ with a different constant of proportionality by using an approximate form of a non-percolative argument due to Mott[36] which is discussed below. Hamilton[37] uses an extension of

Mott's argument and Pollak[11] uses a percolative argument to discuss the case $E_A = 0$. The density of states function is then artificial but both authors find that conductivity exponent is proportional to the (q+4)th root of β^{q+1} in agreement with eqn. (119).

Ambegaokar et al[10] discuss hopping in states on either side of the chemical potential and take the density of states to be a constant ρ_o. We may apply the above formulae to that case by putting $E_A = q = o$ and inserting factors of 2 and 3 into eqns. (116) and (118) respectively because the states extend on either side of the chemical potential. The factor of 2 is obviously correct, the factor of 3 may be verified as correct by direct evaluation of the integral in eqn. (115). With these modifications we find that eqn. (111) yields: $s_p = |40N_p\beta\alpha^3/\pi\rho_o|^{\frac{1}{4}}$. In their original paper, Ambegaokar et al[10] obtain the same result using a different percolation criterion and their best estimate of N_p is 1.26. Seager and Pike[34] estimate $N_p = 2.11$ from a percolation calculation for hyperspheres. They have also made numerical calculations of $\langle\sigma(0)\rangle$ using an expression for g_{mn} taken from the paper by Miller and Abrahams[4]. The low temperature results are consistent with s_p as given above and $N_p = 2.24$, 2.97 and 3.59 when $\rho_o = 10^{19}$, 10^{20} and 10^{21} $cm^{-3}(eV)^{-1}$ respectively. The dependence of N_p on ρ_o presumably arises because s does not have the simple form assumed here[31].

We may readily verify that neither $\langle f\rangle$ nor $dn/d\zeta$ involve an exponential factor in the case currently under discussion. It follows from eqn. (105) that the behaviour of $\langle d^2\rangle$ is also dominated by a factor $\exp(-s_p)$ when α and β tend to zero. We see, therefore, that the $T^{\frac{1}{4}}$ law originally predicted by Mott[36] and verified experimentally in several materials[3], arises directly from the behaviour of $\langle d^2\rangle$ in this case. Mott derives the law in a different way. He considers an exponent $s = 2\alpha r+\beta\Delta$, where $\Delta = |E_m-E_n|$ and minimizes s subject to the constraint $4\pi r^3\Delta\rho_o/3 = 1$. The minimum value of s is

$$s_M = \left[(8\alpha)^3\beta/9\pi\rho_o\right]^{\frac{1}{4}} \tag{120}$$

and it occurs at values of r and Δ given by $r_M = 3s_M/8\alpha$ and $\Delta_M = s_M/4\beta$. Mott's calculation apparently makes no appeal to percolation theory. However, we proceed to show that it may be re-interpreted as providing an approximation to s_p. When the integral (114) is evaluated for the case $s = 2\alpha r+\beta\Delta$ we find that $N(E_m, s_p) = \rho_o 2\pi s_p^4/3\beta(2\alpha)^3$ which is independent of E_m because of the assumed form of s. Hence, $N(s_p) = N(E_m, s_p)$ and eqn. (111) gives $s_p = (12N_p\beta\alpha^3/\pi\rho_o)^{\frac{1}{4}}$ which reduces to s_M if we put $N_p = 128/27$.

To see how this value of N_p arises we notice that eqn. (114) may be rewritten in the form

$$N(s_p) \equiv N(E_m, s_p) = 2\rho_o \int d\Delta \int 4\pi r^2 dr \qquad (121)$$

where the integration is over all values of r and Δ for which $s<s_p$. Instead of evaluating the right-hand side of eqn. (121) exactly (as we did above) let us approximate it by the lower bound $2\rho_o 4\pi r^3\Delta/3$, where $2\alpha r+\beta\Delta = s_p$, and choose r so as to maximize the lower bound. The maximum occurs when $r = r_M$ and $\Delta = \Delta_M$ and is smaller than the exact value by 27/64. By substituting this approximate result into eqn. (111) with $N_p = 1$ we obtain for s_p a value which differs from s_M only by the replacement of ρ_o by $2\rho_o$ because we are concerned with states both above and below the chemical potential. Mott's optimization procedure proceeds by a different route but it leads to the same point in the (r,Δ)-plane. Thus the value $N_p = 128/27$ required to obtain s_M from the exact formula for s_p consists of the factor of 2 associated with the density of states in eqn. (121) and a factor 64/27 associated with the under-estimation of the integral in that equation.

7.5 Approximations to the Conductivity Prefactor

The prefactor $\sigma_o(\mu)$ in the dc conductivity is defined by eqn. (108). The crucial factor in the integrand is D_{mn} which is given by eqns. (109) and (110). We saw in Section (7.3) that the behaviour of D_{mn} is dominated by an exponential decay on either side of the critical percolation surface $s = s_p$. When r is large the boundary condition associated with Kirchoff's equations (99) show that $V_m-V_n \simeq -E(x_m-x_n)$. Hence eqns. (109) and (110) yield

$$D_{mn} = g_o \exp[\mu(s_p-s)]E^2r^2/3 \qquad (122)$$

To obtain a crude estimate of $\sigma_o(\mu)$ we use eqn. (122) for all $s>s_p$ and write

$$D_{mn} = g_o \exp[\mu(s-s_p)]E^2 r_p^{\,2}/3 \qquad (123)$$

for all $s<s_p$ where r_p is the value of r on the critical percolation surface at the energy values E_m and E_n. The expression (123) is the simplest approximation to D_{mn} for $s<s_p$ which exhibits the behaviour discussed in Section 7.3 and is continuous with the expression (122) at the critical percolation surface. When eqns. (122) and (123) are substituted into eqn. (108) we find that the

integral over r may be carried out immediately when $\mu\to\infty$ by changing to s as the integration variable. In all the cases that we consider s involves r through a term $2\alpha r$ and we find that

$$\sigma_o(\mu) = (2\pi g_o/3\alpha\mu)\int dE_m \rho(E_m)\int \rho(E_n) r_p{}^4 dE_n \tag{124}$$

For the energy-independent model with $s = 2\alpha r$, the energy integrals in eqn. (124) yield a factor $n_s{}^2$. Now, we saw in Section 7.4 that n_s is proportional to $r_p{}^{-3}$ in this case. Hence $\sigma_o(\mu)$ is proportional to $g_o r_p^{-1}(\alpha\mu r_p)^{-1}$. Pollak[11] uses a different argument and obtains a prefactor which is not explicitly dependent on r_p. However, the numerical results of Kirkijärvi[33] suggest that $\sigma_o(\mu)$ is proportional to $g_o r_p{}^{-1}(\alpha\mu r_p)^{-\nu}$ with $\nu = 0.6\pm0.25$. For the energy-dependent model with $s = 2\alpha r + 0.5\beta(|E_m|+|E_n|+|E_m-E_n|)$ and a constant density of states ρ_o we find that eqn. (124) reduces to

$$\sigma_o(\mu) = 4\pi\rho_o{}^2 g_o s_p{}^6/\left[15(2\alpha)^5\beta^2\mu\right] \tag{125}$$

Since s_p is proportional to $\beta^{\frac{1}{4}}$ in this case we see that $\sigma_o(\mu)$ is proportional to $g_o\beta^{-\frac{1}{2}}$. Pollak[11] obtains a similar result and Kirkijärvi estimates from his numerical results that $\sigma_o(\mu)$ is proportional to $g_o\beta^{-\lambda}$ with $\lambda = 0.5\pm0.06$. We have retained a factor g_o in these formulae because, from eqn. (100), it is proportional to β. There is little experimental information on the behaviour of the prefactor but the results of Allen and Adkins for heavily doped germanium[38] are consistent with these formulae. When the density of states has the form assumed by Grant and Davis[35] (see Section 7.4) we find that $\sigma_o(\mu)$ is proportional to g_o times the (q+4)th root of $(k_BT)^{2(q+1)}$.

7.6 Approximate Formula for the Mean Square Displacement per Hop

It remains for us to bring together the results obtained in Sections 7.3, 7.4 and 7.5 to obtain an approximate formula for $\langle d^2\rangle$ in the scaled system[39]. Adequate generality is achieved by writing $s = 2\alpha r + U$ where U is a positive function of E_m and E_n. We see from Eqns. (29), (84) and (89) that the denominator in eqn. (105) is given by

$$e^2\langle f\rangle d\eta/d\zeta = e^2\beta\Omega^{-1}N_s^2\langle\Gamma_{mn}\rangle \tag{126}$$

We may use eqn. (100) to eliminate Γ_{mn} from this equation in favour of g_{mn} which takes the form $g_o\exp(-\mu s)$ in the scaled system. The integration over r involved in the system average may be carried out immediately to yield

$$e^2\langle f\rangle dn/d\zeta = 8\pi g_o(2\alpha\mu)^{-3}\int\rho(E_m)dE_m\int\rho(E_n)\exp(-\mu U)dE_n \tag{127}$$

The numerator in eqn. (105) is obtained from eqn. (107) where $\sigma_o(\mu)$ is given by eqn. (124) with $r_p = (s_p-U)/2\alpha$. Thus we have the final result

$$\langle d^2\rangle = \Delta_o^2(\mu)\exp(-\mu s_p) \tag{128}$$

where

$$\Delta_o^2(\mu) = \frac{1}{3}\left(\frac{\mu}{2\alpha}\right)^2 \frac{\int\rho(E_m)dE_m\int\rho(E_n)(s_p-U)^4dE_n}{\int\rho(E_m)dE_m\int\rho(E_n)\exp(-\mu U)dE_n} \tag{129}$$

All our previous results are subsumed in this simple formula. When U=0 we are concerned with the energy-dependent case and the integrals in eqn. (129) are trivial. When $U = 0.5\beta(|E_m|+|E_n|+|E_m-E_n|)$ we have the energy-dependent case with the energies measured from the chemical potential. Provided that the density of states is non-zero at the chemical potential, the integral in the denominator of eqn. (129) is dominated by contributions from near $E_m=E_n=0$ and is in the order of μ^{-1} when $\mu\to\infty$. In that case $\Delta_o^2(\mu)$ is a slowly-varying prefactor. However, when the density of states vanishes below a band-edge at E_A, the denominator of eqn. (129) contains a factor $\exp(-\mu\beta E_A)$ which cancels out the corresponding contribution to $\exp(-\mu s_p)$ in eqn. (128). This case is not really an omalous. The maximum available value of g_{mn} is $g_o\exp(-\beta E_A)$ instead of g_o as we assumed. When a factor of $\exp(-\beta E_A)$ is absorbed in g_o, all three of the quantities s, s_p and U are reduced by βE_A and Δ_o^2 becomes slowly-varying in eqn. (129).

8. CONCLUSION

The treatment of ac conductivity for electrons in localized states in a disordered semiconductor is relatively simple. This is because the high frequency behaviour of $\langle\sigma(\omega)\rangle$ is determined by the short time behaviour of the mean square displacement of an electron. The pair approximation developed in Section 4 gives an adequate description of the frequency dependence which is characteristic of this type of transport. The randomness of the system plays an essential role in the analysis. When it is ignored the frequency dependence disappears. Tunaley[40] has recently given a simple analysis of a random walk which yields a frequency-independent conductivity. He concludes that random walk theory is not applicable to hopping transport problems. However, inspection of Tunaley's analysis shows that he has not taken proper account of the randomness of the system in which the random walk takes place.

The treatment of dc hopping conductivity presents a much more complicated problem because $<\sigma(0)>$ is determined by the characteristics of infinitely long hopping paths. We saw in Section 6 that the mean hopping frequency f in such a path does not show any very interesting features. The controlling influence is exerted by the mean square x-displacement per hop, d^2, which is exponentially smaller than the mean square nearest neighbour intersite separation. The physical reason for this is clear when we consider the nature of a typical hopping path. The electron spends most of its time reciprocating between sites which are anomalously close in the sense that the exponent s is small. These are just the processes which control f and the ac conductivity. However, they do not carry the electron a long way through the material and therefore do not contribute to d^2. The reciprocating hops are interspersed with occasional hops for which s is small but not anomalously so. On the other hand, hops involving large values of s are by-passed. The critical percolation exponent determines the boundary between "small" and "large" values of s and hops with $s=s_p$ dominate the behaviour of d^2.

Since Mott's original non-percolative treatment[36], several authors have made alternative calculations of the exponent in the dc conductivity which do not involve percolation theory[35-37,41,42] (see also the earlier work cited in reference 32). The results obtained are not very different from s_p but the non-percolative treatments do not properly reflect the physical characteristics of the long range hopping paths.

The transition region between ac and dc conductivity is particularly difficult to study. Any development of the pair approximation to include larger clusters of sites is bound to yield zero dc conductivity unless the cluster size is allowed to diverge. Scher and Lax[13,14] have given an illuminating treatment of a continuous-time random walk which gives simple formulae at all frequencies including zero. They consider an energy-independent model and obtain results which are identical to those for a random walk in which every site except the one currently occupied is re-randomized immediately after every hop[8]. Equations (30) to (32) remain valid (with F_m a constant) but in performing the system average we now take the relative site locations available immediately after <u>every</u> hop as statistically independent of one another. This so greatly simplifies the system averaging that it may be carried out explicitly without any further approximations. The final formulae at high frequencies and low site densities are not very different from those obtained in the pair approximation[7]. When ω=o we again have $D(0) = 0.5fd^2$ but, because of the re-randomization, f reduces to the reciprocal of the mean time before the first hop from an occupied site and d^2 reduces to the mean square distance moved in any one hop. In a low density system d^2 is equal to one third of the mean square nearest neighbour site separation because the electron always hops to the nearest of the re-randomized sites and f involves a factor $\exp(-2\eta^{-\frac{1}{2}}/3)$ where $\eta = 4\pi n_s(2\alpha)^{-3}$ because the staying time is dominated by contributions from unfavourable configurations of re-randomized sites[8]. Thus site re-randomization has a drastic effect on the behaviour of both f and d^2. This is because it removes the reciprocating hops which are a characteristic feature of random walks in random systems of fixed sites. In hopping transport we are concerned with fixed sites and the elucidation of the transition region between ac and dc conductivity must await further development of the theory for that case.

Different hopping mechanisms are reflected in the theory developed here in the formulae used for the transition rates. Existing experimental data on both ac and dc hopping conductivity shows little structure against which to test different models. The elementary models used here account for the gross features of the data. High precision data are required on both the frequency and temperature dependence of hopping conductivity in different materials to provide a means of discriminating between one hopping mechanism and another.

REFERENCES

1. E.M. Conwell, Phys.Rev. 103, 51 (1956).
2. N.F. Mott, Canadian J.Phys. 34, 1356 (1956).
3. N.F. Mott and E.A. Davis, "Electronic Processes in Non-crystalline Materials", Clarendon Press, Oxford (1971).
4. A. Miller and E. Abrahams, Phys.Rev. 120, 745 (1960).
5. W. Brenig, G. Döhler and P. Wolfe, Z.Physik. 2, 1 (1971).
6. P.N. Butcher, J.Phys.C: Solid St.Phys. 5, 1817 (1972).
7. P.N. Butcher, J.Phys.C: Solid St.Phys. 7, 879 (1974).
8. P.N. Butcher, J.Phys.C: Solid St.Phys. 7, 264 (1974).
9. M. Pollak and T.H. Geballe, Phys.Rev. 122, 1742 (1961).
10. V. Ambegaokar, B.L. Halperin and J.S. Langer, Phys. Rev. B4, 2612 (1971).

11. M. Pollak, J.Non-cryst.Solids, 11, 1 (1972).

12. R. Jones and W. Schaich, J.Phys.C: Solid St.Phys. 5, 43 (1972).

13. H. Scher and M. Lax, Phys.Rev. B7, 4491 (1973).

14. H. Scher and M. Lax, Phys.Rev. B7, 4502 (1973).

15. M. Pollak, Phys.Rev. 133A, 564 (1964).

16. M. Pollak, Phys.Rev. 138A, 1822 (1965).

17. M. Pollak, Phil.Mag. 23, 519 (1971).

18. P.N. Butcher and P.L. Morys, J.Phys.C: Solid St.Phys. 6, 2147 (1973).

19. A.H. Wilson, "The Theory of Metals", Cambridge University Press, Cambridge (1936).

20. P.N. Butcher and P.L. Morys, Proc. 5th Int.Conf. on Amorphous and Liquid Semiconductors, p.153, Taylor and Francis, London (1974).

21. S. Golin, Phys.Rev. 132, 178 (1963).

22. H.K. Rockstad, J.Non-cryst.Solids, 2, 192 (1970).

23. K.L. Chopra and S.K. Bahl, Phys.Rev. B1, 2545 (1970).

24. A.I. Lakatos and M. Abkowitz, Phys.Rev. B3, 1791 (1971).

25. G.E. Pike, Phys.Rev. B6, 1572 (1972).

26. H. Fröhlich, "Theory of Dielectrics", Clarendon Press, Oxford (1949).

27. A.E. Owen, Prof.Ceramic Sci. 3, 77 (1963).

28. A.K. Jonscher, Nature, 253, 717 (1975).

29. H. Margenau and G.M. Murphy, "The Mathematics of Physics and Chemistry", Van Nostrand, New York (1956).

30. V.K.S. Shante and S. Kirkpatrick, Adv.Phys. 20, 325 (1971).

31. K. Maschke, H. Overhof and P. Thomas, Phys.Stat.Sol. B62, 113 (1974).

32. V. Ambegaokar, S. Cochran and J. Kirkijärvi, Phys.Rev. B8, 3682 (1973).

33. J. Kirkijärvi, Phys.Rev. B9, 770 (1974).

34. C.H. Seager and G.E. Pike, Phys.Rev. B10, 1421,1435 (1974).

35. A.J. Grant and E.A. Davis, Solid St.Comm. 15, 563 (1974).

36. N.F. Mott, Phil.Mag. 19, 835 (1969).

37. E.M. Hamilton, Phil.Mag. 26, 1043 (1972).

38. F.R. Allen and C.J. Adkins, Phil.Mag. 26, 1027 (1972).

39. P.N. Butcher, to be published (1975).

40. J.K.E. Tunaley, Phys.Rev.Letts. 33, 1037 (1974).

41. N. Apsley and H.P. Hughes, Phil.Mag. 30, 963 (1975).

42. R.M. Hill, to be published (1975).

HOPPING TRANSPORT IN HIGH ELECTRIC FIELDS

I. G. Austin

Department of Physics

The University, Sheffield S3 7RH

ABSTRACT

In this lecture, we discuss the use of high electric fields to study phonon and photon assisted tunnelling processes between localised states in solids. Illustrations will be given, based on recent work on transition metal ion (TMI) compounds and chalcogenide glasses.

Many TMI glasses show a field dependent dc mobility of the form $\mu = \mu_o [\sinh \beta F]/\beta F$, where $\beta = ea/2\,kT$, as expected for simple hopping. But the jump distance (a) is much larger than expected. Several possible explanations are discussed, including the problem of averaging in a disordered system, local field corrections, barriers etc. In contrast, chalcogenide glasses show a non-linear mobility of the form $\mu = \mu_o \exp eaF/kT$, which extends to very low fields; this is probably a property of localised states near a mobility edge.

Optical hopping, or photon assisted tunnelling has been studied in charge transfer bands in TMI compounds, and also in connection with models of the Urbach edge in amorphous semiconductors. High field studies give a test of these models and will be briefly discussed.

1. INTRODUCTION

Hopping, or phonon assisted tunnelling between localised states, is a basic mode of electron transport in a wide range of low mobility semiconductors[7]. Typical examples are (i) amorphous solids, (ii) compounds containing transition metal ions (TMI) like ferrites and phosphate glasses, and (iii) molecular crystals like orthorhombic sulphur. Localisation is due to disorder, small polaron formation, or a combination of both. Recently, the high-field conductivity has been studied in detail - especially in disordered systems, where there is a technological interest in switching. Two kinds of non-linear behaviour have been identified; (a) those in which the free carrier density is enhanced by field emission from traps and charged centres, and (b) those in which the mobility becomes field dependent. In this lecture, we shall concentrate on category (b), and to illustrate the phenomena, we discuss in detail recent experiments on TMI compounds and chalcogenide glasses. More complete surveys are given in the articles by Mott[1], Marshall and Miller[2], Austin and Sayer[3]. Switching (which we omit) has been reviewed by Fritsche[4], and field emission processes by Jonscher and Ansari[5], and Hill[6].

Optical hopping, or photon assisted tunnelling between localised states may also occur (Figure 1b). Processes of this type have been studied in various TMI systems[7], e.g. Fe^{2+} - Fe^{3+} transfer in ferrites; they are also invoked in current models of the Urbach edge in amorphous materials. High electric fields modify the optical hopping process and their use in testing various models will be briefly discussed.

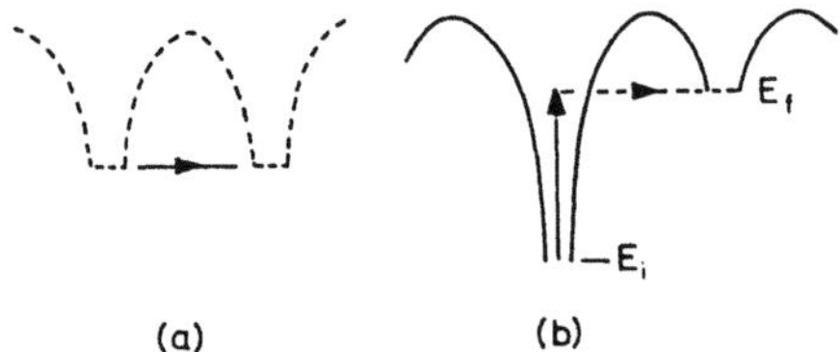

Figure 1. D.c. and optical hopping. (a) Phonon-assisted tunnelling, (b) photon-assisted tunnelling.

2. LOCALISED STATES

First we define more closely the nature of the localised states.

2.1 Transition Metal Ion (TMI) Crystals and Glasses

Many simple 3d oxides, ferrites and garnets show hopping

transport in the d levels when the transition metal ion is present in more than one valence state, e.g. $Mn^{2+} - Mn^{3+}$ transfer in MnO doped with Li. The d overlap is small in these materials, and there is good evidence that the carriers form dielectric of Jahn-Teller small polarons, with hopping energies in the range 0.1 - 0.4 eV[7].

The transition metal oxides also form glasses with P_2O_5 in which the TMI is a major constituent. Comprehensive studies have been made on these phosphate glasses, especially the vanadates ($V_2O_5 - P_2O_5$) and to a lesser extent glasses based on Fe, Cu, Ti etc. Again a "mixed valence" principal applies; thus in the vanadates we require V^{4+} and V^{5+} ions and the ratio is controlled by reduction in the melt. In the 3d glasses, we expect the 3d bands to form localised states in the Anderson sense (i.e. due to disorder), and any polaron hopping energy (W_H) will be increased by a disorder term (W_D) to give a mobility of the form[17,8].

$$\mu = \mu_o \exp - (W_H + \tfrac{1}{2} W_D)/kT \qquad 2.1$$

There is good evidence for this type of model, but it is difficult to determine the size of the W_D term with certainty[8,9]. In the vanadates $W_D \sim 0.1$ eV and in the other glasses it is somewhat larger.

It is important to note that in the 3d glasses and certain ferrites, there is strong evidence from thermopower and esr data that the carrier concentration is independent of temperature above ~ 150 K. Thus we make the assumption (section 3.1) that changes in the conductivity above ~ 150 K due to temperature or electric field are due to variations in the mobility.

2.2 Chalcogenide Glasses

Materials like As_2Se_3, As_2S_3, Se etc. form well defined bulk glasses with semiconducting properties and band-gaps of order 2 - 4 eV. It is generally accepted[10,11] that disorder in these materials gives rise to tails of states at the band edges, which are localised below E_c in the conduction band, and above E_v in the valence band (Figure 2b). Above E_c (or below E_v), the carriers move in extended states with a mobility $\sim 10^2$ cm^2/v sec, determined by scattering; below E_c (or above E_v), transport is by hopping and the mobility is $\lesssim 10^{-1}$ cm^2/volt sec. The extent of tailing is uncertain, but is probably ~ 0.1 eV in simpler glasses like Se and somewhat larger in complex glasses. There is also firm evidence of localised states which are deeper in the energy gap, due to defects such as broken bonds, but the density and distribution in the chalcogenides is uncertain[12].

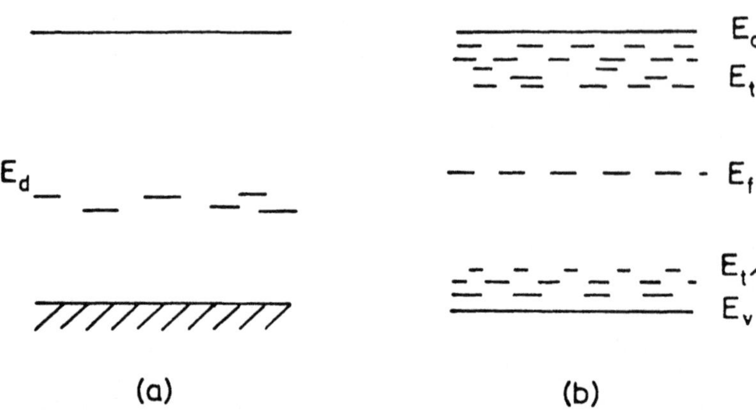

Figure 2. Localised states in (a) a 3d semiconductor and (b) a chalcogenide glass; E_c and E_v are mobility edges in the conduction and valence bands, E_t and $E_{t'}$ are localised tail states and E_f the Fermi energy.

In amorphous Si[13], the transport mechanism changes at a rather well defined temperature ($T_n \sim 400$ K), from hopping between localised tail states at $T < T_n$, to conduction in extended states above T_n. The situation in the chalcogenide glasses is less certain, but the work of Nagels et al.[14] strongly suggests that in at least some of these materials, both mechanisms act in parallel over a substantial temperature range (200 - 400 K). We note that in contrast, there is no indication that a mobility edge plays a part in the d glasses, and the evidence implies that all the d states involved in dc conduction are localised[8,17].

2.3 Hopping in High Fields - simple considerations

For the simplest type of hopping model, the field dependence of the mobility can be calculated as follows. We assume that hopping occurs by activation over a potential barrier of height W_o, between two localised sites separated by a. Then the barrier height for intersite jumps in applied field F is

$$W_{\pm} = W_o \pm \tfrac{1}{2} eaF \qquad 2.2$$

where the $\pm$ signs refer to jumping up or down the field gradient. The net intersite jump probability is proportional to

$$\exp\left[-\frac{W_+}{kT}\right] - \exp\left[-\frac{W_-}{kT}\right] \qquad 2.3$$

giving

$$\mu(F) = \mu(0)\ \frac{\sinh \beta\ F}{\beta F} \qquad 2.4$$

where

$$\beta = ea/2kT \qquad 2.5$$

The sinh term implies ohmic behaviour up to $F \sim kT/ea$ and thereafter an increase in mobility; thus if $a = 5$ Å, $T = 300$ K we require $F \gtrsim 5.10^5$ v cm^{-1} for non-linear effects. Equation 2.2 assumes a classical symmetrical barrier. Phonon assisted tunnelling models give a similar result (see later); but in some models, the downhill jump rate at low temperature is due to spontaneous phonon emission, and F only affects the uphill jump rate[3].

3. EXPERIMENTAL ASPECTS - d COMPOUNDS

High field conductivity studies are usually carried out on thin samples, using dc or pulse conditions and a two electrode configuration. In practice, the true behaviour of the bulk high-field conductivity may be masked by (i) Schottky barriers and other contact effects, (ii) space charge injection, and (iii) heating effects. Effects (i) and (ii) can usually be detected by systematic studies of the way in which the current varies with sample thickness, and different contacts. Pulse techniques enable the mean power dissipation to be kept low, but the high impedance and dielectric response of low mobility semiconductors often prevents the use of very short pulses, as in conventional semiconductors. Even so, a number of materials have been studied at fields up to $\sim 10^6$ v cm^{-1}, although in some glassy semiconductors switching may intervene at lower fields.

In a few chalcogenide glasses, the mobility and its field dependence has been measured directly[2] using the time of flight method developed by Spear and others. Charge carriers are generated on one side of the sample by using a fast electron or light pulse, and drawn across by an external field.

3.1 TMI Glasses and Crystals

Detailed studies of the high-field conductivity in a range of TMI phosphate glasses have been reported by Austin and Sayer[3], using dc pulse techniques. Figure 3 shows a representative set of

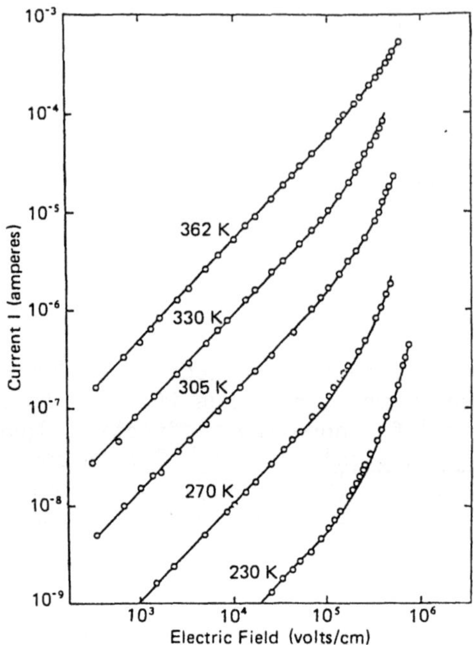

Figure 3. Current against electric field at various temperatures for a glass of composition 38 mol% Fe_2O_3 : 62 mol% P_2O_5. Full lines show fit to equation (2.4). From reference 3.

current-field characteristics at various temperatures for a glass of composition 38 mol% Fe_2O_3; 62 mol% P_2O_5. The behaviour is ohmic up to $\sim 10^5$ v cm^{-1} and at higher fields the current increases up to $\sim$ 20 times that expected from the low field I-E characteristic. Fields approaching $\sim 10^6$ v cm^{-1} could be applied without evidence of breakdown or switching. Measurements on different samples showed a conductivity independent of thickness, thus eliminating space charge conduction and Schottky emission.

For various hopping models (see §4 below) the ratio of the high field mobility to low field mobility should scale as FT^{-1}, as in equations 2.4 and 2.5. Figure 4 shows that the data for the iron glass scale accurately as FT^{-1} over a thickness range 1.9 to 5.9 microns, and a temperature range 230 - 430 K. Also (Figures 3 and 4), the field dependence is of the form (sinh βF)/βF. Excellent FT^{-1} scaling and sinh behaviour was also observed in vanadate glasses and other d glasses (Figure 5). Some deviations were observed at temperatures below $\sim$ 200 K; here again the data fitted a sinh law, but the jump parameter (a) decreased by $\sim$ 30% at low temperatures.

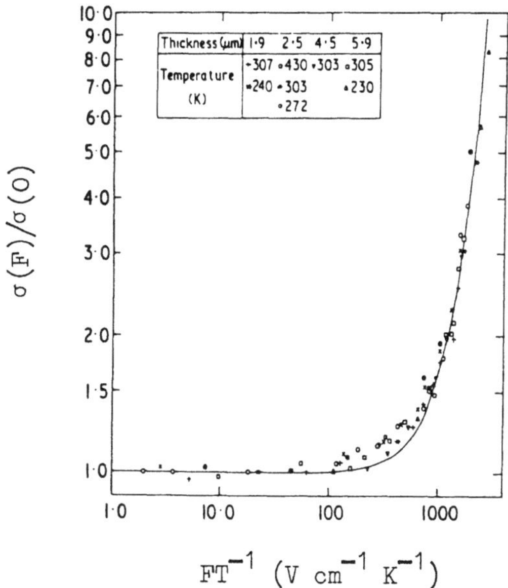

Figure 4. σ (F)/σ (O) against FT^{-1} for a glass of composition 38 mol% Fe_2O_3 : 62 mol% P_2O_5. Full line represents equation (2.4).

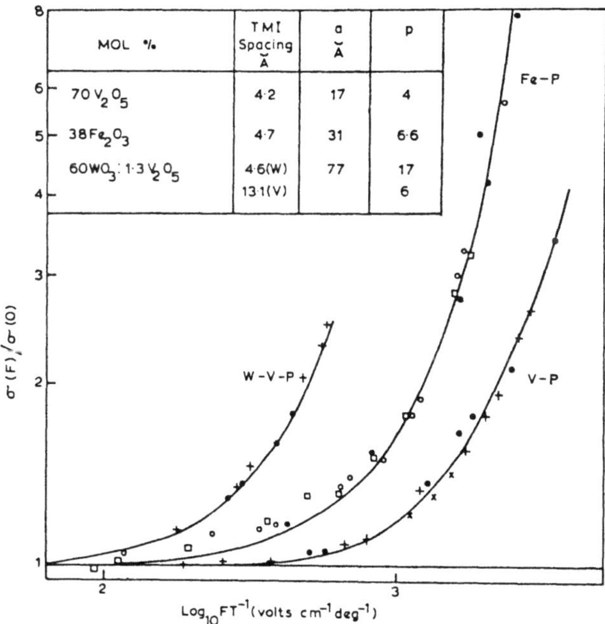

Figure 5. Log σ(F)/σ(O) versus log FT^{-1} for various phosphate glasses. Temperature range 200 – 450°K, thickness range 0.7 – 6 microns. Full line represents equation (2.4). From reference 37.

An analysis of the scaled results using 2.4 and 2.5 gives the jump distance (a). Some representative figures are given in Figure 5 (inset), and show that the apparent jump distance is always a factor p larger than the average spacing between TMI ions, where p = 4 – 7 and is a characteristic of individual glasses. It is interesting to note that a similar field dependence has been observed in spinel ferrite crystals, which have a disordered substitution on octahedral sites[7], and also for ionic diffusion in glasses[15]. In both cases the jump parameters were 20 – 50 Å, corresponding to p values in the range 5 – 10.

We recall from section 2.1 that the evidence from thermopower etc., implies that the field dependence of σ in the TMI glasses is a mobility effect. We turn therefore to the various hopping models and the extent to which they account for the sinh behaviour and large p values.

4. MODELS OF HIGH-FIELD HOPPING

4.1 Small Polarons

For a recent review of the concept, see reference (16). Consider a small polaron hopping from site i to site j, with a polaron hopping energy W_H. If the potential between the sites is $E_i - E_j = \Delta$, then the total activation energy W for hopping at high and intermediate temperatures is (cf equation 10), Austin and Mott[17]

$$W_{\pm} = W_H \pm \tfrac{1}{2}\Delta + \frac{\Delta^2}{16\,W_H} \qquad 4.1$$

assuming that $W_D < W_H$. Thus the jump rates for uphill and downhill hops are

$$\gamma^o_{ij} = \omega_o \exp\{-\beta[W_H + \tfrac{1}{2}(E_j - E_i)]\} \qquad 4.2$$

$$\gamma^o_{ji} = \omega_o \exp\{-\beta[W_H - \tfrac{1}{2}(E_j - E_i)]\} \qquad 4.3$$

Here $\beta = 1/kT$ and the terms in Δ^2 have been omitted for brevity. ω_o is a frequency factor and for non-adiabatic hopping, ω_o varies as $\exp(-2a_{ij}\alpha)$, where α measures the fall-off in overlap with distance. In the absence of an external field, detailed balancing gives

$$\frac{\gamma^o_{ij}}{\gamma^o_{ji}} = \exp \beta \, (E_i - E_j) \ . \qquad 4.4$$

When an external field is applied we can write, using 4.1

$$\Delta = W_D \pm e a_{ij} F \qquad 4.5$$

where a_{ij} is the site separation, and W_D a disorder potential. From the net jump probability along the field, we obtain[3]

$$\frac{\sigma(F)}{\sigma(0)} = \left[\exp - \frac{e^2 a^2 F^2}{16 W_H kT}\right] \frac{\sinh \beta e a' F/2}{\beta e a' F/2} \qquad 4.6$$

where

$$a' = (1 + W_D/4W_H) a \qquad 4.7$$

Equation 4.6, with $W_D = 0$, is identical with the results from Kubo theory[18,19,20]. The presence of the disorder term increases the effective jump parameter (4.7), but this effect is probably small ($\sim$ 20%) since $W_D < W_H$. The exponential term in (4.6) is only significant at high fields, where $eaF \sim 4(W_H kT)^{1/2}$.

The field dependence of formula (4.6) (with $W_D = 0$ and the exponential term equal to unity), is identical with the result for activation over a barrier[21] and for ionic diffusion.

4.2 Variable Range Hopping

If ω_o varies as $\exp(-2\alpha a_{ij})$, and W_D is large ($>$ kT), the carrier may tunnel a distance p times the nearest-neighbour distance (a_o) to a site at a lower energy. Following Mott[22], we put $W_D \propto p^{-3}$; the optimum value of p is then

$$p^4 \simeq \frac{W_D}{kT} (\alpha a_o)^{-1} \ . \qquad 4.8$$

Little is known about the size of the W_D term in the d-glasses, but for the vanadates there is evidence that $W_D \sim kT$[17,9]. From the variation γ^o_{ij} with total vanadium concentration, we have argued elsewhere[3] that $\alpha^{-1} \lesssim 3$ Å, implying $p \simeq 1$ and nearest-neighbour hopping in the vanadates[8]. If we make the plausible assumption that $W_D \lesssim 10$ kT in the other d glasses, then values of $\alpha^{-1} > 100$ Å would be required to account for the p values in figure 5 (inset). Such large tunnel factors are extremely unlikely for the atomic like d orbitals in these glasses, and for comparison we note that $\alpha^{-1} \approx 10$ Å for s states in Ge[10]. We conclude that p values as large as 4 – 6 cannot be explained in this way, although we do not rule out the possibility of p values somewhat greater than unity. Aspley and Hughes[38] have recently shown that $\sigma(F)$ varies as exp (AF^2) in the variable range region.

4.3 Averaging

Formula (4.6) gives the conductance between two sites and must be averaged over all configurations. In their low-field theory Miller and Abrahams (MA)[23] shows that the hopping model is equivalent to a random resistance network, with conductance links G_{ij} which are proportional to the net hopping rate between sites. The average impedance is then calculated assuming that at each hop the electron always chooses the easiest jump. This condition leads to some difficult hops and these dominate the mean impedance. Various authors have criticised the MA method of averaging[24,11]. In a percolation treatment[24,25] the difficult hops are bypassed and the conductivity is assumed to be proportional to the probability that a site lies on a complete conducting path. But present percolation theories seem to overestimate the conductivity, since many of the percolation paths are not fully connected[26,27]. Also, the percolation models neglect any correlation between the energies or occupation numbers at different sites, and any variation of the density of states or tunnel factor with energy.

No formal extension of these averaging theories to high fields has been given, but from (4.6) we infer that the two-site conductance links in the MA network will be replaced by field-dependent links of the form

$$G_{ij} \frac{\sinh ea_{ij}F/2}{ea_{ij}F/2} \qquad 4.9$$

Thus the conductance is most enhanced for the longest hops and if the carrier tunnels to a nearest-neighbour site, these are probably the hardest hops. However, the fluctuations in a_{ij} are far too small ($\sim \pm 30\%$) to account for the large p values in table 1. At higher fields ($eaF \sim kT$), the conductance link varies as exp $[-(2\alpha + \beta eF \cos\theta)\, a_{ij}]$ where θ is the angle between the jump and F – thus the path taken by the carrier may be modified with a preference for long hops in the direction of the applied field. Again however, it is difficult to envisage p factors of the order of 5 – 10 arising from this source.

To summarize, we see that several mechanisms can increase the effective jump parameter in the sinh formula (4.6), but in each case (4.7, 4.8) the increase is too small to account for the p values in Figure 5. An alternative way of viewing our data is to regard p as a field enhancement factor, that is, we assume that hopping does occur across distances which are comparable with nearest-neighbour distances, but the effective field is increased by a factor p. We consider two possibilities below.

4.4 Local Field Corrections

The local field (F_{loc}) acting on an ion in a solid is the sum of the applied field and the field (F_{pol}) due to polarization (P) induced in the medium. According to Lorentz, F_{pol} is given by the surface polarization on the walls of a spherical cavity centred on the ion, and $F_{loc} = F + LP$, where $L = 4\pi/3$.

In electron hopping transport, the local field acting on a carrier before and after hopping will be of the above form, but the question of whether the local field has any effect on the transit from one localised state to another has only recently been examined rigorously[28]. Munn[29] introduces a Lorentz type term in his calculations of carrier mobilities in anthracene, and finds $F_{loc} = 1.7\ F$.

For classical activation over a barrier, as in ionic conduction, Lidiard[30] has argued that no Lorentz correction should be applied, since the Lorentz cavity must always be centred on the hopping carrier: thus the Lorentz field can do no work as the carrier moves from one site to another. This view is strongly supported by the close agreement between ionic diffusion coefficients obtained from conductivity data, NMR relaxation and isotope diffusion[30,31].

A similar argument should apply for electron tunnelling through a barrier, although in this case the electron spends a short time in the barrier and may be momentarily displaced from the centre of its Lorentz cavity. A recent theory based on small polaron formalism shows however[28], that all local field effects cancel in the hopping mobility, because the jump probability between two sites depends only on the energy <u>difference</u> between them, and the Lorentz dipole contributions at the two sites cancel. Thus there is no local field contribution to p in the d-glasses.

4.5 Microscopic Barriers

An alternative possibility is that energy correlations between sites exist in the glasses over regions of order 5 Å upwards, due to (a) structural features in the glass, or (b) long range Coulomb potentials due to donor defects[32]. If the spatial scale of these correlated regions is large enough, then the fluctuations can be described in terms of a local variation in the Fermi level. In this way, we obtain regions of high and low conductivity and higher local fields will appear across the regions of lower conductivity. Clearly this is true in the limit of barriers separated by macroscopic distances in crystalline Se[33] and it is plausible that it remains true on a scale extending down to some tens of Å. In all these cases, the field dependence of a macroscopic sample will be determined by the effective local field which exists in the regions of low conductivity or at difficult hops. We now develop a simple model which incorporates some of these features.

We assume that we have conduction down chains and we neglect any variation in branching caused by the high field. For simplicity we assume a one-dimensional chain with a single hard hop, which may be a potential barrier (figure 6), or tunnelling over a

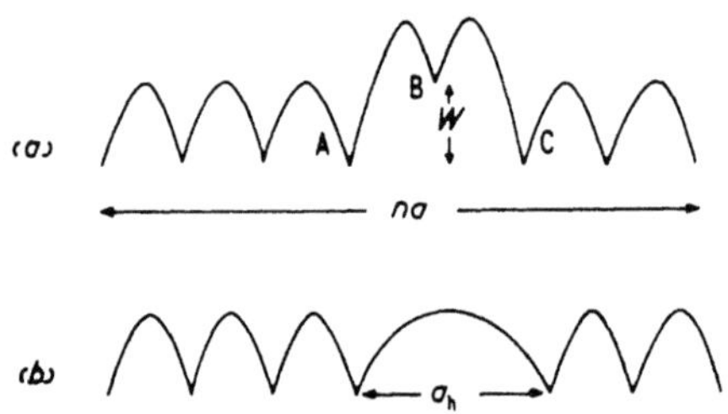

Figure 6. Chain of sites containing a hard hop. (a) Potential barrier W, due to disorder or structure. (b) Large jump distance, a_h, (tunnelling).

long distance (figure 6). n is the average number of links in a chain containing a single hard hop, and a the mean spacing. The impedance of the hard hop is Z in either direction, and in figure 6a the impedance of the "easy links" B to C and B to A in reverse are neglected. All other sites are equivalent with impedance links z. In the SP model a T dependence of the form $\exp(-W_H/kT)$ is contained in all links.

Consider now the onset of the high-field behaviour. The highest field is developed across the high impedance links, and we neglect the field dependence of the other links. Thus the impedance of the high impedance links is $Z\,\phi^{-1}(F_h')$, where

$$\phi(F_h') = [\sinh F_h']/F_h' \simeq 1 + \frac{1}{6}(F_h')^2. \qquad 4.11$$

Here $F_h' \equiv \beta e a F_h/2$, and F_h is the local field across the high-impedance link.

In a field F, the average conductivity per link is of the form[3]

$$\frac{\langle\sigma(F)\rangle}{\langle\sigma(0)\rangle} = 1 + \frac{1}{6}\,(p\,F_{av})^2 \qquad 4.12$$

where

$$p = n(1 + A)^{-3/2} \qquad 4.13$$

and $A = (n - 2)z/Z$. If W is the barrier height, then $Z/z = \exp W/kT$, and p is temperature dependent through A. We note that as $A \to 0$, $p \to n$. When $A < 0.25$, p is relatively insensitive to variations in A, but z_{av}/z begins to rise appreciably. Under these conditions, the main voltage drop occurs across the high impedance links.

For the barrier model (figure 6a) we estimate $n \sim 5$, $W \gtrsim 2$ kT, for the vanadates, and $n \sim 10$, $W \gtrsim 4$ kT for the iron glasses. These numbers are necessarily rough, but the main point we wish to make is that relatively small values of W (0.05 – 0.1 eV) are sufficient to give appreciable field enhancement. Also, these small W values are consistent with independent estimates of the disorder potentials in TMI glasses[3]. A similar analysis can be made in terms of the tunnel model (figure 6b) but this leads to rather large fluctuations in jump distance, and it seems unlikely that the hard hops can be explained in this way[3].

In all the TMI glasses, donor or acceptor defects are present due to reduction or oxidation of the melt. However, the d-carrier concentrations are high ($\sim 10^{19}$ cc^{-1}), implying Debye screening lengths $\sim 5 - 10$ Å, so that the range of any energy correlations due to Coulomb fields should be small. It seems more plausible therefore that structural features are responsible for any inhomogeneities.

One factor which must be considered is phase separation. The present evidence on these glasses rules out separation on a gross scale, but not on scale of a few tens of Å. An alternative possibility is that inhomogeneities of a molecular nature are present. ESR and NMR data on the vanadates show that two types of vanadium site are present, and it has been inferred that VO_5 chains or rings of one type of site; thus the second vanadium site could act as a barrier or hard hop[3].

To summarise, we see that the TMI glasses show a high-field conductivity of the type expected for hopping, but the effective field in formulae like 4.6 is enhanced by a factor p, which is characteristic of individual systems, and probably arises from structural features. Our crude model shows that the macroscopic high-field behaviour is sensitive to inhomogeneities or local fluctuations in conductivity; this could explain the large p values, but independent proof is needed. Also, direct measurements of the high-field mobility would be valuable, by transit-time methods if possible, to check the assumption that the carrier density is independent of F.

5. CHALCOGENIDE GLASSES

Marshall and co-workers[2,34] have recently shown that the field dependence of the conductivity and drift mobility (from transit time experiments) is of the form

$$\sigma(F) = \sigma(0) \exp eaF/kT \qquad 5.1$$

and

$$\mu(F) = \mu(0) \exp eaF/kT \qquad 5.2$$

in a wide range of amorphous chalcogenides. Thus all the field dependence is associated with the mobility. Here (a) is an activation length, of order 10 - 40 Å, which decreases with rise of temperature. Figures 7 and 8 show representative data.

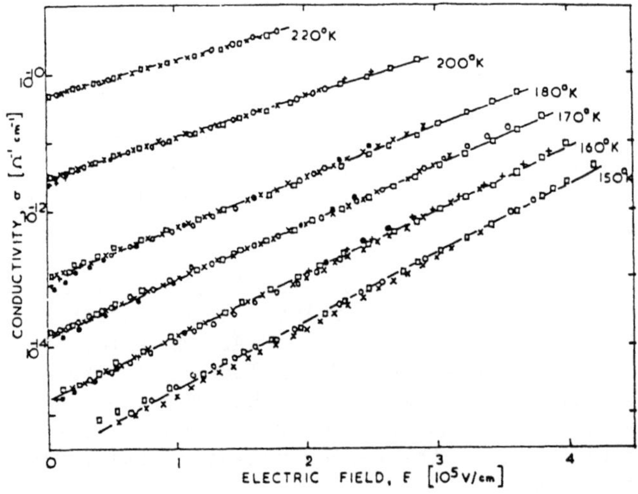

Figure 7. Electric field dependence of the d.c. conductivity of vitreous $As_{30}Tr_{48}Ge_{12}Si_{10}$ (after Marshall and Miller, reference 2).

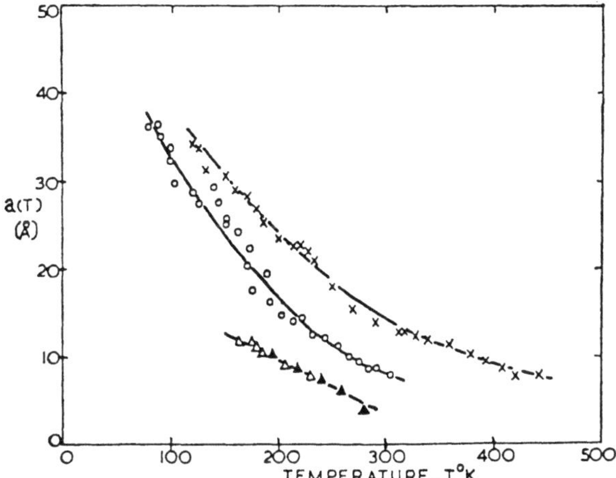

Figure 8. Temperature dependence of the parameter a(T) in several vitreous semiconductors: (Δ) vitreous selenium, hole drift mobility data; (▲) vitreous selenium, electron drift mobility data; (×) vitreous $As_{30}Te_{48}Ge_{12}Si_{10}$, steady and pulsed field conductivity data; (O) vitreous $Ge_{15}Te_{85}$, steady and pulsed field conductivity data. After Marshall et al., reference 34.

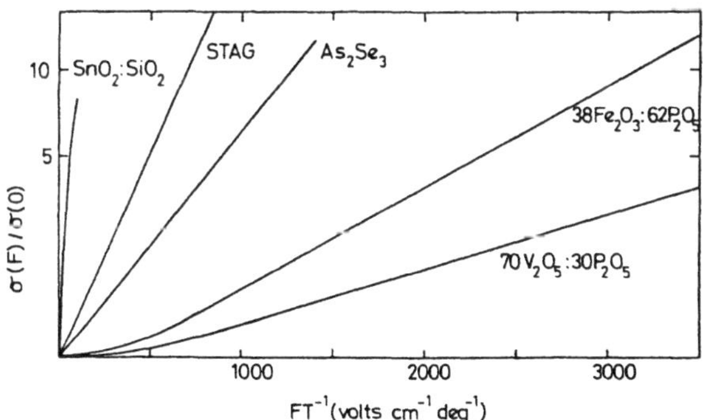

Figure 9. Log σ(F)/σ(O) versus FT^{-1} for various amorphous semiconductors. (i) SnO_2–SiO_2 glaze (two-phase), (ii) STAG glass, (iii) As_2Se_3, (iv) vanadate and iron phosphate glass. See reference 37 for details.

The most significant point is that the mobility is non-ohmic down to the lowest fields at which accurate measurements were possible ($\sim 10^3$ v cm^{-1}). This behaviour resembles that observed in two-phase systems where field emission occurs at internal barriers, but is in striking contrast with the TMI glasses, which are ohmic below $\sim 10^5$ v cm^{-1}. Figure 9 shows a comparison. All the mechanisms we have considered in section 4 predict ohmic behaviour as $F \to 0$, and therefore seem inadequate to explain equation 5.2.

Transit time measurements on these materials show that the low mobility is due to repeated trapping and release in tail states and deep traps. The release time varies exponentially with trap depth (E_{tr}) and theory shows that in a high field, both E_{tr} and the capture cross-section are modified[35], giving a field dependent mobility. But again ohmic behaviour is predicted as $F \to 0$; moreover, Marshall et al.[2,34] show that the parameter (a) in 5.2 is rather insensitive to changes in the distribution of deep traps. Thus they exclude mechanisms of this kind and argue instead that the field dependence is due to a modification of transport processes near the mobility edge.

5.1 Mobility Edge

Mott[1] has pointed out that in a high field, carriers trapped in shallow states just below the mobility edge can always tunnel into extended states (Figure 10), and this mechanism will persist to low fields. Following Marshall et al.[34], the process can be roughly analysed as follows.

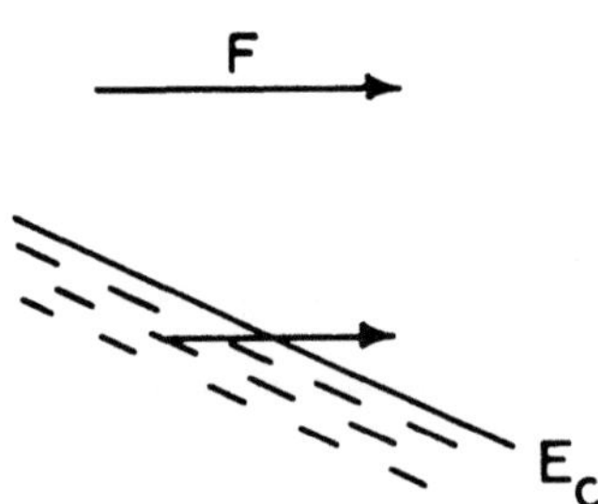

Figure 10. Tunnelling at a mobility edge E_c.

Consider a trap [localisation parameter $\alpha(E)$], which is at a depth E below the mobility edge, and empties into extended states thermally or by tunnelling a distance R (figure 10). The ratio of emission rates is $\sim \exp(-E/kT)/\exp(-\alpha R)$, and field emission will dominate for all localised states above an energy E_o, given approximately by

$$\alpha(E_o) \sim eF/kT \tag{5.3}$$

Thus the total number of free carriers in a field F is $N_t(E_o) f(E_o)\, kT$, compared with $N(E_c)\, f(E_c)\, kT$ at zero field, and

$$\sigma(E)/\sigma(0) = \frac{N_t(E_o)}{N(E_c)} \exp(E_o/kT) \tag{5.4}$$

If we assume an exponential distribution of tail states, $N_t/N = \exp - c\, E_o$, and

$$\sigma(E)/\sigma(0) = \exp a'\, E_o/kT \tag{5.5}$$

where

$$a' = 1 - kT\, c \tag{5.6}$$

A comparison with 5.2 shows that E_o should vary as F, and hence from 5.3, $\alpha(E)$ should be a linear function of E. This is plausible, although one recent theory[34] gives a variation of $E^{0.6}$. Equation 5.6 also predicts the correct temperature trend in the activation length (figure 8).

Clearly this simple model could be improved in a number of ways, but it does account for the basic high-field features, especially the non-ohmic behaviour at low fields. Recent work by Nagels et al.[14] strongly suggests that in the chalcogenides, hopping through tail states occurs in parallel with extended-state conduction. It seems likely however, that the inclusion of this process would only make minor modifications in the behaviour of $\sigma(F)$ at low fields.

6. OPTICAL HOPPING

6.1 Theory

Consider an electron localized at site i, with energy E_i, and an adjacent empty site at an energy E_j (figure 1(b)). The oscillator strength (f) for an optical transition which takes the electron from site i to site f can be calculated using the simple theory of Mulliken[37]. If ϕ_1 and ϕ_2 are the wavefunctions for orbitals at the two sites, then the initial and final wavefunctions are

$$\phi_i = \phi_1 + \alpha_{12}\,\phi_2 \qquad 6.1$$

and

$$\phi_f = \phi_2 + \alpha_{21}\,\phi_1 \qquad 6.2$$

The matrix element for the optical transition at $h\nu = E_f - E_i$ is $\langle\phi_f^* \, ex \, \phi_i\rangle$ giving for the oscillator strength

$$f \simeq 10^{-5}\,\nu\,M^2 \qquad 6.3$$

where

$$M \simeq ea\,\alpha_{12} \qquad 6.4$$

From first order perturbation theory, $\alpha_{12} \simeq -\alpha_{21} \approx J/(E_f - E_i)$, where J is the overlap integral between sites. Thus f gives a measure of J; typically values of $f \sim 10^{-1}$, $a \sim 4$ A, $h\nu \sim 1$ eV correspond to $J \simeq 0.1$ eV.

In the general case, we can write

$$(E_f - E_i) = 2\,W_p + W_D + E' \qquad 6.5$$

where W_p is a polaron binding energy, W_D a disorder potential and E' any electronic energy difference, arising (say) from a change of spin state at one site[7]. If the electron-phonon coupling is strong (small polarons), or W_D is large, the line will be broadened but f is unchanged.

Optical charge transfer (CT) bands of this type have been studied in various d compounds[7], but they are difficult to identify in the presence of internal d-d crystal field bands. Recent work shows however[20,36] that the CT bands are more easily perturbed by an external electric field than d-d transitions at a single site, and thus may be separated using modulation spectroscopy. Theory predicts[36] a symmetric field broadening of the CT band, if optical jumps occur up and down the applied field with equal probability. If the optical hops are constrained to one direction of the applied field, the CT band is shifted to a higher or lower frequency.

6.2 d-Compounds

Figure 11 shows some data for an iron phosphate glass, using blown films 1 - 10 microns thick[37]. The absorption has three general features: (a) a steep edge at $h\nu \gtrsim 2.5$ eV, (b) a long tail extending to low energies (> 1 eV), and (c) broad bands superimposed on the tail, some of which are due to crystal field transitions. Region (a) we interpret as the main absorption edge due to transitions from the oxygen p-band into s or d states of the cations. Broad and unresolved crystal field and CT bands may contribute to the background in the long tail (b), but the origin of this feature is uncertain. Similar tails appear in crystals containing 3d ions, suggesting that the tail is not an exclusive property of amorphous materials.

The peak at ∿ 1.1 eV in the iron glass is a well-known crystal field transition ($f \simeq 10^{-4}$) for Fe^{2+} ions in distorted octahedral sites. Also, the peak near 2.2 eV could be a Fe^{3+} crystal field band, although our measured f-value is somewhat high (∿ 10^{-4}), and the evidence below suggests an overlapping CT band. When the Fe^{2+} concentration is altered by reduction, the tail absorption shows a broad change in the region 1 – 2.5 eV (figure 11).

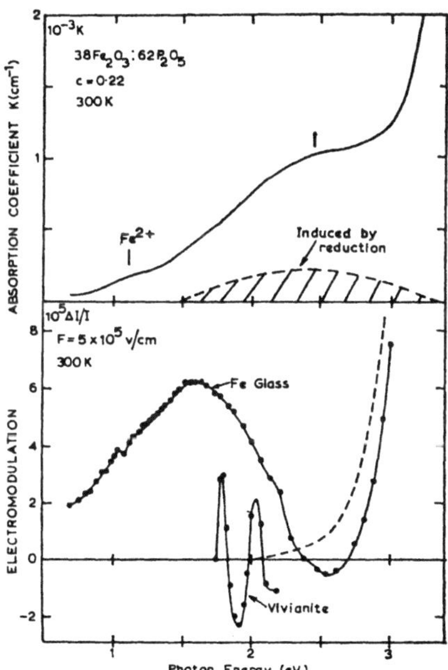

Figure 11. Absorption and electro-transmission (ET) spectra for iron phosphate glass. 38 mol.% Fe_2O_3, Fe^{2+}/Fe total = 0.11. ET spectrum shows fractional change of transmission in an external field, plotted as minus $\Delta I/I$; dashed curve shows extrapolated ET effect for Urbach tail. From reference 37.

A typical modulation spectrum for the iron glass (Fe^{2+}/Fe^{3+} = 0.28) is shown in the lower half of figure 11. Studies at various temperatures and compositions indicate that two distinct features are present: (a) a rapid rise in $\Delta I/I$ near the edge at $h\nu > 2.5$ eV, and (b) a positive to negative swing at lower energies, passing through zero at ~ 2.3 eV. The negative swing is partly obscured by the edge behaviour.

It is useful to compare this glass with vivianite, a crystalline iron phosphate in which there is strong evidence of a $Fe^{2+} - Fe^{3+}$ CT band at 1.9 eV ($f \sim 10^{-3}$). A preliminary ET spectrum for vivianite is shown in figure 11(a), and this is consistent with a field broadening of the CT band at 1.0 eV. Vivianite has a complex crystal field spectrum and it is striking that the only significant ET signals are observed near 1.9 eV. This, and other ET data we have obtained on pure crystals is good evidence that CT bands are much more easily modulated than crystal field bands.

We speculate therefore, that the positive-negative feature in the ET spectrum of the iron glass is associated with an $Fe^{2+} - Fe^{3+}$ CT band centred near 2.2 eV; but the band in the glass is much broader than in the crystal, and under field modulation it is shifted to lower energies. A possible explanation for the latter, is that for carriers trapped at barriers or "hard hops", optical hopping down the applied field will dominate, and such carriers will experience the highest local field. Using the absorption change on reduction (figure 11(a)), we estimate $f \lesssim 2 \times 10^{-3}$ for this band and $J \lesssim 0.02$ eV from equation 6.3.

6.3 Urbach edges in glasses

Many amorphous semiconductors show an Urbach absorption edge of the form

$$K = K_o \exp - (h\nu_o - h\nu)/\sigma \qquad 6.6$$

The detailed mechanisms are uncertain[39], but most models are based on (i) electric field broadening of an edge by internal microfields (with or without exciton effects), or (ii) band-gap variations due to density fluctuations. Models of type (i) can be viewed as a process of photon assisted tunnelling through a barrier at a band edge, the slope of the barrier depending on the local electric microfield. To calculate the net absorption coefficient K, it is necessary to average over some distribution of microfields. The change in absorption induced by an applied <u>external</u> field can also be calculated, and compared with experiment to test the various models[40,41].

Figure 12 shows some recent data on amorphous As_2Se_3 at 300 K.

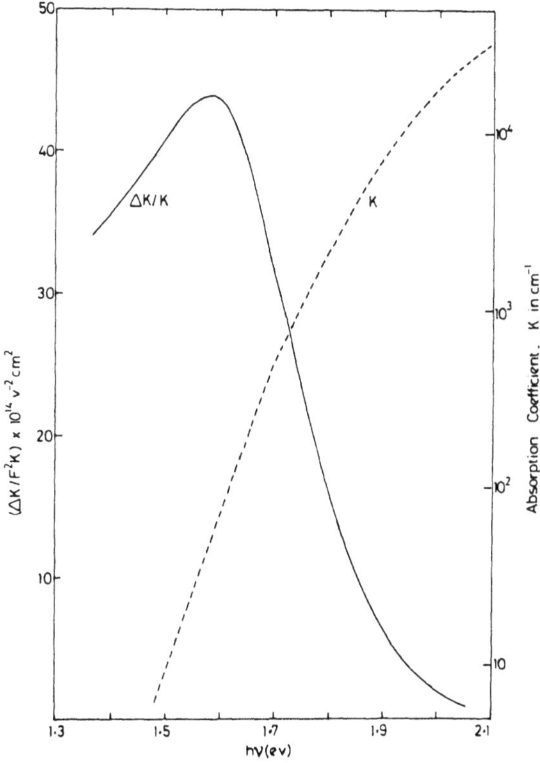

Figure 12. Absorption (K) and electro-absorption ($\Delta K/K$) spectra for amorphous As_2Se_3 at 300 K. From reference 41.

The Urbach edge, which lies below 1.7 eV has a slope of 0.049 eV, which decreases ∿ 20% on cooling to 80 K. Also shown is the electro-absorption spectrum, or ($\Delta K/K$), which varies as the square of the applied field (F) at all wavelengths. ($\Delta K/K$) shows a turnover in the lower part of the Urbach tail in the region $h\nu < 1.6$ eV, implying that in this region the external field decreases σ in equation 6.6. We have observed similar characteristics in amorphous Se and other chalcogenide glasses[41].

According to the Dow and Redfield model[42], the absorption of an exciton in a uniform field F is proportional to

$$\exp h\nu/F \qquad 6.7$$

Thus the logarithmic slope should change in an external field. If an external d.c. field F is applied to such a system, then the local field at some point will be $(\bar{F}_s + \bar{F})$, where $\bar{F}_s$ is the random component. When an average is taken over all sites, we obtain pairs of terms of the form

$$\exp A/(F_s + F) + \exp A/(F_s - F) \qquad 6.8$$

at sites where $F \| F_s$. Here $A = C(h\nu_o - h\nu)$, and we neglect any change in ν_o, since this gives a frequency independent contribution to $\Delta K/K$. For sites where $F \perp F_s$, the local field is $\sqrt{(F^2 + F_s^2)}$. Assuming a weighting factor of $\frac{1}{3}$ for (6.8) and $\frac{2}{3}$ for the case where $F \perp F_s$, we obtain (for $F \ll F_s$)

$$\Delta K/K = A'(2 + \tfrac{1}{2} A')\, F^2/3F_s^2 \qquad 6.9$$

where $A' = (h\nu_o - h\nu)/\sigma$.

Formulae like 6.7 and 6.9 must be averaged over the microfield distribution $p(F_s)$. We have examined the predicted behaviour of $(\Delta K/K)$ for the various microfield models using analytical and computational methods[40,41]. In general

$$\Delta K \propto (h\nu_o - h\nu)^n K \qquad 6.10$$

in the tail region, where $0 < n < 2$, depending on the type of distribution $p(F_s)$ and degree of electron-hole correlation. Thus for a gaussian $p(F_s)$, $n = 2/3$ for the Dow and Redfield exciton model and $n = 1$ for the Franz-Keldysh case. Both models give a reasonably good exponential edge for K after averaging over $p(F_s)$. However, in order to get a turnover in $(\Delta K/K)$ as in Figure 12, we need n to be negative in equation 6.10. If $p(F_s)$ is a coulombic distribution than $n = -2$, but K is no longer a good exponential. A mixed gaussian and coulombic $p(F_s)$ gives a turnover in $(\Delta K/K)$[40], and implies an rms microfield $\sim 5.10^6$ v cm^{-1}; but inevitably there is some departure from exponential behaviour in K. Models based on elastic broadening also give $n \geq 0$. We conclude therefore, that all these microfield models are inadequate in their present form to explain the maximum in the $(\Delta K/K)$ spectrum, and give at the same time, an accurate exponential for K.

An alternative point of view is suggested by our work on Se-As alloys[41]. Namely, that in the Se rich alloys there is a second absorption component in the tail region, which is much less sensitive to electric field broadening. This hypothesis explains the turnover in $(\Delta K/K)$, and a plausible physical origin is that the local electro-absorption response in these materials is anisotropic[41]. Thus the chain structure of Se is preserved in the glass, and it is known from modulation measurements on trigonal selenium crystals that the electro-absorption response is different for fields parallel and perpendicular to Se chains. A similar explanation could apply to As_2Se_3, where the glass retains a considerable amount of the local order of the sheet structure, which forms the crystal.

Acknowledgements

The author is indebted to Dr. L. Friedman and Dr. J. M. Marshall for valuable discussions, and access to unpublished work on the local field question (reference 28).

REFERENCES

1. Mott, N. F., 1971, Phil. Mag. 24, 911.
2. Marshall, J. M. and Miller, G. R., 1973, Phil. Mag. 27, 1151.
3. Austin, I. G. and Sayer, M., 1974, J. Phys. C. (Solid State) 7, 905.
4. Fritsche, H., 1972, Scottish Universities Summer School on Amorphous Semiconductors (London: Academic Press), p. 557.
5. Jonscher, A. K. and Ansari, A. A., 1971, Phil. Mag. 23, 205.
6. Hill, R. J., 1971, Phil. Mag. 24, 1307.
7. Austin, I. G. and Gamble, R. G., 1971, Second Int. Conf. on Conduction in Low Mobility Materials (London: Taylor and Francis) p. 1.
8. Austin, I. G. and Garbett, E. S., 1972, Scottish Universities Summer School on Amorphous Semiconductors (London: Academic Press) p. 393.
9. Sayer, M. and Mansingh, A., 1972, Phys. Rev. 6B, 4629.
10. Mott, N. F. and Davis, E. A., 1971, Electronic Processes in Non-cryst. Materials (London: UP).
11. Brenig, W., 1973, Fifth Int. Conf. on Amorphous and Liquid Semiconductors, (London: Taylor and Francis) p. 31.
12. Spear, W. E., 1973, Fifth Int. Conf. on Amorphous and Liquid Semiconductors, (London: Taylor and Francis) p. 1.
13. Le Comber, P. G., Madan, A. and Spear, W. E., 1972, J. Non-cryst. Solids 11, 219.
14. Nagels, P., Callaerts, R. and Denayer, M., 1973, Fifth Int. Conf. on Amorphous and Liquid Semiconductors (London: Taylor and Francis) p. 867.
15. Vermeer, J., 1956, Physica 22, 157.
16. Emin, D., 1972, Scottish Universities Summer School on Amorphous Semiconductors (London: Academic Press) p. 261.
17. Austin, I. G. and Mott, N. F., 1969, Adv. Phys. 18, 41.
18. Efros, A. L., 1967, Sov. Phys. (Solid State) 9, 901.

19. Klinger, M. I., J. Non-cryst. Solids 4, 463.

20. Reik, R. G. 1970, Solid St. Communic. 8, 1737.

21. Bagley, B. G., 1970, Solid St. Communic. 8, 345.

22. Mott, N. F., 1968, J. Non-cryst. Solids 1, 1.

23. Miller, A. and Abrahams, E., 1960, Phys. Rev. 120, 745.

24. Jones R. and Schaich, W., 1972, J. Phys. C. (Solid St.) 5, 43

25. Ambegaokar. V., Halperin, B. I. and Langer, J. S., 1971, Phys. Rev 4B, 2612.

26. Last, B. J. and Thouless, D. J., 1971, Phys. Rev. Lett. 27. 1719.

27. Adler, D., Flora, L. P. and Santuria, S. D. 1973, Solid St. Communic. 12, 9.

28. Saglam, M. and Friedman, L. 1975, J. Phys. C. (Solid State) 8L, 245.*

29. Munn, R. W., 1972, Chem. Phys. Lett. 16, 429.

30. Lidiard, A. B., 1957, Hanb. Phys. 20, 1, 246.

31. Hoodless, I. M., Strange, J. A. and Wylde, L. E., 1971, J. Phys. C. (Solid State) 4, 2742.

32. Fritsche, H., 1971, J. Non-cryst. Solids 6, 49.

33. Lemercier, C., 1971, Second Int. Conf. on Conduction in Low Mobility Materials, (London: Taylor and Francis p. 251.)

34. Marshall, J. M., Fisher, F. D. and Owen, A. E., 1973, Fifth Int. Conf. on Amorphous and Liquid Semiconductors (London: Taylor and Francis) p 1305.

35. Dussel, G. A. and Boer, K. W., 1970, Phys. Stat. Solidi 39, 375.

36. Austin, I. G., 1972, J. Phys. C. (Solid State) 5, 1687.

37. Austin, I. G., Sayer, M. and Sussman, R. S., 1973, Fifth Int. Conf. on Amorphous and Liquid Semiconductors. (London: Taylor and Francis) p. 1343.

38. Aspley, N. and Hughes, H. P., 1974, Phil. Mag. 30, 963.

39. Dow, J. D., 1972, Comments on Solid State Physics 4B, 35.

40. Street, R. A., Searle, T. M., Austin, I. G. and Sussmann, R. S., 1974, J. Phys. C. (Solid State) 7, 1582.

41. Sussmann, R. S., Austin, I. G. and Searle, T. M., 1975, J. Phys. C. (Solid State), in press.

42. Dow, J. D. and Redfield, D., 1970, Phys. Rev. B1, 3358.

43. Lingelbach, W., Stuke, J. and Weiser, G., 1972, Phys. Rev. 5B, 243.

* See also Munn RW, 1975, J. Phys. C. (Solid State) 8, 2721

THE FORMATION AND MOTION OF SMALL POLARONS*

David Emin

Sandia Laboratories

Albuquerque, New Mexico 87115

I. INTRODUCTION

In this paper I shall endeavor to qualitatively discuss some essential aspects of the theory of the formation and motion of small polarons. This theory has been developed with the view of applying it to materials which are often classified as low-mobility insulators. Typically, the mobilities involved are substantially less than 1 cm^2/V-sec. In these circumstances electrical transport is, for the most part, discussed in terms of phonon-assisted hops (tunneling events) rather than in terms of scattering events. Specific situations to which the theory has been applied include the motion of holes in the ionic crystals MnO,[1] NbO_2,[2,3] and UO_2,[4] and the motion of electrons in the molecular crystals orthorhombic sulfur[5] (S_8) and realgar (As_4S_4).[6] In addition, the theory may also be applied to noncrystalline systems such as the vanadate glasses. However, the following discussion will be concerned solely with the generic features of the theory of small polarons and not with the detailed application to specific systems.

II. DEFINITION OF A SMALL POLARON

Let us begin by defining a polaron in general and a small polaron in particular. Consider the situation illustrated in Fig. 1. If a stationary excess electron, depicted by the small dot, is placed in a solid, the atoms of the material will generally experience a force due to the added carrier. As a result of this interaction, the electron-lattice interaction, the atoms of the material

*Work supported by the U.S. Energy Research and Development Administration.

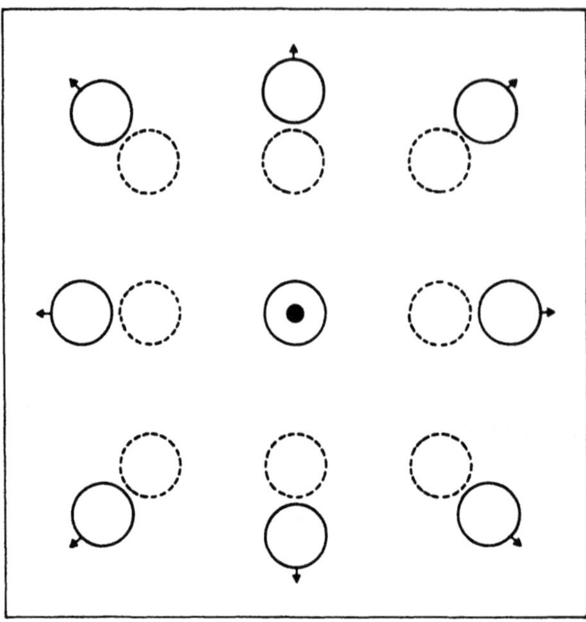

Fig. 1: A stationary carrier (black dot) induces a shift in the atomic equilibrium positions (open circles).

will assume new equilibrium positions consistent with the presence of the added carrier. The solid circles of Fig. 1 represent the atoms in these new equilibrium positions while the dashed circles depict their carrier-free equilibrium positions. These carrier-induced atomic displacements are always such as to lower the energy of the electron. In other words, the carrier-induced atomic displacements produce (or deepen) the potential well in which the electron resides. If this carrier-induced potential well is sufficiently deep, the carrier may occupy a "bound state," being unable to escape completely from the site upon which it was originally placed without an alteration of the positions of the surrounding atoms. In this case, since it is the potential well resulting from the carrier-induced displacements themselves that acts to trap the carrier, the carrier is said to be self-trapped. The unit comprising the self-trapped carrier and its associated lattice deformation is termed a polaron. Furthermore, if the spatial extent of the wavefunction of the self-trapped carrier is less than or comparable to a lattice constant, the polaron is said to be a small polaron.

It should be stressed that although the coining of the term "polaron" had its origin in studies of self-trapping in polar materials, the argument that a stationary carrier will induce a lattice distortion, and hence a potential well, which, if sufficiently deep, will self-trap the carrier is not at all restricted to polar materials. Small-polaron formation can occur in both polar and nonpolar materials and can be associated with a carrier's interaction with acoustic-mode as well as optical-mode lattice vibrations.

III. MOLECULAR CRYSTAL MODEL

An easily generalized model which provides a simple way to conceive of aspects of the small-polaron problem is the molecular crystal model.[7] In this model one considers a single excess electron placed in a periodic array of deformable molecules. Associated with each molecule in the lattice is a configurational coordinate which represents a distortion of the molecule from its carrier-free equilibrium configuration. For instance, the configurational coordinate of a molecule at site $\underset{\sim}{g}$, denoted by $x_{\underset{\sim}{g}}$, may be thought of as the molecular-distortion variable related to the breathing mode of that molecule. In the absence of any coupling between molecules, these local configurational coordinates are taken to vibrate harmonically about their equilibrium positions with the frequency ω_o. Furthermore, coupling between the vibrational motion of one molecule and its neighbors is included in this model, thereby providing the mechanism for the transport of vibrational energy through the crystal; it is this intersite coupling which gives rise to dispersion of the lattice-vibrational frequencies. The Hamiltonian corresponding to this carrier-free molecular crystal is simply

$$H_L = \sum_{\underset{\sim}{g}} \left\{ \frac{\hbar^2}{2M} \frac{\partial^2}{\partial x_g^2} + \frac{M\omega_o^2 x_{\underset{\sim}{g}}^2}{2} + \sum_{\underset{\sim}{h}} M\omega_o \omega_b x_{\underset{\sim}{g}} x_{\underset{\sim}{g}+\underset{\sim}{h}} \right\} , \tag{1}$$

where $\underset{\sim}{h}$ is a nearest neighbor position vector, M is an appropriate reduced atomic mass, and $6\omega_b$ is the optical-phonon bandwidth. Finally, as in most treatments of the electron-lattice interaction, the energy of an excess carrier placed at a site in the lattice is taken to be a linear function of the lattice displacements, the $x_{\underset{\sim}{g}}$'s. In particular, the energy of a carrier placed on site $\underset{\sim}{g}$ is given by

$$E_{\underset{\sim}{g}} = E_o - \sum_{\underset{\sim}{g}'} f(\underset{\sim}{g}' - \underset{\sim}{g})\, x_{\underset{\sim}{g}'} \tag{2}$$

where $f(\underset{\sim}{g}' - \underset{\sim}{g})$ is a weighting factor. It is this factor which characterizes the range of the electron-lattice interaction. For simplicity, most small-polaron studies have taken the electron-lattice interaction to be short-ranged: $f(\underset{\sim}{g}' - \underset{\sim}{g}) = A\delta_{\underset{\sim}{g}', \underset{\sim}{g}}$. In this situation, the carrier-induced molecular deformations are in the main restricted to the vicinity of the carrier itself; it is the presence of finite vibrational coupling between neighboring molecules which is responsible for the deformation pattern not being strictly limited to the occupied site. Figure 2 provides a schematic description of the equilibrium separations (the vertical lines) of a chain of diatomic molecules when a stationary electron (the black dot) resides on one of the molecules.

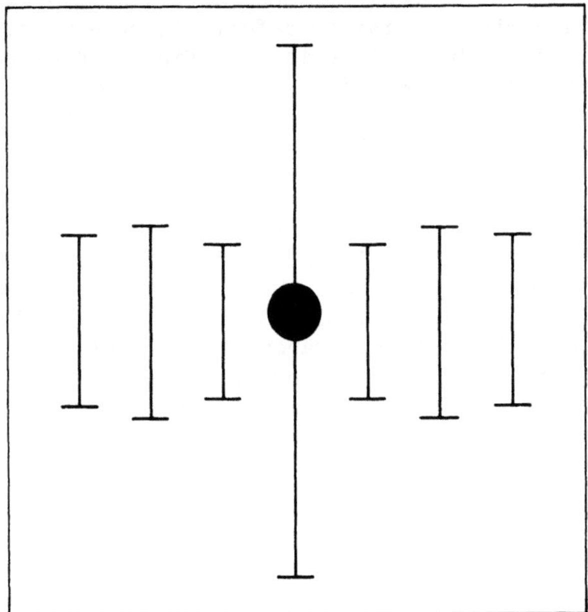

Fig. 2: A schematic representation of equilibrium separations of the molecules of a linear molecular-crystal when a carrier (black dot) resides on the central molecule.

IV. SMALL-POLARON ENERGY SPECTRUM

In Fig. 3 a portion of the energetic situation which prevails in the case of an excess electron forming a small polaron is illustrated. Namely, as a result of a stationary electron deforming the molecule upon which it resides, its energy is reduced by an amount $2E_b$ while the strain energy of the deformed molecules

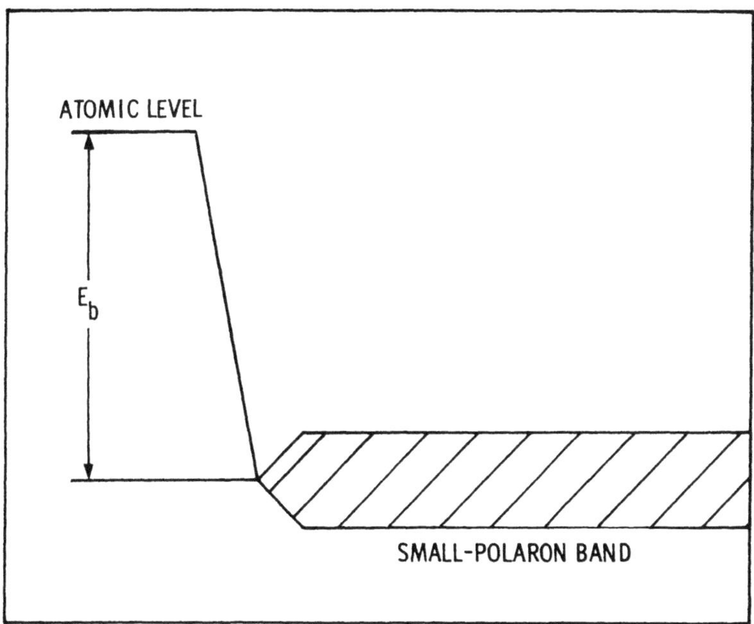

Fig. 3: The energy of the system composed of a stationary carrier in a molecular crystal is reduced by an amount E_b, the small-polaron binding energy, as a result of the molecules deforming in response to the presence of the carrier. The degeneracy inherent in a crystalline system leads to the formation of a small-polaron band.

is increased by an amount E_b. This yields a net reduction of the energy of the system, relative to that of an electron in an undeformed (rigid) molecular crystal, by the amount E_b; this energy is termed the small-polaron binding energy. In the limit of vanishing vibrational dispersion, the small-polaron binding energy is essentially the ratio of the square of the force A exerted by the electron on the configurational coordinate of the site it occupies to twice the stiffness constant of the molecule, $M\omega_o^2$:

$$E_b = A^2/2M\omega_o^2 \quad . \tag{3}$$

Acknowledging the fact that the small polaron (the carrier and its lattice distortion) may equally well reside on any of the geometrically equivalent sites in the crystal, one finds that the eigenstates of the system form a small-polaron band. In particular, pursuing a modified tight-binding approach, one finds that the eigenstates of a small polaron in a simple cubic lattice are

$$E_{\underset{\sim}{k}} = -2J\ \exp(-S)\ \left[\cos(k_x a) + \cos(k_y a) + \cos(k_z a)\right] - E_b\ , \quad (4)$$

where $\underset{\sim}{k}$ is the small-polaron wavevector, a is the lattice constant, J is an electronic transfer integral, and exp(-S) is a vibrational overlap factor. The vibrational overlap factor results from the fact that as a small polaron tunnels from site to site the atomic displacement pattern must be carried along with it. Due to the relatively heavy mass of the atoms compared with electrons, this factor, related (in some sense) to atomic tunneling, is extremely small. Thus the small polaron band is generally viewed as being extremely narrow; typically its width is taken to be very much smaller than even vibrational energies.

V. SMALL-POLARON FORMATION

Having defined a small polaron, one must now specify the conditions under which small polarons are or are not formed. An important consideration in approaching this issue is the range of the electron-lattice interaction. For instance, one may consider the so-called Fröhlich Hamiltonian,[8] in which the carrier interacts with the lattice solely via the long-range electrostatic interaction of a carrier with the electric dipoles induced by longitudinal-optical-mode displacements. Another model is that in which a carrier interacts with the lattice vibrations solely via a short-range interaction with either acoustic phonons (the Bardeen-Shockley deformation potential model[9]) or optical phonons (Holstein's molecular crystal model[7]). Of course, one may envision admixtures of these models. Within the adiabatic approximation, one can demonstrate in the first instance that (for a continuum model) the carrier always forms a polaron (is bound in its carrier-induced well) the radius of which continuously decreases as the electron-lattice coupling strength increases.[10,11] In this case, the carrier-induced well is coulombic (r^{-1}) and hence supports a bound state for all finite values of the coupling strength. On the other hand, it can be shown that for a purely short-range interaction, within the adiabatic approximation (again for a continuum model) the ground state of the system will either be one in which the carrier forms a small polaron or one in which the carrier does not form a polaron at all.[10,11,12,13]

In the case of a short-range interaction, one can gain further insight into the question of when a carrier will form a small polaron and when it will remain nonpolaronic by considering a tight-binding vibrational determination of the eigenstates of an excess electron in a deformable molecular crystal.[13,14] In these calculations, the variational parameters determine the magnitude and spatial extent of the carrier-induced atomic displacements. In particular, by minimizing the free energy of the system with respect to these variational parameters one can find the dynamically stable states of an electron within Holstein's molecular crystal model. Taking the conditions for the existence of the $\underset{\sim}{k} = 0$ nonpolaronic and small-polaron solutions respectively, to be appropriate to the existence of the entire (nonpolaronic) conduction and small-polaron bands, one obtains the conditions under which these two types of eigenstates are self-consistent. Figure 4 portrays the corresponding energy spectrum at absolute zero as a function of the electron-lattice coupling parameter, $E_b/\hbar\omega_o$. Nonpolaronic solutions only

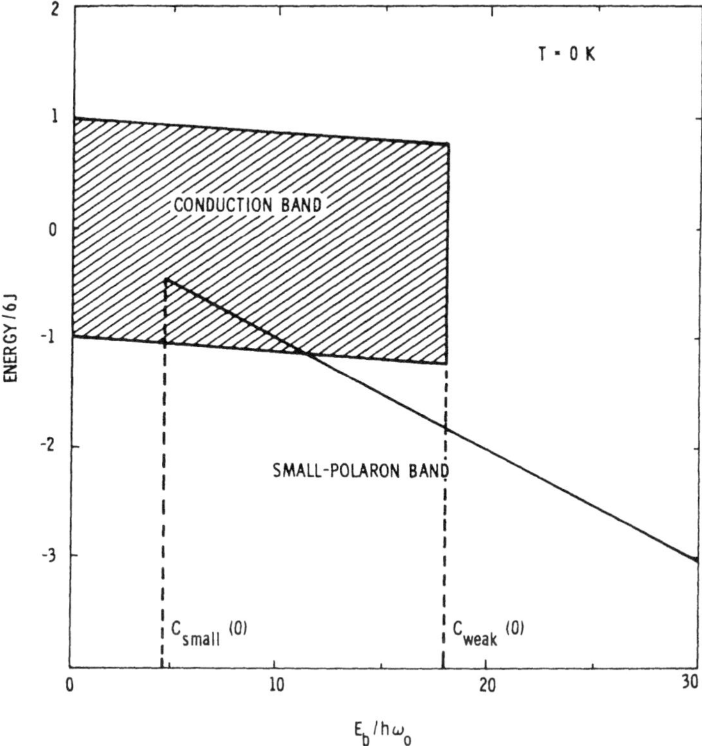

Fig. 4: The one-electron energy spectrum is plotted as a function of the electron-lattice coupling strength, defined as $E_b/\hbar\omega_o$, for $6J/\hbar\omega_o = 10$ at $T = 0$ K. The energy of the coupled system minus the zero-point vibrational energy, $n\hbar\omega_o/2$, is measured in units of the rigid-lattice band halfwidth, 6J.

exist for values of $E_b/\hbar\omega_o$ less than some máximum value, C_{weak}, and small-polaron solutions only exist for values of $E_b/\hbar\omega_o$ greater than a minimum value, C_{small}. Between these two limiting values both types of states may coexist. With a decrease in the adiabaticity parameter, $6J/\hbar\omega_0$, the coexistence regime is decreased until for extremely narrow (rigid-lattice) bands, $6J \leqslant \hbar\omega_0$, only one solution occurs; in this instance the magnitude of the carrier-induced lattice deformation increases continuously as the electron-lattice coupling strength is increased. In the opposite limit, the adiabatic limit, $\hbar\omega_0 \to 0$ with $M\omega_0^2$ finite, the coexistence regime expands to include all nonzero values of the electron-lattice coupling strength.

Focusing attention on the regime in which the coexistence region exists and extends over a finite range of the coupling strength, $E_b/\hbar\omega_0$, one can obtain formulae for the two existence conditions. The weak-coupling existence condition may, apart from a minor numerical factor. be obtained from the following argument.[13] To begin, it is noted that a carrier associated with an energy band of halfwidth 6J can only be confined to a single site for a time $\approx \hbar/6J$. It is this amount of time that is available for the carrier to force a substantial alteration of the positions of the atomic constituents associated with the occupied site. The force that the carrier exerts on the atoms of the occupied molecule is simply the electron-lattice interaction constant A. Thus, during the time that a site is occupied the atoms surrounding the carrier are subjected to the carrier-induced acceleration A/M. The electron-lattice interaction will only be ineffective in producing a substantial deviation of the motion of the surrounding atoms from the motion which characterizes them in the absence of a carrier, if the typical carrier-induced change in the "velocity" of these atoms is small compared with their carrier-free "velocity":

$$\left(\frac{A}{M}\right)\left(\frac{\hbar}{6J}\right) < \sqrt{\frac{\hbar\omega_0(N+\frac{1}{2})}{M}} \quad , \tag{5}$$

where N is the Bose factor, $[\exp(\hbar\omega_0/\kappa T)-1]^{-1}$. Since the interaction between a stationary electron and the atoms surrounding it is typically great enough to produce a large displacement of the atoms surrounding the electron, it is only the ability of an electron to move between sites sufficiently rapidly so as to preclude atomic displacements from occurring which permits one to obtain nonpolaronic solutions. As the temperature is raised, the atomic velocities increase and the presence of an electron on a site poses less of a perturbation to the atomic vibratory motion. Concomitantly, as illustrated in Fig. 5, with a rise in temperature the weak-coupling states remain dynamically stable up to an increased value of the electron-lattice coupling strength.

A mechanism to produce an abrupt conductivity transition emerges from the fact that an electron which at low temperatures could not exist in a nonpolaronic state, is presented with the possibility

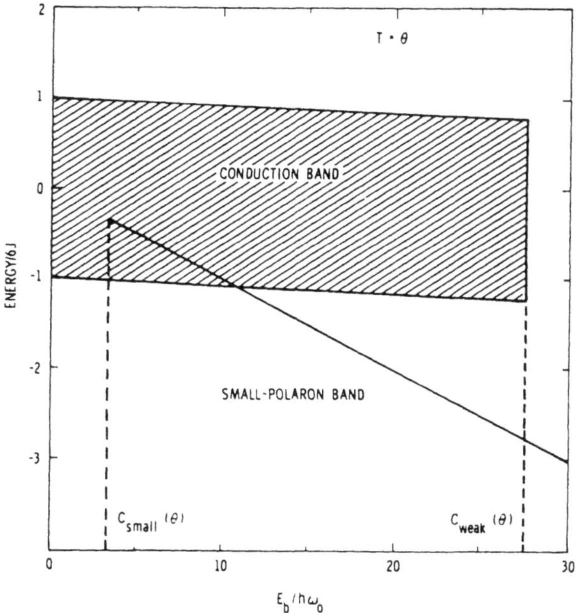

Fig. 5: The one-electron spectrum is plotted as a function of $E_b/\hbar\omega_0$ at the temperature $T = \hbar\omega_0/K \equiv \theta$ with $6J/\hbar\omega_0$ taken, as in Fig. 4, to be 10. The energy of the coupled electron-lattice system minus the purely vibrational contribution to the energy, $n\hbar\omega_0(N+\frac{1}{2})$, is measured in units of the rigid-lattice conduction band half-width, 6J.

of occupying such a state as the temperature is raised.[13,14,15] Assuming that the excited states of the system which are missing from Figs. 4 and 5 do not dominate the transport properties on the low-temperature side of the transition, the low-temperature transport properties will be those characteristic of small-polaron motion. At temperatures above the transition three possibilities present themselves. First, the weak-coupling states may lie so much higher in energy than the small-polaron levels that their addition to the carrier's energy spectrum will have a negligible effect on the transport properties of the system; small-polaron transport will dominate both above and below the transition temperature. Second, the gap between the small-polaron levels and the weak-coupling (conduction-band) states may be sufficiently small so that while the system will be characterized by small-polaron transport properties below the transition, above the transition it will manifest the (presumably) high-mobility transport properties typical of wide-band semiconductors; in this case the transition is from a small-polaron type semiconductor at low temperatures to a high-mobility type semiconductor above the transition. Presently in the literature there is some experimental evidence of semiconductor-to-semiconductor transitions which (with the limited data available) may be interpreted in these terms.[2,3,16,17,18] The third possibility is that the conduction-band levels may lie below the small-polaron levels. In this case, the gap between the conduction band and the valence band may be sufficiently small (or nonexistent) so that the high conductivity state above the transition will manifest metal-like behavior. Thus the abrupt temperature-dependent change in a carrier's energy spectrum provides a mechanism for driving a conductivity transition in which the carriers on the low-conductivity side of the transition are small polarons. Furthermore, it is noted that a slight change in the values of the physical parameters, such as $E_b/\hbar\omega_o$, can produce situations characterized by qualitatively distinct transport properties. These features may prove useful in understanding the transport properties of transition metal oxides.[13]

VI. SMALL-POLARON BAND AND HOPPING MOTION

Having addressed the question of the formation of a small polaron, it is now appropriate to discuss the transport properties associated with small-polaron motion. To begin, note that in an ideal crystal a small polaron may move from one site to its neighbor via two distinct types of process.[19] The first involves the tunneling of a small polaron between adjacent lattice sites with no change in the population of any phonon mode:

$$N_{\underset{\sim}{q}} = N'_{\underset{\sim}{q}}, \text{ for all } \underset{\sim}{q} \quad , \tag{6}$$

where $N_{\underset{\sim}{q}}$ and $N'_{\underset{\sim}{q}}$ are respectively the initial and final-state phonon occupation numbers of the $\underset{\sim}{q}$th vibrational mode associated with an electronic transition (transfer) which carries an electron from one site to another. These so-called <u>diagonal</u> processes involve simply translating the carrier and its self-induced atomic-displacement pattern between neighboring sites without any change in the atomic vibratory motion (motion about the atomic equilibrium positions); such a transition gives rise to small-polaron band motion. The complementary class of process is one in which the phonon occupation does change with a site-to-site transfer:

$$N_{\underset{\sim}{q}} \neq N'_{\underset{\sim}{q}}, \text{ for some } \underset{\sim}{q} \quad . \tag{7}$$

These processes, termed <u>nondiagonal</u> processes, correspond to the phonon-assisted tunneling (hopping) of the carrier between adjacent sites. Thus, the small-polaron mobility is a sum of two contributions: one associated with small-polaron band motion and the other with small-polaron hopping motion.[20]

In an ideal crystal, the small-polaron band mobility will predominate at very low temperatures. With rising temperature the band contribution to the mobility will fall while the hopping component will increase until, at sufficiently high temperatures, small-polaron motion proceeds predominantly via phonon-assisted hopping.[19,20] The actual temperature of the changeover depends on the details of the model [the details of the coupling of the electron to both the optical and acoustic phonons of the crystal and the model for the "scattering" associated with small-polaron band motion]. However, the extreme narrowness of the small-polaron band leads one to expect that the band motion typically will be washed out by whatever disorder exists in a real crystal. Thus, in the following discussion, only small-polaron hopping motion will be discussed. Examples of small-polaron hopping motion appear to be found in the hopping of Ni^{3+} holes around Li^{+} impurities in NiO[21] and in the unbound hopping of Mn^{3+} holes in MnO.[1] In the first case the observed high-frequency conductivity is associated with small-polaron hopping motion. In the second example, the dc conductivity is attributed to the hopping of hole-like small polarons.

VII. SMALL-POLARON JUMP RATE

In the following discussion most of our concern will be focused on calculations in which the electronic transfer integral associated with a small-polaron hop is sufficiently small so as to treat it perturbatively; to lowest nonvanishing order the elemental phonon-assisted site-to-site jump rate is then proportional to the second power of the relevant electronic transfer integral. Physically, this regime corresponds to a situation in which the electron cannot adiabatically follow an alteration of the atomic positions. Thus this is termed <u>nonadiabatic</u> small-polaron motion. The complementary regime is one in which the electron can always adjust to the atomic state. It is termed the <u>adiabatic</u> regime and will be discussed here subsequently.

The fundamental quantity with which we shall now be concerned is the rate which characterizes a phonon-assisted transition that takes a small polaron from one site to a neighboring site. It is this rate which determines the carrier's mobility. If one considers each hop of a carrier to be uncorrelated with its prior hops (or other hops of other carriers) then the relevant jump rate is calculated by placing a small polaron on a particular site at some initial time and then computing the rate at which it moves to a neighboring site. In this picture the lattice vibrations act as a thermal bath with which the electron can exchange energy. It is clear that a necessary condition for the lattice vibrations to fulfill this function is that vibrational energy be capable of being transferred from lattice site to lattice site, i.e., there must be adequate dispersion of the vibrational frequencies.

The nonadiabatic jump rate may be calculated exactly.[12,22] In Fig. 6 the phonon-assisted jump rate associated with a small-polaron hop, in units of $2\pi J^2/\hbar^2\omega_o$ [J is the electronic transfer integral and ω_o is the mean optical-mode frequency], is plotted versus reciprocal temperature, in units of the optical-mode temperature, $\theta = \hbar\omega_o/\kappa$, for two values of the phonon-dispersion parameter, ω_b; the width of the optical band is $6\omega_b$. Focusing

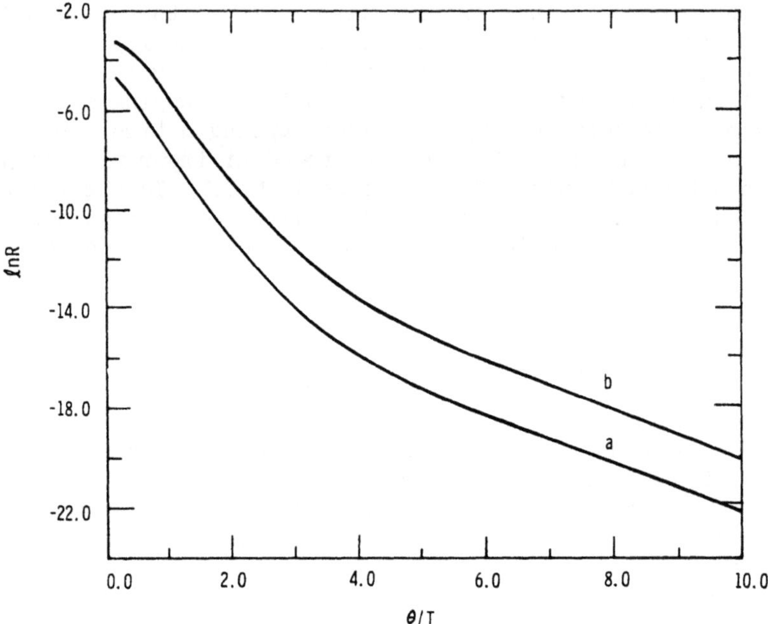

Fig. 6: The natural logarithm of the optical-phonon assisted jump rate in dimensionless units is plotted against θ/T ($\theta = \hbar\omega_o/\kappa$) for $E_b/\hbar\omega_o = 10$ and (a) $2\pi\omega_b/\omega_o = 0.5$, and (b) $2\pi\omega_b/\omega_o = 0.05$.

attention on either one of the two curves of Fig. 6, it is seen that two distinct temperature dependences are manifested. At sufficiently low temperatures, $(2E_b/\hbar\omega_o)\ \mathrm{csch}\ (\hbar\omega_o/2\kappa T) \ll 1$ [in the case of small-polaron hopping $(2E_b/\hbar\omega_o) \gg 1$], multiphonon processes are frozen out and the jump rate is dominated by the phonon-assisted process which involves the absorption of the minimum amount of vibrational energy consistent with the requirement of energy conservation. In the case at hand, a carrier's interaction with only optical phonons, this process is a two-phonon process. Namely, a single phonon of energy $\sim \hbar\omega_o$ is absorbed and another is emitted. The low-temperature jump rate is concomitantly activated with the activation energy $\hbar\omega_o$ that is associated with the probability of absorbing a phonon of energy $\hbar\omega_o$. In the complementary high-temperature regime multiphonon processes are no longer frozen out. Then the jump rate manifests a thermally activated behavior with the activation energy $\epsilon_2 = E_b/2$. It should be noted that this high-temperature activation energy is not associated with phonon energies but simply with the electron-lattice coupling strength and the stiffness of the material, the parameters involved in E_b. The fact that the activation energy depends on no quantum-mechanical quantities suggests that a semiclassical interpretation of this high-temperature activation energy is possible. Later, we shall see that this is, in fact, the case. It is interesting to note that in the typical example illustrated in Fig. 6, the transition between the low-temperature and high-temperature regimes is characterized by a temperature range in which the phonon-assisted jump rate is not simply activated.

Finally, some comment should be made about the role of vibrational dispersion in this calculation. The upper curve of the two on Fig. 6 differs from the lower curve solely in that the dispersion in this case is smaller. The dispersion dependence may be seen to be somewhat greater as the temperature is lowered. If one proceeds to the limit of zero vibrational dispersion the rate increases and becomes undefined. Alternatively, as the dispersion is increased, the dispersion dependence of the jump rate becomes less and less. An interesting aspect of the optical-phonon-assisted jump rate (which will not be dwelt on here) is that it is undefined for a one-dimensional system: An attempt to calculate it yields a divergent result for all values of the optical bandwidth; Fig. 6 is for a three-dimensional lattice.

If a carrier interacts with the atoms of a lattice via a short-range interaction, such as when a carrier on a transition-metal ion displaces adjacent ligands as in, say, MnO,[1] the interaction between the carrier and the acoustic phonons is, a priori, comparable to that between the carrier and optical phonons. Thus one is led to consider the jump rate in the situation in which the carrier interacts solely with acoustic phonons.[23,24] This situation differs from the optical-phonon problem in a fundamental manner. Namely, since the acoustic phonons with which a carrier can interact extend from the Debye energy down to zero energy, at no finite temperature are all acoustic phonons frozen out, i.e., $KT > \hbar\omega_q$ from some q at all finite temperatures. Because there is no special "freezing-out" energy for all the phonons, a low-temperature thermally activated temperature dependence of the jump rate will not be observed. This is illustrated by the curve of Fig. 7 in which the logarithm of the phonon-assisted jump rate, in units of $J^2/\hbar^2\omega_m$, is plotted versus θ_m/T, where θ_m is the temperature corresponding to the energy of the maximum-energy phonon with which a carrier can interact, $\hbar\omega_m = K\theta_m$. At low temperatures the jump

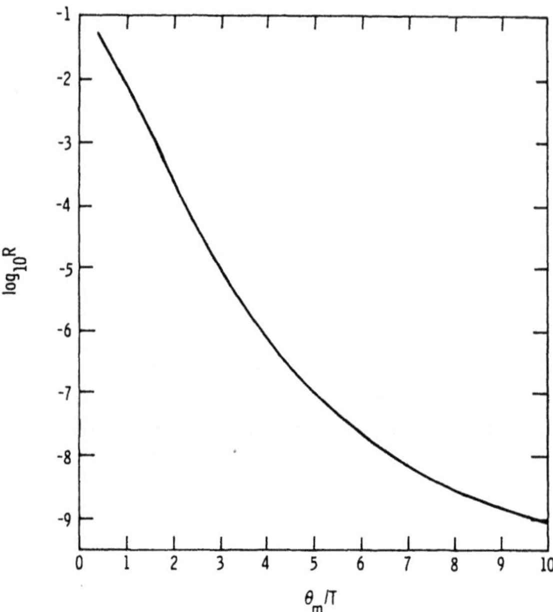

Fig. 7: The logarithm of the acoustic-phonon-assisted jump rate in dimensionless units is plotted against θ_m/T ($\theta_m \equiv \hbar\omega_m/\kappa$) for a typical value of the electron-lattice coupling strength.

rate is nonactivated. As in the optical-phonon calculation, when the temperature is raised above the temperature corresponding to the maximum-energy phonon with which the carrier can interact, the jump rate becomes thermally activated. In fact, as in the optical-phonon problem, this high-temperature behavior can be understood via a semiclassical picture.

VIII. SEMICLASSICAL APPROACH

In the main, the remainder of this summary will be devoted to studies of small-polaron hopping motion within the high-temperature semiclassical regime. A fundamental concept characteristic of this regime is that of a coincidence event. Specifically, taking the electronic energy level associated with a carrier occupying any site in a crystal to be a function of the "instantaneous" positions of the atoms of the crystal, it may be seen that, since the positions of the atoms are constantly changing (associated with the vibratory motion of the lattice), the electronic energy associated with a carrier occupying any given site is also changing. Amidst the myriad of distortional configurations which are assumed by the vibratory atoms, occasionally a situation is encountered in which the electronic energy of an electron at a given site "momentarily" equals that which it would have if it occupied an adjacent site. Such an occurrence, termed a coincidence event, is depicted in Fig. 8. Figure 8a depicts the situation of a carrier residing on a site with the atoms about it being displaced; Figure 8b illustrates the formation of a coincidence event and Fig. 8c represents the situation in which the carrier has hopped and the lattice has relaxed about it. While a coincidence event is viewed as instantaneous in terms of classical physics, in quantum mechanics it is associated with a finite duration. If this time interval is

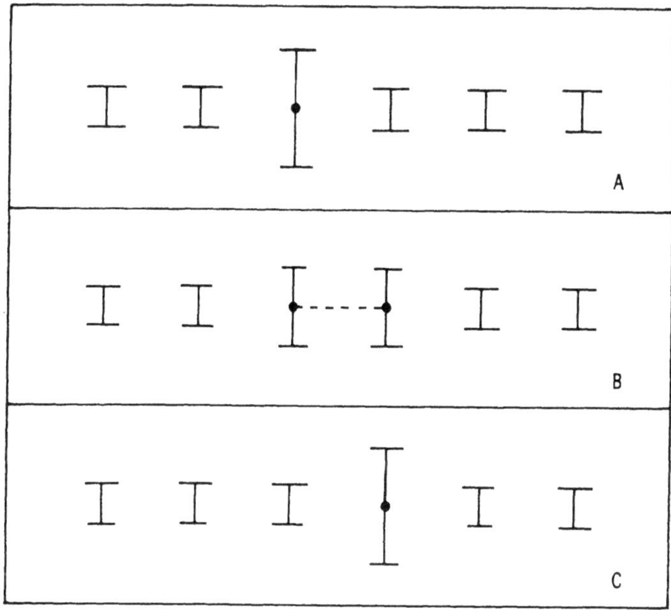

Fig. 8: A schematic representation of the small-polaron jump process in the high-temperature semiclassical regime. Nonessential "thermal" displacements and vibrational dispersion effects are suppressed for clarity.

long compared with the time it takes an electron to transfer between coincident sites, $\sim \hbar/J$, then the electron can always "follow the lattice motion" and avail itself of the opportunity to make a hop. This situation is characteristic of the so-called adiabatic regime. Alternatively, the time required for an electron to hop may be long compared with the duration of a coincidence. Then an electron will not always "follow the lattice motion" and hop when a coincidence event presents itself; this situation corresponds to the nonadiabatic domain. In this case, the jump rate, and hence the drift mobility, is reduced from what it is in the adiabatic regime by a factor P, $P < 1$, where P is the probability that given a coincidence event the carrier will hop.[12,19] Finally, it is noted that the minimum energy required to produce a lattice deformation associated with establishing a coincidence event is just the activation energy characteristic of the high-temperature regime.[25,26]

IX. CORRELATED SMALL-POLARON HOPPING MOTION

In thinking about small-polaron hopping motion, one typically considers successive hops of a carrier to be uncorrelated with one another. However, the preceeding discussion of a semiclassical view of a small-polaron hop is suggestive of a mechanism via which small-polaron motion may be highly correlated. In particular, if a small-polaron's hop is to be considered independent of its previous hop, then the distortion associated with creating the coincidence event of the first hop must relax, with an amount of energy comparable to the hopping-activation energy being dispersed away from the involved sites, in a time which is much shorter than the mean time between small-polaron hops.[27] If the carrier has a substantial probability of hopping (either to a different neighbor or back to the site it occupied previously) before the lattice relaxes, then its motion will be highly correlated. In this case the effective activation energy characterizing small-polaron hopping motion will be substantially reduced from that associated with uncorrelated hopping motion, since much of the distortion needed to form a coincidence event is present residually from the prior hop. In fact, in the highly correlated situation the carrier can be viewed as frequently hopping back and forth between two coincident sites. In this circumstance, a contribution to the net diffusion of the particle occurs when it alters its back and forth jumping motion to hop to a third site. Time is not adequate in the present review to develop a detailed discussion of this phenomenon; for this the reader is referred to the literature.[27,28,29,30] However, as illustrated in Fig. 9, in this type of correlated small-polaron hopping situtation the high-temperature small-polaron hopping mobility need not manifest a clear thermally activated behavior. In fact, the mobility in this domain may even fall slowly with increasing temperature. It is stressed that these correlation effects will only manifest themselves when the jump rate is greater than or comparable to the relaxation rate. Thus for arbitrarily small values of the electronic transfer integral,

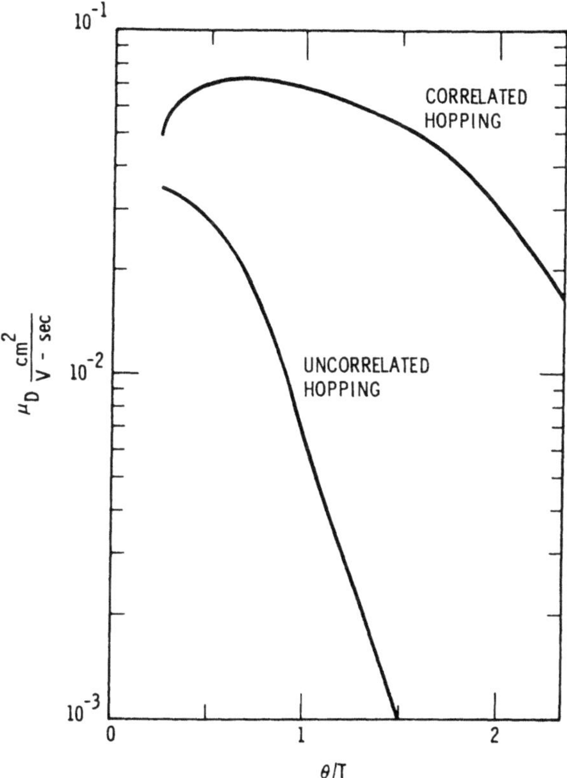

Fig. 9: The semiclassically calculated small-polaron drift mobility is plotted versus θ/T for $E_b/\hbar\omega_o = 10$ with correlation effects being included (upper curve) and with them being ignored (lower curve).

J, the jump rate will also become arbitrarily small and the correlation effects will become insignificant. While most calculations of the phonon-assisted jump rate begin by assuming arbitrarily small values of the electronic transfer integrals, in many systems of physical interest the electronic transfer integrals are sufficiently large so as to require one to contend with the issue of correlated hopping motion.

X. AC CONDUCTIVITY

Small-polaron hopping motion is characterized by a rather distinctive high-frequency ac conductivity.[12,22,31-35] In particular, this phenomenon, associated with the participation of a photon in conjunction with phonons in producing a photon-phonon-assisted hop, generates a characteristic frequency dependence of the small-polaron mobility. In Fig. 10 the natural logarithm of the effective jump rate (related to the mobility by the factor $ea^2/\kappa T$), in dimensionless units, is plotted versus reciprocal temperature in units of the optical-phonon temperature for several values of the external frequency, ω.[22] This plot is for the case of an electron interacting solely with optical phonons. In the high-temperature regime the thermally activated behavior associated with semiclassical hopping is evident. The $\omega \neq 0$ low-temperature regime is characterized by a temperature-independent mobility, which results from the carrier hopping between sites with the absorption of a photon of energy $\hbar\omega$ and the spontaneous emission (but not absorption) of phonons. While this plot shows the mobility rising with increasing frequency, at higher frequencies and temperatures than those illustrated in Fig. 10, the mobility will fall with an increase in frequency. The former feature, indicative

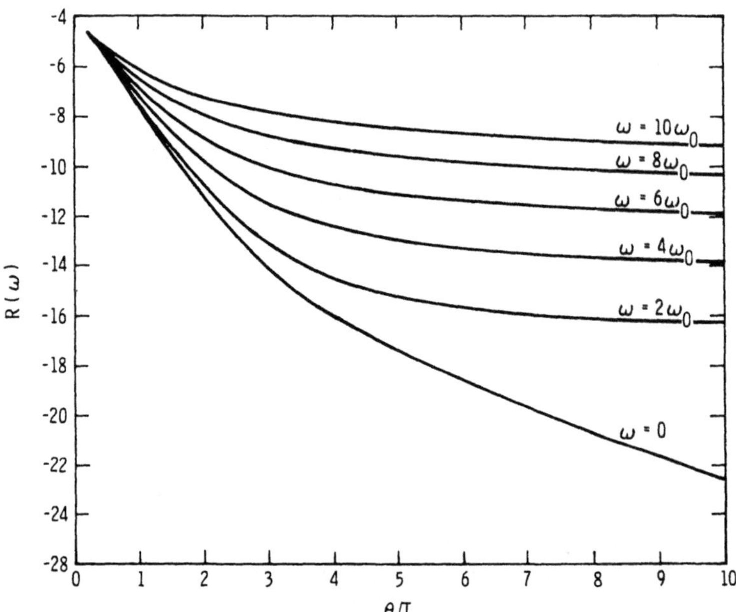

Fig. 10: The natural logarithm of the two-site jump rate plotted versus θ/T for various frequencies of the external field. Here $E_b = 10\ \hbar\omega_0$ and $2\pi\omega_b/\omega_0 = 0.5$, and R is expressed in units of $2\pi J^2/\hbar^2\omega_0$.

of the fact that the probability of emitting phonons ultimately diminishes when the number of emitted phonons, n, demanded by the requirement of energy conservation becomes excessively large ($n > 2E_b/\hbar\omega_o$, i.e., $\omega > 2E_b/\hbar$), is illustrated in Fig. 11. Here the above-described frequency-dependent rate is plotted against frequency, in units of the optical-phonon frequency, for three values of the temperature. The smooth curves of this figure are

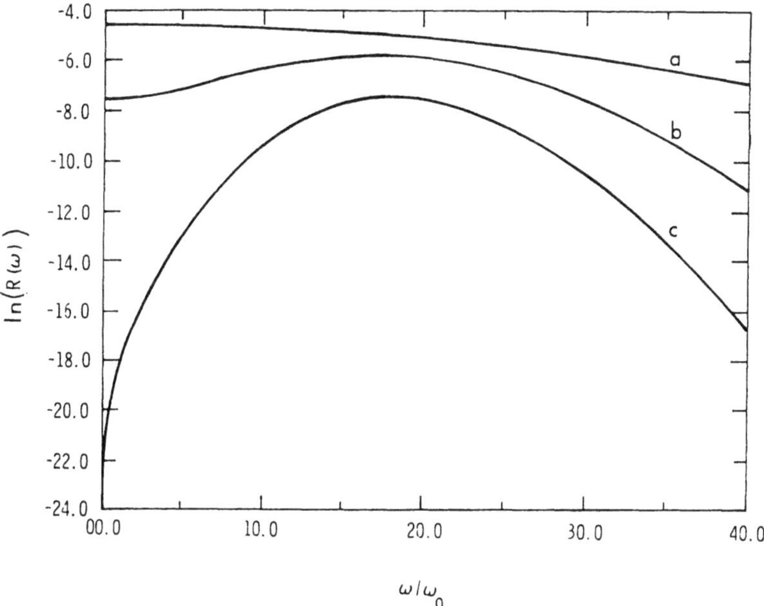

Fig. 11: The natural logarithm of the two-site jump rate, $R(\omega)$, is plotted versus ω/ω_o for (a) $\kappa T = 5\hbar\omega_o$, (b) $\kappa T = \hbar\omega_o$, and (c) $\kappa T = \hbar\omega_o/10$.

drawn through points calculated at integer values of ω/ω_o. Curves a and b refer to the high-temperature semiclassical regime while curve c refers to the low-temperature regime. In the case of a carrier which interacts solely with a narrow band of optical phonons, the frequency dependence of the low-temperature mobility (absorption) comprises a series of peaks centered at integer values of ω/ω_o; curve c is the envelope of this multi-peaked curve. In the high-temperature semiclassical regime these peaks typically broaden sufficiently so as to yield the smooth curves (a and b) shown in the figure. However, in the case of a carrier's interaction with acoustic phonons even the low-temperature absorption is a smooth function of frequency; this is illustrated in Fig. 12.[24] As is evident from Figs. 11 and 12, these curves are often characterized by a noticeable asymmetry. Finally, it should be mentioned that a small-polaron conductivity (absorption) peak may be associated with small-polaron hopping about a defect as well as with "free" small-polaron motion.

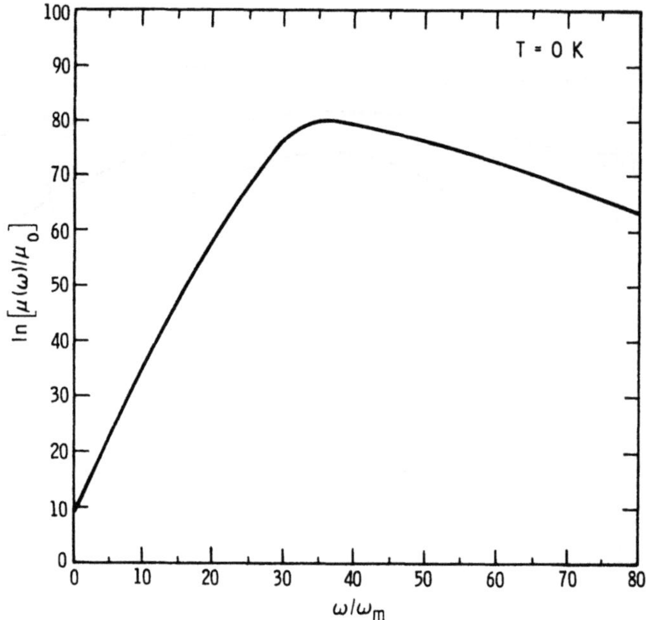

Fig. 12: A typical plot of the natural logarithm of the T = 0 K mobility associated with acoustic-phonon-assisted hopping motion versus the external frequency (expressed in units of the frequency of the highest energy acoustic phonon with which the carrier interacts).

XI. THERMOELECTRIC POWER

A useful transport experiment which is often performed on low-mobility solids is that of measuring the thermoelectric power. In effect, such a measurement allows one to determine the amount of heat transported with a carrier as it moves through a material. In studying hopping motion one is led to ask how much of the energy absorbed and emitted in making a hop is actually carried along with the particle. In the case of the commonly assumed linear electron-lattice interaction, the answer is particularly simple for noninteracting particles in a crystal. Namely, for carriers moving in (nearly) degenerate states, a band narrower than κT, their thermoelectric power contribution, S, is given by

$$S = (\kappa/q) \ln \left[(2 - c)/c \right], \tag{8}$$

where q is the charge of the carrier, and c is the concentration of charge carriers (carriers per site). The factor two arises if particles of either spin can equivalently (singly or doubly) occupy any lattice site; if particles are constrained to not doubly occupy a site then $[(2-c)/c]$ is replaced by $[2(1-c)/c]$.[36] An alternative way of writing this formula is,

$$S = (-1/|q|T)(E_e - E_F), \text{ for electron-like small polarons, } \tag{9a}$$

and

$$S = (1/|q|T)(E_F - E_h), \text{ for hole-like small polarons.} \tag{9b}$$

The above formulae assume the absence of the previously discussed correlation effects. If such correlated hops dominate the (semiclassical) transport, one expects the hopping energy, W_H, to be carried in a direction opposite to the direction of a hop. This is because, in this instance, the hopping energy is primarily absorbed at the "final" site and emitted at the "initial" site. Equations (9a) and (9b) would then be replaced with

$$S = (-1/|q|T)(E_e - E_F - W_H) \tag{10a}$$

and

$$S = (1/|q|T)(E_F - E_h - W_H), \tag{10b}$$

for noninteracting electrons and hole-like carriers, respectively.

Ignoring such correlation effects, one finds that, beyond a linear electron-lattice interaction, some fraction of the hopping energy will typically be carried along with the carrier. Ad hoc studies of a situation in which the occupation of a site produces a carrier-induced change in the lattice stiffness associated with the atoms surrounding that site yields a modification of Eqs. (9a) and (9b):[37]

$$S = (-1/|q|T) \left[E_e - E_F + (r-1)/(r+1)\, E_T \right], \tag{11a}$$

and

$$S = (1/|q|T) \left[E_F - E_h + (r-1)/(r+1)\, E_T \right], \tag{11b}$$

where r is the ratio of the local stiffness at the unoccupied site to that at the occupied site, and E_T is an energy which in the high-temperature semiclassical limit is $W_H - \kappa T/2$. As the temperature approaches zero, E_T tends to zero for a carrier interacting with acoustic phonons, but tends to $\hbar\omega_0$ for a carrier interacting solely with optical phonons.

XII. HALL MOBILITY

The calculation of the Hall mobility associated with small-polaron hopping motion is a much more complex undertaking than the calculation of the drift mobility. This is because the Hall mobility measurement probes a carrier's response to combined small electric and magnetic fields, whereas the drift mobility is only concerned with the carrier's response to electric-field-induced motion. To calculate the Hall mobility one must consider the alteration of the probabilities of hopping to several "final" sites as a result of the application of the magnetic field. This requires one to consider a basic unit of an "initial" site and at least two possible "final" sites; in the simplest calculation of the drift mobility only a pair of sites need be considered.

Generally speaking, the magnetic field only manifests itself when the lattice distorts in such a way as to provide a carrier in a zero-field situation with equal probabilities of hopping to anyone of several sites. In these instances, when the carrier may be said to have a choice of "final" sites, a magnetic field can affect the hopping motion so as to provide a Hall current.[26] Not surprisingly, the Hall mobility associated with hopping motion differs qualitatively from that which results from a free-electron-like picture.[25,26,38] For instance, the Hall mobility will generally differ very substantially in both magnitude and temperature dependence from the mobility associated with a conductivity measurement; concomitantly the Hall coefficient does not simply measure the carrier number. In fact, it has been demonstrated that when the drift mobility is low ($\ll 10^{-1}$ cm^2/V-sec) and thermally activated, the mobility measured in a Hall experiment can be much higher ($\sim 10^{-1}$ cm^2/V-sec) and less activated; it can even be a decreasing function of increasing temperature.[25,26,38] In addition, the Hall mobility can be nontrivially dependent on the lattice geometry. Figure 13, depicting the results of an adiabatic calculation[26] of the drift and Hall mobilities in a triangular lattice structure, illustrates some of these features. Furthermore, although the sign of the Hall effect for free-electron-like behavior by itself determines the sign of the charge carrier, in a hopping situation this is no longer the case.[39] For instance, the hopping motion of "holes" will yield a hole-signed Hall effect in a square lattice but not in a hexagonal (triangular) lattice. Observation of these effects has been reported in a number of materials.[1,2,3,12,21]

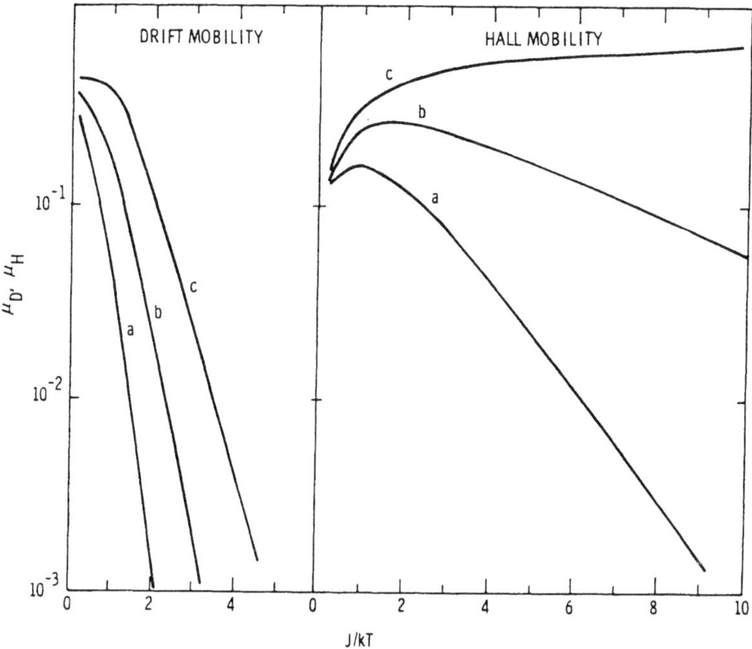

Fig. 13: The drift and Hall mobilities (cm^2/V-sec) associated with adiabatic small-polaron motion in a triangular lattice are plotted versus reciprocal temperature, in units of J^{-1}, for three successively decreasing values of the electron-lattice coupling strength.

REFERENCES

1. C. Crevecoeur and H. J. de Wit, J. Phys. Chem. Solids 31, 783 (1970).
2. I. K. Kristensen, J. Appl. Phys. 40, 4992 (1970).
3. G. Belanger, J. Destry, G. Perluzzo and P. M. Raccah, Can. J. Phys. 52, 2272 (1975).
4. J. Devreese, R. DeCominck and H. Pollak, Phys. Stat. Sol. 17, 825 (1966).
5. D. J. Gibbons and W. E. Spear, J. Phys. Chem. Solids 27, 1917 (1966).
6. G. B. Street and W. D. Gill, Phys. Status Solidi 18, 601 (1966).
7. T. Holstein, Ann. Phys. (N.Y.), 8, 325 (1959).
8. H. Fröhlich, "Polarons and Excitons," edited by C. G. Kuper and G. D. Whitfield, Plenum, New York, 1963.
9. J. Bardeen and W. Shockley, Phys. Rev. 80, 72 (1950).
10. Y. Toyozawa, Prog. Theor. Phys. 26, 29 (1963).
11. D. Emin and T. Holstein, to be published.
12. D. Emin, "Electronics and Structural Properties of Amorphous Semiconductors," edited by P. G. LeComber and J. Mort, Academic, New York, 1973.
13. D. Emin, Adv. Phys. 22, 57 (1973).
14. D. Emin, Phys. Rev. Lett. 28, 604 (1972).
15. D. Emin, J. Solid State Chem. 12, 393 (1975).
16. K. Sakata, J. Phys. Soc. Japan 26, 867 (1964).
17. R. F. Janninck and D. H. Whitmore, J. Phys. Chem. Solids 27 1183 (1966).
18. G. Villeneuve, A. Bordet, A. Casalot, J. P. Pouget, H. Lavnois and P. Ledever, J. Phys. Chem. Solids 33, 1953 (1972).
19. T. Holstein, Ann. Phys. (N.Y.) 8, 343 (1959).
20. L. Friedman, Phys. Rev. 135, A233 (1964).
21. A. J. Bosman and H. J. van Daal, Advan. Phys. 19, 1 (1970).
22. D. Emin, Adv. Phys. 24, 305 (1975).
23. D. Emin, Phys. Rev. Lett. 32, 303 (1974).
24. D. Emin, Adv. Phys., to be published.
25. L. Friedman and T. Holstein, Ann. Phys. (N.Y.) 21, 494 (1963).
26. D. Emin and T. Holstein, Ann. Phys. (N.Y.) 53, 439 (1969).
27. D. Emin, Phys. Rev. Lett. 25, 1751 (1970).
28. D. Emin, Phys. Rev. B3, 1321 (1971).
29. D. Emin, Phys. Rev. B4, 3639 (1971).
30. D. Emin, J. Non-Cryst. Solids 8, 511 (1972).
31. M. Klinger, Phys. Lett. 7, 102 (1963).
32. H. G. Reik and D. Heese, J. Phys. Chem. Solids, 28, 581 (1967).
33. V. N. Bogomolov, E. K. Kudinov, D. N. Mirlin, and Y. A. Firsov, Sov. Phys.-Solid State 9, 1630 (1968).
34. V. N. Bogomolov and D. N. Mirlin, Phys. Stat. Sol. 27, 443 (1968).
35. E. K. Kudinov, D. N. Mirlin, and Yu. A. Firsov, Sov. Phys. Solid State 11, 2257 (1970).
36. D. Emin, Phys. Rev. Lett. 35, 882 (1975).

37. D. Emin, unpublished.
38. D. Emin, Ann. Phys. (N.Y.) 64, 336 (1971).
39. T. Holstein, Phil. Mag. 27, 225 (1973).

ELECTRONIC PROPERTIES OF AMORPHOUS SEMICONDUCTORS

P. Nagels

Materials Science Department

S.C.K./C.E.N., B-2400 MOL (Belgium)

1. INTRODUCTION

In recent years there has been an increasing interest in non-crystalline materials from both fundamental and technological point of view. One of the major reasons for the intensive research in this field was the discovery of non-destructive fast electrical switching in thin amorphous films of certain chalcogenide alloys. This phenomenon, which deals with current-voltage characteristics in amorphous films, can be described as follows. When low electric fields are applied across the film the current is ohmic and the material is in an almost non-conductive state, in which the resistance is typically in the range 10^5-10^7 ohm at room temperature. Above a certain critical voltage the resistance drops drastically and the material switches from a non-conductive to a conductive state, which is maintained at much lower voltages. Two principal forms of switching devices have been developed which are called threshold and memory switches. In the threshold switch the conductive state returns back to its original high resistance state when the current falls below a critical value. In contrast, the memory switch remains in the conductive state even when the applied electric field is reduced to zero. The high resistive state can be reobtained by passing a brief pulse of high current through the film. The physical basis underlying their operation is not completely clear at the present time. Nevertheless, these devices can now be made with a high degree of reproducibility.

The most important application of an amorphous solid is that of selenium which is currently used in large amounts as a photoreceptive material in the copying process of xerography. The successful commercial application of this process stimulated extensive fundamental research in the field of the optical and electrical properties of this element and of other selenium compounds.

From fundamental point of view the non-crystalline materials are of considerable intrinsic interest. Indeed it can be expected that they will show many special features, as compared to crystalline materials, arising from the absence of long range periodicity in the atomic positions. Recently many efforts have been made in order to find a correlation between the structure of non-crystalline materials and their electrical and optical properties. Some models have been developed describing e.g. the electronic density of states in non-crystalline materials but many of these ideas are rather speculative and as yet not firmly based on theoretical work. Among non-crystalline materials one generally considers liquid and solid amorphous metals and semiconductors. In this review we shall mainly be concerned with amorphous semiconductors, which are typical examples of highly disordered materials. In the literature the terms amorphous, glassy and vitreous are commonly used to describe the same state.

Amorphous materials can be prepared in different ways. A widely used technique, which yield glasses in bulk form, exists in a rapid quenching of the element or the alloy from the molten to the solid state. Those materials for which the glass-forming ability is small cannot be prepared by quenching the liquid state but can often be obtained in the form of thin films by vacuum evaporation or sputtering.

Amorphous solids are classified in the same manner as crystalline solids depending on the type of chemical bonding. Two main groups can be distinguished which contain on the one hand ionic and on the other hand covalent materials. The ionic materials that have been studied most extensively are those based on glass-forming oxides such as SiO_2, P_2O_5 and B_2O_3. Electrical conduction in these glasses may be ionic when significant amounts of alkali metal ions are present or electronic by electron transfer between metal ions of different valency state. Typical examples of semi-conducting glasses are the ones in which the major constituent is a transition metal oxide. For example, V_2O_5 samples can be prepared containing V^{5+} ions as well as V^{4+} ions. Conduction is due to hopping of an electron from a V^{4+} ion to a V^{5+} ion.

The covalent amorphous semiconductors can be divided into two main groups. The first class contains materials with tetrahedral coordination such as the elements Si, Ge and the III-V compounds. These materials can only be prepared in the amorphous phase by thin film deposition. The second group contains the so-called chalcogenide glasses which are based on the elements S, Se and Te and to which other elements, e.g. Si, Ge and As can be added. A glass containing four components Si, Te, As and Ge is widely used to prepare switching devices.

In the next chapter the different models, describing the electronic density of states, will be presented. Here the main question arises whether the features of the band structure of the crystalline material are preserved when going from the crystalline to the disordered state.

The third section is concerned with some electrical properties of amorphous semiconductors, particularly with the conductivity, thermoelectric power and Hall coefficient. On the basis of the models discussed in chapter 2, the formulae describing the three transport coefficients will be presented.

The last section deals with some typical experimental results which have been obtained on chalcogenide glasses and amorphous $CdGe_xAs_2$. These materials have been selected because it is generally accepted that they are the best suited to test the band models proposed for amorphous semiconductors. Indeed these covalent alloys may show compositional disorder in addition to the translational disorder encountered in elemental and compound amorphous semiconductors.

2. ELECTRONIC STRUCTURE OF AMORPHOUS SEMICONDUCTORS

The main characteristics of the density of electronic states N(E) of crystalline solids are the sharp structure in valence and conduction bands and the abrupt terminations at the valence band maximum and the conduction band minimum. The sharp edges in the density of states produce a well-defined forbidden energy gap. These features of the band structure are consequences of the perfect short-range and long-range order of the crystal. In an amorphous solid the long-range order is destroyed, whereas the short-range order, i.e. the interatomic distance and the valence angle, is only slightly changed. The concept of the density of electronic states is also applicable to non-crystalline solids. The basic problem in amorphous semiconductors is to know then how the energy distribution of electronic states is changed when the long-range periodicity in the atomic positions is lost. At present, there appears to be widespread agreement that, if the short-range order is the same in the amorphous state as in the crystalline state, some basic features of the electronic structure of the crystal are preserved. In particular, it is generally accepted that the energy distribution of the density of electronic states remains essentially the same. Theoretical support to this idea has been put forward by Klima and McGill [1], who calculated the density of states for a random array of disordered clusters.

Based on earlier ideas, Mott [2] and Cohen, Fritzsche and Ovshinsky [3] (further called CFO) have developed a model for the electronic structure of amorphous semiconductors. The main feature of the model is the existence of localized states at the extremities of the valence and conduction bands, arising from the absence of long-range order. The densities of localized states decrease gradually into the forbidden energy gap, destroying in this way the sharpness of the valence and conduction bands. Figure 1(a) gives a schematic representation of the density of states in an amorphous semiconductor illustrating the tails of localized states. Until now opinions vary as to the extent of tailing. Cohen et al. [3] have suggested that in amorphous alloys, such as the chalcogenide glasses, the disorder is sufficiently great so that the tails of the conduction and valence bands overlap, leading to an appreciable density of states in the middle of the gap. This model ensures self-compensation and hence pins the Fermi level close to the middle of the gap, a feature required by the electrical properties of these materials. The CFO model is represented in figure 1(b). According to Davis and Mott [4] the tails of localized states should be rather narrow and should extend a few tenths of an electron volt into the forbidden gap. They proposed furthermore the existence of a band of compensated levels near the middle of the band, which will also pin the Fermi level. The levels should arise from unsatisfied or dangling bonds. This picture is illustrated in figure 1(c).

Both models are equal in the sense that they both use the concept of localized states in the band tails. In either band a well-defined energy separates the ranges of energy where the states are localized and extended. Mott suggested that at the transition from extended to localized states, represented by E_C and E_V in figure 1(a), the mobility drops by several orders of magnitude, producing a mobility shoulder. The interval between the energies E_C and E_V acts as a pseudo-gap and is defined as the mobility gap. Cohen [5] proposed a slightly different picture for the energy dependence of the mobility. He suggested that there should not be an abrupt but rather a continuous drop of the mobility occurring in the extended states just inside the mobility edge. In this intermediate range the mean free path of the carriers becomes of the order of the interatomic spacing so that the ordinary transport theory based on the Boltzmann equation cannot be used. Cohen

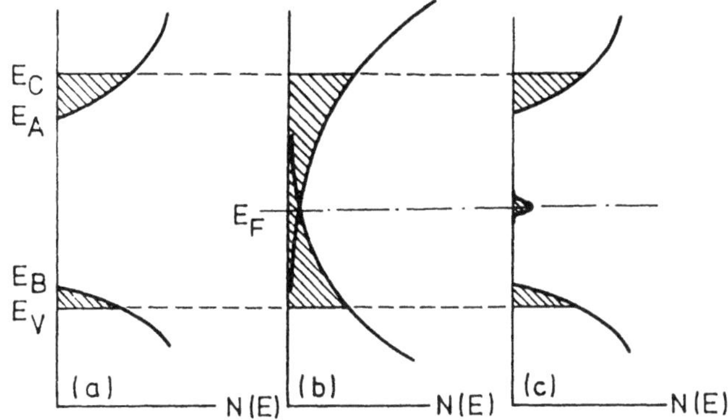

FIGURE 1

Density of states as a function of energy in amorphous semiconductors

described the transport as a Brownian motion in which the carriers are under the influence of a continuous scattering.

On the basis of the model of Davis and Mott, there can be three processes leading to conduction in amorphous semiconductors. Their relative contribution to the total conductivity will change markedly in different temperature ranges. At very low temperatures conduction can occur by thermally assisted tunneling between states at the Fermi level. At higher temperatures charge carriers are excited into the range of localized states. Carriers in the localized states can take part in the electric charge transport by hopping only. At still higher temperatures carriers are excited across the mobility edge into the extended states. The mobility in the extended states is much higher than in the localized states. From this it follows that electrical conductivity measurements over a wide temperature range are needed to study the electronic structure of amorphous semiconductors. The formulae of d.c. conductivity, thermoelectric power and Hall coefficient dealing with these three types of conduction processes will be discussed in the next section.

A different approach to the understanding of the electrical properties of amorphous semiconductors, in particular of the chalcogenide glasses, has been put forward by Emin [6]. He suggested that the charge carriers in some amorphous semiconductors may be small polarons. In support of this hypothesis he argued that the presence of disorder in a non-crystalline material tends to slow down a carrier. This slowing down may lead to a localization of the carrier and, if the carrier stays at an atomic site sufficiently long enough for atomic rearrangements to take place, it may induce displacements of the atoms in its immediate vicinity, causing small polaron formation. Since the small polaron is local in nature, the absence of long-range order in non-crystalline solids may be expected to have no significant influence on its motion. In support of the validity of his model, Emin was able to analyse experimental data of d.c. conductivity, thermopower and Hall mobility obtained on some chalcogenide glasses in the framework of the existing small polaron theories developed for crystalline solids.

3. ELECTRICAL PROPERTIES

The essential feature of the Mott-CFO model for the band structure of amorphous semiconductors is the existence of tails of localized states at the extremities of the valence and conduction bands. This leads to two basically different channels for conduction:

1) conduction in extended states above the mobility edge;
2) hopping conduction in localized states below the mobility edge.

To calculate the electron density in each energy range one has to know the density of electronic states as a function of energy. Whatever the nature of the states, the number of electrons in a band lying within the energies E_1 and E_2 is given by:

$$n = \int_{E_1}^{E_2} N(E)f(E)dE \tag{3.1}$$

where f(E) is the Fermi-Dirac distribution function:

$$f(E) = 1/\{\exp\,(E-E_F)/kT + 1\} \tag{3.2}$$

We proceed now to calculate the electron densities in the two different energy regimes. In order to do this, we are obliged to use very simplified models for the energy distribution of the electron states.

3.1. Electron Density

a) In extended states. In the Mott-CFO model the Fermi level E_F is situated near the middle of the gap and thus sufficiently far from E_C, the energy which separates the extended from the localized states, so that Boltzmann statistics can be used to describe the occupancy of states:

$$f(E) = \exp\,\{-(E-E_F)/kT\} \tag{3.3}$$

Furthermore it will be assumed that the density of states in the extended states remains constant within kT. The number of electrons is then given by:

$$n = \int_{E_C}^{\infty} N(E_C) \exp \{-(E-E_F)/kT\} \, dE$$
$$= N(E_C) \, kT \exp\{-(E_C-E_F)/kT\} \qquad (3.4)$$

where $N(E_C)$ is the density of states at E_C.

b) In band tails. As mentioned earlier, various opinions exist about the basic question whether the band tails should exist in amorphous semiconductors and how far they should extend in the band gap. Experimental data on optical absorption, photoemission and photoconductivity seem to indicate that the tails should be rather narrow. These findings favour Davis and Mott's ideas about the band structure of amorphous solids and therefore we shall consider a model as illustrated in figure 1(a) in which there is a range $\Delta E = E_C-E_A$ of localized states at the bottom of the band. It will be assumed that the density of localized states behaves like some power of the energy E:

$$N(E) = \frac{N(E_C)}{(\Delta E)^s} (E-E_A)^s \qquad (3.5)$$

For s = 1 the density of states various linearly with E. The bottom E_A of the energy interval is sufficiently removed from E_F that, as in the former case, Boltzmann statistics can be applied. The number of electrons can then be calculated as follows [7]:

$$n = \frac{N(E_C)}{(\Delta E)^s} \int_{E_A}^{E_C} (E-E_A)^s \exp \{-(E-E_F)/kT\} \, dE \qquad (3.6)$$

Putting $x \equiv (E-E_A)/kT$ we get:

$$n = \frac{N(E_C)}{(\Delta E)^s} (kT)^{s+1} \exp \{-(E_A-E_F)/kT\} \int_0^{\frac{\Delta E}{kT}} e^{-x} x^s \, dx \qquad (3.7)$$

The integral in eq. (3.7), which we shall denote by C, is an incomplete gamma function. It can be calculated in the following way:

$$C = \int_0^{\frac{\Delta E}{kT}} e^{-x} x^s dx = \int_0^{\infty} e^{-x} x^s dx - \int_{\frac{\Delta E}{kT}}^{\infty} e^{-x} x^s dx$$

$$= \Gamma(s+1) - \int_{\frac{\Delta E}{kT}}^{\infty} e^{-x} x^s dx$$

By subsequent partial integrations we get for the integral $\int e^{-x} x^s dx$:

$$\int e^{-x} x^s dx = -x^s e^{-s} \left[1 + \frac{s}{x} + \frac{s(s-1)}{x^2} + \frac{s(s-1)(s-2)}{x^3} + \ldots \right]$$

so that:

$$C = s! - \left(\frac{\Delta E}{kT}\right)^s \exp\left(-\frac{\Delta E}{kT}\right) \left[1 + s\left(\frac{kT}{\Delta E}\right) + s(s-1)\left(\frac{kT}{\Delta E}\right)^2 + \ldots \right] \quad (3.8)$$

3.2. d.c. Electrical Conductivity

The conductivity for any semiconductor can be expressed in the form:

$$\sigma = e \int N(E)\mu(E)f(E)(1-f(E))\, dE \quad (3.9)$$

where f(E) is the Fermi-Dirac distribution function. According to Mott's view, the mobility drops sharply at the critical energy E_C (or E_V) but at present it is not known how the mobility depends on energy in both conduction regimes. It will be assumed that μ does not vary too rapidly with energy so that one can use an average value.

a) Extended state conduction of electrons above E_C. In the non-degenerate case and under the assumption of a constant density of states and constant mobility, the conductivity due to electrons excited beyond the mobility edge into the excited states is given by:

$$\sigma = eN(E_C)\, kT\, \mu_C \exp\{-(E_C-E_F)/kT\} \quad (3.10)$$

where μ_C is the average mobility.

Measurements of drift mobility and Hall mobility in amorphous semiconductors have shown that the mobility is very low, even smaller than 1 $cm^2V^{-1}s^{-1}$. This corresponds to a mean free path comparable or less than the interatomic distance. Cohen [5] has suggested that conduction in this case would be more properly described as a diffusive or Brownian-type motion. In this regime the mobility can be obtained with the help of the Einstein relation:

$$\mu = eD/kT \tag{3.11}$$

The diffusion coefficient D may be written as:

$$D = 1/6\ \nu a^2 \tag{3.12}$$

where ν is the jump frequency and a the interatomic separation.

The mobility in the Brownian motion regime is then given by:

$$\mu = 1/6\ \frac{ea^2}{kT}\ \nu \tag{3.13}$$

We therefore expect that the expression for the conductivity would be of the form:

$$\sigma = \sigma_o \exp\ \{-(E_C-E_F)/kT\} \tag{3.14}$$

Optical absorption measurements made on amorphous semiconductors have shown that the band gap decreases with increasing temperature. The energy distance E_C-E_F will show a similar behaviour, and, if one assumes a linear temperature dependence:

$$E_C - E_F = E(0) - \alpha T \tag{3.15}$$

then the expression for the conductivity becomes:

$$\sigma = \sigma_o \exp\ (\alpha/k)\ \exp\ \{-E(0)/kT\} \tag{3.16}$$

Here E(0) is the energy distance at T = 0.
We can write this formula in the form:

$$\sigma = C_o \exp\ \{-E(0)/kT\} \tag{3.17}$$

where $C_o = eN(E_C)\ kT\ \mu_C \exp\ (\alpha/k)$.
As seen before μ_C is proportional to 1/T, so that the pre-exponential C_o is temperature independent. A plot of $\ln\sigma$ versus 1/T will yield a straight line. The slope of this line yields E(0), whereas the intercept at 1/T = 0 gives $\sigma_o \exp\ (\alpha/k)$. Mott [2] has made an estimate of the pre-exponential σ_o. In general σ_o may lie between 10 and 10^3 $ohm^{-1}cm^{-1}$ in most amorphous semiconductors. An estimate of α can be obtained from the temperature dependence of the optical gap. In chalcogenide glasses the temperature

coefficient of the optical gap generally lies between $4\text{-}8\times10^{-4}$ eV deg^{-1}. As the Fermi level is situated near the middle of the gap, values of α of approximately half these magnitude are expected and hence values of $\exp(\alpha/k)$ in the range 10-100 seem to be most probable.

A different approach based on a so-called random phase model was used by Hindley [8] and Friedman [9] to calculate the mobility in the extended states near E_C or E_V. The basic feature of this model, used to describe an idealized situation of the amorphous state, was first put forward by Mott [2] and Cohen [5] and is related to the character of the extended state wave function. It is assumed that the phase of the probability amplitude for finding an electron on a particular atomic site varies randomly from site to site. On the basis of this model, Hindley and Friedman derived the following expression for the d.c. conductivity in the extended states:

$$\sigma = \frac{2\pi e^2}{3\hbar a}\{za^6J^2[N(E_C)]^2\}\exp\{-(E_C-E_F)/kT\} \qquad (3.18)$$

Here a is the interatomic spacing, z the coordination number and J the electronic transfer integral. The d.c. conductivity within the random phase model follows a simple exponential law:

$$\sigma = \sigma_o \exp\{-(E_C-E_F)/kT\} \qquad (3.19)$$

The conductivity mobility can be derived from eq. (3.18) by dividing by ne, where the electron density is given by:

$$n = \int_{E_C}^{\infty} N(E_C)f(E)dE \simeq N(E_C)\; kT \exp\{-(E_C-E_F)/kT\} \qquad (3.20)$$

This yields:

$$\mu_C = \frac{2\pi ea^2}{3\hbar}\, z\, \left[\frac{J}{kT}\, a^3 J\, N(E_C)\right] \qquad (3.21)$$

The mobility decreases with increasing temperature as T^{-1}. This is in agreement with the formula derived by Cohen using the arguments that charge transport occurs via a Brownian-type motion.

<u>b) Conduction in the band tails.</u> Conduction in the localized states of the band tails will be by hopping and hence it may be expected that the mobility will be thermally activated:

$$\mu_{hop} = \mu_o \exp\{-W(E)/kT\} \qquad (3.22)$$

The pre-exponential μ_o has the form:

$$\mu_o = 1/6\, \nu_{ph}\, eR^2/kT \qquad (3.23)$$

where ν_{ph} is the phonon frequency and R the distance covered in one hop.

The conductivity being an integral over all available energy states will depend on the energy distribution of the density of localized states. We shall consider now two simplified models for the density of states and calculate the corresponding formulae for the conductivity assuming again that the mobility μ_{hop} is energy independent [7]. The first model implies a linear variation of N(E) in the band tails:

$$N(E) = \frac{N(E_C)}{\Delta E}(E-E_A) \tag{3.24}$$

with $\Delta E = E_C - E_A$.
The second one implies a quadratic variation of N(E):

$$N(E) = \frac{N(E_C)}{\Delta E^2}(E-E_A)^2 \tag{3.25}$$

It may be mentioned that Mott has presented some simple arguments to demonstrate that a linear variation of N(E) in the band tails might be a good approximation.

In the next section we shall present measurements of electrical conductivity on some chalcogenide glasses. We shall see that the experimental data can be analysed in terms of parallel conduction of holes in extended and localized states. In order to perform this analysis, it is necessary to adopt a model for the energy distribution of N(E) in the band tails. The two models, where we have taken s = 1 and s = 2 in eq. (3.5), will be used for this purpose.

For a given s value the conductivity σ_{hop} is easily found with the help of eq. (3.9) and (3.7) :

$$\sigma_{hop} = \sigma_{ohop}\left(\frac{kT}{\Delta E}\right)^s C \exp\{-(E_A-E_F+W)/kT\} \tag{3.26}$$

where $\sigma_{ohop} = eN(E_C)\mu_o'$.
μ_o' is defined by:

$$\mu_o = 1/6\ \nu_{ph}\ eR^2/kT = \mu_o'/kT \tag{3.27}$$

and:

$$C = s! - \left(\frac{\Delta E}{kT}\right)^s \exp\left(-\frac{\Delta E}{kT}\right)\left[1 + s\left(\frac{kT}{\Delta E}\right) + s(s-1)\left(\frac{kT}{\Delta E}\right)^2 + \ldots\right] \tag{3.28}$$

For the two specific cases s=1 (linear variation) and s=2 (quadratic variation) the conductivity is given by:

$$s=1 \qquad \sigma_{hop} = \sigma_{ohop} \frac{kT}{\Delta E} C_1 \exp\{-(E_A-E_F+W)/kT\} \qquad (3.29)$$

with:

$$C_1 = 1 - \exp(-\Delta E/kT)\left[1 + (\Delta E/kT)\right] \qquad (3.30)$$

The expression $N(E_C)kT(kT/\Delta E)C_1$ represents the effective density of states for a given temperature T.

$$s=2 \qquad \sigma_{hop} = \sigma_{ohop} \left(\frac{kT}{\Delta E}\right)^2 C_2 \exp\{-(E_A-E_F+W)/kT\} \qquad (3.31)$$

with:

$$C_2 = 2 - \exp(-\Delta E/kT)\left[2 + 2(\Delta E/kT) + (\Delta E/kT)^2\right] \qquad (3.32)$$

c) Conduction in localized states at the Fermi energy. If the Fermi energy lies in a band of localized states, as predicted by the Davis-Mott model, the carriers can move between the states via a phonon-assisted tunneling process. This is the transport analogous to impurity conduction observed in heavily doped and highly compensated semiconductors at low temperatures. To calculate the tunneling conduction at E_F we shall follow Mott's treatment [2]. The transition from one atomic site to another one, between which there exists an energy difference W, has a probability given by:

$$p = \nu_{ph} \exp(-2\gamma R) \exp(-W/kT) \qquad (3.33)$$

where R is the interatomic spacing and γ a quantity which is representative for the rate of fall-off of the wave function at a site.

By making use of the Einstein relation the conductivity can be written as:

$$\sigma = N(E_F)\, kT\, e\mu = N(E_F)\, e^2 D \qquad (3.34)$$

where $N(E_F)$ is the density of states at the Fermi level. The diffusion coefficient D can be expressed as:

$$D = 1/6\, pR^2 \qquad (3.35)$$

and hence:

$$\sigma = 1/6\, e^2R^2\, \nu_{ph}\, N(E_F) \exp(-2\gamma R) \exp(-W/kT) \qquad (3.36)$$

As the temperature is lowered the number and energy of phonons decreases and the more energetic phonon-assisted hops will become progressively less favourable. Carriers will tend to hop to larger distances in order to find sites which lie energetically closer than the nearest neighbours. Mott's treatment of this variable range tunneling process leads to a temperature dependence for the conduction of the form:

$$\sigma = \text{const.}\ \exp\left[-\left(\frac{A}{T}\right)^{1/4}\right]$$

3.3. Thermoelectric Power

In this section we shall derive the formulae for the thermopower associated with the two main processes of conduction in an amorphous semiconductor, i.e. extended states conduction and hopping conduction in the band tails.

A general expression for the thermopower is given by:

$$S = \frac{k}{e\sigma} \int e\ \mu(E)\ N(E)\ kT\ (E-E_F)/kT\ \frac{\partial f}{\partial E}\ dE \tag{3.37}$$

with a corresponding formula for the conductivity:

$$\sigma = -\int e\ \mu(E)\ N(E)\ kT\ \frac{\partial f}{\partial E}\ dE \tag{3.38}$$

Thus we have:

$$S = -\frac{k}{e} \frac{\int e\ \mu(E)\ N(E)\ kT\ (E-E_F)/kT\ \frac{\partial f}{\partial E}\ dE}{\int e\ \mu(E)\ N(E)\ kT\ \frac{\partial f}{\partial E}\ dE} \tag{3.39}$$

f is the Fermi-Dirac distribution function. When we use the relationship:

$$\frac{\partial f}{\partial E} = -\ f(1-f)/kT \tag{3.40}$$

then S can be written as:

$$S = -\frac{k}{e} \frac{\int e\ \mu(E)\ N(E)\ (E-E_F)/kT\ f(1-f)\ dE}{\int e\ \mu(E)\ N(E)\ f(1-f)\ dE} \tag{3.41}$$

For a non-degenerate semiconductor, classical Boltzmann statistics are appropriate. In this case, i.e. $E-E_F \gg kT$, the factor $f(1-f)$ in eq. (3.41) reduces to a Boltzmann factor $f = \exp\left[-(E-E_F)/kT\right]$

<u>a. Conduction in extended states.</u> In order to simplify the calculation of the conductivity arising from carrier excitation in

the extended states we have assumed that the mobility is energy independent and that the density of states remains constant over kT, so that $N(E) = N(E_C)$. Under the same circumstances the thermopower can readily be found by integrating eq. (3.41).
This gives:

$$S = -\frac{k}{e}\frac{\int_0^\infty (E-E_F)/kT \exp\left[-(E-E_F)/kT\right] dE}{\int_0^\infty \exp\left[-(E-E_F)/kT\right] dE} \tag{3.42}$$

Writing $(E-E_F)$ as the sum $(E-E_C) + (E_C-E_F)$ and knowing that the integral:

$$\int_0^\infty e^{-x}x^n dx = \Gamma(n+1)$$

we obtain for the thermopower an expression of the familiar form:

$$S = -\frac{k}{e}\left[\frac{E_C-E_F}{kT} + 1\right] \tag{3.43}$$

For a crystalline semiconductor the kinetic term A, which is equal to unity in the case considered here, depends on the scattering mechanism. If E_C-E_F varies linearly with temperature, so that:

$$E_C - E_F = E(0) - \alpha T \tag{3.44}$$

eq. (3.43) can be written as:

$$S = -\frac{k}{e}\left[\frac{E(0)}{kT} - \frac{\alpha}{k} + 1\right] \tag{3.45}$$

A plot of S as a function of 1/T yields a straight line showing exactly the same slope as the corresponding $\ln \sigma$ versus 1/T curve. The intercept of the S curve on the S axis at 1/T = 0 enables us to calculate the temperature coefficient α, at least if the assumption of a constant N and μ is a good approximation.

<u>b. Conduction in localized states.</u> The thermopower built up by carriers conducting in the localized states of the band tails will be given by:

$$S = -\frac{k}{e}\frac{\int (E-E_F)/kT \exp\left[-(E-E_F)/kT\right] N(E)\, dE}{n} \tag{3.46}$$

or:

$$S = -\frac{k}{e}\{\frac{E_A - E_F}{kT} + \frac{1}{n}(\int \frac{E - E_A}{kT} \exp\left[-(E-E_A)/kT\right] N(E)\, dE)$$

$$\exp\left[-(E_A - E_F)/kT\right]\} \tag{3.47}$$

For the "s-model" this becomes:

$$S = -\frac{k}{e}\left[\frac{E_A - E_F}{kT} + \frac{C^*}{C}\right] \tag{3.48}$$

where:

$$C^* = \int_0^{\frac{\Delta E}{kT}} e^{-x} x^{s+1} dx \tag{3.49}$$

and C is defined by eq. (3.8).
For a linear variation of N(E), S can be represented by:

$$S = -\frac{k}{e}\left[\frac{E_A - E_F}{kT} + \frac{C_1^*}{C_1}\right] \tag{3.50}$$

with:

$$C_1^* = 2 - \{\exp(-\Delta E/kT)\left[2 + 2(\Delta E/kT) + (\Delta E/kT)^2\right]\} \tag{3.51}$$

and:

$$C_1 = 1 - \{\exp(-\Delta E/kT)\left[1 + (\Delta E/kT)\right]\} \tag{3.52}$$

Assuming a quadratic variation we have:

$$S = -\frac{k}{e}\left[\frac{E_A - E_F}{kT} + \frac{C_2^*}{C_2}\right] \tag{3.53}$$

where:

$$C_2^* = 6 - \{\exp(-\frac{\Delta E}{kT})\left[6 + 6(\frac{\Delta E}{kT}) + 3(\frac{\Delta E}{kT})^2 + (\frac{\Delta E}{kT})^3\right]\} \tag{3.54}$$

and:

$$C_2 = 2 - \{\exp(-\frac{\Delta E}{kT})\left[2 + 2(\frac{\Delta E}{kT}) + (\frac{\Delta E}{kT})^2\right]\} \tag{3.55}$$

If both electrons and holes contribute to the conductivity then the thermopower is the algebraic sum of the individual contributions S_e and S_h but each weighed according to the ratio of its conductivity to the total conductivity. Thus we have:

$$S = \frac{S_e\sigma_e + S_h\sigma_h}{\sigma} \tag{3.56}$$

and:

$$\sigma = \sigma_e + \sigma_h \tag{3.57}$$

At very low temperature charge transport can arise from electrons tunneling between localized states at E_F. Cutler and Mott [10] have suggested that the thermoelectric power in this regime should be identical to the equation used for metals:

$$S = \frac{\pi^2 k^2 T}{3e} \left[\frac{d(\ln\sigma)}{dE}\right]_{E=E_F}$$

since the Fermi energy lies in a region where the density of states is finite.

3.4. Hall Effect

The basic transport properties usually measured in crystalline semiconductors are the conductivity σ and the Hall coefficient R_H. Measurements of R_H provide a reliable guide to the charge carrier concentration. For n-type semiconductors the Hall coefficient is negative and is given by the general formula:

$$R_H = -\frac{r}{ne} \tag{3.58}$$

Here r is the scattering factor. It is usually not much greater than one.

In p-type semiconductors the Hall coefficient is positive. From σ and R_H we may determine the Hall mobility:

$$\mu_H = |R_H|\,\sigma = r\,\mu \tag{3.59}$$

Thus the Hall mobility is greater than the conductivity mobility μ by the scattering factor r.

Interpretation of the Hall coefficient on this basis is valid for materials in which the mean free path is long compared to the interatomic spacing. In amorphous semiconductors the mobilities are found to be very low so that the carriers will move with a mean free path comparable to the interatomic distance. As a consequence the ordinary transport theory based on the Boltzmann equation cannot be used anymore. So far measurements of the Hall coefficient have only been made on a limited number of amorphous semiconductors due to the high resistivities and the low carrier mobilities. In most cases the Hall coefficient was found to be negative, in contrast with the thermopower which showed a positive sign.

Using the random phase model, describing the crystal wave function as a linear superposition of atomic wave functions with coefficients which have no phase correlation from site to site, Friedman [9] was able to find an expression for the Hall mobility. This approach, which was used by Hindley [8] to calculate the conductivity, may be applicable to the conduction regime in the extended states near the mobility edge. Friedman's formula is:

$$\mu_H = 4\pi \left(\frac{ea^2}{\hbar}\right) \left[a^3 J\, N(E_C)\right] \eta \left(\frac{\bar{z}}{z}\right) \tag{3.60}$$

Here J is the overlap energy integral between neighbouring sites; a is the interatomic spacing; z is the coordination number and $\bar{z}$ is the number of interacting sites (a minimum of three sites is necessary in order to obtain a Hall effect); η is a parameter usually of the order of 1/3.

From eq. (3.60) it follows that the Hall mobility within the random phase model is temperature independent. Friedman made an estimate of the magnitude of μ_H and obtained $\mu_H \simeq 10^{-1}$ $cm^2V^{-1}s^{-1}$. The Hall mobility is related to the conductivity mobility by a simple relation:

$$\frac{\mu_H}{\mu_C} \simeq \frac{kT}{J} \tag{3.61}$$

J is a quantity of the order of 1 eV (2 zJ is the band width obtained from a tight binding calculation), so that J>>kT. The Hall mobility will be considerably smaller than the conductivity mobility, the ratio being at least 1/10. An important feature, resulting from Friedman's treatment, is connected with the sign of the Hall coefficient. For a configuration of three interacting sites the Hall coefficient is negative, even if holes are responsible for the conductivity. This result is consistent with the p-n sign anomaly between the thermopower and the Hall effect commonly encountered in amorphous semiconductors.

A satisfactory theory of the Hall mobility when conduction takes place by hopping of charge carriers in the localized states of the band tails does not exist at the present time. Experimental data, which might give some insight in this problem, are also lacking.

4. CHALCOGENIDE GLASSES

In the past a considerable amount of experimental work has been published on the transport properties of covalent amorphous semiconductors. This section will mainly be devoted to a brief description of some results of d.c. conductivity, thermopower and Hall effect which have been obtained on systems containing one or more of the chalcogenide elements S, Se and Te. We shall not attempt to summarize all the published data on the chalcogenide glasses but we shall focus our attention to some results which have contributed to get some insight in the electronic structure of these materials. Therefore the main purpose will be to find out whether the experimental results may yield convincing evidence for the band structure of amorphous semiconductors as described in section 2.

The binary systems As-S and As-Se, which include the compounds As_2S_3 and As_2Se_3, have been studied extensively. Ternary and quaternary systems, containing elements such as S, Se, Te, As, Si, Ge, Tl and others can also be prepared in the glassy state. The multicomponent systems, e.g. $As_{3.0}Te_{4.8}Si_{1.2}Ge_{1.0}$, are of great practical importance because of their switching and memory properties. Most of the chalcogenide glasses can be prepared by quenching from the melt. In the next section we shall discuss in detail the electrical properties of the three following systems: a binary As-Se, a ternary As-Te-Si and a quaternary As-Te-Si-Ge system.

4.1. As-Se Alloys

In the As-Se system much work has been devoted to the stoichiometric composition As_2Se_3, mainly because of the existence of the corresponding crystalline compound. It is supposed that the nearest neighbour configuration is the same in the glass as in the crystal, allowing a direct comparison of the properties of the amorphous state with those of the crystalline one. Measurements of d.c. electrical conductivity on amorphous As_2Se_3 have been reported by many authors. The room temperature conductivity of this material is extremely low (about 10^{-12} $ohm^{-1}cm^{-1}$). The $\ln\sigma$ versus 1/T curve gives a straight line with a single slope in the whole of the investigated temperature range. The activation energy calculated from the slope is in the range of 0.9 to 1.0 eV, which represents about half of the optical gap. Thermopower measurements on an As_2Se_3 glass have been reported by Callaerts et al. [11]. The thermopower, which is positive, can be fitted to an expression of the form $S = (k/e)(E_S/kT + B)$ where $E_S = 1.01 \pm 0.05$ eV and $B = -7.1$. The conductivity is of the form $\sigma = C_o \exp(-E_\sigma/kT)$ with $E_\sigma = 0.98$ eV and $C_o = 6.6 \times 10^3$ $ohm^{-1}cm^{-1}$. Measurements of d.c. conductivity and thermopower for vitreous As_xSe_{100-x} alloys containing from 30 % to 50 % As have recently been published by Hurst and Davis [12]. Their results for the stoichiometric composition As_2Se_3 are in complete agreement with the data obtained by Callaerts et al. The temperature dependence of the conductivity and the thermopower for five As-Se alloys is shown in figures 2 and 3. Across the whole composition range the conductivity obeys a single

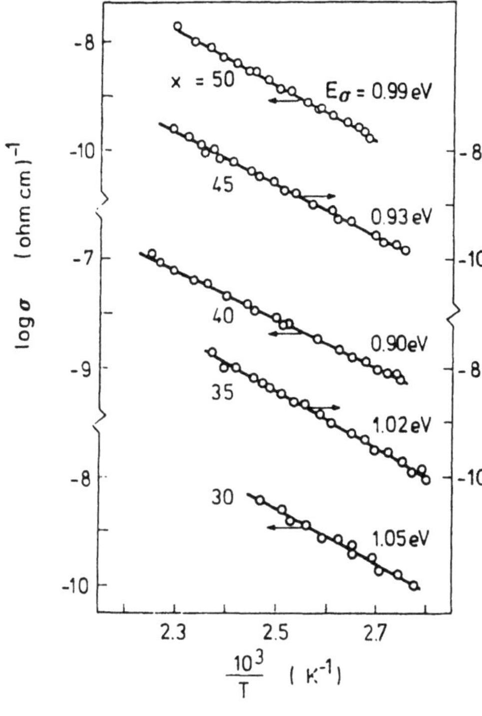

FIGURE 2

Log conductivity versus $10^3/T$ for five As_xSe_{100-x} alloys. E_σ is the conductivity activation energy. After Hurst and Davis (1974).

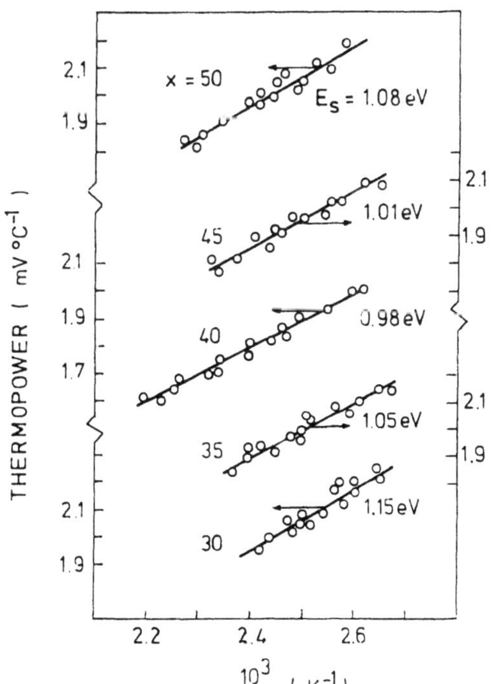

FIGURE 3

Thermopower versus $10^3/T$ for five As_xSe_{100-x} alloys. E_S is the thermopower activation energy. After Hurst and Davis (1974).

exponential relationship. The pre-exponential term C_o had values between 10^3 and 10^4 $ohm^{-1}cm^{-1}$. For each composition, E_S was equal to E_σ within experimental error. Nagels et al. [13] have reported Hall mobility measurements in vitreous As_2Se_3. The sign of the Hall coefficient was negative. The Hall mobility was temperature independent and had a value of 7.6×10^{-2} $cm^2V^{-1}s^{-1}$.

In the light of the concepts about the different ways of current transfer in amorphous semiconductors (see section 2), the major feature of these results is the equal value of the activation energy associated with the conductivity and thermopower (~ 1.0 eV). This feature strongly suggests that in the investigated temperature range conduction occurs via delocalized states. Indeed, if the carriers move by hopping in the localized states, the conductivity varies nearly exponential with temperature [apart from a small temperature dependent pre-exponential term, as shown by eq. (3.29) and (3.31)] and the measured activation energy is the sum of the activation energies for carrier creation and for hopping. Therefore, one expects a difference in slope between the conductivity and thermopower curve, which has not been observed in the As-Se alloys. The high value of the pre-exponential term ($C_o \approx 10^4$ $ohm^{-1}cm^{-1}$) in the conductivity expression also points to conduction in the extended states. Furthermore, the magnitude of the Hall mobility, its temperature independence and the negative sign of the Hall coefficient are in complete agreement with Friedman's theoretical considerations within the random phase model.

We shall discuss now in somewhat more detail the results obtained on the stoichiometric As_2Se_3 compound. In the general case of electron and hole contribution, the thermopower for conduction in the extended states is given by the equation:

$$\sigma S = \sigma_e S_e + \sigma_h S_h \tag{4.1}$$

with:

$$S_e = -(k/e)\left(\frac{E_C - E_F}{kT} + 1\right); \quad S_h = (k/e)\left(\frac{E_F - E_V}{kT} + 1\right) \tag{4.2}$$

and:

$$\sigma_e = C_{oe} \exp\left[-(E_C - E_F)/kT\right]; \quad \sigma_h = C_{oh} \exp\left[-(E_F - E_V)/kT\right] \tag{4.3}$$

where E_C, E_V separate the energies of localized from extended states. For glassy As_2Se_3 the thermopower has a positive sign which indicates that holes are the most numerous carriers ($E_F - E_V < E_C - E_F$) and/or $C_{oh} > C_{oe}$. By assuming that E_F lies in the middle of the gap, but that the tail of localized states is wider for the conduction band than for the valence band, Mott [14] favoured the possibility $E_F - E_V < E_C - E_F$. For a higher number of holes and on the assumption of a linear variation of the Fermi level with temperature the conductivity and the thermopower can be represented by eq. (3.16) and (3.45):

$$\sigma = \sigma_o \exp(\alpha/k) \exp\left[-(E_F-E_V)_{T=0}/kT\right]$$

and:

$$S = (k/e)\left[(E_F-E_V)_{T=0}/kT - \alpha/k+1\right]$$

The temperature coefficient α obtained from the thermopower data, as reported by Callaerts et al., is 6.3×10^{-4} eV K^{-1}. The temperature coefficient of the optical gap found by Kolomiets and Pavlov [15] is 11×10^{-4} eV K^{-1}, whereas Hurst and Davis [12] reported a somewhat smaller value equal to 9×10^{-4} eV K^{-1}. For As_2Se_3 the activation energies E_σ and E_S are equal within the limits of precision, which amounts to about 5 %. It results that $\sigma_h \geq 50\, \sigma_e$. If $C_{oe} = C_{oh}$ in eq. (4.3), then:

$$E_C - E_F \geq E_F - E_V + kT \ln 50 = E_F - E_V + 0.14 \text{ eV}$$

at 150°C.

From the ratio of the Hall to the drift mobility, obtained with the aid of eq. (3.21) and (3.60), one is able to calculate the conductivity mobility μ_C if an estimate of the overlap integral J can be made:

$$\frac{\mu_H}{\mu_C} = \frac{6kT}{J} \eta \frac{\bar{z}}{z^2} \qquad (4.4)$$

J can be eliminated in favour of the bandwidth W = 2zJ. Then we get:

$$\frac{\mu_H}{\mu_C} = \frac{12kT}{W} \eta \frac{\bar{z}}{z} \qquad (4.5)$$

The bandwidth W can be evaluated from:

$$W = \frac{2ze^{-1}\hbar^2}{2m^* a^2} \qquad (4.6)$$

where m^* is the effective mass.
Kolomiets et al. [16] have determined the reduced effective mass of holes and electrons for a number of chalcogenide glasses using the method of electroabsorption. For glassy As_2Se_3 they reported $m^*/m_o = 2.9$. Taking z = 6 and a ≃ 3 Å [17] one finds W = 1.7 eV. With the values η = 1/3, $\bar{z}$ = 3 one obtains from eq. (4.5) $\mu_H/\mu_C \simeq 0.04$ and $\mu_C \simeq 2$ $cm^2V^{-1}s^{-1}$. This very rough estimate of μ_C lies within the limits 10^{-2} - 5 $cm^2V^{-1}s^{-1}$ estimated by Cohen [5] for the mobility in the Brownian motion regime.

The density of states at the mobility edge $N(E_V)$ can be evaluated from the pre-exponential constant $C_o = \sigma_o \exp(\alpha/k)$, where σ_o is given by the equation:

$$\sigma_o = e\, N(E_V)\, kT\, \mu_C \quad (4.7)$$

This yields $N(E_V) \simeq 4 \times 10^{20}$ $cm^{-3}eV^{-1}$ at 150°C. This value, which is an extremely rough approximation because of the uncertainties in the choice of the parameters, is consistent with Mott's [2] estimate of the density of states at E_C or E_V [$N(E_{C,V}) \simeq 10^{21} cm^{-3}eV^{-1}$].

It can be concluded that the transport data obtained on amorphous As_xSe_{100-x} alloys (with x ranging between 30 and 50) strongly suggest conduction by holes in extended states.

4.2. Ternary and Multicomponent Chalcogenide Glasses

Numerous papers on the electrical properties of mixed systems of binary chalcogenides and of multicomponent glasses have been published in the literature. In a number of the chalcogenide glasses an exponential variation of the conductivity as a function of temperature has been reported. In the majority of the cases it has been suggested that conduction occurs in extended states near the mobility edge. Detailed measurements of the thermopower as a function of temperature are less frequent, mainly because of the difficulty of measuring small voltages in high-ohmic material. The thermopowers are normally positive. For most of the glasses a linear variation of S with 1/T was reported. The activation energies of the thermopower curves were often found to be less than the corresponding slopes of the conductivity. To account for this difference in activation energies, typically being between 0.1 and 0.2 eV, various explanations have been suggested: (a) two carrier intrinsic conduction (b) one carrier conduction in the localized states, involving an activation energy for hopping (c) phonon assisted hopping of small polarons [18, 19].

Careful measurements of electrical conductivity and thermopower over a wide temperature range have been made by Nagels et al. [20, 7] on ternary glasses in the system As-Te-Si and on a quaternary Si-Te-As-Ge (STAG) glass. Their results are presented in figures 4 and 5. The conductivity of the different samples cannot be represented by one single exponential of the form $\sigma = C \exp(-E/kT)$. The data are best fitted by smooth curves with gradually increasing activation energy. The deviation from a purely exponential behaviour is small, thus requiring measurements over many decades in order to detect the slight curvature in a log σ versus 1/T plot. The thermopower data can also be best fitted by slightly curved lines. They exhibit a less steep temperature dependence than the conductivity curves. Measurements of the Hall mobility made on the same samples are shown in figure 6. The Hall

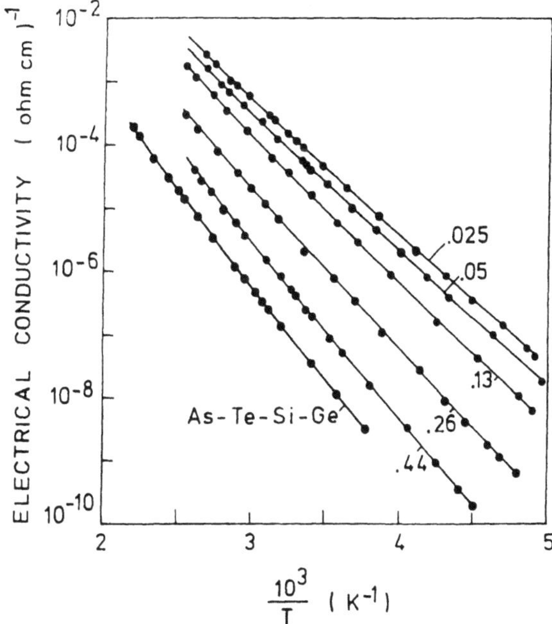

FIGURE 4

Electrical conductivity versus $10^3/T$ for five $AsTe_{1.5}Si_x$ glasses (with x = 0.44, 0.28, 0.13, 0.05 and 0.025) and a $As_{3.0}Te_{4.8}Si_{1.2}Ge_{1.0}$ glass. After Nagels et al. (1973).

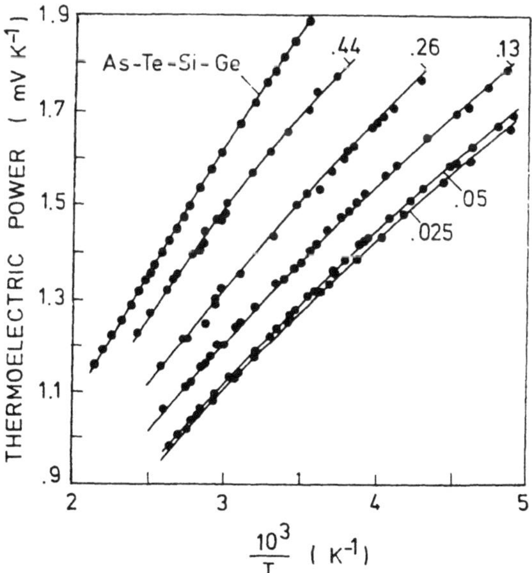

FIGURE 5

Thermopower versus $10^3/T$ for five $AsTe_{1.5}Si_x$ glasses (the Si content is again indicated) and a $As_{3.0}Te_{4.8}Si_{1.2}Ge_{1.0}$ glass. After Nagels et al. (1973).

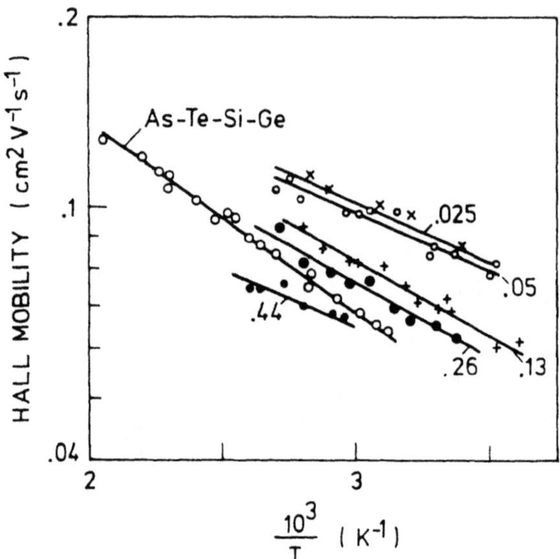

FIGURE 6

Hall mobility versus $10^3/T$ for various $AsTe_{1.5}Si_x$ glasses and a $As_{3.0}Te_{4.8}Si_{1.2}Ge_{1.0}$ glass. After Nagels et al. (1973).

mobility increases very slightly with increasing temperature, the activation energy being in the range 0.03 to 0.05 eV. An exponentially activated Hall mobility with energies around 0.05 eV has also been observed in some other chalcogenide glasses. This behaviour, together with the difference in slopes between σ and S, has been used by Seager et al. [19] as an argument to describe the conduction by phonon-assisted tunneling of small polarons. Nagels et al. analyse their results on the basis of parallel conduction of holes in extended and localized states. As seen in section 3, both conduction mechanisms can be described by an exponential temperature dependence of σ. The slight curvature observed in the experimental conductivity lines indicates that the difference in the rate constants, i.e. the activation energies, of the two exponentials is small, so that there is a gradual change from one process to the other and no kink is observed. In this transition region the slope of the thermopower versus 1/T curve has no particular significance.

As pointed out in chapter 3, the contribution of the conduction in localized states depends on the energy distribution of the density of states in the tail of the valence band. The conductivity and the thermopower, calculated for two simplified models assuming a linear and quadratic variation of N(E) in the band tails, are given by eqs. (3.29), (3.31), (3.50) and (3.53). Using these formulae, combined with the corresponding expressions for conduction in the extended states, a detailed numerical analysis has been made for two glasses of composition $AsTe_{1.5}Si_{0.13}$ and $As_{3.0}Te_{4.8}Si_{1.2}Ge_{1.0}$ [7]. The total conductivity and thermopower is given by:

$$\sigma = \sigma_{ext} + \sigma_{loc} \tag{4.8}$$

and:

$$S = (S_{ext}\sigma_{ext} + S_{loc}\sigma_{loc})/\sigma \tag{4.9}$$

The main purpose was to find out whether the conductivity and thermopower curves can be decomposed into a sum of two contributions and particularly whether some preference can be given to one of the models for the density of states in the localized states. For both models fitting to the σ and S data within the limits of experimental accuracy turned out to be possible, yielding physically meaningful values for the activation energies and the pre-exponential terms. This is not surprising since the exponential term in both the eqs. (3.29) and (3.31) plays by far the predominant role and the magnitude of the thermopower is nearly completely determined by the activation energies deduced from the exponentials. Measurements to much lower temperatures would be necessary in order to find a pronounced difference between the two models. Nevertheless the theoretical fitting of the experimental conductivity and thermopower curves demonstrated that the proposed picture of combined conduction of holes in extended and localized states is a very plausible one.

The values of the different parameters, resulting from the curve fitting procedure, are shown in Table 1. Here $C_{ext} = e\, N(E_V)\mu_{C_{ext}} \exp(\alpha/k)$, $C_{loc} = e\, N(E_V)\, \mu_{C_{loc}} \exp(\alpha/k)$ and ΔE is the width of the valence band tail. Fitting of the thermopower curves was only possible for an activation energy $E_F - E_B$ close to the total activation energy for conduction $E_F - E_B + W$, thus yielding a small value for W ($\simeq$ 0.03 eV). The analysis gives evidence for the existence of a narrow tail of localized states in both materials. Indeed, from Table 1 it can be seen that for the linear model (s=1), favoured by Mott, the width of the tail is equal to about 0.15 eV in the STAG glass and 0.13 eV in the As-Te-Si glass.

The ratio C_{ext}/C_{loc} enables us to calculate the decrease in mobility when passing from the extended to the localized states. It may be remembered here that it has been assumed that the mobility is energy independent in both regimes and drops suddenly at E_V. Considering again the linear model, the decrease in mobility amounts to a factor 21 in the STAG glass and 17 in the As-Te-Si glass. The effective density of states at E_B is given by $N(E_V)C_1(kT^2/\Delta E)$ with $C_1 \simeq 1$. This leads to a decrease in $N(E)_{eff}$ between E_V and E_B equal to 6.1 for the Ge-Si-As-Te alloy and 5.1 for the As-Te-Si alloy at room temperature (again for s=1).

TABLE 1

Composition	Model	C_{ext} $(ohm\ cm)^{-1}$	C_{loc} $(ohm\ cm)^{-1}$	$(E_F-E_V)_{T=0}$ eV	ΔE eV
$As_{3.0}Te_{4.8}Si_{1.2}Ge_{1.0}$	s=1	3900	580	0.683	0.154
	s=2	3900	940	0.683	0.177
$AsTe_{1.5}Si_{0.13}$	s=1	4000	730	0.520	0.128
	s=2	4000	1218	0.520	0.148

The experimental Hall mobility has been found to increase with temperature (see figure 6). This behaviour can be interpreted by supposing that the Hall mobility in the extended states near E_C is given by Friedmann's theory, and furthermore that the Hall mobility in the hopping regime is much smaller than in the extended states. Indeed, for carriers conducting in both regimes the measured Hall mobility will be given by:

$$\mu_H = (\mu_{H,ext}\sigma_{ext} + \mu_{H,loc}\sigma_{loc})/\sigma \qquad (4.10)$$

Under the assumption $\mu_{H,loc} << \mu_{H,ext}$, the Hall mobility in the extended states can be calculated from $\mu_{H,ext} = \mu_H \cdot \sigma/\sigma_{ext}$. This leads to temperature independent values equal to 1.2×10^{-1} $cm^2V^{-1}s^{-1}$ for As-Te-Si and 1.9×10^{-1} $cm^2V^{-1}s^{-1}$ for As-Te-Si-Ge.

5. $CdGeAs_2$ GLASSES

$CdGeAs_2$ belongs to a ternary system based on $CdAs_2$ which is able to form amorphous material when elements such as Ge, Tl, Si are added to it. Cervinka and coworkers [21] have studied the structure of amorphous $CdGe_xAs_2$ by X-ray diffraction. By comparing the radial distribution curves of the atomic density they concluded that the structure of amorphous $CdGe_xAs_2$ ressembles that of crystalline $CdAs_2$. This compound has a tetragonal lattice and is composed of tetrahedra formed from one Cd atom and four As atoms. The ternary system $CdGe_xAs_2$ is of particular interest since it is one of the few semiconductors with tetrahedral coordination which can be prepared in the glassy state by quenching the melt.

The temperature dependence of the conductivity of $CdGe_xAs_2$ glasses has been determined by Cervinka et al. [22]. A detailed study of the transport properties, comprising conductivity, thermopower and Hall effect measurements, has been made by Nagels et al.

[23, 24]. Their results of electrical conductivity obtained on $CdGe_xAs_2$ glasses with x varying between 0.1 and 1.0 mol are shown in figure 7. The temperature dependence of the thermopower is represented in figure 8. With increasing Ge content the thermopower changes from p-type to n-type. The thermopower of the 0.1 and 0.2 mol compositions is positive in the whole temperature range. For x = 0.3 and 0.4 a change in sign occurs near 320 K. The thermopower of the glasses with Ge contents higher than 0.5 is always negative. Measurements of the Hall coefficient yield a negative sign for all samples except for the stoichiometric composition $CdGeAs_2$. The behaviour of the thermopower gives evidence for a conduction process in which both electrons and holes contribute to the total conductivity and the thermopower. Nagels et al. have discussed the electrical data in terms of the commonly accepted band scheme of amorphous semiconductors. As a result of a quantitative analysis based on a theoretical fitting of the experimental conductivity and thermopower curves they were able to show that the transport properties can be explained by parallel conduction of electrons and holes in the localized and extended states of both the conduction and the valence band. A model for the density of states, resulting from this analysis, is schematically illustrated in figure 9 for a $CdGe_{0.1}As_2$ and a $CdGe_{1.0}As_2$ glass. In samples with low Ge content (x = 0.1 and 0.2) the Fermi level lies closer to the mobility edge of the valence band, yielding p-type conduction in the whole temperature range. In $CdGe_{0.1}As_2$ the Fermi level is located at 0.02 eV below the middle of the mobility gap at zero degrees K. The Fermi level shifts to the middle of the gap with increasing Ge content and in the sample with x = 0.5 it adopts an energy exactly intermediate between E_C and E_V. In the latter case n-type conduction arises from a higher contribution of electron conduction in the extended states due to a higher value of its preponential constant ($C_{ext,e}$ = 9000 $ohm^{-1}cm^{-1}$ for $C_{ext,h}$ = 5200 $ohm^{-1}cm^{-1}$). For samples with compositions in the range x = 0.6 to 1.0 the Fermi level lies closer to the mobility edge of the conduction band, giving rise to n-type conduction. The mobility gap $E_V - E_C$ increases slightly from 1.13 eV for $CdGe_{0.1}As_2$ to 1.26 eV for $CdGe_{1.0}As_2$, in agreement with observations made by Cervinka et al. [22] on the optical properties of $CdGe_xAs_2$. These authors reported a shift of the optical absorption edge to slightly higher energies with increasing Ge content. The energy range of the localized states of both the conduction and valence band is about 0.12 eV wide. The addition of Ge does not seem to affect the width of this energy range. All the values used to construct the band scheme (represented in figure 9) and those mentioned above result from computing conductivity and thermopower curves with the help of four contributions. Hence, it is clear that, because of the high number of parameters involved in the calculations and the accuracy of the measurements, a variety of sets of slightly different values can be obtained which also would fit the experimental results in a satisfactory way. However, this will not affect the essential feature

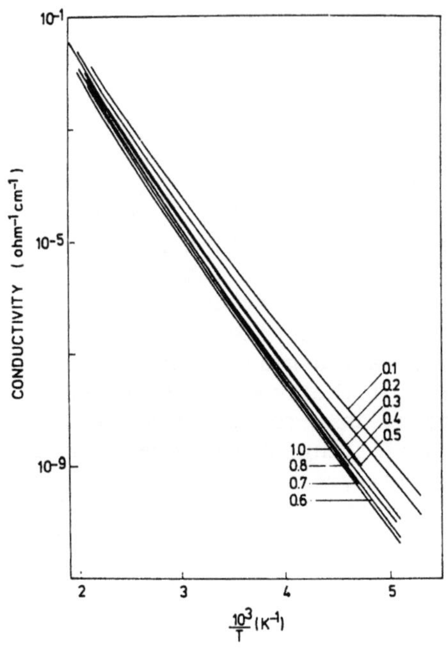

FIGURE 7

Electrical conductivity versus $10^3/T$ for nine $CdGe_xAs_2$ glasses containing from 0.1 to 1.0 mol of germanium. After Nagels et al. (1973).

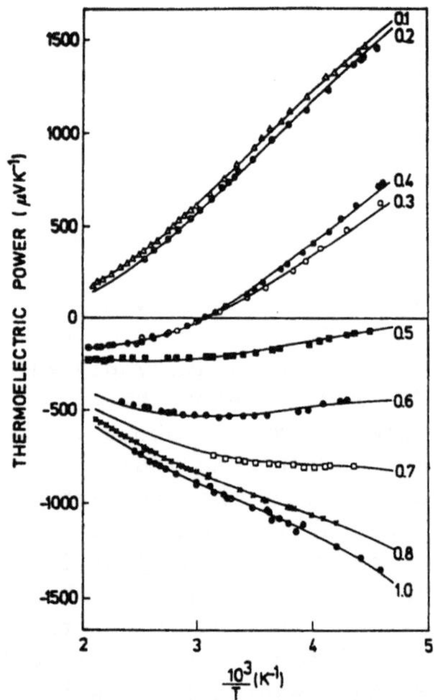

FIGURE 8

Thermoelectric power versus $10^3/T$ for nine $CdGe_xAs_2$ glasses containing from 0.1 to 1.0 mol of germanium. After Nagels et al. (1973).

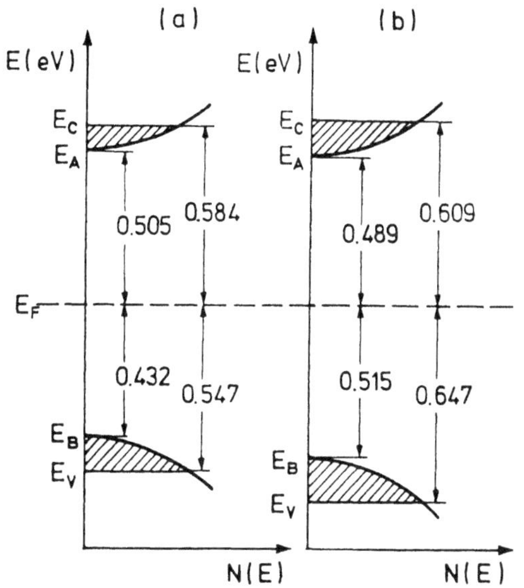

FIGURE 9

Schematic representation of the band structure in (a) p-type $CdGe_{0.1}As_2$ and (b) n-type $CdGe_{1.0}As_2$. After Nagels et al. (1973).

of the analysis, i.e. the parallel conduction of electrons and holes in both the localized and extended states.

6. CONCLUSIONS

Until now much contradiction exists about the fundamental question whether the salient characteristic of the band structure of crystalline solids, i.e. the sharp edges in the density of states forming a well-defined energy gap between valence and conduction bands, is also valid for the amorphous case. Mott and Cohen-Fritzsche-Ovshinsky have developed a model for the non-crystalline state, which assumes that, due to the absence of long-range order, there are tails of localized states at the band extremities. There arc sharp transitions at the energies E_C and E_V where the wave function changes its nature, varying from localized to extended. Because of this change, the mobility should fall at E_C and E_V by several orders of magnitude. The electrical conductivity of many amorphous semiconductors appears to depend exponentially on temperature. According to the Mott-CFO model this dependence should arise from a mobility gap formed between the mobility edges E_C and E_V. The existence of the mobility edges has been a matter of considerable speculation. Nevertheless the theoretical situation on the energy dependence of μ near E_C is unsettled. Obviously more theoretical work needs to be done on this subject. The extent of tailing of the bands also needs further investigation. In general, all of the essential features contained in the Mott-CFO model require more theoretical foundation. It is clear that this is an extremely ambitious task since the amorphous state is more difficult to approach than the perfect crystal. Once the band structure of non-crystalline solids is settled on a firm basis, it will become possible to treat the experimental results with more confidence.

In this article we have reviewed some experimental results on the transport properties of amorphous semiconductors which, in our opinion, can be interpreted in the framework of Mott's model. The temperature dependence of the electrical conductivity, thermopower and Hall effect measured on vitreous $CdGe_xAs_2$ and particularly on some chalcogenide glasses lends some support for the existence of narrow tails of localized states at the band extremities (of the order of 0.1 eV). However, it must be emphasized that other explanations, e.g. the plausibility of small polaron formation, cannot be completely ruled out. It follows that much more experimental work is required to investigate the nature of conduction in non-crystalline solids.

References

[1] J. Klima, T.C. McGill and J.M. Ziman, Discuss. Faraday Soc. 50, 20 (1970)
[2] N.F. Mott, Phil. Mag. 22, 7 (1970)
[3] M.H. Cohen, H. Fritzsche and S.R. Ovshinsky, Phys. Rev. Lett. 22, 1065 (1969)
[4] E.A. Davis and N.F. Mott, Phil. Mag. 22, 903 (1970)
[5] M.H. Cohen, J. Non -Cryst. Solids 4, 391 (1970)
[6] D. Emin, Scottish Universities Summer School in Physics, Aberdeen, Scotland 1972 (P. Le Comber, editor), Academic Press, London and New York
[7] P. Nagels, R. Colson, M. Denayer and R. Gevers, Annual Meeting Belgian Phys. Soc. (1974), Mons, Belgium
[8] N.K. Hindley, J. Non-Cryst. Solids 5, 17 (1970)
[9] L. Friedman, J. Non-Cryst. Solids 6, 329 (1971)
[10] M. Cutler and N.F. Mott, Phys. Rev. 181, 1336 (1969)
[11] R. Callaerts, P. Nagels and M. Denayer, Phys. Lett. 38A, 15 (1972)
[12] C.H. Hurst and E.A. Davis, J. Non-Cryst. Solids 16, 343 (1974)
[13] P. Nagels, M. Denayer and R. Callaerts, Mat. Res. Bull. 6, 1247 (1971)
[14] N.F. Mott, Phil. Mag. 24, 1 (1971)
[15] B.T. Kolomiets and B.V. Pavlov, Sov. Phys.-Semiconduct. 1, 350 (1967)
[16] B.T. Kolomiets, T.F. Mazets and Sh.M.Efendiev, Sov. Phys. - Solid State 12, 514 (1970)
[17] A.A. Vaipolin and E.A. Porai-koshits, Sov. Phys. - Solid State 2, 1500 (1960)

[18] C.H. Seager, D. Emin and R.K. Quinn, J. Non-Cryst. Solids 8-10, 341 (1972)
[19] C.H. Seager, D. Emin and R.K. Quinn, Phys. Rev. B 8, 4746 (1973)
[20] P. Nagels, R. Callaerts and M. Denayer, Proc. Fifth Int. Conf. Amorphous and Liquid Semicond. (1973), Garmisch, Germany, p. 867, Taylor & Francis Ltd., London
[21] L. Cervinka, R. Hosemann and W. Vogel, J. Non-Cryst. Solids 3, 294 (1970)
[22] L. Červinka, A. Hrubý, M. Matyáš, T. Šimeček, J. Skácha, L. Štourač, J. Tauc, V. Vorlíček and P. Höschl, J. Non-Cryst. Solids 4, 258 (1970)
[23] R. Callaerts, M. Denayer, F.H. Hashmi and P. Nagels, Discuss. Faraday Soc. 50,27 (1970)
[24] P. Nagels, R. Callaerts and M. Denayer, Proc. Symposium Amorphous and Vitreous Systems (1973), Charleroi, Belgium, p. N1.

Part IV
Seminars

FUNCTIONAL INTEGRALS

G.J. Papadopoulos

University of Leeds, Leeds LS29JT,UK and
E.S.I.S., Universitaire Instelling Antwerpen
2610 Wilrijk, Belgium

I. INTRODUCTION

It has by now become abundantly clear, from the lectures of Professor Devreese [1] and Professor Thornber [2], that the propagators associated with certain Hamiltonians are of crucial importance in handling transport theory problems.

In dealing e.g. with the polaron problem we need the propagators associated with Hamiltonians H and H', where:

$$H = \frac{1}{2m}\vec{p}^2 + \sum_{\vec{k}} \hbar\Omega a^+_{\vec{k}} a_{\vec{k}} + \sum_{\vec{k}} (V_{\vec{k}} e^{i\vec{k}\cdot\vec{x}} a_{\vec{k}} + V^*_{\vec{k}} e^{-i\vec{k}\cdot\vec{x}} a^+_{\vec{k}}) \tag{1a}$$

$$H' = H - e\vec{E}\cdot\vec{x} \tag{1b}$$

A very good way of obtaining these propagators (and to a great extent explicitly) is by the method of path or functional integration. On this method we intend to expound in this lecture. In fact this will complement Professor Devreese's introductory lecture on Path Integrals. However, here we shall concentrate on the operational side of the method rather than on any other aspect.

Before we work with the problem of the polaron Hamiltonian, involving first and second quantization, it would be helpful if we dealt with simpler situations, and this for the purpose of establishing certain notions.

II. FEYNMAN PATH INTEGRALS

Let us consider a system with Hamiltonian

$$H(\vec{x}) = -\frac{\hbar^2}{2m}\frac{\partial^2}{\partial \vec{x}^2} + U(\vec{x}) \tag{2.1}$$

The problem we are faced with is to find the evolution in time of a wavefunction, which at a given time (say t = 0) is given through $\Psi_0(\vec{x})$.

The answer to the problem is obtained by solving Schrödinger's equation:

$$\{i\hbar\frac{\partial}{\partial t} - H(\vec{x})\}\Psi(\vec{x},t) = 0 \tag{2.2a}$$

with the initial condition:

$$\Psi(\vec{x},0) = \Psi_0(\vec{x}) \tag{2.2b}$$

In terms of the evolution, $\exp(-iHt/\hbar)$, the solution is given by:

$$\Psi(\vec{x},t) = \exp\{-\frac{i}{\hbar}H(\vec{x})t\}\Psi_0(\vec{x}) \tag{2.3}$$

This clearly satisfies the Schrödinger equation, together with the appropriate initial condition.

In (2.3) the evolution operator acts locally on the initial wavefunction, which it transforms in time, so that it obeys the Schrödinger equation. If we wish, now, to disengage the process of propagation from the particular content of the wavefunction we rewrite (2.3) as follows:

$$\Psi(\vec{x},t) = \int \exp\{-\frac{i}{\hbar} H(\vec{x})t\}\delta(\vec{x}-\vec{x}')\Psi_0(\vec{x}')d\vec{x}'$$

$$= \int K(\vec{x}t|\vec{x}'0)\Psi_0(\vec{x}')d\vec{x}' \qquad (2.4a)$$

The kernel:

$$K(\vec{x}t|\vec{x}'0) = \exp\{-\frac{i}{\hbar} H(\vec{x})t\}\delta(\vec{x}-\vec{x}') \qquad (2.4b)$$

is the propagator in coordinate representation. It supplies the wavefunction at time t (not necessarily later) from the information carried by the wavefunction at t = 0.

Since

$$\delta(\vec{x}-\vec{x}') = \sum_{\vec{k}} \langle\vec{x}|\vec{k}\rangle\langle\vec{k}|\vec{x}'\rangle = \langle\vec{x}|\vec{x}'\rangle$$

where all $\langle\vec{x}|\vec{k}\rangle$ form a complete set of wavefunctions, (2.4b) can take the form:

$$K(\vec{x}t|\vec{x}'0) = \exp\{-\frac{i}{\hbar} H(\vec{x})t\}\langle\vec{x}|\vec{x}'\rangle = \langle\vec{x}|\exp\{-\frac{i}{\hbar} H(\vec{x})t\}|\vec{x}'\rangle \qquad (2.4c)$$

Now, one way to obtain an explicit expression for the propagator is to operate as follows:

$$K = \{1 - \frac{i}{\hbar} Ht + \frac{1}{2!} (-\frac{i}{\hbar})^2 H^2 t^2 + \ldots\} \delta(\vec{x} - \vec{x}')$$

This is an additive procedure based on a series expansion of the evolution operator.

Another way to go about it is via a multiplicative process based on a product expansion of the evolution operator as follows:

$$K \simeq \underset{1}{(1 - \frac{i}{\hbar} H\frac{t}{N})} \underset{2}{(1 - \frac{i}{\hbar} H\frac{t}{N})} \ldots \underset{N}{(1 - \frac{i}{\hbar} H\frac{t}{N})} \delta(\vec{x} - \vec{x}') \qquad (2.5)$$

If we take the limit as $N \to \infty$ we have the required propagator, since:

$$\lim_{N \to \infty} (1 - \frac{i}{\hbar} H\frac{t}{N})^N = \exp\{-\frac{i}{\hbar} Ht\}$$

The multiplicative procedure is the one employed in path integral constructions.

Next, before we go any further notice that by adding a term of $O\{(\Delta t)^2\}$ $(\Delta t = t/N)$ to any one of the factors $(1 - \frac{i}{\hbar} H\Delta t)$ in (2.5) the limit of the product as $N \to \infty$ is not affected. So we can obtain our propagator as a limiting case of the matrix element:

$$\langle \vec{x} | (1 - \frac{i}{\hbar} H\Delta t)(1 - \frac{i}{\hbar} H\Delta t) \ldots (1 - \frac{i}{\hbar} H\Delta t) | \vec{x}' \rangle$$

Inserting complete sets of wavefunctions between the various operators we have:

$$K_N(\vec{x}t|\vec{x}'0) = \int <\vec{x}|1-\frac{i}{\hbar}H(\vec{x})\Delta t|\vec{x}_{N-1}><\vec{x}_{N-1}|1-\frac{i}{\hbar}H_{N-1}\Delta t|\vec{x}_{N-2}>\dots$$

$$\dots<\vec{x}_2|1-\frac{i}{\hbar}H_2\Delta t|\vec{x}_1><\vec{x}_1|1-\frac{i}{\hbar}H_1\Delta t|\vec{x}'>d\vec{x}_1 d\vec{x}_2\dots d\vec{x}_{N-1} \tag{2.6}$$

We have in (2.6) N short-time propagators of the form:

$$<\vec{x}_{j+1}|1-\frac{i}{\hbar}H_{j+1}\Delta t|\vec{x}_j> = (1-\frac{i}{\hbar}H_{j+1}\Delta t)\delta(\vec{x}_{j+1}-\vec{x}_j) \tag{2.7}$$

but (N-1) 3D integrations.

To obtain the form of the Feynman path integral we employ the plane-wave decomposition of the identity transformation, i.e.:

$$\delta(\vec{x}_{j+1}-\vec{x}_j) = \frac{1}{(2\pi)^3}\int d\vec{k}\,\exp\{i\vec{k}\cdot(\vec{x}_{j+1}-\vec{x}_j)\} \tag{2.8}$$

The decomposition is done using essentially the eigenfunctions of tho kinetic energy operator.

Putting the Hamiltonian (2.1) into (2.7), we have with the aid of (2.8) an approximate expression for the short-time propagator, as:

$$<\vec{x}_{j+1}|1-\frac{i}{\hbar}\{-\frac{\hbar^2}{2m}\frac{\partial^2}{\partial\vec{x}^2_{j+1}}+U(\vec{x}_{j+1})\}\Delta t|\vec{x}_j>$$

$$= \frac{1}{(2\pi)^3}\int d\vec{k}[1-\{i\frac{\hbar\ \vec{k}^2}{2m}+\frac{i}{\hbar}U(\vec{x}_j)\}\Delta t]\exp\{i\vec{k}\cdot(\vec{x}_{j+1}-\vec{x}_j)\} \tag{2.9}$$

Notice that on the r.h.s. of (2.9) $\vec{x}_{j+1}$ in $U(\vec{x}_{j+1})$ was replaced by $\vec{x}_j$, on account of the δ-function accompanying it.

As pointed out earlier we can add to the short-time operators of (2.6) any terms of order higher than Δt, without this affecting the limit; as long as our short-time propagators are correct to first order in Δt we are o.k. So, we can replace in (2.9) the term in the angular brackets by an exponential, which has the same expansion to order Δt.

i.e.: $\langle\vec{x}_{j+1}|1 - \frac{i}{\hbar} H_{j+1}\Delta t|x_j\rangle$ is replaced by:

$$\frac{1}{(2\pi)^3}\int d\vec{k}\,\exp\{-i\frac{\hbar\,k^2}{2m}\Delta t - \frac{i}{\hbar}U(\vec{x}_j)\Delta t\}\exp\{i\vec{k}\cdot(\vec{x}_{j+1}-\vec{x}_j)\}$$

$$= \left(\frac{m}{2\pi i\hbar\Delta t}\right)^{3/2}\exp\left[\frac{i}{\hbar}\left\{\frac{m}{2}\left(\frac{\vec{x}_{j+1}-\vec{x}_j}{\Delta t}\right)^2 - U(\vec{x}_j)\right\}\Delta t\right] \tag{2.10}$$

Making the replacements (2.10) in (2.6) we have for the approximate propagator the expression:

$$K_N'(\vec{x}t|\vec{x}'0) = \int \exp\left[\frac{i}{\hbar}\sum_{j=0}^{N-1}\left\{\frac{m}{2}\left(\frac{\vec{x}_{j+1}-\vec{x}_j}{\Delta t}\right)^2 - U(\vec{x}_j)\right\}\Delta t\right]$$

$$\times\left(\frac{m}{2\pi i\hbar\Delta t}\right)^{3/2}\prod_{j=1}^{N-1}\left(\frac{m}{2\pi i\hbar\Delta t}\right)^{3/2} d\vec{x}_j \tag{2.11}$$

with: $\vec{x}_0 = \vec{x}'$, $\vec{x}_N = \vec{x}$.

Notice that in the (2.11) we have the Lagrangian (in discrete form) of our problem revealed.

The product accompanying the exponential on the r.h.s. of (2.11) is the path differential $D[\vec{x}_1,\vec{x}_2,\ldots,\vec{x}_{N-1}]$. It contains the right normalizing factors (measure of integration) for obtaining the propagator, through the multiple process of integration, in the limit of infinite subdivision of the interval [0,t]. In other words, in the limit as $N\to\infty$, the sequence $\{K'_N\}$ goes to the required propagator

$$K'_N(\vec{x}t|\vec{x}'0) \to K(\vec{x}t|\vec{x}'0)$$

The above method of path integral construction is essentially Hamiltonian based, and the steps (2.5) - (2.11) originated from Abé [3].

In the limit of infinite refinement of the partitions of the time interval [0,t] we make use of a notation, naturally emanating from (2.11), and indicating the multiple integrations (integration over paths). We have:

$$K(\vec{x}t|x'0) = \int \exp[\frac{i}{\hbar}\int_0^t \{\frac{m}{2}\dot{\vec{x}}^2(\tau)-U(\vec{x}(\tau))\}d\tau]D[\vec{x}(\tau)] \tag{2.12a}$$

with: $\vec{x}(0) = \vec{x}'$, $\vec{x}(t) = \vec{x}$, and where:

$$D[\vec{x}(\tau)] = (\frac{m}{2\pi i\hbar d\tau})^{3/2} \prod_{0<\tau<t} (\frac{m}{2\pi i\hbar d\tau})^{3/2} d\vec{x}(\tau) \tag{2.12b}$$

Examples of this type of integration have been dealt with by Professor Devreese, and we shall not proceed in this direction.

Here we just like to draw a picture depicting the axes of our multidimensional space of integration, appropriate for a one-dimensional space per variable.

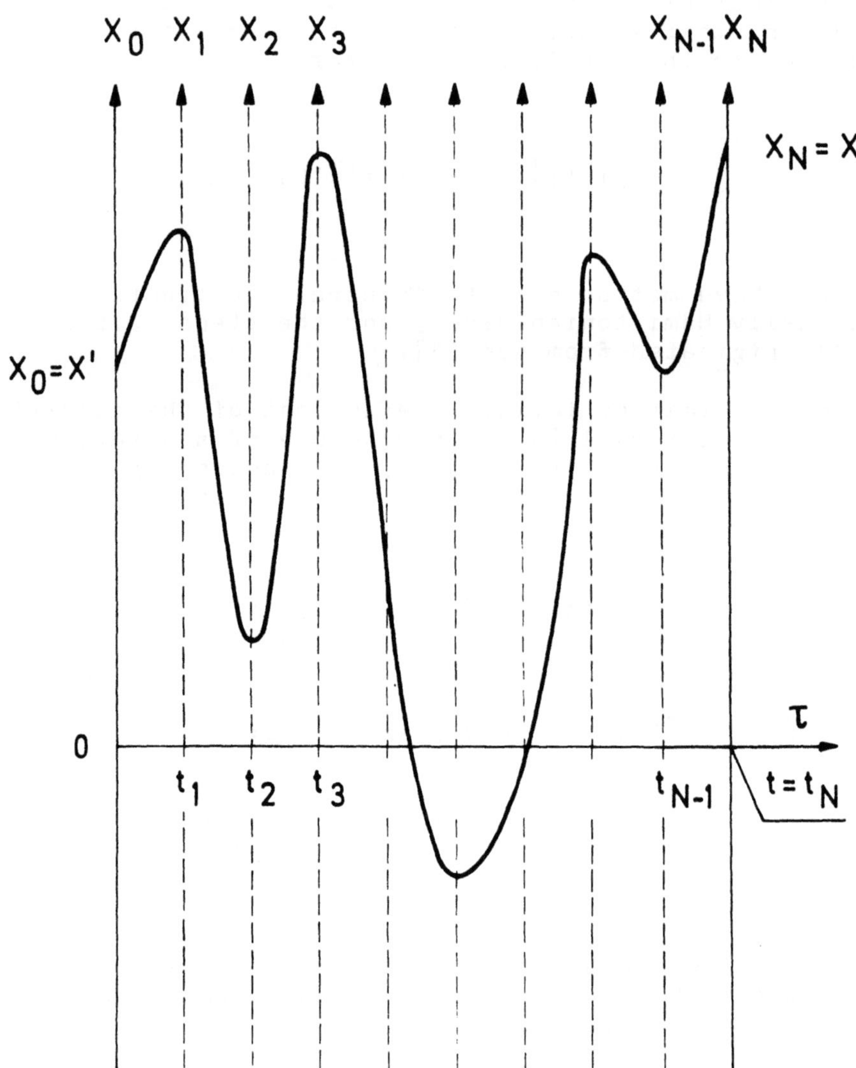

Fig.: 1 Shows the multidimensional space of integration for one-dimensional motion.

The one-dimensional motion analogue of (2.11) is an (N-1)-fold integral and the range of each of its variables $x_1, x_2, \ldots, x_{N-1}$ is from $-\infty$ to $+\infty$; these variables take their values independently of each other. Let us then freeze one particular set of these values and see what picture we get. We get a set of points each of which lies on a different axis of the set of axes $x_0, x_1, x_2, \ldots, x_N$, as shown in the figure.

If we connect these points by a continuous curve we get an idea of the continuous paths entering the process of multiple integration. Actually, this is somewhat illusive, for the continuous paths are not by any means all the paths involved in this integration; for imagine a sequence of denser and denser partitions of [0,t], which implies that any two consecutive axes are separated from each other by a smaller and smaller distance, tending to zero. Now since the variables of integration take any values from $-\infty$ to $+\infty$ (which are represented by points of the various axes) this means that the representative points on two consecutive axes (which tend to coalesce) can be separated by an appreciable distance. This sort of situation produces in the limit of infinite, time interval, subdivision a broken path. In fact our integration involves all imaginable paths which cross once only each of the axes $x(\tau)$.

Notice that we managed to give the essentials for path integral constructions, keeping sophistication at a low key; we have not even mentioned the functional, though by rights it is unquestionably a highly related notion. You see things are not so complicated, after all !

III. FUNCTIONAL INTEGRALS WITH BOSE OPERATORS

Our Fröhlich Hamiltonian was partly second quantized (Boson operators of the polarization field) and partly Schrödinger (first) quantized (electron position and momentum operators).

In the previous section we have dealt with the Feynman path integral giving the propagator associated with a first quantized Hamiltonian. In this section we shall develop functional integrals for second quantized Hamiltonians with Bose operators. To obtain an expression for the propagator associated with the Fröhlich Hamiltonian we have to employ both types of functional integrals.

For the purpose of developing our functional integral we shall simplify matters and deal with the following simple Hamiltonian

$$H = \hbar\Omega a^{+}a \tag{3.1}$$

with the annihilation, and creation operators a, a^{+} obeying the commutation relation

$$[a,a^{+}] = 1$$

In the occupation number representation, the effects of these operators on the state with n particles, $|n\rangle$, as is well known, are:

$$a|n\rangle = n^{1/2}|n-1\rangle, \quad a^{+}|n\rangle = (n+1)^{1/2}|n+1\rangle$$

from which we have the number operator $a^{+}a$, through

$$a^{+}a|n\rangle = n|n\rangle$$

Talking about occupation numbers, this is the natural representation system for our Hamiltonian (3.1), for in this system H becomes diagonal, and so the required propagator can be most easily obtained, in this representation, as:

$$\langle m|\exp(-\frac{i}{\hbar}\hbar\Omega a^{+}at)|n\rangle = \exp(-i\Omega nt)\delta_{nm}$$

but we shall pretend that we have no knowledge of this simple fact here, and proceed on purpose to make things a little complicated.

We shall use the system of coherent states

$$|\alpha\rangle = \frac{1}{\sqrt{\pi}} \sum_{n=0}^{\infty} \frac{\alpha^n}{\sqrt{n!}} e^{-\frac{1}{2}|\alpha|^2} |n\rangle \qquad (3.2a)$$

for our representation.

Where α labelling the state $|\alpha\rangle$ is complex. Before we go any further it might be an idea to assemble together a few needed facts about these states. More about coherent states and functional integrals is included in references [4].

The bra form of the coherent state $|\alpha\rangle$ is given by:

$$\langle\alpha| = \frac{1}{\sqrt{\pi}} \sum_{m=0}^{\infty} \frac{\bar{\alpha}^m}{\sqrt{m!}} e^{-\frac{1}{2}|\alpha|^2} \langle m| \qquad (3.2b)$$

where $\bar{\alpha}$ denotes the complex conjugate of α.

The coherent state $|\alpha\rangle$ is an eigenfunction of the annihilation operator a with eigenvalue α:

$$a|\alpha\rangle = \alpha|\alpha\rangle \quad \text{or} \quad \langle\alpha|a^+ = \langle\alpha|\bar{\alpha} \qquad (3.2c)$$

The scalar product of two coherent states is given by:

$$\langle\alpha|\alpha'\rangle = \frac{1}{\pi} \exp\{\bar{\alpha}\alpha' - \frac{1}{2}|\alpha|^2 - \frac{1}{2}|\alpha'|^2\} \qquad (3.2d)$$

which means that two different such states are not orthogonal.

The coherent states obey the completeness relation:

$$\int \langle\lambda|\alpha\rangle\langle\alpha|\mu\rangle \, d^2\alpha = \langle\lambda|\mu\rangle \tag{3.2e}$$

where: $d^2\alpha = d(\mathrm{Re}\alpha)d(\mathrm{I_m}\alpha)$

Finally, we shall need the following integral identity:

$$\int \exp(-\gamma|\alpha|^2+\lambda\bar{\alpha}+\mu\alpha)\frac{1}{\pi} \, d^2\alpha = \frac{1}{\gamma} \exp(\frac{\lambda\mu}{\gamma}) \tag{3.2f}$$

where $\mathrm{Re}\gamma > 0$.

<u>The propagator</u>: We are looking for the propagator

$$\langle\alpha|\exp(-\frac{i}{\hbar} Hs)|\alpha'\rangle$$

in terms of the coherent states. The parameter s could be the time t or $-i\hbar\beta$ to get matrix elements of the Boltzmann factor, $\exp(-\beta H)$, or could be a complex parameter a in Professor Langreth's lectures [5]. We shall, nevertheless, call it time.

For our simple example we have, by analogy to (2.6):

$$\langle\alpha|\exp(-i\Omega a^+ as)|\alpha'\rangle \approx \int \langle\alpha|1-i\Omega a^+ a\Delta s|\alpha_{N-1}\rangle$$

$$\times \langle\alpha_{N-1}|1-i\Omega a^+ a\Delta s|\alpha_{N-2}\rangle \ldots$$

$$\ldots\langle\alpha_2|1-i\Omega a^+ a\Delta s|\alpha_1\rangle\langle\alpha_1|1-i\Omega a^+ a\Delta s|\alpha'\rangle$$

$$\times \, d^2\alpha_1 d^2\alpha_2 \ldots d^2\alpha_{N-1} \tag{3.3}$$

where $\Delta s=s/N$ of the line joining 0 with s.

In (3.3) we have N short-time propagators, but N-1 integrations.

With the aid of (3.2c) a typical short-time propagator takes the form:

$$<\alpha_{j+1}|1-i\Omega a^+ a\Delta s|\alpha_j> = (1-i\Omega\bar{\alpha}_{j+1}\alpha_j\Delta s)<\alpha_{j+1}|\alpha_j>$$

$$= \frac{1}{\pi}\exp\{(1-i\Omega\Delta s)\bar{\alpha}_{j+1}\alpha_j - \frac{1}{2}|\alpha_{j+1}|^2 - \frac{1}{2}|\alpha_j|^2\}+O\{(\Delta s)^2\} \tag{3.4}$$

We have added $O\{(\Delta s)^2\}$ on the far r.h.s. of (3.4) in order to remove the excess terms of higher order in Δs (and thus maintain the equality with the l.h.s.) introduced through the replacement of $(1-i\Omega\bar{\alpha}_{j+1}\alpha_j\Delta s)$ by the exponential $\exp(-i\Omega\bar{\alpha}_{j+1}\alpha_j\Delta s)$.

As pointed out earlier on the typical short-time propagator, appearing in the (approximate) composition law (3.3), which is exact to first order in Δs can be replaced by another one again exact to first order in Δs without this affecting the limit as $\Delta s\to 0$, which gives the exact propagator $<\alpha|\exp(-i\Omega a^+ as)|\alpha'>$.

With the above in mind let us replace the short-time propagators on the r.h.s. of (3.3) by the exponentials on the far r.h.s. of (3.4). We have:

$$<\alpha|\exp(-i\Omega a^+ as)|\alpha'>$$

$$\approx \int\exp\{-\frac{1}{2}|\alpha'|^2-|\alpha_1|^2+(1-i\Omega\Delta s)\bar{\alpha}_1\alpha'-|\alpha_2|^2+(1-\Omega\Delta s)\bar{\alpha}_2\alpha_1 +$$

$$\ldots-|\alpha_{N-1}|^2+(1-i\Omega\Delta s)\bar{\alpha}\alpha_{N-1} -\frac{1}{2}|\alpha|^2\}\frac{1}{\pi}\prod_{j=1}^{N-1}\frac{d^2\alpha_j}{\pi} \tag{3.5a}$$

The integrations over the α_j's can be performed one after the other, with the aid of the integral identity (3.2f), and the result is:

$$\langle\alpha|\exp(-i\Omega a^+ as)|\alpha'\rangle$$

$$\approx \frac{1}{\pi} \exp\{-\frac{1}{2}|\alpha|^2 + (1-i\Omega \frac{S}{N})^N \bar{\alpha}\alpha' - \frac{1}{2}|\alpha'|^2\} \qquad (3.5b)$$

which in the limit as $N\to\infty$ leads to our propagator in terms of coherent states:

$$\langle\alpha|\exp(-i\Omega a^+ as)|\alpha'\rangle = \frac{1}{\pi} \exp\{-\frac{1}{2}|\alpha|^2 + e^{-i\Omega s} \bar{\alpha}\alpha' - \frac{1}{2}|\alpha'|^2\} \qquad (3.5c)$$

Let us now work out a simple example by obtaining the average occupation number for a system of free phonons or photons.

The density matrix ρ in terms of the coherent states is given by:

$$\langle\alpha|\rho|\alpha'\rangle = \{\int\langle\alpha|\exp(-i\Omega a^+ as)|\alpha\rangle d^2\alpha\}^{-1}\langle\alpha|\exp(-i\Omega a^+ as)|\alpha'\rangle$$

$$= \{1-\exp(-i\Omega s)\}\frac{1}{\pi}\exp\{-\frac{1}{2}|\alpha|^2 + e^{-i\Omega s} \bar{\alpha}\alpha' - \frac{1}{2}|\alpha'|^2\} \qquad (3.6)$$

with: $s = -i\hbar\beta = -i\hbar/kT$

The average occupation number is obtained as:

$$\langle a^+ a\rangle = \mathrm{Tr}\rho a^+ a$$

$$= \int\langle\alpha|\rho|\alpha'\rangle\langle\alpha'|a^+ a|\alpha\rangle d^2\alpha' d^2\alpha \qquad (3.7a)$$

which with the aid of (3.6) and $<\alpha'|a^+a|\alpha> = \bar{\alpha}'\alpha<\alpha'|\alpha>$ leads to:

$$<a^+a> = \frac{1}{\exp(\frac{\hbar\Omega}{kT})-1} \tag{3.7b}$$

the well-known Bose function.

Before we deal with more general Hamiltonians we wish to rearrange the terms in the exponent of (3.5a). We shall initially concentrate our attention on the expression deriving from the product of the scalar products of the form $<\alpha_{j+1}|\alpha_j>$ (see(3.4)).

We have the following three equivalent expressions:

$$<\alpha_N|\alpha_{N-1}><\alpha_{N-1}|\alpha_{N-2}>\ldots<\alpha_3|\alpha_2><\alpha_2|\alpha_1><\alpha_1|\alpha_0> \prod_{j=1}^{N-1} d^2\alpha_j$$

$$= \exp[\frac{1}{2}\sum_{j=0}^{N-1}\{\alpha_j(\bar{\alpha}_{j+1}-\alpha_j)-\bar{\alpha}_{j+1}(\alpha_{j+1}-\alpha_j)\}]\frac{1}{\pi}\prod_{j=1}^{N-1}\frac{1}{\pi}d^2\alpha_j \tag{3.8a}$$

$$= \exp\{-\frac{1}{2}|\alpha_0|^2 - \sum_{j=0}^{N-1}\bar{\alpha}_{j+1}(\alpha_{j+1}-\alpha_j) + \frac{1}{2}|\alpha_N|^2\}\frac{1}{\pi}\prod_{j=1}^{N-1}\frac{1}{\pi}d^2\alpha_j \tag{3.8b}$$

$$= \exp\{\frac{1}{2}|\alpha_0|^2 + \sum_{j=0}^{N-1}\alpha_j(\bar{\alpha}_{j+1}-\bar{\alpha}_j) - \frac{1}{2}|\alpha_N|^2\}\frac{1}{\pi}\prod_{j=1}^{N-1}\frac{1}{\pi}d^2\alpha_j \tag{3.8c}$$

In the limit infinite subdivision of the interval [0,t] (hope you have no objection to reverting back to straight time!) the notation suggested by these expressions is respectively:

$$\exp[\frac{1}{2}\int_0^t\{\alpha(\tau)\frac{\partial\bar{\alpha}(\tau)}{\partial\tau} - \bar{\alpha}(\tau)\frac{\partial\alpha(\tau)}{\partial\tau}\}d\tau]\frac{1}{\pi}\prod_{0<\tau<t}\frac{1}{\pi}d^2\alpha(\tau) \tag{3.9a}$$

$$\exp\{-\frac{1}{2}|\alpha(0)|^2-\int_0^t \bar{\alpha}(\tau)\frac{\partial\alpha(\tau)}{\partial\tau}d\tau + \frac{1}{2}|\alpha(t)|^2\}\frac{1}{\pi}\prod_{0<\tau<t}\frac{1}{\pi}d^2\alpha(\tau) \tag{3.9b}$$

$$\exp\{\frac{1}{2}|\alpha(0)|^2+\int_0^t \alpha(\tau)\frac{\partial\bar{\alpha}(\tau)}{\partial\tau}d\tau - \frac{1}{2}|\alpha(t)|^2\}\frac{1}{\pi}\prod_{0<\tau<t}\frac{1}{\pi}d^2\alpha(\tau) \tag{3.9c}$$

Although integration by parts transforms the expressions (3.9a,b,c) one from the other, it would be dangerous to begin with them the game of functional integration. The trouble comes when writing back discrete forms from the 'limiting' expressions, for these forms could be different from those given by (3.8a,b,c).

Consider for example (3.9b): The integral, $\int_0^t \bar{\alpha}\frac{\partial\alpha}{\partial\tau}d\tau$, can be interpreted in a different way from the one in (3.8b) namely: $\sum_{j=0}^{N-1}\bar{\alpha}_j(\alpha_{j+1}-\alpha_j)$.

This will certainly lead to erroneous evaluations. The moral is that really, as far as the process of multiple integration is concerned, we have no other choice than using (3.8a,b,c) for interpreting (3.9a,b,c). The reason for having to adhere rigidly to the discrete forms given by (3.8a,b,c) lies in the fact that these expressions constitute the various δ-functions of our problem and which we are not allowed to spoil. However, we have some freedom in the choice of the discrete forms which derive from the Hamiltonian of the problem, but in this case we have to adjust the normalizing factors accordingly.

Consider now a more general Hamiltonian $H(a^+,a)$ which is expressed in its normal form (the creation operators preceeding the annihilation operators) we have:

$$\langle\alpha_{j+1}|H(a^+,a)|\alpha_j\rangle = H(\bar{\alpha}_{j+1},\alpha_j)\langle\alpha_{j+1}|\alpha_j\rangle \tag{3.10}$$

The propagator associated with such a Hamiltonian, as a functional integral, can take the form:

$$\langle\alpha|\exp\{-\frac{i}{\hbar}\int_0^t H(a^+,a)d\tau\}|\alpha'\rangle$$

$$= \int \exp[-\frac{1}{2}|\alpha'| - \int_0^t \{\bar{\alpha}(\tau)\frac{\partial\alpha(\tau)}{\partial\tau} + \frac{i}{\hbar} H(\bar{\alpha}(\tau),\alpha(\tau))\}d\tau + \frac{1}{2}|\alpha|]\ \frac{1}{\pi} \prod_{0<\tau<t} \frac{1}{\pi}\ d^2\alpha(\tau) \tag{3.11}$$

with: $\alpha(0) = \alpha'$, $\alpha(t) = \alpha$.

Again in writing the integral $\int_0^t \bar{\alpha}\dot{\alpha}d\tau$ in (3.11) we have in mind the discrete form (3.8b); the discrete form of the integral $\int_0^t H(\bar{\alpha}(\tau),\alpha(\tau))d\tau$ is obtained by use of (3.10) and is $\sum_{j=0}^{N-1} H(\bar{\alpha}_{j+1},\alpha_j)\Delta\tau$.

<u>Quadratic Hamiltonian with linear terms</u>: Let us now proceed to evaluating the propagator associated with the following Hamiltonian:

$$H_1 = \hbar\Omega a^+a+\bar{\varphi}(t)a^+ +\varphi(t)a \tag{3.12}$$

The solution to this problem can be used for handling the vibrational degrees of freedom in the polaron Hamiltonian (1.1).

To obtain the propagator associated with the Hamiltonian H_1 given in (3.12) we make use of the functional integral (3.11) with H replaced by H_1. With this replacement the integral appearing in (3.11) will be:

$$\int_0^t \{\bar{\alpha}(\tau)\frac{\partial\alpha(\tau)}{\partial\tau} + \frac{i}{\hbar} H_1(\bar{\alpha}(\tau),\alpha(\tau))\}d\tau$$

$$= \int_0^t [\bar{\alpha}(\tau)\frac{\partial\alpha(\tau)}{\partial\tau} + i\Omega\bar{\alpha}(\tau)\alpha(\tau) + \frac{i}{\hbar}\{\bar{\varphi}(\tau)\bar{\alpha}(\tau)+\varphi(\tau)\alpha(\tau)\}]d\tau \tag{3.13}$$

This differs from the corresponding integral of the previous evaluation, where the Hamiltonian were $\hbar\Omega a^+a$, by the linear functional in $\bar{\alpha}$ and α.

A linear transformation which reduces the present evaluation to the previous one can be formed using a particular solution of the Euler-Lagrange equations, obtained, as per usual, by setting the first variational derivative of (3.13) w.r.t. $\bar{\alpha}$ equal to zero, i.e.:

$$\frac{\delta}{\delta\bar{\alpha}(\tau)} \int_0^t \{\bar{\alpha}(\tau')\frac{\partial\alpha(\tau')}{\partial\tau'} + \frac{i}{\hbar}H_1(\bar{\alpha}(\tau'),\alpha(\tau'))\}d\tau' = 0 \tag{3.14a}$$

This leads to the following (classical) equation of motion:

$$\dot{\alpha}(\tau)+i\Omega\alpha(\tau) + \frac{i}{\hbar}\bar{\varphi}(\tau) = 0 \tag{3.14b}$$

Or equivalently its complex conjugate

$$\dot{\bar{\alpha}}(\tau)-i\Omega\bar{\alpha}(\tau) - \frac{i}{\hbar}\varphi(\tau) = 0 \tag{3.14c}$$

We, now, pick up a particular solution of (3.14b), say,

$$A(\tau) = -\frac{i}{\hbar}\int_0^\tau \exp\{-i\Omega(\tau-\tau')\}\bar{\varphi}(\tau')d\tau' \tag{3.14d}$$

and form the linear transformation:

$$\alpha(\tau) = A(\tau)+\eta(\tau) \tag{3.15a}$$

In order to preserve the end conditions on the $\alpha(\tau)$ variables $(\alpha(0)=\alpha'\ ,\ \alpha(t)=\alpha)$, the $\eta(\tau)$ variables must obey the end conditions:

$$\eta(0) = \alpha'\ ,\ \eta(t) = \alpha-A(t) \tag{3.15b}$$

If we put now the transformation (3.15a,b) into (3.13) and use the resulting expression in (3.11) the functional integral giving the required propagator takes (in terms of the new variables, $\eta(\tau)$) the form:

$$\langle\alpha|\exp\{-\frac{i}{\hbar}\int_0^t H_1(a^+,a)d\tau\}|\alpha'\rangle$$

$$= \exp\{-\frac{1}{2}|\alpha'|^2-\bar{A}(t)\eta(t)-\frac{i}{\hbar}\int_0^t \varphi(\tau)A(\tau)d\tau\}$$

$$\times\int \exp[-\int_0^t \{\bar{\eta}(\tau)\dot{\eta}(\tau)+i\Omega\bar{\eta}(\tau)\eta(\tau)\}d\tau]\frac{1}{\pi}\prod_{0<\tau<t}\frac{1}{\pi}\,d^2\eta(\tau) \tag{3.16}$$

For the derivation of (3.16) we have made use of the equation of motion given by (3.14b) and its equivalent complex conjugate form (3.14c), satisfied by A and $\bar{A}$. Furthermore, we did not bother ourselves with the Jacobian of the transformation (3.15), for this is just unity.

The procedure followed for obtaining (3.16) has essential analogies with the one presented in Feynman and Hibbs (6) for the treatment of the forced harmonic oscillator. In fact here the mathematical manipulations are somewhat simpler, for the classical equations of motion are only of first order w.r.t. the time.

To complete the evaluation of our propagator we need to perform the functional integration on the r.h.s. of (3.16), which by now is something we really know how to do. We recall first of all that the integral $\int_0^t \bar{\eta}\dot{\eta}d\tau$ appearing in the functional integral of (3.16) corresponds to the discrete form given by the sum over j in (3.8b) (the role of the α_j is now taken by the η_j). Furthermore, remembering that $i\Omega\bar{\eta}\eta$ comes from the Hamiltonian $\hbar\Omega a^+a$, then by adding the fixed quantity $\{-\frac{1}{2}|\eta(0)|^2+\frac{1}{2}|\eta(t)|^2\}$ to the exponent in our functional integral, we make, according to (3.11) the propagator for the Hamiltonian $\hbar\Omega a^+a$, known from (3.5c). We do not want to alter our functional integral and so we must subtract from the exponent what we have added to it, and in this way we have established for ourselves the formula:

$$\int \exp[-\int_0^t \{\bar{\eta}(\tau)\dot{\eta}(\tau)+i\Omega\bar{\eta}(\tau)\eta(\tau)\}d\tau] \prod_{0<\tau<t} \frac{1}{\pi} d^2\eta(\tau)$$

$$= \exp\{\bar{\eta}(t)\eta(0)\exp(-i\Omega t)-|\eta(t)|^2\} \tag{3.17}$$

It is now a matter of routine substitutions to be made in (3.16) by use of (3.17) and the formulae for A(t), and $\eta(0)$, $\eta(t)$, from (3.14d), and (3.15b) for obtaining the propagator associated with the Hamiltonian H_1 given by (3.12). We have:

$$\langle\alpha|\exp\{-\frac{i}{\hbar}\int_0^t H_1(a^+,a)d\tau\}|\alpha'\rangle$$

$$= \frac{1}{\pi}\exp[-\frac{1}{2}|\alpha'|^2+\bar{\alpha}\alpha'\exp(-i\Omega t)-\frac{1}{2}|\alpha|^2-\frac{i}{\hbar}\bar{\alpha}\int_0^t \exp\{-i\Omega(t-\tau)\}\bar{\varphi}(\tau)d\tau$$

$$-\frac{i}{\hbar}\alpha'\int_0^t \exp(-i\Omega\tau)\varphi(\tau)d\tau-\frac{1}{\hbar^2}\int_0^t\int_0^\tau \exp\{-i\Omega(\tau-\tau')\}\varphi(\tau)\bar{\varphi}(\tau')d\tau d\tau'] \tag{3.18}$$

This completes the lecture. More about these wonderful integrals on another occasion.

REFERENCES

1. J.T. DEVREESE, An Introduction to the Use of Path Integrals in Transport Theory, Proc. Antwerp Summer School (1975), present volume. Eds.: J.T. Devreese and V.E. Van Doren.

2. K.K. THORNBER, Aspects of Linear and Nonlinear Electronic Conduction in Dissipative Media, Proc. Antwerp Summer School (1975), present volume. Eds.: J.T. Devreese and V.E. Van Doren.

3. R. ABE, Busseiron Kenk-yu, 79, 101 (1954)
The original formulation of the Lagrangian based path integrals given by Feynman, can be found in:
R.P. FEYNMAN and A.R. HIBBS, "Quantum Mechanics and Path Integrals", McGraw-Hill (New York, 1965).

4. S.S. SCHWEBER, J. Math. Phys. 3, 831 (1962)
M. REVZEN, Phys. Rev. 185, 337 (1969)
M. REVZEN, A.J.P., 38, 611 (1970)
A. CASHER, D. LURIE and M. REVZEN, J. Math. Phys., 9, 1312 (1969).
R.J. GLAUBER, Phys. Rev. 136, 2766 (1963)
Y. TAKAHASHI and F. SHIBATA, J. Phys. Soc. Japan, 38, 656 (1975)
J.R. KLAUDER, Ann. Phys. (N.Y.), 11, 123 (1960)
B.R. MOLLOW, Phys. Rev., 162, 1256 (1967)
W.H. LOUISSELL, "Radiation and Noise in Quantum Electronics", McGraw-Hill (New York, 1964)
S.F. EDWARDS and K.F. FREED, J. Phys. C, 3, 739 (1970).

5. D. LANGRETH, Linear and Nonlinear Response Theory and Applications, Proc. Antwerp Summer School (1975), present volume. Eds.: J.T. Devreese and V.E. Van Doren.

SOME RECENT FINDINGS IN NOISE THEORY AND THEIR IMPLICATIONS FOR TRANSPORT PROCESSES

K. K. Thornber

Bell Laboratories

Murray Hill, New Jersey 07974 U. S. A.

ABSTRACT

We review the Langevin approach, the neglect of correlations, microscopic sources of noise and their macroscopic approximations, the eindrif method and spontaneous and induced fluctuations, diffusion noise, and the Einstein relation and effective temperatures. By focusing on the underlying physics, we stress the necessity of carefully checking results of noise theories based on mathematical or engineering models. The eindrif method, which provides a means of modelling noise based on the actual physical processes which give rise to the noise, is outlined in some detail. We also indicate how the fluctuating forces which play such a natural role in classical noise theory can be applied to quantum-mechanical treatments of transport.

I. INTRODUCTION

Noise and transport are two very closely related phenomena. Indeed, the Einstein relation[1] and the more general fluctuation-dissipation theorem[2-6] indicate how inseparable these processes really are.[7] It is the purpose of this review to show how the connection between the theory of noise and the theory of transport has been used to enhance our understanding of both.[8-14] We shall focus on only a small portion of each type; however, in the case of noise, where the emphasis is on microscopic fluctuations, we shall find ourselves at the very heart of the problem.[15-18]

Recently a number of results appeared which suggest that a critical review of noise theory as applied to physical problems is in order: The Langevin approach[19] of random forces was found to be inconsistent[6,13] with the fluctuation-dissipation theorem;[2-6] some correlations among induced fluctuations were found to have been ignored;[11] the Einstein relation was found in some cases to have been improperly extended to finite fields and currents;[6,11] the microscopic origin of noise was found to be untreated by the two principal methods used to calculate device noise;[18] noise sources and responses to them were found to be sometimes improperly separated;[6,11,13] some mathematical models and engineering shortcuts were found to be applied to problems without due regard for the laws of nature.[6] To be sure, careful physicists had avoided these pitfalls, mathematicians had concentrated on the stochastic nature of their models and engineers on the device implications and simplicity of theirs. Nonetheless, when these three disciplines were united to solve a problem in solid-state device noise, it was the physical portion of the problem which was usually subordinated. The Langevin approach is an excellent example of this. The physicist, however, must shun such widely used approximations as the Langevin *Ansatz*, the Boltzmann equation, the Markov process, etc., if their use is not justified by the physics of the problem; that is, due regard must be given to the nature of the actual physical processes from which the noise arises.

In discussing the above problems, we base our evaluation ultimately on the results of quantum mechanics, either explicitly as with the fluctuation-dissipation theorem or implicitly as with the eindrif method. From the simplicity of our findings it will be clear that the physicist must give careful consideration to the applicability of any result of standard noise theory derived without careful regard for the underlying physical processes giving rise to the noise and to the physical properties which couple these fluctuations to measurable quantities.

II. AN INCONSISTENCY IN LANGEVIN'S *ANSATZ*

In this section we summarize discussions presented elsewhere[6,13] which explicitly exhibit internal inconsistencies in the Langevin approach.[19] We do this to emphasize the very basic problem associated with treating noise problems too macroscopically, and to make a connection with transport coefficients.

A charged particle in a dissipative medium responds to a probe electric field according to

$$v_e(\omega) = \mu(\omega)e(\omega) \tag{1}$$

where v_e is the expected velocity response to the field e, and μ is the mobility, all at frequency ω. [The function $\mu(\omega)$ is in principle the ac mobility obtained from an exact quantum-mechanical calculation of the system of interest.] According to Langevin we can use (1) to write the equation of motion of the particle including velocity fluctuations in the form

$$j\omega m v(\omega) = -m\zeta(\omega)v(\omega) + qe(\omega) + X(\omega) = F(\omega) \tag{2}$$

where the resistive factor $\zeta(\omega)$ satisfies

$$\mu(\omega) = (q/m)(j\omega + \zeta(\omega))^{-1} \tag{3}$$

and where we have introduced the random force $X(\omega)$ to produce fluctuations in v. The quantity F is the total force on the particle of mass m. In the absence of the probe field e we can solve for the spectral density of X and F in terms of that of v using (2). The results are

$$S_X = |j\omega m + m\zeta(\omega)|^2 S_v, \tag{4}$$

$$S_F = |j\omega m|^2 S_v. \tag{5}$$

However, we know independently from the fluctuation-dissipation (FD) theorem that in thermal equilibrium

$$S_v = (4kT/q)\mathrm{Re}(\mu(\omega)) \tag{6}$$

for $\omega \ll kT/\hbar$. Inserting (6) into (4) and (5) we obtain

$$S_X = 4kTm\mathrm{Re}(\zeta(\omega)) = 4kTq\mathrm{Re}(\mu^{-1}(\omega)), \tag{7}$$

$$S_F = m\omega^2(4kT/q)\mathrm{Re}(\mu(\omega)). \tag{8}$$

From (7) we conclude that unless $\zeta(\omega)$ is independent of frequency, the Langevin force $X(\omega)$ will not be white, and, therefore, cannot be assumed to be so. A similar difficulty arises if in using Boltzmann methods, the scattering rate depends upon particle velocity.[6]

Such difficulties as these arise because of lumping together too much detail into the phenomenological Eq. (2) and then attempting to introduce fluctuations by merely superimposing a random force. Considering the essence of the FD theorem, it is not surprising that in the Langevin sources we must incorporate spectral details characteristic of the dynamical system itself. One must, therefore, treat the problem more microscopically. In the next section we outline the eindrif method[6,13] which avoids the above difficulties while preserving the utility of driving forces in flucutation calculations.

Turning to transport coefficients, we note that (7) and (8) suggest that S_X, the spectral density of the fluctuating forces, is closely related to resistivity, while, except for zero frequency, S_F, that of the total force, is closely related to conductivity. At zero frequency $S_F = 0$, and more sophisticated techniques must be used to determine transport coefficients from force-force correlation functions, as discussed by March.[20] We shall return to the problem of calculating resistivity from S_X calculated quantum mechanically in §6. Here we merely point out as does March for his results, how in the spirit of Langevin such relations can be motivated. Ballentine and Heaney[21] provide a similar motivation for expressions similar to (7) and (8), but do not note the above-mentioned difficulty with the Langevin idea nor its implications for noise theory.

III. THE EINDRIF METHOD

The method introduced in Ref. 6 can be summarized as follows: If we treat a fluctuating system sufficiently microscopically, we find that the origin of the fluctuations is the uncertainty of the times at which the particles comprising the system transfer from one state to another. The origin of statistical correlations between fluctuations in the occupancies of these states is then seen to arise from the dependence of the dynamics governing the transfer on the occupancies of the states. What we assume, therefore, is that we understand the physical processes which govern the behavior of the device in sufficient microscopic detail that all statistical correlations which enter can be understood as arising from the interrelation of the dynamics of these processes. (Indeed it is hard to see how one could understand even the noiseless operation of a device without such knowledge.) We then transfer correlations from the statistical portion of the problem to the dynamical portion, which can be understood without a detailed knowledge of probability theory. In the statistical portion we are left with a number of statistical driving terms, one for each elementary transfer process which enters into the equations of motion. The statistics of these driving terms are simple, independent, and can be determined <u>a priori</u>. Their physical origin is easily understood. We thus avoid not only the complications of the Langevin approach, but some of the difficulties as well. In addition, non-Markovian and nonstationary problems can be treated. The five features of the Langevin method which render it more flexible than the Markov method, as given by Lax,[22] are also applicable to our method. The primary difference between the eindrif method and those of others is the emphasis placed on the physical features of the problem at a microscopic level.[18]

The elementary independent driving force (eindrif) method was discussed in some detail in Ref. 6 and summarized briefly in Ref. 13. For our purposes here let us outline the essence of the approach. While this method originally arose from the desire to treat linear nonstationary noise in a simple manner,[12,15,16] we shall see that it might well have arisen from the desire to avoid the inconsistency presented in the previous section. We stress at the outset that we are only concerned with linear noise; however, we cover the entire spectrum from Markov problems where fluctuations occur very rapidly to quasi-static problems where they occur very slowly. The extention to nonlinear problems, while possible, is nonobvious. We consider only classical noise; however, the method applies equally well to systems either in thermal equilibrium or far from equilibrium, steady-state or transcient.

Focusing on statistical processes at a microscopic level greatly simplifies two problems associated with calculating noise. On the one hand the physical origin of the noise is readily apparent, and hence the statistical properties of the fluctuations are most easily determined. On the other hand, and this is the heart of the matter, it becomes possible to distinguish and hence separate the spontaneous portion of the fluctuations from the induced portion. Such a separation is both convenient and crucial. Convenient because being spontaneous, this portion is independent of all other fluctuations; being induced, this portion depends upon the spontaneous portion and would be zero in the latter's absence. Since the spontaneous portions can be separated out, they can be represented as driving terms for the fluctuations. Being spontaneour, each driving term will be independent of all others, substantially simplifying the problem of statistical correlations, which are now transformed to the dynamical part of the problem from the statistical part. Being elementary, the statistics of the driving terms are readily calculated. The problem of expressing flucuations in terms of these elementary, independent driving functions (eindrifs) is considered below. Being able to distinguish between the spontaneous and induced portions of fluctuations permits a critical assessment of the treatment of correlations in other methods.[11] Once these portions are clearly distinguished and separated, it becomes possible to treat some device noise at a more macroscopic level.[23,24]

The eindrif method was motivated and derived in Ref. 6. Let us reivew the main points in the derivation. We regard the particles, usually electrons and/or holes, to be in a number of states. From the statistics of the occupation of these states, all noise properties can be derived in much the same way that noiseless properties can be derived from the noiseless occupation of these states.

States can be trapping sites, elements of position-velocity, phase space, etc. The occupation of a given state fluctuates because of the stochastic nature of the flow of particles out to other states and in from other states. Consider the flow into state i from state j. Then, if N_i is the occupancy of state i, the kinematic equation expressing this flow is simply

$$\dot{N}_i = \Sigma_k f_{ij}(t-t_{ijk}) + \ldots \tag{9}$$

where the function $f_{ij}(t)$, often just a delta function $\delta(t)$, characterizes the change in occupation of state i due to transfers from j. f_{ij} can be stochastic, its statistics being determined by the noiseless operation of the device. The "..." refers to kinematic terms of a similar form for transfers from, and with a minus sign to, other states. The t_{ijk} are the times at which a particle transfers from state j to state i and are stochastic variables. Treating one term will suffice as the others are treated similarly.

We now transform (9) into a dynamical expression from which the noise can be extracted. We do this by writing (9) in the form

$$\dot{N}_i = [R_{ij}(\underset{\sim}{N}) + d_{ij}(t)] + \ldots \tag{10a}$$

where we define $d_{ij}(t)$ by

$$d_{ij}(t) \equiv \Sigma_k f_{ij}(t-t_{ijk}) - R_{ij}(\underset{\sim}{N}). \tag{10b}$$

Here $R_{ij}(\underset{\sim}{N})$ is a functional of $\underset{\sim}{N}$ governing the rate of transfer of particles to i from j. The ith component of the vector $\underset{\sim}{N}$ is, of course, N_i. While (10a,b) may appear to be somewhat tautological, in fact it is just the decomposition which we seek. For suppose the d_{ij} were all zero. Then the (10a) all reduce to equations solved by the noiseless solutions $\underset{\sim}{N}^o(t)$. Thus, writing $\underset{\sim}{N}(t) = \underset{\sim}{N}^o(t) + \underset{\sim}{n}(t)$, where $\underset{\sim}{n}(t)$ is the fluctuation in occupancy from the noiseless occupation, we note that $R_{ij}(\underset{\sim}{N})$ can contain only the induced portion of the fluctuations, along with the noiseless part of the transfer. (If it contained any spontaneous noise sources, the noise in $\underset{\sim}{N}$ would not vanish with the d_{ij}.) But the first term on the right-hand side of (10b) contains both the spontaneous and induced portions of the fluctuations, as well as the noiseless part of the transfer. On taking the difference in (10b), therefore, we are left with only the spontaneous portion of the fluctuations for our $d_{ij}(t)$. Thus the $d_{ij}(t)$ are the eindrifs which we seek: by the arguments given above they are all <u>independent</u> driving terms, each d_{ij} being uncorrelated statistically with any other $d_{k\ell}$. From a knowledge of the statistics of the $d_{ij}(t)$ we can determine from (10a) the statistics of $\underset{\sim}{n}(t)$.

The final step in setting up the eindrif method is to determine the statistics of the $d_{ij}(t)$. Here we make an approximation valid rigorously only for linear noise. We assume that the statistics of the $d_{ij}(t)$ can be approximated by the statistics of another stochastic variable $d^{o}_{ij}(t)$ which we define by

$$d^{o}_{ij}(t) \equiv \Sigma_k f_{ij}(t-t^{o}_{ijk}) - R^{o}_{ij}(t) \tag{11}$$

where $R^{o}_{ij}(t) \equiv R_{ij}(\tilde{N}^{o})$. Here the R^{o}_{ij} is a specific function of time, not a stochastic function as in (10b), and consequently the t^{o}_{ijk} are independent, as in pure shot noise. This error in $\langle d_{ij}(t_1)d_{ij}(t_2)\rangle$, $\langle d_{ij}(t_1)d_{ij}(t_2)\rangle - \langle d^{o}_{ij}(t_1)d^{o}_{ij}(t_2)\rangle$, is then of the order of $(R_{ij}-R^{o}_{ij})$, that is of the order of the noise. For linear noise this is negligible by definition. The Ansatz (11) can be motivated physically.[6,13] That the correlation functions among the $\tilde{n}$ are identical to those obtained by standard methods for stationary, linear, Markov processes[25,26] indicates the validity of this approach for linear noise. Equation (11) can be used to calculate the complete statistics of the d_{ij}.[6] The most important expressions, however, are these.

$$\langle d_{ij}(t)\rangle = 0 \tag{12a}$$

$$\langle d_{ij}(t_1)d_{k\ell}(t_2)\rangle = R^{o}_{ij}(t_1)\delta_{ik}\delta_{j\ell}\int_{-\infty}^{\infty}\langle f_{ij}(\tau)f_{ij}(\tau+t_2-t_1)\rangle d\tau \tag{12b}$$

Using (12a,b) it is straightforward to determine $\langle\tilde{n}(t)\rangle$ and $\langle\tilde{n}(t_1)\tilde{n}(t_2)\rangle$ by solving (10a) linearized about $\tilde{N}^{o}$ for $\tilde{n}$.

Before proceeding we must call attention to certain features of noise calculations brought out by the eindrif method and stress how this method differs from the Langevin approach and the Markov approach. First note that the $\tilde{\tilde{d}}$-correlation function (12b) depends upon the noiseless solution through $R_{ij}(\tilde{N}^{o})$. Thus the $\tilde{n}$-correlation function will depend upon not only on the ac response function arising from linearizing (10a) but also on the noiseless solution $\tilde{N}^{o}$ through the $\tilde{\tilde{d}}$-function. Thus in general one would not expect to be able to express the noise solely in terms of either ac or noiseless quantities alone.[13] Indeed diagonalizing the linearized equations for $\tilde{n}$ in terms of $\tilde{\tilde{d}}$ does <u>not</u> lead to independent Langevin-like driving terms among the equations.[13] This is because the ac-response functions and the noiseless results do not have the same symmetry properties and hence are not simultaneously diagonalized. This points to the basic difference between the eindrif method and that of Langevin. In the former an eindrif enters in a well-determined physical manner for each microscopic transfer process, it is uncorrelated with all other eindrifs, its statistics are in general nonstationary, and when stationary its spectral density is in general nonwhite; both are predetermined. In the latter, by contrast, a Langevin force enters each macroscopic equation of motion ad hoc, it is correlated with all other such forces, one per equation, its statistics are assumed to be stationary with white spectral density, and its strength and degree of correlation must be calculated by solving the entire set of equations of motion. With respect to the Markov approach, the eindrif method avoids the cusp in the autocorrelation function at $t = 0$:[25] indeed, the first derivative of this function at $t = 0$ is zero as required by microscopic reversibility.[13] In this way a more realistic expression for the retrogression of a fluctuation is obtained. Another distinction for the Markov approach is that the calculation of conditional probabilities is bypassed.

As mentioned above Lax[22] lists five features of the Langevin method which render it more flexible than the Markov method. As these features apply even more strongly to the eindrif method as well, it is worth summarizing. (1) Deterministic coupling between two dissipative systems does not alter the noise sources but can alter considerably the responses to these sources. (2) Finite correlation times which arise in non-Markov processes can be treated. (3) Transforming from one set of variables to another is easier. (4) Only M rather than M^2 equations need be solved. (5) The method is usable even in unstable situations, e.g., in amplifiers and oscillators. The origin of this greater flexibility results from transforming the correlations between various processes out of the statistics and into the dynamics of the problem where they can be handled more easily.

As stressed above, caution must be exercised in treating a noise problem on a level more macroscopic than that of the elementary, physical noise processes. As an example, a situation was presented in a previous paper[13] in which, if one treated the problem macroscopically, one would completely miss the internal noise intrinsic to each microscopic subsystem treated macroscopically as a single state. The concept of internal noise is quite important and easily handled with the eindrif method.[13] Internal noise does not always affect the measured noise, but this occurs only under very severe restrictions (approximations). A feel for this can be obtained from the following example. Consider a system of subsystems, each subsystem consisting of a number of states. Let N_i^I be the occupancy of state i in subsystem I and let each state be coupled to all others in the system. Then we can proceed as follows,

$$\dot{N}_i^I = \Sigma_{Jj}[\Sigma_r \delta(t-t_{ij}^{IJ}(r)) - \Sigma_s \delta(t-t_{ji}^{JI}(s))] \tag{13a}$$

$$= \Sigma_{Jj}[R_{ij}^{IJ}(t) + d_{ij}^{IJ}(t) - R_{ji}^{JI}(t) - d_{ji}^{JI}(t)], \tag{13b}$$

from which it follows at once that

$$\dot{N}_i^{Io} = \Sigma_{Jj}[R_{ij}^{IJo}(t) - R_{ji}^{JIo}(t)] \tag{14a}$$

$$\dot{n}_i^I = \Sigma_{Jj}(r_{ij}^{IJ}n_j^J + d_{ij}^{IJ} - s_{ij}^{IJ}n_j^J - d_{ji}^{JI}) \tag{14b}$$

where

$$r_{ij}^{IJ} \equiv \Sigma_{Kk}\delta R_{ik}^{IK}/\delta N_j^J \tag{15a}$$

$$s_{ij}^{IJ} \equiv \Sigma_{Kk}\delta R_{ki}^{KI}/\delta N_j^J. \tag{15b}$$

[The r and s defined in (15a,b) are in general linear operators on the n.] To cast (14a,b) into macroscopic expressions relating the subsystems we must somehow replace the microscopic occupancies by macroscopic ones. Let $n^I = \Sigma_i n_i^I$. Then we note from (14b) that

$$\dot{n}^I = \Sigma_J(\Sigma_{Kk}r_k^{IJK}n_k^K + d^{IJ} - \Sigma_{Kk}r_k^{JIK}n_k^K - d^{JI}) \tag{16a}$$

$$r_k^{IJK} \equiv \Sigma_{ij}\delta R_{ij}^{IJ}/\delta N_k^K \tag{16b}$$

$$d^{IJ} \equiv \Sigma_{ij}d_{ij}^{IJ}. \tag{16c}$$

If now the elements of a given subsystem are sufficiently uniform so that r_k^{IJK} can be taken to be independent of k for each IJK, then (16a) becomes

$$\dot{n}^I = \Sigma_J(\Sigma_K r^{IJK}n^K + d^{IJ} - \Sigma_K r^{JIK}n^K - d^{JI}) \tag{16d}$$

which would have been obtained by treating each subsystem as a single state. The statistics for the d^{IJ}, obtainable from those of the d_{ij}^{IJ}, are also those which would have been obtained from the same macroscopic model. However, if the r_k^{IJK} do in fact depend on k, then in passing to a macroscopic equation similar to (16d), internal noise[13] will enter and the necessity of starting microscopically is clear. Equation (16d) also implies that if a given state is subdivided so that transfer rates to all other states from each subdivision remain the same but internally arbitrary transfers are permitted between subdivisions, the noise obtained by treating the subdivided state as a single state is unaltered, as physically it must be.

IV. DIFFUSION NOISE, THERMAL EQUILIBRIUM

In Appendix C of Ref. 6 we derived an extension of the Einstein relation, $\mu = qD/kT$, for mobility μ and diffusion constant D which were functions of the electric field, $\mu(E) \equiv v(E)/E$ and $D(E)$. The result was the relation[11]

$$dv(E)/dE = qD(E)/kT, \tag{17}$$

which in the limit of constant μ and D yields the familiar Einstein relation. While the derivation given was correct and quite rigorous, it was presented in a manner not readily understood. Considering the possible significance of the above result, it was felt some clarification was in order. It is the purpose of this section to provide this clarification for thermal equilibrium problems: non-equilibrium problems are treated in the next section. We shall also use this opportunity to generalize a very important result of microscopic noise theory.

The Einstein relation is very important for noise theory since it provides a link between the spectral density of the spontaneous current-current fluctuations within a small volume of the device and the mobility of the carriers. Since this spectral density, denoted by $S_{\underset{\sim}{\eta}}(\underset{\sim}{r}_1,\underset{\sim}{r}_2)$, is given by[6,27-30]

$$S_{\underset{\sim}{\eta}}(\underset{\sim}{r}_1,\underset{\sim}{r}_2) = 4qD(\underset{\sim}{r}_1)\rho(\underset{\sim}{r}_1)\delta(\underset{\sim}{r}_1-\underset{\sim}{r}_2), \tag{18}$$

for thermal equilibrium we can use (17) to obtain

$$S_{\underset{\sim}{\eta}}(\underset{\sim}{r}_1,\underset{\sim}{r}_2) = 4kT\rho(\underset{\sim}{r}_1)\delta(\underset{\sim}{r}_1-\underset{\sim}{r}_2)dv(E)/dE. \tag{19}$$

In (18) q is the elementary carrier charge, $D(\underset{\sim}{r})$ is the diffusion constant at $\underset{\sim}{r}$, and $\rho(\underset{\sim}{r})$ is the charge density at $\underset{\sim}{r}$. Equations (18) and (19) can be used in macroscopic noise expressions which relate the noise induced at the outputs of a device to the microscopic current fluctuations which give rise to this noise. Expression (19) is particularly convenient since the mobility is often much easier to measure than the diffusion constant.

In Section 3 of Ref. 6 very simple derivations of (18) and (19) were presented in order to illustrate our method of treating fluctuations in noise theory. van Vliet has treated microscopic fluctuations in a more rigorous manner;[30] however, his derivation of (18), despite its generality, nonetheless, depends for its validity on the assumption of Markov processes and the validity of the Boltzmann equation. In addition, electron-electron scattering is awkward to treat under these restrictions. We can relax these assumptions by treating the problem entirely quantum mechanically.

As with van Vliet's derivation, no assumption of thermal equilibrium will be made.

Although our derivation is very brief, let us carefully define the problem. One method of calculating device noise is to determine the density matrix for the states of the entire device under the operating conditions of interest. If one could do this readily for devices of interest, (18) would be superfluous. Unfortunately this is usually not possible, so one makes what is probably an excellent approximation for most problems, and one which we shall make as well. In classical language we assume that the spontaneous current fluctuations in one microscopic region are stochastically independent from the spontaneous current fluctuations in all other regions. To be sure induced fluctuations in one region will be coupled to induced fluctuations in other regions. It is only their spontaneous sources which we take to be independent (uncorrelated). In quantum-mechanical language, we assume that we can divide the device up into regions in such a way that all quantum mechanical features can be adequately and relatively easily treated in each region independent of what is going on in other regions. The interactions between regions are then treated macroscopically, usually in terms of Maxwell's equations. One thus divides the problem into a microscopic and a macroscopic part[17,18] and treats each as well as is necessary (or possible). We shall now turn our attention to the microscopic part, the derivation of (18).

To derive (18) we must do two things. We must calculate the spectral density of the current-current fluctuations, and we must express our result in terms of the diffusion constant. If $\hat{\underset{\sim}{x}}_i$ is the position operator of the ith carrier in the region of volume Ω and q is its charge, then the current density operator is

$$\hat{\underset{\sim}{j}}(\underset{\sim}{r}) = \frac{1}{\Omega}\Sigma_i q\dot{\hat{\underset{\sim}{x}}}_i \tag{20}$$

where Σ_i is over all the electrons in the region at $\underset{\sim}{r}$. Thus the correlation function of $\hat{\underset{\sim}{j}}(\underset{\sim}{r})$ is given by

$$\langle\hat{\underset{\sim}{j}}(\underset{\sim}{r}_1)_{t_1}\hat{\underset{\sim}{j}}(\underset{\sim}{r}_2)_{t_2}\rangle = \frac{q^2}{\Omega^2}\Sigma_i(\Sigma_j\langle\dot{\hat{\underset{\sim}{x}}}_{it_1}\dot{\hat{\underset{\sim}{x}}}_{jt_2}\rangle) \qquad \underset{\sim}{r}_1 = \underset{\sim}{r}_2 \tag{21a}$$

$$= 0 \qquad \underset{\sim}{r}_1 \neq \underset{\sim}{r}_2 \tag{21b}$$

where (21b) follows by the assumption discussed above. The meaning of the averaging $\langle\ \rangle$ used here and below is discussed in the following section. Because of the identity of the carriers, the sum over j in (21a) is independent of i, even for interacting particles. Also, in the macroscopic limit where (18) applies, one factor of Ω^{-1} becomes $\delta(\underset{\sim}{r}_1-\underset{\sim}{r}_2)$. Thus (21a,b) reduce to

$$\langle \hat{j}(\underset{\sim}{r}_1)_{t_1} \hat{j}(\underset{\sim}{r}_2)_{t_2} \rangle = q\rho(\underset{\sim}{r}_1)\delta(\underset{\sim}{r}_1 - \underset{\sim}{r}_2)\Sigma_j \langle \hat{\dot{\underset{\sim}{x}}}_{1t_1} \hat{\dot{\underset{\sim}{x}}}_{jt_2} \rangle . \tag{22}$$

The current-current spectral density can be obtained from (22) according to the relations[6]

$$\underset{\approx}{S}_{\underset{\sim}{\eta}}(\underset{\sim}{r}_1, \underset{\sim}{r}_2, \omega) \equiv 4 \operatorname{Re}(\underset{\approx}{K}_{\underset{\sim}{j}\underset{\sim}{j}}(\omega)) \tag{23a}$$

$$\underset{\approx}{K}_{\underset{\sim}{j}\underset{\sim}{j}}(\omega) \equiv \int_0^\infty d\omega \, \underset{\approx}{k}_{\underset{\sim}{j}\underset{\sim}{j}}(t) e^{i\omega t} \tag{23b}$$

$$\underset{\approx}{k}_{\underset{\sim}{j}\underset{\sim}{j}}(t) \equiv \frac{1}{2} \langle [\underset{\sim}{j}_{t_1}, \underset{\sim}{j}_{t_2}]_+ \rangle . \tag{23c}$$

Having determined $\underset{\approx}{S}_{\underset{\sim}{\eta}}$, we now turn to the diffusion constant.

Shockley, et al.,[27] define a frequency-dependent, diffusion constant, which they show gives the usual diffusion constant of microscopic diffusion theory $[d\langle x_t^2 \rangle / dt = 2Dt]$ and of macroscopic transport theory $[J_{(\text{diffusion})} = \partial(Dn)/\partial x]$ (see their Section 6). For interacting carriers, it is straightforward to generalize their definition to obtain a diffusion constant per particle. Quantizing this expression yields the definition

$$\underset{\approx}{D}(r,\omega) \equiv \frac{1}{N(\underset{\sim}{r})} \operatorname{Re} \int_0^\infty dt \; e^{i\omega t} \frac{1}{2} \langle [\Sigma_i \hat{\dot{\underset{\sim}{x}}}_{it}, \; \Sigma_j \hat{\dot{\underset{\sim}{x}}}_{j0}]_+ \rangle \tag{24}$$

where $N(\underset{\sim}{r})$ is the number of carriers at $\underset{\sim}{r}$. [The corresponding classical expression for independent carriers

$$\underset{\approx}{D}(\omega) \equiv \int_0^\infty dt \; e^{i\omega t} \langle \dot{\underset{\sim}{x}}_t \dot{\underset{\sim}{x}}_0 \rangle \tag{25}$$

is equivalent to the definition (Eq. 42) used previously, if in that expression $t_1 \to \infty$ and (t_1+t) remains constant.] Arguments similar to those used by Shockley indicate that $\underset{\approx}{D}(\underset{\sim}{r},0)$ defined in (24) corresponds to the usual constant of microscopic diffusion. We have not determined whether or not (24) reduces to the zero-frequency definition of $\underset{\approx}{D}$ used by van Vliet[30] in the Markov-Boltzmann limit.

It is now straightforward to write the current-current spectral density (23a) in terms of the diffusion constant defined in (24). Inserting (22) into (23c), (23c) into (23b), and (23b) into (23a), and comparing with (24), we obtain

$$\underset{\approx}{S}_{\underset{\sim}{\eta}}(\underset{\sim}{r}_1,\underset{\sim}{r}_2,\omega) = 4q\rho(\underset{\sim}{r}_1)\underset{\approx}{D}(\underset{\sim}{r}_1,\omega)\delta(\underset{\sim}{r}_1-\underset{\sim}{r}_2). \tag{26}$$

This expression is valid for any stationary noise problem whether the device be in thermal equilibrium with no current flow or in a state of nonequilibrium with arbitrary applied fields and currents, and dissipation. (In the latter case velocities and currents introduced above are taken relative to their expected values.) In the limit of zero frequency we recover (18).

We can now turn to the principal purpose of this section - the clarification of the thermal-equilibrium portion of Appendix C of Ref. 6. As stated above, the primary goal of that appendix was to derive (17) in a fully quantum mechanical manner for thermal equilibrium. The key point in that derivation [Eq. (100)] was to note that the dissipation term in the fluctuation-dissipation (FD) theorem was proportional to $dv/dE = d(\mu E)/dE$ and not to $v/E = \mu$ for $\mu = \mu(E)$. The point requiring clarification is that the FD theorem [Eq. (98)] as proven is exact only for thermal equilibrium. It is valid, for example, if the solid is placed in an externally applied, dc electric field such that no current flows.[31] For such a case the mobility $\mu(E)$ would be defined by

$$\mu(E) \equiv \frac{1}{E}\int_0^E \left.\frac{dv}{de}\right|_{E'} dE'$$

where v is the velocity response to the low frequency probe field e.

In writing Appendix C of Ref. 6, however, we had a different problem in mind. There is a tendency to attribute a field-dependent mobility $\mu(E)$ to electron heating, and vice versa. Yet we are acquainted with no rigorous argument requiring such a correspondence. Indeed, experimental results indicating the absence of electron heating while $\mu = \mu(E)$ have already been made known,[32] and the existence of substantial heating while μ remains constant is certainly not inconceivable. Thus suppose that, even though a current is flowing, conditions are such that only a small error would be made if we treated the problem as if thermal equilibrium applied. This is, after all, the same qualification implicitly made, but seldom stated, about all noise results of driven systems which are based in some significant manner on thermal equilibrium expressions. This was the spirit in which Appendix C was written. Thus, when we stated[6] that "we shall rederive the FD theorem here primarily for the purpose of calling attention to its application to cases of steady-state but nonequilibrium conditions," we intended to convey the idea that while we were concerned with what is strictly speaking a nonequilibrium problem ($I \neq 0$), if deviations from equilibrium could be neglected, then our result follows.

Probably more misleading, however, was our use of the words "effective temperature". [See just preceding Eqs. (94) and (98b), and just following (109).] Referring to the third place we used this concept, it should be clear that we required the density matrix to be of the form $\exp(-\beta H_s)$ where $\beta \equiv 1/kT$. Thus to us, effective temperature refers to a genuine temperature: it is "effective" in the sense that it might not be the same as the lattice temperature. Thus we were including the possibility of heated but still approximately thermally distributed carriers. We did not in using "effective" in this sense realize the confusion which would naturally arise with what is more commonly referred to as effective temperature:

$$qD(E) = kT_{eff}\mu(E),$$

an expression used even when electron heating cannot be ignored. The derivation of the FD theorem given in Appendix C does <u>not</u> refer to such problems. On the other hand, in the next section we argue that under certain nonequilibrium conditions we can write

$$qD(E) = kT_{eff}dv(E)/dE$$

where T_{eff} is well defined.

A minor algebraic oversight also detracts from the presentation. Define S(K) and A(X) as follows:

$$S(K_{ij}(\omega)) \equiv \frac{1}{2}\,(K_{ij}(\omega) + K_{ji}(-\omega)) \tag{27a}$$

$$A(X_{ij}(\omega)) \equiv \frac{1}{2i}\,(X_{ij}(\omega) - X_{ji}(-\omega)). \tag{27b}$$

Then replace $\mathrm{Re}(K_{ij}(\omega))$ in (98a), (106), and (115a), and $\mathrm{Im}(X_{ij}(\omega))$ in (98a), (104), and (115b), by $S(K_{ij}(\omega))$ and $A(K_{ij}(\omega))$, respectively. Since Re ≡ S and Im ≡ A for i = j, our derivation of (17) is still valid. Finally, the former treatment[6] assumed the electrons did not interact. Replacing $\hat{r}_i = \hat{x}_i$ with $\hat{r}_i = \Sigma_j\hat{x}_j$ yields (17) including electron-electron interaction.

V. DIFFUSION NOISE, NONEQUILIBRIUM

In the two concluding paragraphs of Appendix C, Ref. 6, we very briefly considered a genuinely nonequilibrium problem. Unfortunately too little attention was given to this discussion, and one almost obvious result was yet to be realized. We commence with a definition of the averaging which we used in deriving (26) and then consider some aspects of nonequilibrium problems. Lax[25] has also considered systems driven away from equilibrium using a macroscopic, Markovian technique.

Although usually associated with quantum statistical mechanics, the concept of the density matrix is actually much more general.[33] If we know that a quantum system is in a certain state $|i\rangle$, then to find the average or expectation value of any physical quantity A, one calculates

$$A = \langle\hat{A}\rangle \equiv \langle i|\hat{A}|i\rangle \tag{28}$$

where $\hat{A}$ is the operator corresponding to A. If, however, the system of interest interacts with an external system, e.g., a heat bath, then in place of (28), we usually define $\langle\hat{A}\rangle$ in terms of the density matrix ρ:

$$A = \langle\hat{A}\rangle \equiv \Sigma_{\alpha\beta} A_{\beta\alpha}\, \rho_{\alpha\beta}/\Sigma_{\alpha}\, \rho_{\alpha\alpha} \tag{29}$$

$$= \mathrm{Tr}(\hat{A}\rho)/\mathrm{Tr}(\rho). \tag{30}$$

The operator-like quantity ρ can be calculated in principle by starting with a known $\rho(t_1)$ at time t_1 and propagating to the time of interest $t_2 > t_1$ according to the dynamical relation

$$d\rho/dt = i[\rho, \hat{H}_s]/\hbar \tag{31}$$

where $\hat{H}_s$ is the Hamiltonian of the system. For statistical mechanical problems, $\rho = \exp(-\beta\hat{H}_s)$ where $\beta = 1/kT$, and T is the temperature of the system. However, we see from the above that we do not have to assume thermal equilibrium conditions in order to calculate expectation values. The basic problems, therefore, of nonequilibrium situations are first to calculate the density matrix ρ and second to evaluate the expectation values of interest.[8-10,34] Examples can be found in the literature.

The foregoing should not only have clarified what we meant in §IV by $\langle\ \rangle$, but also make the last paragraph in Appendix C of Ref. 6 more understandable. All we did there was to explore the consequences of a density matrix $\rho_{\alpha\beta}$ which depends only on the eigenenergy E_α of the state α:

$$\rho_{\alpha\beta} = \delta_{\alpha\delta} f(E_\delta). \tag{32}$$

Again, we are not assuming that this form of $\rho_{\alpha\beta}$ applies in all cases. We only ask what happens if $\rho_{\alpha\beta}$ has the form of (32), either exactly or to a good approximation. As shown previously,[6] it follows at once that

$$\frac{2}{\hbar} S(K_{ij}(\omega)) = \frac{1}{2\pi i} \int_{\sigma-i\infty}^{\sigma+i\infty} ds\ G(s,\omega)(1+e^{s\hbar\omega}) \tag{33a}$$

and

$$A(X_{ij}(\omega)) = \frac{1}{2\pi i} \int_{\sigma-i\infty}^{\sigma+i\infty} ds\ G(s,\omega)(1-e^{s\hbar\omega}) \tag{33b}$$

where

$$G(s,\omega) \equiv \pi F(s)\Sigma_{\ell m} e^{sE_\ell} \langle \ell|\hat{r}_j|m\rangle\langle m|\hat{r}_j|\ell\rangle \delta(\hbar\omega - (E_m - E_\ell)) \tag{33c}$$

and

$$F(s) = \int_0^\infty dE\ f(E)e^{-sE}, \tag{33d}$$

$f(E)$ being zero for $E < 0$. These are genuine nonequilibrium results. In Ref. 6 we stopped here. Now we push on.

If in the above we let $r_i = r_j = \Sigma_i \hat{\dot{x}}_j \equiv \hat{\dot{x}}$ and take the limit $\omega \to 0$ ($\hbar\omega \ll kT_{lattice}$ is really all that is necessary), then we find, using the same manipulations which led to (94) of Ref. 6 [(17) of the present paper], that we can write

$$dv(E)/dE = qD(E)/kT_{eff} \tag{34}$$

and, therefore,

$$S_{\underset{\sim}{\eta}}(\underset{\sim}{r}_1,\underset{\sim}{r}_2) = 4kT_{eff}\rho(r_1)\delta(\underset{\sim}{r}_1-\underset{\sim}{r}_2)dv(E)/dE \tag{35a}$$

where

$$\frac{1}{kT_{eff}} \equiv \beta_{eff} \equiv \frac{\frac{1}{2\pi i}\int_{\sigma-i\infty}^{\sigma+i\infty} ds\ G(s,0)(-s)}{\frac{1}{2\pi i}\int_{\sigma-i\infty}^{\sigma+i\infty} ds\ G(s,0)}\ . \tag{35b}$$

With some further manipulation (35b) becomes

$$\beta_{eff} = \left. \frac{\int_0^\infty dt \; \Sigma_\ell \langle \ell | -\frac{df(E)}{dE} [\hat{\dot{x}}_t, \hat{\dot{x}}_0]_+ | \ell \rangle}{\int_0^\infty dt \; \Sigma_\ell \langle \ell | f(E) [\hat{\dot{x}}_t, \hat{\dot{x}}_0]_+ | \ell \rangle} \right|_{E=\hat{H}_s} . \tag{35c}$$

Note that if $f = \exp(-\beta E)$ then $\beta_{eff} = \beta$. This provides a well-defined T_{eff} in those cases where (32) holds.

The above result, (35c), enables us to say something else about nonequilibrium systems. We set $\hat{r}_i = \hat{r}_j = \hat{\dot{x}}$ in deriving (35c); however, it is clear from our derivation of (33a), (33b) and (33c) above and the FD theorem that we could choose $\hat{r}_i = \hat{r}_j = \hat{r}$ to be <u>any</u> operator whatever. In thermal equilibrium we would find that

$$\beta = \underset{\omega \to 0}{L} \frac{\mathrm{Im}(X_{ii}(\omega))/\omega}{\mathrm{Re}(K_{ii}(\omega))} \tag{36}$$

while away from equilibrium we could define a $\beta_{eff}(\hat{r})$ according to

$$\beta_{eff}(\hat{r}) \equiv \left. \frac{\int_0^\infty dt \; \Sigma_\ell \langle \ell | -\frac{df(E)}{dE} [\hat{r}_t, \hat{r}_0]_+ | \ell \rangle}{\int_0^\infty dt \; \Sigma_\ell \langle \ell | F(E) [\hat{r}_t, \hat{r}_0] \ell \rangle} \right|_{E=\hat{H}_s} \tag{37}$$

so long as (32) holds. It would be a rather remarkable coincidence if now $\beta_{eff}(\hat{r})$ were independent of $\hat{r}$, the operator used to define it. If this is the case, that indeed the effective temperature resulting from the measure of one physical quantity were not that resulting from the measurement of another, the utility of the concept of effective temperature would be greatly restricted. In addition, $\beta_{eff}(\hat{r})$ will in general be a function of position $\tilde{r}$ in the device as well. For systems in which (32) does not hold, the density matrix is even more general, and the possibility of a "universal" effective temperature would seem to be remote indeed.

Returning to (35c), let us examine our expression for β_{eff}. For simplicity let us assume that the carriers are independent, and that they relax in a characteristic time $\tau(E)$, a function of the energy E of the carrier. Then (35c) simplifies to

$$\beta_{eff} = \frac{-\Sigma_\ell \langle \ell | \dot{x}^2 | \ell \rangle \tau(E_\ell) df(E_\ell)/dE_\ell}{\Sigma_\ell \langle \ell | \hat{\dot{x}}^2 | \ell \rangle \tau(E_\ell) f(E_\ell)} \tag{38}$$

where now $\langle \ell | \hat{\dot{x}}^2 | \ell \rangle$ is (2/m) times the kinetic energy K_ℓ of a carrier in the state ℓ with eigenenergy E_ℓ. Let us also assume that $K_\ell = K(E_\ell)$, and let n(E) be the number of states per unit energy. Then (35c) simplifies further, becoming

$$\beta_{eff} = \frac{-\int_0^\infty dE\ n(E)\tau(E)K(E)df(E)/dE}{\int_0^\infty dE\ n(E)\tau(E)K(E)f(E)} . \tag{39}$$

Integrating the numerator of (39) by parts and taking n(0) = 0, again for simplicity, one obtains

$$\beta_{eff} = \frac{\int_0^\infty dE\ f(E)d(n(E)\tau(E)K(E))/dE}{\int_0^\infty dE\ f(E)n(E)\tau(E)K(E)} . \tag{40}$$

We remark that had $\beta_{eff}(\hat{r})$ been simplified for some other operator $\hat{r}$ than the velocity, the kinetic energy K(E) in (40) would be replaced by the expectation value of the mean square of the operator of interest.

The effective temperature given by (40), $\beta_{eff} = 1/kT_{eff}$, has several interesting properties. If (i) the $\tau(E)$ is independent of E so that $\tau(E) = \tau$, a simple collision or relaxation time, (ii) $n(E) \propto E^\delta$, and (iii) $K(E) \propto aE$, then

$$kT_{eff} = a^{-1}(\delta+1)^{-1}\overline{K(E)} = (\delta+1)^{-1}\overline{E} \tag{41a}$$

where we use for the average of a function of E, A(E),

$$\overline{A(E)} \equiv \frac{\int_0^\infty dE\ f(E)n(E)A(E)}{\int_0^\infty dE\ f(E)n(E)} . \tag{41b}$$

For free (or nearly free) electrons, $\delta = -1/2$ in one dimension, $\delta = 0$ in two dimensions, $\delta = 1/2$ in three dimensions, and $a = 1$. Thus, for constant τ, we recover the more intuitive definition of effective temperature, namely that $3/2\ kT_{eff}$ equals the average kinetic energy. (For harmonic oscillators, $\delta = 0,1,2$ for one, two, and three dimensions, respectively, and $a = 1/2$.) We emphasize, however, that if τ is indeed a function of energy, and if $n(E)$ and $K(E)$ are more complicated than assumed, then such a simple definition of effective temperature cannot be expected to be valid.

VI. FLUCTUATING FORCES IN QUANTUM MECHANICS

Based on the classical ideas presented in §§2 and 3, it is straightforward to define operators for the corresponding fluctuating forces in transport problems treated quantum mechanically. If we define the impedance $\underset{\approx}{Z}$ according to

$$\int_{-\infty}^{\infty} dt\ \underset{\approx}{Z}_{t_1 t}\underset{\approx}{Y}_{tt_2} \equiv \underset{\approx}{I}\delta(t_1-t_2), \tag{42}$$

then the operator for the driving force $\hat{\underset{\sim}{d}}_t$ can be defined in the Heisenberg representation according to

$$\hat{\underset{\sim}{d}}_{t_2} \equiv \int_{-\infty}^{\infty} dt_1 \underset{\approx}{Z}_{t_2 t_1}\hat{\underset{\sim}{x}}_{t_1}. \tag{43a}$$

Depending on our definition of the admittance $\underset{\approx}{Y}$, and hence of the impedance the following relations arise: for $\underset{\approx}{Y}_{t_2 t_1}$ antisymmetric in (t_1, t_2)

$$\underset{\approx}{Y}_{t_2 t_1} \equiv i\langle[\hat{\underset{\sim}{x}}_{t_2}, \hat{\underset{\sim}{x}}_{t_1}]\rangle/\hbar, \tag{44a}$$

$$i\langle[\hat{\underset{\sim}{d}}_{t_2}, \hat{\underset{\sim}{d}}_{t_1}]\rangle/\hbar = \tilde{\underset{\approx}{Z}}_{t_1 t_2} \tag{44b}$$

(where $\tilde{\underset{\approx}{Z}}$ is the transpose of $\underset{\approx}{Z}$), and for $\underset{\approx}{Y}_{t_2 t_1}$ causal

$$\underset{\approx}{Y}_{t_2 t_1} \equiv i\langle[\hat{\underset{\sim}{x}}_{t_2}, \hat{\underset{\sim}{x}}_{t_1}]\rangle/\hbar\ u(t_2-t_1), \tag{45a}$$

$$i\langle[\hat{\underset{\sim}{d}}_{t_2}, \hat{\underset{\sim}{d}}_{t_1}]\rangle/\hbar\ u(t_1-t_2) = \tilde{\underset{\approx}{Z}}_{t_1 t_2}, \tag{45b}$$

an <u>anti</u>causal relation as expected from the anticausal nature of driving forces relative to responses. Both (44a,b) and (45a,b) illustrate the close connection between the usual Kubo form for the admittance and the corresponding form for the impedance, an inverse transport quantity. In the path integral representation, defining $\underset{\sim}{d}_t$ as

$$\underset{\sim}{d}_{t_2} \equiv \int_{-\infty}^{\infty} dt_1 \underset{\approx}{Z}_{t_2 t_1} \underset{\sim}{x}_{t_1}, \tag{43b}$$

then we obtain the pair

$$\underset{\approx}{Y}_{t_2 t_1} \equiv i\langle \underset{\sim}{x}_{t_2} (\underset{\sim}{x}_{t_1} - \underset{\sim}{x}'_{t_1}) \rangle / \hbar, \tag{46a}$$

$$i\langle \underset{\sim}{d}_{t_2} (\underset{\sim}{d}_{t_1} - d'_{t_1}) \rangle / \hbar = \tilde{\underset{\approx}{Z}}_{t_1 t_2}, \tag{46b}$$

as we must considering (45a,b). We note in passing that a transformation of the form (43b) in a path integral over $(\underset{\sim}{x},\underset{\sim}{x}')$ will change the integral to one over $(\underset{\sim}{d},\underset{\sim}{d}')$, permitting the entire transport problem to be treated in the world of fluctuating forces. For completeness we note that the total force operator $\hat{\underset{\sim}{F}}_t$ satisfies

$$\hat{\underset{\sim}{F}}_{t_2} = \int_{-\infty}^{\infty} dt_1 \underset{\approx}{G}^{o}_{t_2 t_1} \hat{\underset{\sim}{x}}_{t_1} \tag{47a}$$

$$\underset{\approx}{G}^{o}_{t_2 t_1} \equiv m\underset{\approx}{I}\delta''(t_2 - t_1) \tag{47b}$$

from which it follows that

$$i\langle [\hat{\underset{\sim}{F}}_{t_2}, \hat{\underset{\sim}{F}}_{t_1}] \rangle / \hbar \equiv \int_{-\infty}^{\infty} dt'_2 \int_{-\infty}^{\infty} dt'_1 \underset{\approx}{G}^{o}_{t_2 t'_2} \underset{\approx}{Y}_{t'_2 t'_1} \tilde{\underset{\approx}{G}}^{o}_{t'_1 t_1}, \tag{48}$$

again expressing the relation between total force and admittance.

It should be noted that the above applies to almost any system whatever, be it in thermal equilibrium, or well out of equilibrium under the influence of strong electric and magnetic fields. In the latter, one replaces $\hat{\underset{\sim}{x}}_t$ by $\hat{\underset{\sim}{y}}_t$ where $\underset{\sim}{x}_t = \underset{\sim}{x}^o_t + \hat{\underset{\sim}{y}}_t$ and $\underset{\sim}{x}^o_t$ is the expected motion of the particle of interest, that is the expectation value of the displacement operator of the particle. The impedance is then closely related to the response of the particle to a small ac probe electric field, a very useful quantity in many respects.[9,10]

If we restrict ourselves to thermal equilibrium, then we can make use of the FD theorem to express the impedance in terms of autocorrelation functions of the fluctuating forces. It follows at once that the resistivity satisfies

$$\mathrm{Im}[\underset{\approx}{\tilde{Z}}(\omega)] = \frac{2}{\hbar}\,\frac{1 - \exp(\hbar\omega/kT)}{1 + \exp(\hbar\omega/kT)}\,\mathrm{Re}[\underset{\sim}{\underset{\approx}{K}}_d(\omega)] \tag{49a}$$

$$\underset{\sim}{\underset{\approx}{K}}_d(\omega) \equiv \int_0^\infty dt\, \underset{\sim}{\underset{\approx}{k}}_d(t) e^{i\omega t} \tag{49b}$$

$$\underset{\sim}{\underset{\approx}{k}}_d(t) \equiv \frac{1}{2}\,\langle\{\underset{\sim}{d}(t),\underset{\sim}{d}(0)\}\rangle, \tag{49c}$$

where as expected $\underset{\approx}{k}_d$ is a correlation function of $\underset{\sim}{d}_t$. Ballentine and Heaney use a different method, that of superoperators, to derive similar relations between correlation functions of the fluctuating forces and the resistivity at thermal equilibrium.[21] It should be pointed out that in the presence of finite currents, thermal equilibrium is no longer strictly maintained, and then the results (42) - (48) rather than (49a,b,c) should be used.

It would be out of place here to develop these ideas further. Suffice it to say that the fluctuating-force concept is useful both in classical noise theory and in quantum mechanics, and has even found some application in simplifying certain problems in irreversible thermodynamics.[13] We expect further use of this concept to be made in these fields in the future.

VII. CONCLUSIONS

In the foregoing we have discussed some recent developments in noise theory and indicated how these relate to similar problems in transport theory. Indeed, these two fields are so closely related that it would be hard to imagine a new finding in one field not giving rise to new results in the other. In particular we have focused on the following recent results:

(1) The <u>eindrif method</u>,[6,13] which simultaneously generalizes and greatly simplifies the characterization of microscopic, statistical fluctuations in noise theory, which, by focusing attention on the elementary fluctuations accompanying all transfer processes, separates these fluctuations in the transfer currents into an induced portion and a spontaneous portion, and which enables correlation to be treated in the dynamical portion of the problem, reserving for the statistical portion only well-defined, uncorrelated random processes whose statistics are readily calculated, provides a means of including all correlation and relaxation effects at a

microscopic level, thus filling a gap in the Langevin and the impedance-field methods of calculating noise in devices.[17,18] Related findings include:

(a) An internal inconsistency[6,13] in the Langevin Ansatz,

(b) The importance of spatial correlation[11] in device noise calculations,

(c) The presence of internal noise[13] in macroscopic treatments of microscopic problems, and

(d) Some simplifications in treating fluctuations in irreversible thermodynamics.[13]

(2) The noise accompanying diffusion processes has been reinvestigated and a very general relation between the current spectral density and the frequency-dependent diffusion coefficient was given. Related findings include:

(a) The Einstein relation has been rederived[6,11] from the FD theorem with particular attention to its form when finite electric fields are present. The relation $eD = kT\mu$, $\mu \equiv v/E$, is no longer valid if the mobility μ and the diffusion constant D are functions of E, even if conditions are such that only a small error is made in treating the problem as if thermal equilibrium applied (zero noiseless current), as is usually assumed in using $eD = kT\mu$ in problems in which currents flow in response to applied fields. Instead the correct relation is

$$eD(E) = kT\, dv(E)/dE, \tag{50}$$

an intuitively obvious result considering the nature of the linear ac response to fluctuating forces in terms of which velocity and hence current fluctuations are usually described.

(b) In some restricted cases, if the carriers become heated, the T in (50) can be replaced by a well-defined $T_{eff}(E)$, a field-dependent effective temperature.

(3) The application of the FD theorem and the concept of fluctuating forces to problems in quantum-mechanical transport theory has enabled certain aspects of such problems to be viewed in a new light. For example, a linear transformation enables one to calculate in the fluctuating-force domain instead of in the position domain, which results in certain simplifications.

Our primary message is this. Noise problems have been treated by mathematicians and engineers as well as by physicists.

Mathematicians usually focus on model problems which need not correspond to any physical situation while engineers tend to rely on very simple, macroscopic characterizations of noise sources often scaled down uncritically to microscopic dimensions. Unlike the mathematician, the physicist is bound not only by the rules of mathematics but also by the laws of nature. For example, if the Boltzmann equation is not valid for the problem of interest, or if the processes giving rise to the noise are non-Markovian, assumptions to the contrary cannot be made, however interesting the resulting mathematical problem may seem. Unlike that of the engineer, the world of the physicist is usually microscopic, obeying the rules of quantum mechanics. Considering the simplicity of the recent findings summarized in this paper, it is clear that the physicist must be extremely critical of "standard" results and methods when confronting noise problems in physics.

REFERENCES

1. A. Einstein, Ann. d. Phys. (4) 17, 549 (1905).
2. H. B. Callen and T. A. Welton, Phys. Rev. 83, 34 (1951).
3. H. B. Callen, Phys. Rev. 111, 367 (1958).
4. W. Bernard and H. B. Callen, Rev. Mod. Phys. 31, 1017 (1959).
5. H. B. Callen, in D. ter Haar, ed., Fluctuation, Relaxation and Resonance in Magnetic Systems, New York: Plenum (1962).
6. K. K. Thornber, BSTJ 53, 1041 (1974).
7. K. K. Thornber, Phys. Rev. B 9, 3489 (1974).
8. K. K. Thornber and R. P. Feynman, Phys. Rev. B 1, 4099 (1970); Errata, Phys. Rev. B 4, 674 (1971).
9. K. K. Thornber, Phys. Rev. B 3, 1929 (1971); Errata Phys. Rev. B 4, 675 (1971).
10. K. K. Thornber, in J. T. Devreese, ed., Polarons in Ionic Crystals and Polar Semiconductors, Amsterdam: North-Holland (1972).
11. K. K. Thornber, Solid-State Electronics 17, 95 (1974).
12. K. K. Thornber, BSTJ 53, 1211 (1974).
13. K. K. Thornber, J. Appl. Phys. 46, 2781 (1975).
14. K. K. Thornber, in this volume.
15. K. K. Thornber, Proc. IEEE 60, 1113 (1972).
16. K. K. Thornber and M. F. Tompsett, IEEE Trans. Elect. Dev. ED-20, 456 (1973).
17. T. C. McGill, M-A. Nicolet and K. K. Thornber, Solid-State Electronics 17, 107 (1974).
18. K. K. Thornber, T. C. McGill and M-A. Nicolet, Solid-State Electronics 17, 587 (1974).
19. P. Langevin, Compt. Rend. 146, 503 (1908).
20. N. H. March, in this volume.
21. L. E. Ballentine and W. J. Heaney, J. Phys. C: Solid-State Phys. 7, 1985 (1974).

22. M. Lax, Rev. Mod. Phys. 38, 541 (1966).
23. K. M. van Vliet, A. Friedmann, R. J. J. Zijlstra, A. Gisolf and A. vanderZiel, J. Appl. Phys. 46, 1804 (1975).
24. K. M. van Vliet, A. Friedmann, R. J. J. Zijlstra, A. Gisolf and A. vanderZiel, J. Appl. Phys. 46 1814 (1975).
25. M. Lax, Rev. Mod. Phys. 32, 25 (1960).
26. K. M. van Vliet and J. R. Fassett, in R. E. Burgess, ed., Fluctuation Phenomena in Solids, New York: Academic (1965).
27. W. Shockley, J. A. Copeland and R. P. James, in P. O. Löwdin, ed., Quantum Theory of Atoms, Molecules and the Solid State, New York: Acedemic (1962).
28. K. M. van Vliet, Solid-State Electronics 13, 649 (1970).
29. K. M. van Vliet, J. Math. Phys. 12, 1981 (1 71).
30. K. M. van Vliet, J. Math. Phys. 12, 1998 (1971).
31. K. M. van Vliet, private communication.
32. J. L. Tandon, H. R. Bilger and M-A. Nicolet, Solid-State Electronics 18, 113 (1975).
33. R. P. Feynman, Statistical Mechanics, a Set of Lectures, New York: Benjamin (1972) Ch. 2.
34. K. K. Thornber, Ph.D. Thesis Part II, California Institute of Technology (1966), unpublished.

LOW FREQUENCY FLUCTUATIONS IN ELECTRONIC TRANSPORT PHENOMENA*

Peter H. Handel

Physics Department

University of Missouri-St. Louis

St. Louis, Mo 63121, USA

1. INTRODUCTION

Various fluctuations can be present in the output of solid state electronic devices. The study of fluctuation phenomena is important both for practical applications and theoretical considerations. Electrical fluctuations are known as noise in general, even if their frequency is not in the acoustic domain, and no sound waves are generated.

A noise measurement [1-6] is always limited to a certain frequency domain. An important measure of a noise process is the variance $<(\delta I)^2>$, and its spectral density $<(\delta I)^2>_f$ which is a function of frequency f . The electric current fluctuations $\delta I = I - <I>$ generate voltage fluctuations $\delta U = R\delta I$ in a series resistance R . Voltage (or current) fluctuations are measured in general by amplifying them first with a wide band preamplifier. The output is amplified and squared, yielding the variance $<(\delta I)^2>$ after low-pass filtering. In order to obtain

*This paper was written in the framework of the joint project ESIS (Electronic Structure in Solids) of the Universities of Antwerpen and Liège, Belgium, while the author was a visiting Professor at the International Center for Theoretical Solid State Physics in Antwerpen.

the spectral density, a tunable band-pass filter is inserted between the preamplifier and the main amplifier, or a multichannel analyser may be used. For frequencies below 10 Hz, the spectrum is measured by recording the noise at low speed, and analysing the output of a subsequent high speed playback. This procedure can be iterated, making very low frequencies accessible.

The purpose of this contribution is to present a theory of low frequency electronic noise. While noise at higher frequencies is well understood theoretically [1-7] on the basis of the concepts introduced by Nyquist and Schottky in the first four decades of our century, low frequency noise (below 10^4-10^5Hz) is usually characterized by a 1/f spectrum down to frequencies as low as measurements have been made, i.e. to 10^{-4}-10^{-6}Hz, and has defied any coherent theoretical interpretation so far [1-6]. I will try to give here a more detailed account of the quantum theory of 1/f noise published recently [8-9].

Specifically, after a short discussion of various forms of electronic noise and of the main experimental facts which characterize 1/f noise, the propagator formalism will be used in Sec.4, in order to construct the wave function of a current carrier in interaction with the electromagnetic field. Although some readers would prefer a complete derivation of the infrared radiative correction factor for the bremsstrahlung matrix element, only a discussion of the basic (uncorrected) bremsstrahlung matrix element with its 1/f dependence will be given here. The occurrence of current fluctuations will be described both in space and in time in Sec.6. Some fundamental properties of 1/f noise are derived in Sec. 7, and elementary considerations concerning a sample with N electrons are presented in Sec.9. Finally, a generalized concept of 1/f noise is developed in Sec.10.

2. Basic forms of Electronic Noise

The measured frequency spectrum of current fluctuations originating from a circuit element of conductance G=ReY has the general form

$$\langle(\delta I)^2\rangle_f = A\frac{\langle I\rangle^2}{f} + \frac{4\tau B\langle I\rangle^\gamma}{1+(2\pi\tau f)^2} + 4kTG, \qquad (1)$$

where A, B, γ and τ are constants, while Y is the complex admittance. The three terms in Eq(1) are known as 1/f noise, shot noise and thermal (nyquist noise. The main contribution comes from 1/f noise at low frequencies, shot noise at intermediate frequencies and thermal noise at high frequencies.

In the absence of electrical bias, in thermal equilibrium at the temperature T only thermal noise is present at any frequency. At very high frequencies, or low temperatures, the Planck factor $hfk^{-1}T^{-1}[e^{hf/kT}-1]^{-1}$ has to be included in the expression of thermal noise .

Shot noise is due to the atomistic corpuscular character of electric charge. Each current carrier yields a current pulse while it passes through the sample or during its lifetime in a semiconductor. In solid state or vacuum diodes $\gamma=1$, and the Schottky formula $<(\delta I)^2>_f = 2|e|<I>$ can be applied at saturation for $f<<\tau^{-1}$, where τ is the duration of a current pulse, i.e. the time of passage for a carrier of charge e . The term "shot noise" remembers the noise from pooring small shot on a drum or from raindrops on a tin roof. In semiconductors $\gamma=2$ and two terms with different relaxation times τ will give the contributions of electrons and holes respectively. Integrated over f from 0 to ∞, the shot noise term in Eq.(1) yields $B<I>^{\gamma}$. The $[1+(2\pi f\tau)^2]^{-1}$ dependence characterizes the spectrum of fluctuation processes with exponential relaxation. This is a consequence of the Wiener-Khintchine theorem [1-7]

$$<\delta I(t)\delta I(t+\tau)> = \int_0^{\infty} <(\delta I)^2>_f \cos 2\pi f\tau \, df, \qquad (2)$$

which states that the spectral density is the Fourier transform of the autocorrelation function $<\delta I(t)\delta I(t+\tau)>$. The bracketts signify ensemble averages everywhere, but can be replaced by time averages for stationary ergodic systems.

3. Experimental Aspects of 1/f Noise

The first term of Eq.(1) has a universal character, because 1/f noise is found at sufficiently low frequencies whenever a current is transported by a small number of charge carriers. Thus, in the eyes of the experimental physicist, 1/f noise appears to be just as universal for low frequency fluctuations of transport phenomena, as is "white" Nyquist noise for thermal equilibrium. 1/f noise is known as flicker noise in vacuum tubes and semiconductors, excess noise in solids, or contact noise.

The upper frequency f_1, below which 1/f noise becomes dominant, depends on the relative magnitude of the 1/f part and of the other noise contributions. f_1 is quite different for different systems. In homogeneous semiconductor samples with ohmic contacts f_1 varies in the range 10^2-10^4Hz, while it is 10^4-10^6Hz for carbon resistors and carbon microphones. In semiconductor film f_1 is generally higher and for granulated metallic films or for contact noise f_1 varies over an even wider frequency interval. This is related - as in the general case - to the inverse proportionality of 1/f noise to the effective number of charge carriers, as we shall see. In semiconductor diodes f_1 is of the order of 10^3-10^4Hz for forward, and 10^7Hz for reverse bias. It can be as high as 10^8-10^9Hz for microwave (point contact) diodes and is lower for diodes with lower high frequency cutoff frequencies. In vacuum tubes with wolfram filements f_1 is 10^2-10^3Hz, while tubes with oxyde cathodes have $f_1 \approx 10^3$-10^5Hz.

In metals with good metallic contacts only thermal noise is found, even for large currents, unless very small samples are considered [10-11]. We consider here that this only means that f_1 is very low for metals, due to the large number of current carriers present in a conventional sample. It is important to note that the low heat capacity C of very small probes leads to large temperature fluctuations at thermal equilibrium $<(\delta T)^2>=kT^2C^{-1}$. These temperature fluctuations lead to resistance fluctuations $<(\delta R)^2>=\beta^2kT^2R^2C^{-1}$, where β is the temperature coefficient of the resistance, and generate current fluctuations if a voltage is applied to the sample. This is the so called "bolometer effect" (sec. e.g. Bittel and Storm [1] p.p. 152-157). The bolometer effect frustrates 1/f measurements on very small samples unless sufficiently low frequencies are used. Although it is not sufficient, in general, to characterize the sample through only one thermal relaxation time constant τ, the general form of the spectrum of the bolometer effect is similar to $(1+\omega^2\tau^2)^{-1}$, i.e. is constant for $\omega<<\tau^{-1}$ if the smallest τ value given by the diffusion equation is used. Unfortunately, a strong bolometer effect is sometimes confused with 1/f noise [13-14].

This brings us to the two most remarcable properties which define 1/f noise : the absence of a lower frequency limit and of a temperature dependence. A 1/f spectrum down to $f=0$ involves a logarithmically divergent variance. This difficulty is not removed completely by the observation that it takes an infinite time to observe what happens at $f=0$. This argument has led many investigators to conclude that the 1/f spectrum must flatten out at some not yet observed, very low, frequency. Gigantic efforts have been made to discover a low-frequency turnover of 1/f noise. Rollin and Templeton [15] have followed the 1/f law down to 2.10^{-4}Hz in carbon resistors and germanium monocrystals while Baker [16] and Firle and Winston [17] have verified the 1/f dependence down to 6.10^{-6}Hz on point contact diodes. For some of these diodes the 1/f dependence has been verified over 13 frequency deacdes. At high frequencies the shot noise contribution has been subtracted. More recent measurements [18] have again verified the 1/f dependence from the 10 kHz region down to $5\ 10^{-5}$Hz . No sign of a low-frequency limit of 1/f noise has been found so far.

Some temperature dependence of 1/f noise is present in cases in which the system undergoes changes with temperature, e.g. adsorbed gases or water being evaporated from the surface of carbon resistors, or strong variation of the carrier concentration in semiconductors. In general [1], however, an approximate result obtained by Hooge [10-11] by comparing many measurements on semiconductors and own measurements on metals, shows that the relative 1/f noise depends only on the total number of carriers in the sample, N :

$$\langle(\delta I)^2\rangle_f/\langle I\rangle^2 = 2.10^{-3}N^{-1}f^{-1}=\nu N^{-1}f^{-1} \qquad (3)$$

Although this relation is only aproximately valid, it generalizes the well known inverse proportionality of 1/f noise to the volume of the current carrying sample. Eq.(3) has also been applied successfully to metallic point-contacts [11]. For ions in water ν has been found [19] to be roughly proportional to the concentration N/V and $\nu=10$ for molar electrolytic solutions. 1/f noise had been found earlier [20] in the voltage across the membranes of Ranvier's nodes in nerve cells.

Often 1/f noise shows small deviations of the exponent of f around -1, of the order of 10%. In addition, the dependence on the average current I can be slightly different; particularly nonlinear circuit elements show a stronger dependence on I. A marked dependence of 1/f noise on the state of the surface is found in semiconductors and is of technical importance in MOSTs (metal-oxide-silicon transistors). This later fact may lead to the misinterpretation of 1/f noise as an effect which is characteristic only to semiconductor surfaces. The above analysis of experimental facts suggests that 1/f noise is a fundamental property of any electrical current.

4. Bremsstrahlung, Turbulence, and 1/f Noise

Starting from the idea formulated in the conclusion of Sec.3 I have formulated first a qualitative [21], then a more quantitative turbulence theory [22] of 1/f noise for intrinsic semiconductors [23-24] and for the case of a semiconductor or of a metal [25]. The turbulence theory was formulated in terms of the magnetic field components and restricted to the case of homogeneous isotropic turbulence which frustrated all attempts to determine the intensity of the noise. Efforts [26-27] to develop a quantized form of the turbulence theory have led to a progressive simplification and schematization of the considered physical model of a 1/f noise measurement, and to the inclusion [26] of the phenomena of bremsstrahlung with infrared radiative corrections as the quantum form of the nonlinear electromagnetic field turbulence. The nonlinearity appears both classically and quantum mechanically through the elimination of the sources of the electromagnetic field, and is in fact the object of the perturbation theory of conventional quantum electrodynamics. At this point it became clear that the exponent ε determined self-consistently by the intensity of turbulence in the classical theory [25] is the coefficient of the logarithmic energy term in the infrared radiative correction exponent of the quantum theory [9] . Eq.(75') of the classical theory [23] corresponds to Eq.(13) of the quantum form [9] with $\varepsilon/2 \leftrightarrow \alpha A$

As a result of the mentioned process of gradual simplification and schematization of a typical noise measurement, I have found that the fundamental cause of 1/f noise is already present in a beam of electrons emerging in vacuum from an arbitrary scattering process. I identify this cause as a peculiar interference effect between the part of the wave function of the scattered beam (and in fact of every electron) which has emitted a bremsstrahlung photon in the (otherwise elastic) scattering process, and the part which did not emit any. Indeed, due to soft photon emission (bremsstrahlung) a monochromatic incident beam will yield a coherent superposition of scattered waves with only slightly different energies, which is not an eigenfunction of the current (or momentum) operator any more, even if we single out the current emerging in an arbitrarily narrow cone of directions. Consequently, the current emerging in the considered narrow solid angle will exhibit quantum fluctuations, although its expectation value is constant in time. We will show in Sec.6 that the spectrum of these current fluctuations at low frequencies is identical with the spectrum of the emitted photons, i.e. also of 1/f type.

The wave function $\Psi(x)$ of the interacting electron can be constructed [28] with the help of the Feynman propagator $S_F(x-y)$ as

$$\Psi(x)=\phi(x)+e\int d^4y S_F(x-y)\, A_\mu(y)\gamma^\mu\Psi(y), \qquad (4)$$

where the spinor $\phi(x)$ represents the incident wave, and the integral is the scattered wave which contains only positive frequencies for $t>0$. Here A_μ is the operator of the four-vector potential. The electromagnetic field is quantized in our description, but the Dirac field is not. We have set $\hbar=c=1$ and x stands for $(x^0, x^1, x^2, x^3)=(t,x,y,z)$. Eq.(4) is the formal solution of the Dirac equation [28]

$$(i\gamma^{\mu}\partial_{\mu}-m)\Psi=eA_{\mu}\gamma^{\mu}\Psi, \qquad (5)$$

in which γ^{μ} are the γ matrices of Dirac. For $t\to\infty$ Eq.(4) is written in terms of the elements of the S matrix (the square brackett)

$$\Psi(x)=\int d^3p \sum_{r=1}^{2}\phi_p^r(x)[-ie\int dy^4 \ \phi_p^r(y)A_{\mu}(y)\gamma^{\mu}\Psi(y)] \quad , \qquad (6)$$

where $\phi_p^r(x)$ are free particle states. The lowest order matrix elements are of second order for bremsstrahlung, as we see on Fig.1. In addition, there are two similar graphs which correspond to absorption of radiation (inverse bremsstrahlung) arising from the negative frequency part of A.

5. The Scattered Wave

For the case of low frequency bremsstrahlung emission (small k) the wave scattered in the direction $\hat{p}\equiv\vec{p}/|\vec{p}|$ is obtained from Eq.(6) in the form [28-29]

$$(\Psi(x)-\phi(x))_{\hat{p}}^{(e)} =$$

$$\sum_{\lambda=1}^{2}\int\frac{\vec{p}^{2}d|\vec{p}|}{E}\iint\frac{V}{(2\pi)^3}\frac{\sqrt{m}}{\sqrt{V}}\ e^{-ip\cdot x}u(p)a^{+}_{\vec{k},\lambda}|o> \ x$$

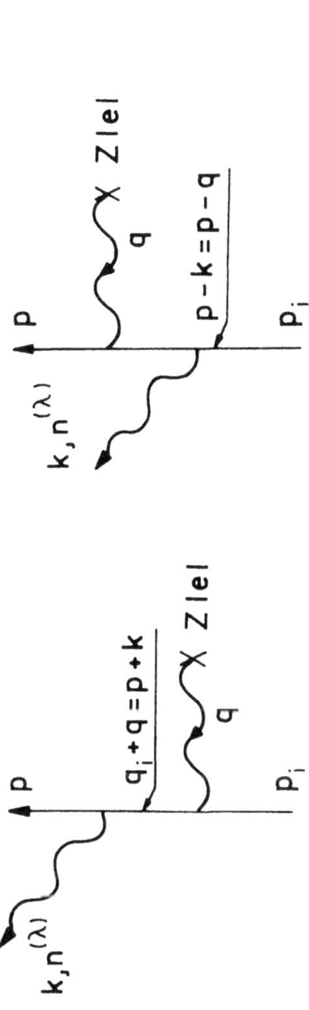

Fig. 1. Lowest order bremsstrahlung of an electron in a fixed Coulomb potential.

$$\times\, e^{i\gamma(\varepsilon)} \frac{V}{(2\pi)^3} \frac{iZe^3}{\sqrt{2\varepsilon}V^{3/2}} \frac{m}{\sqrt{E_i}} \left(\frac{\varepsilon}{E}\right)^{\alpha\frac{A}{2}} \frac{\overline{u}(p)\gamma_o u(p_i)}{|\vec{q}|^2}$$

$$\times \left(\frac{n^{(\lambda)}\cdot p}{k\cdot p} - \frac{n^{(\lambda)}\cdot p_i}{k\cdot p_i}\right) 2\pi\delta(E-\varepsilon-E_i)\varepsilon^2 d\varepsilon d\Omega_{\vec{k}} \ . \qquad (7)$$

This would yield the scattered wave by integration over the solid angle differential $d\Omega_{\vec{p}}$. Here the final state $|\vec{k},\lambda\rangle = a^+_{\vec{k},\lambda}|o\rangle$ of the electromagnetic field, corresponding to the presence of a photon of wave vector $\vec{k}=\varepsilon\vec{k}/|\vec{k}|$ and polarization four-vector $n^{(\lambda)}(\vec{k})$, is put in evidence explicitely, while the "in" state ϕ is considered to include the vacuum of the electromagnetic field in its definition, together with an incoming free electron state $u(p_i)$ of four-momentum p_i . The factors succeeding $|0\rangle$ i.e. the last two rows of Eq.(7) represent the bremsstrahlung matrix element (included in brackets in Eq.(6)), calculated to first order in the fixed Coulomb potential which has been assumed as the cause of the scattering process. q is the momentum transfer. If we ignore $(\varepsilon/E)^{\alpha A/2}$, the matrix element is limited to second order, i.e. to only one real photon vertex in addition to the Coulomb interaction vertex, and with no virtual photons considered, as in Fig.1. $(\varepsilon/E)^{\alpha A/2}$ and part of the phase factor $e^{i\gamma(\varepsilon)}$ are introduced as infrared radiative correction factors [29] with A a constant and α the fine structure constant (See Eq.19). They give the effect of summing an infinite series of diagrams involving an arbitrary number of soft real and virtual photons [29]. By soft photons we understand any photons with energy below the resolution threshold ε_o . In any physical experiment there is a minimal energy ε_o for which photons can be detected [30], all lower energies escape observation and contribute to what will be called the elastic part of the process. Considering in addition to these soft photons the emission of one or more real photons with energies above the detection threshold, one describes bremsstrahlung by summing again over soft photon contributions of all orders, either at the wave function and matrix element level, or only later, by summing the cross sections. We will put the result of the usual summation at the cross section level in the form obtained when the summation is performed at the wave function level. For this purpose, one has to divide the Hilbert space of photons into a subspace containing all soft photons and a subspace formed by the observable photon states. For soft photons, the Fock basis is replaced [31] by a basis of coherent states. The reason for using the coherent states (translated principal vectors) for the soft photons is that the matrix elements would otherwise show infrared divergences which would cancel only at the cross section level.

Returning to Eq.(7) we introduce creation operators for photons of indefinite direction and polarization

$$a_\varepsilon^+ = \varepsilon N_o e^{i\gamma(\varepsilon)} \int \sum_{\lambda=1}^{2} \left(\frac{n^{(\lambda)}\cdot p}{k\cdot p} - \frac{n^{(\lambda)}\cdot p_i}{k\cdot p_i}\right) a^+_{\vec{k},\lambda}\, d\Omega_{\vec{k}} \, , \qquad (8)$$

where $e^{i\gamma(\varepsilon)}$ is a phase factor to be defined later through Eq.(16). From

$$[a_{\vec{k},\lambda}, a^+_{\vec{k}',\lambda'}] = \delta(\vec{k}-\vec{k}')\delta_{\lambda,\lambda'}, \quad [a,a] = [a^+,a^+] = o, \qquad (9)$$

we obtain the commutation relations

$$[a_\varepsilon, a^+_{\varepsilon'}] = \delta(\varepsilon-\varepsilon') \, , \quad [a_\varepsilon, a_{\varepsilon'}] = [a^+_\varepsilon, a^+_{\varepsilon'}] = o \, , \qquad (10)$$

by defining the normalization factor N_o through

$$\varepsilon^2 N_o^2 \int \sum_{\lambda=1}^{2} \left(\frac{n^{(\lambda)}.p}{k.p} - \frac{n^{(\lambda)}.p_i}{k.p_i}\right) d\Omega_{\vec{k}} = 1 \quad ; \quad N_o = (2\pi)^{-1} A^{-1/2}. \tag{11}$$

Using this notation and writing $pdp=EdE$, Eq.(7) becomes

$$(\Psi(x)-\phi(x))^{(e)}_{\hat{p}} =$$

$$\frac{iZe^3m^2}{(2\pi)^5}\int^{\varepsilon_1} \frac{|\vec{p}'|u(p')}{N_o\sqrt{2E}} e^{i\vec{p}'.\vec{r}-i(E_i-\varepsilon)t} a^+_\varepsilon|o\rangle\left(\frac{\varepsilon}{E}\right)^{\alpha A/2}$$

$$\frac{\bar{u}(p')\gamma_o u(p_i)}{|\vec{q}|^2} \frac{d\varepsilon}{\sqrt{\varepsilon}} . \tag{12}$$

The lower integration limit has to be pushed to zero with coherent states used for $\varepsilon<\varepsilon_o$. Here $\vec{p}'$ is the momentum of the scattered electron. Its magnitude $|\vec{p}'|$ differs from $|\vec{p}_i|=|\vec{p}|$ only due to the small energy change ε introduced by bremsstrahlung.

$$|\vec{p}'|=|\vec{p}|-\varepsilon/v , \tag{13}$$

where v is the velocity of the electron We shall put $\vec{p}'=\vec{p}$ in Eq.(12), except for $\varepsilon>\varepsilon_o$ in the exponential. In Eq.(12) the integration over the energy loss ε is limited by the maximal energy of a bremsstrahlung photon ε_1 which equals the kinetic energy of the incoming electron. The integrand is exact only for $\varepsilon_0 < \varepsilon << \varepsilon_1$. The momentum transfer $|\vec{q}|$ is $2|\vec{p}|\sin\theta_q/2$, where θ_0 is the scattering angle of the electron in the Coulomb potential. As ε_0 is the dection threshold for low energy i.e. the lowest frequency registered by the multichannel electrical noise spectrometer, we divide the integration interval in Eq.(12) in (o,ε_0) and $(\varepsilon_0,\varepsilon_1)$. In the interval (o,ε_0) the form of the integrand in Eq.(12) is incorrect. In fact we have to integrate the squared amplitude over this interval and take the square root of the result as the elastic scattering matrix element :

$$E^{-\alpha A/2}\left[\int_0^{\varepsilon_0}\left(\frac{\varepsilon^{\alpha A/2}}{\sqrt{\varepsilon}}\right)^2 d\varepsilon\right]^{1/2} = \frac{1}{\sqrt{\alpha A}}\,(\varepsilon_0/E)^{\alpha A/2}. \qquad (14)$$

The validity of this approach is discussed later in connection with Eq.(21) I denote by $|f>$ the coherent state of the electromagnetic field associated [31] in the Hilbert space of soft photons with the final state of momentum $p=(E,\vec{p})$:

$$|f>=\exp\{-\frac{1}{2}\sum_{\lambda=1}^{2}\int|S_f^{(\lambda)}(k)|^2 d^3k\}$$

$$\exp\{\sum_{2=1}^{2}\int d^3k S_f^{(\lambda)}(k)\, n^{(\lambda)}(k)\, a^+_{\vec{k},\lambda}\}|o>, \qquad (15)$$

where $k=(k_0,\vec{k})$ and

$$S_f^{(\lambda)}(k) \equiv \frac{e}{[2(2\pi)^3\varepsilon]^{1/2}}\;\frac{p.n^{(\lambda)}}{k.p} \;. \qquad (16)$$

We are interested only in the relative magnitude and phase of the elastic and bremsstrahlung contributions, and so we approximate $\bar{u}(p)\gamma_0 u(p_i)$ by unity, neglecting terms of order $(v/c)^2$ in the common factor. This leads from Eq.(12) to a form of the scattered wave containing an elastic contribution and a bremsstrahlung term

$$(\Psi(x)-\phi(x))_{\hat{p}}^{(e)} =$$

$$\frac{iZe^3m^2u(p)e^{i\vec{p}.\vec{r}-iEt}}{(2\pi)^5 4|\vec{p}|N_o\sqrt{2E}\sin^2\theta_o/2}$$

$$\{\frac{1}{\sqrt{\alpha A}}(\frac{\varepsilon_o}{E})^{\alpha A/2}+\int_{\varepsilon_o}^{\varepsilon_1}(\frac{\varepsilon}{E})^{\alpha A/2}e^{i\varepsilon t-i\frac{\varepsilon}{v}\hat{p}.\vec{r}}\,a_\varepsilon^+\frac{d\varepsilon}{\sqrt{\varepsilon}}$$

where $|o\rangle$ is the vacuum state in the Hilbert space of observable photons. The negative frequency part of A_μ in Eq.(6) yields an absorption term $(\Psi(x)-\phi(x))_{\hat{p}}^{(a)}$ similar to Eq.(16), which is included into the final form of the scattered wave operator :

$$\frac{1}{|f\rangle|o\rangle}(\Psi(x)-\phi(x))_{\hat{p}} \equiv \Psi_s =$$

$$a'e^{i\vec{p}.\vec{r}-iEt}\{1+\int_{\varepsilon_o}^{\varepsilon_1}\rho_\varepsilon\,[e^{i\varepsilon(t-\hat{p}.\vec{r}/v)}a_\varepsilon^+ + e^{-i\varepsilon(t-\hat{p}.\vec{r}/v)}a_\varepsilon]\frac{d\varepsilon}{\sqrt{\varepsilon}}\}. \quad (17)$$

Here the notations

$$a' = \frac{iZe^3 m^2 u(p)(\varepsilon_0/E)^{\alpha A/2}}{(2\pi)^4 4|\vec{p}|(2\alpha E)^{1/2} \sin^2\theta_0/2} , \quad \rho_\varepsilon = \sqrt{\alpha A}(\varepsilon/\varepsilon_0)^{\alpha A/2} , \tag{18}$$

have been introduced. The general phase factor is included in $u(p)$ and the relative phase of the elastic and bremsstrahlung terms has been included in the definition of a_ε^+ in Eq.(8). Although Eq.(17) has been derived for electrons, it can be shown that it is valid in fact for charged particles of arbitrary spin, with the same definition of the constant A in the infrared exponent [29]

$$A = \frac{2q \cdot q}{3\pi m^2} \approx \frac{8\beta^2}{3\pi} \sin^2 \frac{\theta_0}{2} . \tag{19}$$

Here $2\beta \sin \frac{\theta_0}{2}$ is the magnitude of the difference between the initial and final velocity vectors in units of c . The last form of Eq.(19) is a nonrelativistic approximation.

6. Current Fluctuations

If S is the cross section of the beam and $a = a'/\sqrt{S}$ represents a new constant, the current operator corresponding to Eq.(17) will be

$$\vec{j} = \frac{-iS}{2m}[\Psi_s^+ \nabla \Psi_s - (\nabla \Psi_s^+)\Psi_s] = \vec{p}\frac{|a|^2}{m}\{1 + 2\int_{\varepsilon_0}^{\varepsilon_1} \rho_\varepsilon (e^{i\varepsilon\theta} a_\varepsilon^+ + e^{-i\varepsilon\theta} a_\varepsilon)\frac{d\varepsilon}{\sqrt{\varepsilon}}$$

$$+\int_{\varepsilon_0}^{\varepsilon_1}\frac{d\varepsilon}{\sqrt{\varepsilon}}\int_{\varepsilon_0}^{\varepsilon_1}\frac{d\varepsilon'}{\sqrt{\varepsilon'}}\rho_\varepsilon\rho_{\varepsilon'}[e^{i(\varepsilon-\varepsilon')\theta}a^+_\varepsilon a_{\varepsilon'}+e^{-i(\varepsilon-\varepsilon')\theta}a_\varepsilon a^+_{\varepsilon'}$$

$$+e^{-i(\varepsilon+\varepsilon')\theta}a_\varepsilon a_{\varepsilon'}+e^{i(\varepsilon+\varepsilon')\theta}a^+_\varepsilon a^+_{\varepsilon'}]\},$$

where $\theta=t-\hat{p}.\vec{r}/v$ is the retarded temporal variable. Denoting by $|o>$ the vacuum state in the Hilbert space of detectable photons $(\varepsilon>\varepsilon_0)$ we obtain for the expectation value of the current a constant expression

$$<\vec{j}> \equiv <o|\vec{j}|o>=\vec{p}\frac{|a|^2}{m}[1+\int_{\varepsilon_0}^{\varepsilon_1}\rho^2_\varepsilon\frac{d\varepsilon}{\varepsilon}]\equiv\vec{p}\frac{|a|^2}{m}(\frac{\varepsilon_1}{\varepsilon_0})^{\alpha A} \quad (21)$$

A comparison with Eq.(18) shows that $<\vec{j}>$ does not depend on ε_0 . Eq.(21) gives in fact the well known cross section for elastic nonradiative and bremsstrahlung scattering. The relative contribution of the nonradiative and bremsstrahlung parts coincides with the result obtained in Eqs.(2.51-2.60) by Yennie, Frautschi and Suura [29] . This shows that the relative magnitude of the two terms in Eq.(17) is correct.

The operator of the current fluctuations $\delta\vec{j}=\vec{j}-<\vec{j}>$ has the form

$$\delta\vec{j}=\vec{p}\frac{|a|^2}{m}\{2\int_{\varepsilon_0}^{\varepsilon_1}\rho_\varepsilon(e^{i\varepsilon\theta}a^+_\varepsilon+e^{-i\varepsilon\theta}a_\varepsilon)\frac{d\varepsilon}{\sqrt{\varepsilon}}-\int_{\varepsilon_0}^{\varepsilon_1}\rho^2_\varepsilon\frac{d\varepsilon}{\varepsilon}$$

$$+\int_{\varepsilon_0}^{\varepsilon_1}\frac{d\varepsilon}{\sqrt{\varepsilon}}\int_{\varepsilon_0}^{\varepsilon_1}\frac{d\varepsilon'}{\sqrt{\varepsilon'}}\rho_\varepsilon\rho_{\varepsilon'}[e^{i(\varepsilon-\varepsilon')\theta}a^+_\varepsilon a_{\varepsilon'}+e^{-i(\varepsilon-\varepsilon')\theta}a_\varepsilon a^+_{\varepsilon'}$$

$$+e^{-i(\varepsilon+\varepsilon')\theta}a_\varepsilon a_{\varepsilon'}+e^{i(\varepsilon+\varepsilon')\theta}a_\varepsilon^+ a_{\varepsilon'}^+]\} \quad . \tag{22}$$

With this operator the autocorrelation function in space and time can be constructed

$$A(\eta)=\frac{1}{2}[<\delta\vec{j}^+(\theta+\eta)\delta\vec{j}(\theta)>+<\delta\vec{j}^+(\theta)\delta\vec{j}(\theta+\eta)>] =$$

$$= 2|a|^4\frac{\vec{p}^2}{m^2}[2\int_{\varepsilon_0}^{\varepsilon_1}\rho_\varepsilon^2\cos\varepsilon\eta\frac{d\varepsilon}{\varepsilon}+\int_{\varepsilon_0}^{\varepsilon_1}\frac{d\varepsilon}{\varepsilon}\int_{\varepsilon_0}^{\varepsilon_1}\frac{d\varepsilon}{\varepsilon'}\rho_\varepsilon^2\rho_{\varepsilon'}^2\cos(\varepsilon+\varepsilon')\eta] \tag{23}$$

where

$$\eta=\tau-\hat{p}.\vec{s}/v \tag{24}$$

is a shift due to a time delay τ and a spatial displacement $\vec{s}$.

The second term on the right hand side of Eq.(23) is small of order αA with respect to the first and can be neglected as a "noise of noise" contribution together with the vector potential term in the current (see the appendix). Applying the Wiener-Khintchine theorem as in Eq.(2), we obtain by fixing $\vec{s}$ at the value $\vec{s}=0$ the spectral density of relative current noise at the frequency f

$$<(\delta j)^2>_f/<j>^2 = \frac{4\rho_f^2 f^{-1}}{[1+\int_{f_0}^{f_1}\rho_{f'}^2 df'/f']^2} = 4\alpha A(ff_0/f_1^2)^{\alpha A} f^{-1} \quad . \qquad (25)$$

In this formula the units have been restored, with $\varepsilon = hf$, but the time scale covariance and invariant character of this relation are evident. If we integrate from f_0 to f_1 we obtain for the total noise the remarcable result

$$<(\delta j)^2> = 4(f_0/f_1)^{\alpha A}[1-(f_0/f_1)^{\alpha A}]<j>^2 \quad . \qquad (26)$$

We know from Eqs.(8) and (21) that $<j>$ is independent of f_0. Eq.(26) shows that the integrated noise will not increase monoton ously for $f_0 \to 0$. We obtain the maximal noise $<(\delta j)^2> = <j>^2$ for the case of "halfway resolution" in which $(f_0/f_1)^{\alpha A} = 1/2$. See also Eq.(29). The most remarcable fact is that the maximum noise $<(\delta j)^2> = <j>^2$ is independent of αA, i.e. of the degree of violence with which the charge carrier has been jolted. I am led to conclude that if one looks at it the right way, the current transported by a charge carrier is entirely a 1/f noise current.

7. Fundamental Properties of 1/f Noise from One Carrier

If we limit ourselves to a fixed frequency interval, or if we consider the spectral density at a given frequency, the independence of αA mentioned above disappears. For a single particle of arbitrary mass carrying one elementary charge, the highest value of αA allowed by Eq.(19) is $8\alpha/(3\pi) \approx 161^{-1}$ formally. The fact, that Eq.(19) is a nonrelativistic approximation does not change much. The extremely relativistic (high energy) approximation [29] provides for a logarithmic energy-dependece of αA above $\alpha A \approx \alpha$.

Although this logarithmic increase may enhance the value of αA by up to an order of magnitude at high energies, where the rest mass of the particle is negligible with respect to the kinetic mass, we shall nevertheless use the value 161^{-1} as a maximal value, in order to gain a first insight. Accordingly, we would have to lower f at least 49 orders of magnitude below f_1, before $(f/f_1)^{\alpha A}$ would even introduce a factor of 1/2 in Eq.(25). We conclude that the half-length of the 1/f spectrum on the negative logarithmic frequency scale is more than 49 decades. Practically, this means that the factor $(f/f_1)^{\alpha A}$ can be neglected in Eq.(25) for all applications, excepting only the calculation of the total noise power.

If we take into account that the age of the universe is estimated to be of the order of 10^{17} sec., we realize how far from "halfway resolution" we are. However, three remarks are in order :

1. If particles of charge ze (or groups of z particles of charge e) are scattered, αA will increase proportional to z^2 and the factor $4\alpha A$ in front of Eq.(25) will pick up an additional z (if the average current is to stay constant), as will be shown in Sec.9 This means that the bulk of 1/f noise will be shifted to higher frequencies. Moreover, the half-lenght will be reduced, and may become accessible to the experiment. On the other hand, for small αA the noise disappears behind the low frequency horizon of our existence.

2. If $f_o \to 0$ in Eq.(26), the integrated noise seems to disappear since no elastic part is left with which the bremsstrahlung part can interfere. But even if we would carve out the stem of elastic scattering completely, the integrated noise would not go to zero as Eq.(26) predicts. A residual noise would be provided in this case by the second term on the right hand side of Eq.(23) and the treatment of the problem has to be modified. The noise will remain finite.

3. If τ is put zero in Eqs.(23-24) we obtain the same result as Eq.(25) with the wavenumber K in the $\hat{p}$ direction replacing f . Thus, along the direction of the beam we have spatial turbulence, with the same 1/k spectrum as the 1/f spectrum in frequencies. As in the classical turbulence model [23-25], we have thus spatiotemporal 1/f noise (which corresponds to the K^{-3}

spectrum in the isotropic three-dimensional case).

The autocorrelation function of $1/f$ noise in time and in the spatial direction of the beam is defined by Eq.(23). If we extend the limits to zero and infinity respectively, the first term on the right hand side can be integrated [32] yielding

$$A(\eta)\approx 4|a|^4\frac{\vec{p}^2}{m^2}\int_0^\infty\left(\frac{\varepsilon}{\varepsilon_0}\right)^{\alpha A}\cos\varepsilon\eta\frac{d\varepsilon}{\varepsilon}=4|a|^4\frac{\vec{p}^2}{m^2}\frac{\Gamma(\alpha A)}{(\varepsilon_0\eta)^{\alpha A}}\cos\frac{\pi\alpha A}{2}\ , \tag{27}$$

where $\Gamma(z)$ is the Γ-function. This simple $\eta^{-\alpha A}$ law has to be modified both at very large ($\gtrsim\varepsilon_0^{-1}$) and very small ($\lesssim\varepsilon_1^{-1}$) correlation times (or distance). The exact form can be given in terms of the incomplete Γ-function [32]

$$A(\eta)=2|a|^4\frac{\vec{p}^2}{m^2}(\varepsilon_0\eta)^{-\alpha A}\{e^{-i\frac{\pi}{2}\alpha A}[\Gamma(\alpha A,i\varepsilon_0\eta)-\Gamma(\alpha A,i\varepsilon_1\eta)]$$

$$+\ e^{i\frac{\pi}{2}\alpha A}[\Gamma(\alpha A,-i\varepsilon_0\eta)-\Gamma(\alpha A,-i\varepsilon_1\eta)]\}\ . \tag{28}$$

On the other hand, we can simplify Eq.(27) even further by writing $\Gamma(\alpha A)\approx(\alpha A)^{-1}$ because of $\alpha A<<1$.

8. Measured 1/f Noise

1. In any measurement of current fluctuations, the current is divided into a constant component and noise. By gradually extending the duration of the experiment and lowering the low frequency limit of the spectral measurement, we gradually redefine the meaning of "constant current", implicitely reducing the d.c. component. In the limit of infinite resolution no d.c.

component is left, and the whole current is a noise current. Due to this process of continuous redefinition of the meaning of "constant", the precision of a $1/f$ noise limited measurement is not improved by simply measuring for a longer time. Indeed, the expectation value of the variance increases logarithmically with time as lower and lower frequencies are included. This has led to the conclusion that $1/f$ noise is not ergodic.

Our results show that the $1/f$ spectrum is shaped and attenuated by infrared radiative corrections so that is is convergent at low frequencies. Furthermore, the normalization of the whole $1/f$ spectrum changes gradually when lower and lower frequencies are included, but remains finite (see Eq. 29 below, and Eqs. 25-26). Thus, in general, time averages will not be equivalent to ensemble averages. However, they are equivalent in the limit of low f_o, i.e. when the whole process is included. We conclude that $1/f$ noise is ergodic in the limit $f_o \to 0$, where f_o is the low frequency threshold of the noise measurement (quasiergodicity).

We may, however, perform noise measurements in a fixed frequency interval with a floating determination of the average current. Thus, if the interval is 10^{-2} -10^3Hz, the average current reference should be floating with a relaxation time of 10^3sec, i.e. should be in every moment the average current over the last 10^3sec. The fluctuations defined with respect to the floating average, with a fixed integration time, will be ergodic for any f_o.

2. So far we have considered only the case of a multichannel noise spectrometer, by which the various spectral components are measured simultaneously with the same value of f_o. Often, however, a spectral measurement is gradually extended to lower frequencies by performing consecutive monochannel measurements over a series of adjacent narrow spectral intervals of monotonously decreasing center frequency. In this case f_o is below, but very close to, any frequency in the given small interval and we have to set $f_o \approx f$ in the last form of Eq.(25). If we integrate again from some lower limit f_{oo} which represents the low frequency limit of the lowest frequency interval measured, we obtain

$$<(\delta j)^2>' = 2[1-(f_{oo}/f_1)^{2\alpha A}]<j>^2 \qquad (29)$$

instead of Eq.(26). This result increases monotonously when f_{oo} is lowered, and reaches a limit of $2<j>^2$ for $f_{oo}=0$. For practical applications, the difference between the two results is usually negligible because of the smallness of the accessible logarithmic frequency intervals compared to the half-length of 1/f noise.

3. The upper frequency limit of 1/f $f_1=\epsilon_1/h$ is essentially the kinetic energy of the carriers, divided by Planck's constant. Generally, the measuring apparatus will register the passage of individual current carriers already at much lower frequencies $f_1'\approx 1/\tau_o$ where τ_o is a characteristic time for the passage of a carrier through the measuring system. Consequently, a shot noise term of the form $4\tau_o(1+\omega^2\tau_o^2)^{-1}$ will be present in the noise spectrum in addition to the 1/f contribution. While shot noise arises from individual pulses, 1/f noise arises from the distribution in time of the same pulses which make up the current. Thus, the 1/f and shot noise contribution express different spects of the same basic phenomenon. The distribution properties of the shot noise pulses, i.e. the noise properties at lower frequencies, are determined as fluctuations of the probability current j . The upper frequency,at which 1/f noise is covered by other types of fluctuations, depends on the relative magnitude of the 1/f and other noise contributions.

At the characteristic frequency of shot noise, and above, 1/f noise will be reduced by a self-interference effect. Indeed, if Δr is the spatial extent of the measuring area (or sample) Eq.(13) shows that "spatial" 1/f noise will be averaged over, if $(\epsilon/v)\Delta r=\epsilon\tau_o\geqslant 1$. This, however, corresponds to the characteristic frequency at which shot noise begins to decrease.

A question left open here is the question of the dependence of 1/f noise on the coupling with the measuring apparatus. The measuring apparatus may increase the density of final states for bremsstrahlung photons. In addition, the consideration of bremsstrahlung and inverse bremsstrahlung at an equilibrium radiation temperature $T \neq 0$ is of interest.

9. 1/f Noise from Macroscopic Samples

Consider, e.g., the current noise of a cylindrical semiconductor sample of length l with ohmic contacts at the ends. Suppose that a nonlocalized electron from a certain point of the Fermi sphere of the metal at one side is scattered into a state which carries the current through the semiconductor to the metal on the other side. If we average this state over the distribution of atomic position coordinates and of atomic and crystal parameters, we obtain the coherent wave [33] which carries the electron through the sample. This averaged wave also includes the effect of multiple scattering in the semiconductor. The coherent wave is defined over the whole sample, and has a form similar to Eq.(17) with the effective mass of the carriers and the value of the phase velocity of the light in the medium substituted in Eqs.(18) and (19). This coherent wave will yield a current operator similar to Eq.(20) and current noise in accord with Eqs.(25) - (26).

If $j_i = \langle j_i \rangle + \delta j_i$ denotes the current carried by the carrier i in the axial direction, the relative current fluctuation through the sample can be written in the form

$$\frac{\delta I}{\langle I \rangle} = \frac{(1/l)\sum_i \delta j_i}{(1/l)\sum_i \langle j_i \rangle} . \qquad (30)$$

Let N be the number of carriers in the semiconductor sample. In terms of the single-carrier noise spectrum $\langle (\delta j)^2 \rangle_f$ we obtain for the whole sample

$$\frac{\langle (\delta I)^2 \rangle_f}{\langle I \rangle^2} = \kappa \frac{\sum_i \langle (\delta j_i)^2 \rangle_f}{(\sum_i \langle j_i \rangle)^2} = \kappa \frac{N \langle (\delta j)^2 \rangle_f}{N^2 \langle j \rangle^2} \qquad (31)$$

where $\kappa \geqslant 1$ is a correction factor which takes into account the correlations between carriers. For $\kappa = 1$ no correlations are present. On the other hand, if the carriers would always scatter in groups of z and would carry the current in groups of z, with N still denoting the number of individual carriers, one would obtain $\kappa = z^3$ provided no correlation is present between the groups, and $\kappa \geqslant z^3$ in general. Considerations of this type should be applicable not only for a semiconductor sample, but also for vacuum tubes, bad contacts, thin sheats, ions in superfluid helium [34], and, in a special way, for electrolytes. In superconductors (weak links, SQIDs) one would expect $\kappa = 2$ for the Cooper pairs. Using Eq.(25) with $(ff_o/f_1^2)\alpha A \approx 1$, one obtains from Eq. (31)

$$<(\delta I)^2>_f = 4\kappa\alpha\bar{A}f^{-1}N^{-1}<I>^2 \tag{32}$$

Here $\bar{A}$ is an average over the scattering angle θ_o in Eq.(19). If we take the average of $\sin^2(\theta_o/2)$ to be 1/3 and set $\beta = 1$ in Eq.(32) we obtain with $\kappa = 1$

$$4\kappa\alpha\bar{A} = 4\frac{8\alpha}{9\pi} \approx 8.10^{-3} \tag{33}$$

This result is of the order of magnitude of the experimental value estimated by Hooge [1-10-11] to be approx. 2.10^{-3}. I do not think this coincidence is accidental. In fact, the relevant phase velocity of electromagnetic waves in metals is quite small, of the order of a few cm/s. It is conceivable that most of the low frequency photons are emitted when the velocity of the carriers approaches the order of magnitude $\beta = 1$ from below. For the present problem of noise from a sample of N electrons it is important to note that the results obtained in Sec. 6 for a pure inititial state $e^{i\vec{p}.\vec{r}-iEt}$ are trivially extended to the case in which the initial state is a mixture described by a density matrix. The 1/f noise contributions of the various components will add up to give the total 1/f noise.

10. The Universality of 1/f Noise

The low frequency 1/f noise spectrum obtained here for the case of the electromagnetic interaction will occur for any interaction which leads to infrared divergences. This includes some of the known elementary excitations in solids as replacements for the photons considered above.

More important is the fact, that the gravitational interaction leads to similar infrared divergences. Consequently, the emission of gravitons from a deflected stream of matter will cause the flow rate (debit) to present low frequency fluctuations with a similar 1/f spectrum as described here. For the case of a jet of water (fire hose) deflected, e.g. by the incidence on a rubber conweyer-belt, the equivalent of the αA constant can be much larger than αA and strong, easily detectable, 1/f noise should be present in the flow rate of the outgoing jet. The fact, that large masses of the order of grams scatter as a whole, overcompensates the smallness of the gravitational constant, allowing for strong quantum-gravitational 1/f noise. Needless to say, astrophysical jets like spiral arms of galaxies or Saturn's belts, and the angular velocity of planets and stars should present similar spatio-temporal 1/f noise, over reciprocal cosmic intervals, as the water-jet considered above. We are led to believe that spatio-temporal 1/f noise is the cause of a hierarchic cluster structure of the universe not only in space, but also in time.

The macroscopic quantum phenomenon of 1/f noise opens a new avenue of quantum physics, the road of very low frequency (secular) fluctuations. It seems that nature has decided to lift the veil of secrecy from the nonlinear quantum theory of matter, time and space, by starting with the limit of low frequencies and wave-numbers.

The author acknowledges many useful encouraging discussions with C. Eftimiu, T.P. Cheng, and P.B. James from the University of Missouri-St. Louis, as well as with A. van der Ziel, D. Halford, S. Teitler, J.J. Brophy, V. Sergiescu and G. Uhlenbeck, who are also insiders and pioniers of the 1/f noise problem. He is indebted to J. Devreese, K.H. Michel and W. Huybrechts from the Universitaire Instelling Antwerpen (Belgium) for interesting, useful, discussions and a careful verification of some basic calculations respectively. Last, but not least, the continuous support of the National Science Foundation is acknowledged with pleasure.

APPENDIX

Evaluation of the Contribution from Vector Potential Terms

The current operator is

$$\vec{j}=\frac{-i}{2m}[\,\Psi_s^+\nabla\Psi_s-(\nabla\Psi_s^+)\Psi_s] +\frac{e}{m}\Psi_s^+\vec{A}\Psi_s\,, \tag{A1}$$

where

$$\vec{A}=\int\frac{d^3k}{\sqrt{2k_o(2\pi)^3}}\sum_\lambda \vec{n}^{(\lambda)}(k)[\,a_{\vec{k},\lambda}e^{-ik.x}+a^+_{\vec{k},\lambda}e^{ik.x}]\,, \tag{A2}$$

and Ψ_s is given by Eq(17) with

$$a_\varepsilon=\frac{\sqrt{\alpha}}{2\pi\sqrt{\alpha A}}\int\sum_\lambda(\frac{n^{(\lambda)}.p}{k.p}-\frac{n^{(\lambda)}.p_i}{k.p_i})a_{\vec{k},\lambda}e^{-i\gamma(\varepsilon)}\varepsilon^2 d\Omega_{\vec{k}}\;. \tag{A3}$$

The explicit form of the operator of the current fluctuation defined by Eqs.(A1-2) is given by

$$|a|^{-2}\delta\vec{j}=\frac{\vec{p}}{m}[\,1+2\int_{\varepsilon_o}^{\varepsilon_1}(a_\varepsilon^+ + a_\varepsilon)\frac{d\varepsilon}{\varepsilon^{1/2}}$$

$$+\int_{\varepsilon_0}^{\varepsilon_1}\int_{\varepsilon_0}^{\varepsilon_1}(a^+_\varepsilon a^+_{\varepsilon'}+a_\varepsilon a^+_{\varepsilon'}+a^+_\varepsilon a_{\varepsilon'}+a_\varepsilon a_{\varepsilon'})\,\frac{d\varepsilon d\varepsilon'}{(\varepsilon\varepsilon')^{1/2}}] \qquad (A4)$$

$$+\ \frac{e}{m}[\ \vec{A}+\int_{\varepsilon_0}^{\varepsilon_1}(a^+_\varepsilon\vec{A}+a_\varepsilon\vec{A}+\vec{A}a^+_\varepsilon+\vec{A}a_\varepsilon)\ \frac{d\varepsilon}{\varepsilon^{1/2}}$$

$$+\int_{\varepsilon_0}^{\varepsilon_1}\int_{\varepsilon_0}^{\varepsilon_1}(a^+_\varepsilon\vec{A}a^+_{\varepsilon'}+a_\varepsilon\vec{A}a^+_{\varepsilon'}+a^+_\varepsilon\vec{A}a_{\varepsilon'}+a_\varepsilon\vec{A}a_{\varepsilon'})\ \frac{d\varepsilon d\varepsilon'}{(\varepsilon\varepsilon')^{1/2}}]-|a|^{-2}<\vec{j}>$$

Here and in the following expression we have to read $\rho_\varepsilon d\varepsilon$ every time we see an energy differential. In addition, we have suppressed all phase factors.

In the corresponding autocorrelation function normal ordering of the two $\vec{A}$ operators in all terms quadratic with respect to the vector potential $\vec{A}$ is assumed, and the bracketts are vacuum expectation values :

$$|a|^{-4}<\vec{j}^+(t)\vec{j}(t+\tau)>=\frac{\vec{p}^2}{m^2}\ [\ 1+4\int<a_\varepsilon a^+_{\varepsilon'}>\frac{d\varepsilon d\varepsilon'}{(\varepsilon\varepsilon')^{1/2}}$$

$$+\iiiint<a_\varepsilon a^+_{\varepsilon'}a_{\varepsilon''}a^+_{\varepsilon'''}+a_\varepsilon a_{\varepsilon'}a^+_{\varepsilon''}a^+_{\varepsilon'''}>\frac{d\varepsilon d\varepsilon' d\varepsilon'' d\varepsilon'''}{(\varepsilon\varepsilon'\varepsilon''\varepsilon''')^{1/2}}]$$

$$+\ \frac{e\vec{p}}{m^2}\cdot\ [\ 4\int<a_\varepsilon\vec{A}+\vec{A}a^+_\varepsilon>\frac{d\varepsilon}{\varepsilon^{1/2}}$$

$$+\ 3\iiint<a_\varepsilon a^+_{\varepsilon'}\vec{A}a^+_{\varepsilon''}+a_\varepsilon a_{\varepsilon'}\ \vec{A}a^+_{\varepsilon''}+h.c.>\frac{d\varepsilon d\varepsilon' d\varepsilon''}{(\varepsilon\varepsilon'\varepsilon'')^{1/2}}$$

(A5)

$$+\iiint \langle a_\varepsilon a_{\varepsilon'} a^+_{\varepsilon''}\vec{A} + a_\varepsilon a^+_{\varepsilon'} a_{\varepsilon''}\vec{A} + h.c.\rangle \frac{d\varepsilon d\varepsilon' d\varepsilon''}{(\varepsilon\varepsilon'\varepsilon'')^{1/2}} \,]$$

$$- |a|^{-4}\langle\vec{j}\rangle^2$$

$$+ \frac{e^2}{m^2}\,[\iint \langle 2\vec{A}a^+_\varepsilon\vec{A}a^+_{\varepsilon'} + 2a_\varepsilon\vec{A}a_{\varepsilon'}\vec{A} + a_\varepsilon\vec{A}\vec{A}a^+_{\varepsilon'} \frac{d\varepsilon d\varepsilon'}{(\varepsilon\varepsilon')^{1/2}}$$

$$+\iiiint \langle a_\varepsilon\vec{A}a^+_{\varepsilon'} a^+_{\varepsilon''}\vec{A}a^+_{\varepsilon'''} + h.c. + a_\varepsilon\vec{A}a_{\varepsilon'} a^+_{\varepsilon''}\vec{A}a^+_{\varepsilon'''}\rangle$$

$$\frac{d\varepsilon d\varepsilon' d\varepsilon'' d\varepsilon'''}{(\varepsilon\varepsilon'\varepsilon''\varepsilon''')^{1/2}} \,] \qquad \text{(A5)}$$

Here h.c. denotes the hermitian conjugate of all preceding terms in the bracket. The vector potential introduces four types of terms :

$$1)\quad \frac{e\vec{p}}{m^2}\cdot\int_{\varepsilon_o}^{\varepsilon_1} \rho_\varepsilon e^{i\varepsilon t}\langle\vec{A}a^+_\varepsilon\rangle\frac{d\varepsilon}{\sqrt{\varepsilon}} =$$

$$\frac{\alpha\vec{p}}{\sqrt{2\alpha A}(2\pi)^{5/2}m^2}\cdot\int_{\varepsilon_o}^{\varepsilon_1} \rho_\varepsilon \varepsilon d\varepsilon \int d\Omega_{\vec{k}} e^{i\vec{k}\cdot\vec{r}+i\gamma(\varepsilon)} \sum_\lambda n^{(\lambda)}(k)$$

$$\times\left(\frac{n^{(\lambda)}\cdot p}{k\cdot p} - \frac{n^{(\lambda)}\cdot p_i}{k\cdot p_i}\right). \qquad \text{(A6)}$$

This type does not contribute to 1/f noise; due to the absence of the energy denominator it has a white spectrum. Its contribution to the low frequency spectrum is negligible as $\varepsilon/\varepsilon_1$. Indeed, for $\varepsilon=\varepsilon_1$ it is of the same order as the main 1/f term.

$$2)\ \frac{e\vec{p}}{m^2}\iiint \rho_\varepsilon \rho_{\varepsilon'} \rho_{\varepsilon''} e^{i(\varepsilon'-\varepsilon+\varepsilon'')t+i\varepsilon''\tau} \langle a_\varepsilon a^+_{\varepsilon'} \vec{A} a^+_{\varepsilon''}\rangle$$

$$\times\frac{d\varepsilon d\varepsilon' d\varepsilon''}{(\varepsilon\varepsilon'\varepsilon'')^{1/2}} = \frac{\alpha(\varepsilon_1-\varepsilon_0)\vec{p}}{\sqrt{2}(2\pi)^{5/2} m^2}\int_{\varepsilon_0}^{\varepsilon_1} \rho_\varepsilon^2 d\varepsilon \int d\Omega_{\vec{k}} e^{i\vec{k}.\vec{r}+i\gamma(\varepsilon)} \sum_\lambda n^{(\lambda)}(k)$$

$$\times\left(\frac{n^{(\lambda)}.p}{k.p} - \frac{n^{(\lambda)}.p_i}{k.p_i}\right) . \qquad (A7)$$

Terms of this type contribute to 1/f noise. However, they are smaller and can safely be neglected, being of order $\alpha\beta|\Delta\vec{\beta}|/(2\pi)^2$, compared to the leading 1/f term obtained by neglecting the vector potential contribution.

$$3)\ \frac{e^2}{m^2}\iint \rho_\varepsilon \rho_{\varepsilon'} e^{i(\varepsilon+\varepsilon')t+i\varepsilon'\tau} \langle \vec{A} a^+_\varepsilon \vec{A} a^+_{\varepsilon'}\rangle \frac{d\varepsilon d\varepsilon'}{(\varepsilon\varepsilon')^{1/2}} =$$

$$\frac{m^2}{\vec{p}^2}\ (\text{type } 1) \ . \qquad (A8)$$

These terms are similar to type 1, do not yield a 1/f spectrum, and are negligible as $(\varepsilon/\varepsilon_1)^2$ at low frequencies.

$$4)\quad \frac{e^2}{m^2}\iiiint \rho_\varepsilon \rho_{\varepsilon'} \rho_{\varepsilon''} \rho_{\varepsilon'''} e^{-i(\varepsilon+\varepsilon')t+i(\varepsilon''+\varepsilon''')(t+\tau)}$$

$$<a_\varepsilon \vec{A} a^+_{\varepsilon'} a^+_{\varepsilon''} \vec{A} a^+_{\varepsilon'''} \frac{d\varepsilon d\varepsilon' d\varepsilon'' d\varepsilon'''}{(\varepsilon\varepsilon'\varepsilon''\varepsilon''')^{1/2}} =$$

$$\frac{\alpha}{m^2}\int \rho^2_{\varepsilon'} e^{i\varepsilon'\tau} \frac{d\varepsilon'}{\varepsilon'} \iint \rho_\varepsilon \rho_{\varepsilon''} e^{i(\varepsilon''-\varepsilon)t+i\varepsilon''\tau} <a_\varepsilon \vec{A}\vec{A} a_{\varepsilon''}>$$

$$\frac{d\varepsilon d\varepsilon''}{(\varepsilon\varepsilon'')^{1/2}}$$

This yields a negligible 1/f noise contribution of order $[\alpha\beta\Delta\vec{\beta}/(2\pi)^2]^2$, compared to the leading term.

Consequently, all vector potential terms can be neglected. Note that the $e^{i\vec{K}\cdot\vec{r}}$ dependence leads also to a $1/r$ (types 1 and 2) or $1/r^2$ dependence on the distance r from the scattering point, which further reduces the terms, but is not important at low frequencies.

REFERENCES

1. H. Bittel and L. Storm : Rauschen (Springer, Berlin, 1971).
2. A. Van der Ziel : Noise, Sources, Characterization, Measurement, (Prentice Hall Inc., Englewood Cliffs, N.J., 1970).
3. A. Van der Ziel : Fluctuation Phenomena in Semiconductors (Butterworths, London, 1959).
4. L.D. Smullin and H.A. Haus: Noise in Electron Devices (J. Wiley & Sons, New York, 1959).
5. D.A. Bell : Electrical Noise; Fundamentals and Physical Mechanism (Van Rostrand, London, 1960).
6. H. Pfeifer : Elektronisches Rauschen (B.G. Teubner, Leipzig, 1959).
7. R.E. Burgess : Fluctuation Phenomena in Solids (Academic Press, New York, 1965).
8. P.H. Handel, Bull. Amer. Phys. Soc. January 1975 (Annual Meeting, Anaheim).
9. P.H. Handel, Phys. Rev. Letters 34, 1492,1495 (1975).
10. F.N. Hooge, Physica 60, 130 (1972); Physics Letters 29A, 139 (1969).
11. F.N. Hooge and A.M.H. Hoppenbrouwers, Physica 45, 386 (1969); Physics Letters 29A, 642 (1969).
12. K. Scheidhauer, Z.Angew.Phys. 13, 380 (1961).
13. J. Clarke and R.F. Voss, Phys. Rev. Letters 33, 24 (1974).
14. J. Clarke and T.Y. Hsiang, Phys. Rev. Letters 34,1217 (1975).
15. B.V. Rollin and I.M. Templeton, Proc. Phys. Soc.(Lond.) B66, 259 (1953); B67,271 (1954).
16. D.K. Baker, J. Appl. Phys. 25, 922 (1954).
17. T.E. Firle and H. Winston, J. Appl. Phys. 26, 716(1955)
18. I.R.M. Mansour, R.J. Hawkins and G.G. Bloodworth,Radio Electronic Engr. 212 (1968); M.A. Caloyannides, J. Appl. Phys. 45, 307 (1974).
19. F.N. Hooge, Phys. Letters 33, 165 (1970).
20. A.A. Verveen and H.E. Derksen, Proc. I.E.E.E. 56,906 (1968).
21. P.H. Handel, Rev. Roumaine de Phys. 7, 407 (1962).
22. P.H. Handel, Physics Letters 18, 224 (1965).
23. P.H. Handel, Z. Naturforschung 21a, 579, 573 (1966).
24. P.H. Handel, Phys. Status Solidi 29, 299 (1968).
25. P.H. Handel, Phys. Rev. A3, 2066 (1971), with minus in both forms of Eq.(4.7).
26. P.H. Handel, Quantum Theory of 1/f Noise (1972), unpublished.
27. P.H. Handel, Single particle Diffusion and 1/f Noise (1973), unpublished.

28. See e.g. J.D. Bjorken and S.D. Drell, Relativistic Quantum Mechanics (McGraw-Hill, New York, 1964) p.p.96, 121-126.
29. D.R. Yennie, S.C. Frautschi, H. Suura, Annals of Physics (N.Y.) 13, 379 (1961).
30. F. Bloch and A. Nordsieck, Phys. Rev. 52, 54 (1937).
31. V. Chung, Phys. Rev. 140B, 1110 (1965).
32. I.S. Gradstein and I.M. Ryzhik : Table of Integrals, Series, and Products (Academic Press, New York, 1965) Eqs. 3.761-9 and 3.7761-7.
33. M. Lax, Rev. Mod. Phys. 23, 387 (1951), Eq.(5.5).
34. H. Hoch, L. Busse and F. Moss, Phys. Rev. Lett. 34, 384 (1975).

ELECTRON-HOLE LIQUID CONDENSATION IN SEMICONDUCTORS

T. L. Reinecke*

Naval Research Laboratory
Washington, D.C. 20375 USA

and

S. C. Ying† ‡

Laboratoire de Physique des Solides
Université Paris-Sud, 91405-Orsay, France

The gas-liquid phase diagram for electron-hole liquid condensation in semiconductors is discussed. Its general features can be understood from straightforward statistical considerations using a knowledge of the ground state energy; particular attention is given to the region $T \gtrsim \frac{1}{2} T_c$.

I. INTRODUCTION

At sufficiently low temperature and density, electrons and holes created in pure Ge and Si form bound excitons. At higher density, interactions between excitons become important, and at sufficiently high density (of the order of 10^{12} to 10^{17} cm^{-3}), a new collective state of the electrons and holes, which is lower in energy than the exciton, appears. This is a "metallic liquid-like" collective state of electrons and holes, and it occurs as micron sized (or larger) droplets of the electron-hole liquid (EHL). In recent years much experimental and theoretical work has been done on this system.[1]

As a function of density and temperature, the carrier system consists of varying densities of excitons, free electrons and holes, and EHL, and the electronic properties of the material depend strongly on the state of this system. In (the indirect gap semiconductors) Ge and Si the carrier lifetimes are sufficiently long that the electron-hole system can be thought of as existing in thermal equilibrium.

An especially full picture of EHL condensation is given by its phase diagram, which is similar to that for a first order classical gas-liquid condensation (see Fig. 1). This diagram is the boundary as a function of density and temperature separating a "gas" phase, composed of excitons, electrons, and holes, from the high density EHL and from a coexistence region (characterized by droplets of EHL) of the two phases. Condensation occurs only for temperatures less than a critical temperature T_C. Thomas et al.[2] have made the most detailed measurements of the phase diagram, and their data are shown in Fig. 1. In these measurements, the gas side of the phase diagram is obtained from the density (or temperature) of the onset of carrier recombination radiation characteristic of the EHL phase, and the density on the liquid side is obtained from the lineshape of the EHL recombination radiation.

It is the purpose of this article to describe the theoretical understanding of the various portions of the EHL phase diagram. We shall show that its general features can be understood using relatively straightforward statistical considerations and a knowledge of the ground state energy. For clarity we shall refer to the case of Ge; similar considerations apply to Si.

We begin by briefly describing the current relatively good understanding of the ground state energy of EHL. For an infinite, homogeneous system at zero temperature and at high density, the electrons and holes move in the semiconductor band structure (four conduction bands and two coupled valence bands for Ge) and interact with one another via the Coulomb interaction screened by the static semiconductor dielectric constant. The ground state energy of ρ electron-hole pairs per unit volume is of the form

$$E(\rho) = E_k(\rho) + E_x(\rho) + E_c(\rho) \qquad (1)$$

where E_k is the positive kinetic energy. The exchange energy, E_x, and the correlation energy, E_c, lower the total energy because like charges are anticorrelated quantum mechanically. E_k and E_x are calculated exactly, and E_c has been calculated by several groups using modifications of the Random Phase Approximation (RPA).[3,4] They find that $E(\rho)$ has a minimum at a density ρ_o where the binding energy per pair is lower than that of the exciton. The density ρ_o lies in the "high density" ($r_s < 1$) regime (r_s = interparticle spacing/exciton Bohr radius), where the RPA is expected to be a good approximation. In Ge (and Si) the electrons and holes are distributed over several bands which lowers their kinetic energy and thus plays an important role in stabilizing EHL relative to the exciton.

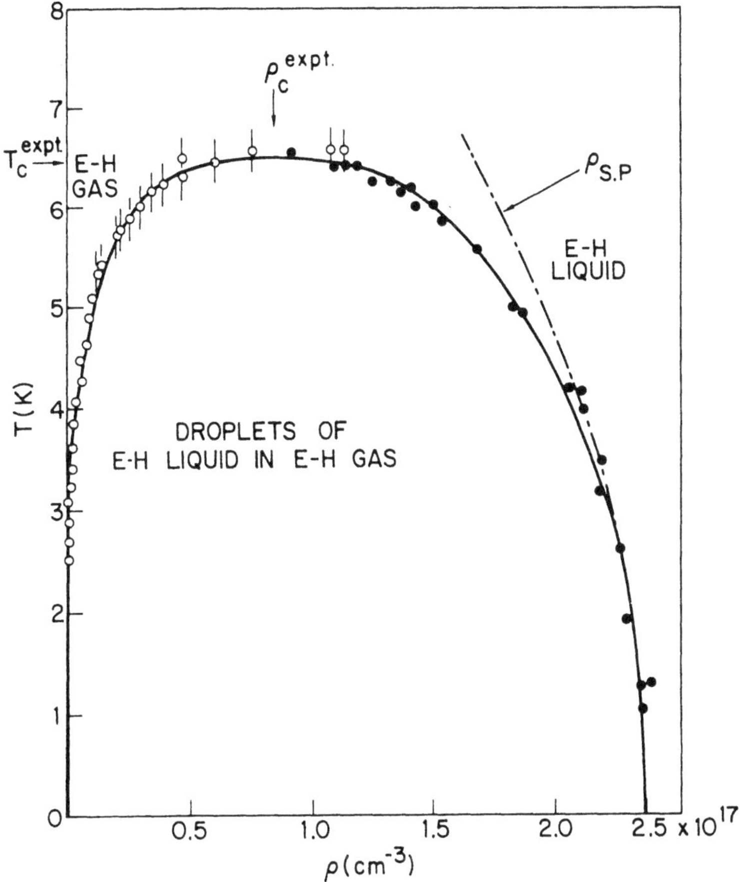

Fig. 1. Phase diagram in temperature and density of electrons and holes in Ge: experimental values are from Ref. 2; solid curve is the result of droplet fluctuation model discussed in text, Sec. III; $\rho_{S.P.}$ gives density variation due to single particle excitations obtained from the experimental values.

II. PHASE DIAGRAM FOR $T \lesssim \frac{1}{2} T_c$

The low temperature ($T \lesssim \frac{1}{2} T_c$) liquid side of the phase diagram can be understood by considering the thermal expansion of the dense EHL.[4,5] At these low temperatures the vapor pressure of the surrounding gas will have negligible effect on the liquid. The free energy of the EHL can be expanded for small T/T_F (the Fermi temperature is 45K for holes and 29K for electrons)

$$F(\rho,T) \cong E(\rho) - \frac{1}{2} \gamma(\rho) T^2 \qquad (2)$$

where the entropy γT arises from single particle excitations. The liquid density is obtained from the minimum of Eq. (2), and its resulting T^2 variation has been observed in detail experimentally.[5] These measurements also show that the free particle value of $\gamma(\rho)$ gives reasonably good agreement for the T^2 coefficient in the density; Rice[6] has pointed out that this is the case because the Coulomb many-body interactions should not significantly change the shape of the bands, but rather lower them rigidly. The liquid density for $T \lesssim \frac{1}{2} T_c$ as obtained from the measurements of Thomas et al. is shown as $\rho_{S.P.}(T)$ in Fig. 1.

The low temperature ($T \lesssim \frac{1}{2} T_c$) gas side of the phase diagram can be understood with a simple picture of a classical gas in equilibrium with the liquid.[7,8] At sufficiently low temperature and density, the gas phase consists of excitons for which the chemical potential is that of a classical gas

$$\mu_g = E_B + k_B T \,\ell n \left[\rho_g \, h^3/d \, (2\pi M k_B T)^{3/2}\right] \tag{3}$$

where ρ_g is the gas density, and E_B, M, and d are the exciton binding energy, mass and degeneracy. On the condensation curve $\mu_g = \mu_\ell$, the chemical potential of the liquid, and the gas density is

$$\rho_{g,coex.} \propto (k_B T)^3 \exp\left[-(E_B - \mu_\ell)/k_B T\right] \tag{4}$$

This dependence of $\rho_{g,coex.}$ on T gives very good agreement with experiment for 10^{12} cm^{-3} < $\rho_{g,coex.}$ < 10^{15} cm^{-3} and 2K < T < 4.2K.[7] It is not presently clear for temperatures as high as 4K to what extent the gas remains in the form of excitons or dissociates into electrons and holes (see further discussion, Section IV). Nonetheless, Eq. (3) appears to be a good approximation to its chemical potential.

III. THE PHASE DIAGRAM FOR $T \gtrsim \frac{1}{2} T_c$

It is clear in Fig. 1 that the phase diagram for $T \gtrsim \frac{1}{2} T_c$ deviates significantly on both the liquid and gas sides from that obtained from the simple descriptions given above. One way to see the reason for this is shown in Fig. 2. In principle, for $T < T_c$, $\rho_{g,coex.}$ and $\rho_{\ell,coex.}$ are obtained from a complete knowledge of μ as a function of ρ at a given T; a Maxwell construction is then used to locate both $\rho_{g,coex.}$ and $\rho_{\ell,coex.}$ At the present time, however, μ is known only at high density, Eqs. (1) and (2), and at low density, Eq. (3); it is not known in intermediate density regime. Thus neither side of the coexistence curve for temperatures $T \gtrsim \frac{1}{2} T_c$ (except very near T_c, see below) can be obtained. The physical meaning of the Maxwell construction can be seen in the case of $\rho_{\ell,coex.}$ by noting that the liquid is in equilibrium with its vapor pressure, which makes an appreciable

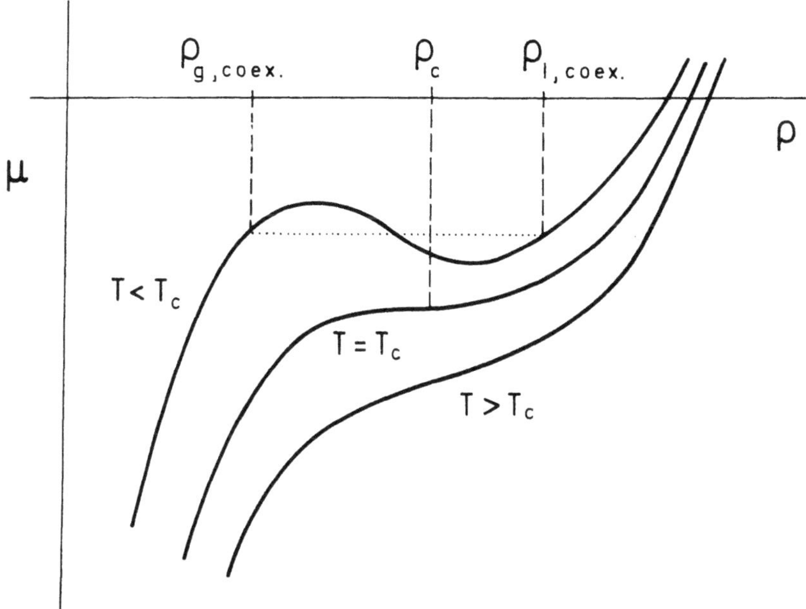

Fig. 2. Sketch of chemical potential μ vs. density ρ showing the coexistence values of the liquid and gas densities and the critical density and temperature ρ_c and T_c.

contribution to the liquid free energy, Eq. (2), at these higher temperatures.

Combescot[8] has pointed out that the critical density ρ_c, which is located by an inflection point in $\mu(\rho)$ (Fig. 2), in fact occurs in the high density regime. Therefore, since no Maxwell construction is required there, the critical point (ρ_c, T_c) can be determined from a knowledge of the dense liquid phase alone. This has been done using the expansion of the EHL free energy in Eq. (2) with several treatments of the correlation energy $E_c(\rho)$ and yields satisfactory values of T_c in the range 5.9 - 8.3K.[2,4,8] The location of the inflection point involves $\partial^3 F(\rho)/\partial\rho^3$, and therefore is very sensitive to details of $E_c(\rho)$ for the dense liquid, which is known only approximately.

In order to treat the phase diagram throughout the region $T \gtrsim \frac{1}{2} T_c$, a droplet fluctuation model[9] is employed. This is a generalization of a model discussed by Fisher and others[10] in the theory of the phase transition and pre-transition phenomena for the classical gas-liquid condensation. The gas phase is taken to consist of electrons, holes and droplet fluctuations of the dense EHL phase. They are assumed to be non-interacting, and then the gas density at constant μ and T is

$$\rho_g = q_o \sum_{n=1}^{\infty} \exp\{-(F_B n + F_S a_o n^{\sigma} + k_B T\tau \ln(n) - \mu n)/k_B T\} \quad (5)$$

where the free energy of a droplet of n electron-hole pairs[12] is separated into a bulk term $F_B n$, a surface term $F_S a_o n^{\sigma}$, where $a_o n^{\sigma}$ is the droplet surface area, and a higher order term in $\ln(n)$; q_o is an overall proportionality constant. Condensation, which involves the onset of a stable liquid phase, is identified in this model by the incipient divergence of the fluctuations to the EHL phase: For $\mu > F_B$ the probability of very large (infinite) size droplet fluctuations diverges, and for $\mu < F_B$, it does not; therefore $\mu(T) = F_B$ gives the condensation point at a given T. Further, the number and size of the fluctuations is controlled by the surface tension of the EHL droplets, $F_S(T)$. As T approaches T_c, $F_S(T)$ decreases, and fluctuations increase. Finally at a temperature for which $F_S(T)$ vanishes, stable droplet formation is never possible, and therefore the condition $F_S(T = T_c) = 0$ determines T_c.

The liquid phase is pictured in this model in a similar way to the above as dense EHL containing non-interacting "bubble" fluctuations of gas. On the coexistence curve ($\mu = F_B$), the liquid density is

$$\rho_{\ell,\text{coex.}} = \rho_{S.P.}(T) - q_o \sum_{n=1}^{\infty} \exp\{-(F_S a_o n^{\sigma} + k_B T \ln(n))/k_B T\} \quad (6)$$

Here $\rho_{S.P.}(T)$ gives the density variation of EHL due to single particle excitations (from Eq. (2)), and the second term gives that due to bubbles. At low temperatures, Eqs. (5) and (6) reduce to the simpler forms used to describe the low temperature regions above.

Eqs. (5) and (6) give the phase diagram for $T \gtrsim \frac{1}{2} T_c$ entirely in terms of $\rho_{S.P.}(T)$ and $F_S(T)$, which are properties of the dense EHL phase alone. For all $T \leq T_c$, the expansion in small T/T_F, as in Eq. (2), can be used

$$\rho_{S.P.} \cong \rho_o (1 - \delta_\rho T^2) \quad (7a)$$

$$F_S(T) \cong F_S(0)(1 - \delta_F T^2) \quad (7b)$$

The phase diagram can now be obtained either "theoretically" by using $\rho_{S.P.}(T)$ and $F_S(T)$ taken from first principles calculations or by using experimental values for them.

Let us consider first whether Eqs. (5), (6) give a good account of the shape of the phase diagram for $T \gtrsim \frac{1}{2} T_c$ by choosing $\rho_{S.P.}(T)$ and $F_S(T)$ from experiment. $\rho_{S.P.}(T)$ from the data of Thomas et al.[2] (at $T << T_c$) is shown in Fig. 1 ($\rho_o = 2.38 \times 10^{17}$ cm^{-3}, $\delta_\rho = 0.0072K^{-2}$); the experimental value of $T_c = 6.5K$ gives δ_F ($\delta_F = T_c^{-2}$). Values for a_o, σ, τ, and q_o are obtained straightforwardly by assuming spherical droplets.[9] Then the solid curve in Fig. 1 is obtained by choosing $F_S(0) = 1.0 \times 10^{-4}$ erg/cm^2. Clearly the agreement with the experimental phase diagram is good on both sides. This value of $F_S(0)$ provides an experimental estimate of EHL surface energy from these measurements, and it is in reasonably good agreement with independent measurements based on supersaturation.[12]

By employing $\rho_{S.P.}(T)$ and $F_S(T)$ obtained from first principles calculations, a "theoretical" phase diagram for Ge which has the same shape as that in Fig. 1 is obtained. In this way, the droplet model, Eqs. (5), (6) also gives a new way to calculate the critical properties ρ_c and T_c, upon which we now comment. ρ_o and δ_ρ have been calculated by several groups[3,4] who obtain good agreement with experimental measurements. The present authors[13] have developed a method for calculating $F_S(T)$ based on the density functional method, and for Ge obtain

$$F_S(T) = (1.1 \times 10^{-4})\ (1 - (T/5.0)^2)\ \mathrm{erg/cm^2}.$$

This gives the estimate $T_c \cong 5K$. From Eqs. (5) and (6), $\rho_c = \frac{1}{2} \rho_{S.P.}(T_c)$; using $\rho_{S.P.}(T)$ calculated in Ref. 4, this gives $\rho_c \cong 0.845 \times 10^{17}$ cm^{-3}. These ρ_c and T_c for Ge compare favorably with the experimental values and also with theoretical estimates[4,8] based on the rather different approach illustrated in Fig. 2.

The phase diagram in Si has not yet been measured in detail. Using Eqs. (5) and (6) with calculated values of $\rho_{S.P.}(T)$ and $F_S(T)$, a complete diagram has been constructed for Si.[9] The values of ρ_c and T_c (= 1.0×10^{18} cm^{-3} and 21.6K) compare favorably with available measurements.[14,15]

IV. INSULATOR-METAL TRANSITION IN THE GAS

Returning to the gas phase, at very low temperatures and low density, it is composed primarily of excitons. As the temperature and density are raised, however, the excitons tend to dissociate into a gas of free electrons and holes. Rough estimates, based on the limits of stability of these two gas phases, give values of the density for this insulator-metal type transition between 10^{13} and 10^{15} cm^{-3}.[2,16] This suggests that such a transition is associated with the gas side of the phase diagram for T $\lesssim$ 4K; therefore, the assumption that the gas phase for T $\gtrsim \frac{1}{2} T_c$ is composed of free electrons and holes, which was used above in the treatment of that region, appears to be justified.

There has been some discussion in the literature concerning the nature of this transition in the gas - - specifically whether it is continuous or discontinuous, at what density and temperature it occurs, and to what extent it may be observable. At the moment the experimental situation on this point is unclear, and theoretical calculations are not adequate to give reliable information. The question of such a transition in the gas remains an important, unsettled problem.

In summary, we have shown that the general features of the EHL phase diagram can be understood quite well using straightforward statistical considerations and a knowledge of the ground state properties of the dense EHL.

REFERENCES

* NAS-NRL Resident Research Associate

† Permanent address: Department of Physics, Brown University, Providence, Rhode Island 02912, USA

†† A. P. Sloan Foundation Research Fellow. Supported in part by the National Science Foundation.

1. Useful reviews include Ya. E. Pokrovskii, Phys. Status Solidi (a) 11, 385 (1972); T. M. Rice, Proceedings of the XII International Conference on the Physics of Semiconductors, Stuttgart, 1974, ed. M. H. Pilkuhn, p. 23; M. Voos, Proceedings of the XII International Conference on the Physics of Semiconductors, Stuttgart, 1974, ed. M. H. Pilkuhn, p. 33.

2. G. A. Thomas, Phys. Rev. Lett. 33, 219 (1974).

3. W. F. Brinkman and T. M. Rice, Phys. Rev. B7, 1508 (1973); M. Combescot and P. Nozières, J. Phys. C: Solid State Physics 5, 2369 (1972); P. Bhattacharyya et al., Phys. Rev. B10, 5127 (1974).

4. P. Vashishta et al., Phys. Rev. Lett. 33, 911 (1974).

5. G. A. Thomas et al., Phys. Rev. Lett. 31, 386 (1973); T. K. Lo, Solid State Commun. 15, 1231 (1974).

6. T. M. Rice, Il. Nuov.Cim. 23B, 1 (1974).

7. J. C. Hensel, Phys. Rev. Lett., 30, 227 (1973); C. Benoit à la Guillaume and M. Voos, Phys. Rev. B7, 1723 (1973); T. K. Lo et al., Phys. Rev. Lett. 31, 224 (1973).

8. M. Combescot, Phys. Rev. Lett. 32, 15 (1974).

9. T. L. Reinecke and S. C. Ying, to be published.

10. M. E. Fisher, Physics 3, 255 (1967); J. Frenkel, Kinetic Theory of Liquids (Oxford Press, Oxford, 1946), Chapt. VII; F. F. Abraham, Homogeneous Nucleation Theory (Academic, New York, 1974).

11. Droplet fluctuations with unequal numbers of electrons and holes give a small contribution which can easily be included, see Ref. 9.

12. A. S. Alekseev et al., Proceedings of the XII International Conference on the Physics of Semiconductors, Stuttgart, 1974, p. 91; T. K. Lo et al., to be published.

13. T. L. Reinecke, F. Crowne, and S. C. Ying, Proceedings of the XII International Conference on the Physics of Semiconductors, Stuttgart, 1974, p. 61; T. L. Reinecke and S. C. Ying, to be published.

14. B. M. Ashkinadze et al., Zh. Eksp. Teor. Fiz. 58, 507 (1970), (Sov. Phys. JETP 31, 271 (1970)).

15. We would like to point out that M. Droz and M. Combescot (to be published) have recently used a quite different approach based on a phenomenological lattice gas model to treat the high temperature region of EHL condensation.

16. W. F. Brinkman and T. M. Rice, Ref. 3; T. M. Rice, Ref. 1.

THERMODYNAMIC AND TRANSPORT PROPERTIES IN THE HUBBARD MODEL*

E. N. Economou

Department of Physics, University of Virginia

Charlottesville, Virginia 22901, USA

I. INTRODUCTION

I present here a progress report on our study of the Hubbard model; in the present work the Hubbard Hamiltonian is approximated by a static, random, self-consistent potential which allows a quantitative treatment over the whole range of values of the relevant parameters.

The present work was undertaken in order to develop a formalism which would allow the quantitative study of systems such as impurity bands in crystalline semiconductors[1] and one dimensional organic conductors.[2] Electron correlations seem to play a very significant role in the physical properties of these systems. Thus one is forced to consider a Hubbard type of Hamiltonian which is actually the simplest model Hamiltonian believed to incorporate the essential physical features resulting from the electron correlations.

The physical systems of interest here exhibit structural disorder in addition to electron correlations. Thus it becomes next to impossible to distinguish the physical origin of the observed properties without a quantitative study of both structural disorder and electron correlations. In this work a one-electron approximation to the electron correlation problem is developed. This approximation simplifies the problem conceptually and it is amenable to detailed quantitative study while, in my opinion, still retains the essential physical properties of the original Hubbard Model.

* Work supported by NSF Grant No. GH-37264

II. THE BASIC IDEAS

The Hubbard Hamiltonian is given by[3]

$$H = \sum_{n\sigma} \varepsilon_o a^+_{n\sigma} a_{n\sigma} + \sum_{nm\sigma} V_{nm} a^+_{n\sigma} a_{m\sigma} + U \sum_n a^+_{n\uparrow} a_{n\uparrow} a^+_{n\downarrow} a_{n\downarrow} , \tag{1}$$

where the sites {n} form a lattice, σ takes two values, down ($\downarrow$) or up ($\uparrow$), ε_o is a constant, V_{nm} is usually taken (for simplicity) as a constant V when n,m are nearest neighbors and zero otherwise, and $a^+_{n\sigma}, a_{n\sigma}$ are creation and anihilation operators respectively to a local state at the site n with spin σ, $|n\sigma)$. The last term of the right-hand side of Eq.(1) describes a local electron-electron repulsion of strength U. Note that interactions between electrons being at different sites are omitted. This omission may be a serious drawback of the model at low values of U/V, where electrons tend to spread around. However, as $U/V \to o$ the model is again satisfactory, since it reduces to the tight binding approximation.

Hubbard[3] had the idea of approximating the electron interaction term by a <u>random</u> one electron potential taking the values U or 0 at each site depending on whether or not an electron of opposite spin is already occupying the site. The probability of the value U is $\frac{n}{2}$ and of the value 0 is $1-\frac{n}{2}$ where n is the number of electrons per site. In what follows we will assume, unless otherwise stated, that $n = 1$. It is the randomness in the above defined approximate potential which accounts for the relative success of this approach.[3] However, as Hubbard pointed out, this approximation has its limitations, the most important of which stems from the dynamic nature of the problem allowing the local potential to change in time back and forth between the values U and 0. If this change is fast enough (as in the case of $U/V \ll 1$) the electron sees an average potential of value U/2 and not U or 0. One way to improve this approximation as to incorporate partially the fluctuating local potential is to introduce a <u>random</u>, <u>static</u>, one electron potential whose <u>probability distribution</u> will be determined by a self-consistent Hartree-Fock type of condition.

A binary type of probability distribution has been chosen; this choice is the simplest one which correctly reproduces the results in the two extreme limits $U/V \to \infty$ and $U/V \to o$; it interpolates for the intermediate values. More explicitly the approximation is as follows

$$U a^+_{n\uparrow} a_{n\uparrow} a^+_{n\downarrow} a_{n\downarrow} \approx \Delta\varepsilon_{n\sigma} a^+_{n\sigma} a_{n\sigma} , \tag{2}$$

where $\Delta\varepsilon_{n\sigma}$ is a random variable having a binary alloy type of distribution, i.e.

$$\left.\begin{aligned} \Delta\varepsilon_{n\uparrow} &= \Delta\varepsilon_{\uparrow}^{A} \equiv Un_{-} \\ \Delta\varepsilon_{n\downarrow} &= \Delta\varepsilon_{\downarrow}^{A} \equiv Un_{+} \end{aligned}\right\} \quad \text{with probability } \chi_A = \frac{1}{2} \quad , \tag{3a}$$

$$\left.\begin{aligned} \Delta\varepsilon_{n\uparrow} &= \Delta\varepsilon_{\uparrow}^{B} \equiv Un_{+} \\ \Delta\varepsilon_{n\downarrow} &= \Delta\varepsilon_{\downarrow}^{B} \equiv Un_{-} \end{aligned}\right\} \quad \text{with probability } \chi_B = \frac{1}{2} \quad , \tag{3b}$$

In Eq.(3) the symmetry for spin up and down has been taken into account. Thus only two quantities, n_+, n_-, determine the assumed probability distribution. These two quantities can be found from a generalized Hartree-Fock type of self-consistency, namely

$$n_+ = \langle a_{n\uparrow}^{+} a_{n\uparrow} \rangle_{n=A} = \int_{-\infty}^{E_F} \rho_{\uparrow n=A}(E)\,dE \quad , \tag{4a}$$

$$n_- = \langle a_{n\downarrow}^{+} a_{n\downarrow} \rangle_{n=A} = \int_{-\infty}^{E_F} \rho_{\downarrow n=A}(E)\,dE \quad , \tag{4b}$$

where the subscript $n=A$ denotes that the random potential $\Delta\varepsilon_{n\sigma}$ at the site n has the fixed value $\Delta\varepsilon_{\sigma}^{A}$; $\rho_{\sigma n=A}(E)$ is the n site contribution to the density of states and E_F is the Fermi level. The quantity $\rho_{\sigma n=A}(E)$ is given by the well-known formula

$$\rho_{\sigma n=A}(E) = -\frac{1}{\pi}\,\mathrm{Im}\left\langle \left(n\sigma \left| \frac{1}{E+is-H_\sigma} \right| n\sigma \right) \right\rangle_{av;n=A} \tag{5}$$

where

$$H_\sigma = \sum_n \left(\varepsilon_o + \Delta\varepsilon_{n\sigma}\right) a_{n\sigma}^{+} a_{n\sigma} + \sum_{nm} V_{nm} a_{n\sigma}^{+} a_{m\sigma} \quad . \tag{6}$$

The brackets $\langle\rangle_{av}$ in Eq.(5) indicate average over the random variables $\{\Delta\varepsilon_{m\sigma}\}$. It follows from Eq.(4) that

$$n_+ + n_- = \sum_\sigma \langle a_{n\sigma}^{+} a_{n\sigma} \rangle = 1 \tag{7}$$

since we have assumed one electron per site.

When $n_+=1$, $n_-=0$ the present scheme reduces to the original Hubbard binary alloy approximation. It is worthwhile to note that the present approach predicts local moments which can take two values: $(\langle a_{n\uparrow}^+ a_{n\uparrow}\rangle - \langle a_{n\downarrow}^+ a_{n\downarrow}\rangle)_{n=A} = n_+ - n_-$ or $(\langle a_{n\uparrow}^+ a_{n\uparrow}\rangle - \langle a_{n\downarrow}^+ a_{n\downarrow}\rangle)_{n=B} = n_- - n_+$. Obviously the fully averaged local moment equals to zero. The partially averaged moment corresponds roughly to the average $\frac{1}{T}\int_0^T(\langle a_{n\uparrow}^+(t)a_{n\uparrow}(t)\rangle - \langle a_{n\downarrow}^+(t)a_{n\downarrow}(t)\rangle)\,dt$; where $T\sim\hbar/U$ and $\langle a_{n\uparrow}^+(o)a_{n\uparrow}(o)\rangle - \langle a_{n\downarrow}^+(o)a_{n\downarrow}(o)\rangle$ is 1 or -1 for cases A or B respectively. Note that the local moment $n_+ - n_-$ is produced without any assumption about the form of the electronic eigenfunctions the latter being determined from the solution of the self-consistent random potential. Thus even itinerant electrons can produce local moments.

Eq.(4) predicts that $n_+ = 1$, $n_- = 0$ when $U/V = \infty$, which provide exact results in this limit, as was pointed out by Hubbard.[3] In contrast to Hubbard's original approximation, the present approach gives exact results in the opposite limit of $U/V = 0$, where $n_+ = n_- = 1/2$ and the system behaves as an ordinary non-magnetic metal. The magnetic moment varies continuously from zero (when $U/V = 0$) to its maximum value one (when $U/V = \infty$).

The present approximate scheme allows also the incorporation of magnetic order in the system by making the random variables $\{\Delta\varepsilon_{n\sigma}\}$ statistically dependent; the simplest way to achieve this is to introduce a parameter $P_{B/A}$ giving the probability of the site $n+1$ being B, given that the site n is A. In other words $P_{B/A}$ gives the probability of an antiparallel arrangement of two neighboring moments while $1 - P_{B/A}$ is the probability of a parallel arrangement. Since the local moments can take only two values (A or B) the magnetic order is equivalent to that of an Ising model with nearest neighbor coupling J. The quantity $P_{B/A}$ is related with J as follows

$$P_{B/A} = \frac{e^{-\beta J}}{e^{-\beta J} + e^{\beta J}} , \tag{8}$$

where $\beta = \frac{1}{k_B T}$ is the inverse temperature. The quantity $P_{B/A}$ can be determined by minimizing the free energy

$$\left(\partial F\left(P_{B/A}\right)/\partial P_{B/A}\right)_\beta = 0. \tag{9}$$

The free energy $F = E - TS$ depends on $P_{B/A}$ because the electronic density of states (and hence the energy E and electronic entropy $S_{e\ell}$) depend on $P_{B/A}$ and because the "lattice" entropy $S_{latt} \equiv S - S_{e\ell}$ depends on $P_{B/A}$ since the latter directly affects the ways you can arrange the local moments on the lattice. Obviously $S_{latt} = k_B N \ell n2$ when $P_{B/A} = 1/2$ (no correlation) and $S_{latt} = k_B \ell n2$ when $P_{B/A} = 1$ or 0; N is the total number of lattice sites.

Since our way of introducing magnetic ordering is equivalent to that of an Ising model (with a self-consistent temperature dependent J) it is clear that the scheme is capable of handling ferromagnetic or antiferromagnetic phase transitions in a rather realistic way: One expects perfect long (as well as short) range order at $T = 0$. As the temperature is raised imperfect long range order is still present until a critical temperature where the long range is destroyed. Beyond this temperature only short range order survives. This latter disappears at a considerably higher temperature.

The present approach can also be used to study transport properties such as conductivity, magnetoresistance etc. It is well known that the Hubbard Model predicts a metal-insulator transition (in two or three dimensions). This transition can be quantitatively studied within the framework of the present scheme. It is worthwhile to mention the one-dimentional case. For any $U \neq 0$ our approach predicts that the system will be insulating (for any random one dimensional system all eigenstates are localized and consequently the conductivity vanishes). This is in agreement with exact results.[4]

Extensions of the scheme already outlined allow the study of various interesting cases: e.g. the role of an external static magnetic field can be examined; in particular negative magnetoresistance is predicted resulting from a reduction of the randomness of the self-consistent potential due to ordering effect of the magnetic field. Note that negative magnetoresistance is a common occurrence in impurity bands in crystalline semiconductors. One can also handle cases where the number of electrons per site is different than one by introducing a three component random potential, where the third component corresponds to the case of double or no occupation of the site.

Cyrot[5] in a series of papers has employed similar ideas to the ones outlines here. Our approach is different with respect to the self-consistency conditions (4). One can prove that our self-consistency conditions give exact results in the atomic limit ($U =$ finite, $V = 0$), in contrast to that of Cyrot. Cyrot has not examined in any detail the question of magnetic ordering. Finally he restricted himself to the thermodynamic properties. Licciardello and Economou[6] have also employed a very similar approach. Magnetic order was introduced in a simpler but less satisfactory way through a long range order parameter.[7]

III. QUANTITIES OF INTEREST

Having outlined the basic ideas, I will present here the quantities which we try to calculate:

1. The magnetic order parameter $P_{B/A}$ (T) as a function of the temperature T. For each lattice there is a critical value of $P_{B/A}$, $P_{B/A} = p_c$, at which the long range order disappears. The equation

$$P_{B/A}(T_c) = p_c \, , \tag{10}$$

defines the critical temperature for the phase transition. One can also find the temperature T_o at which the short range order disappears:

$$P_{B/A}(T_o) = .5 \quad . \tag{11}$$

2. The partially and the fully averaged density of states: $\rho_{n\sigma;n=A \text{ or } B}(E)$ $\rho_{n\sigma}(E) = 1/2\,[\rho_{n\sigma;n=A}(E) + \rho_{n\sigma;n=B}(E)]$. The density of states enters in almost all quantities of physical interest.

3. The mobility $\mu(E)$. This quantity determines together with the density of states the conductivity of the system. The value of μ at $E = E_F$ determines whether or not the system is metal or insulator. Since $\mu(E)$ is temperature dependent [through the parameter $P_{B/A}$ (T)] a reasonable definition of the metal-insulator transition can be obtained by equating the temperature k_BT to the mobility gap $|E_F - E_c|$, where E_c is the mobility edge[8] closest to the Fermi level.

$$k_B T_{MI} = |E_F - E_c| \, . \tag{12}$$

4. The local magnetic moment m(T) defined as

$$m(T) = \left\{1 - 2\left[n_+ \langle a^+_{n\downarrow} a_{n\downarrow}\rangle_{T;n=A} + n_- \langle a^+_{n\uparrow} a_{n\uparrow}\rangle_{T;n=B}\right]\right\}^{\frac{1}{2}} \tag{13}$$

where $\langle a^+_{n\sigma} a_{n\sigma}\rangle_{T;n=A \text{ or } B} = \int_{-\infty}^{\infty} \rho_{n\sigma;n=A \text{ or } B}(E) f(E)\, dE$ and

f(E) is the fermi function. It is easy to see that as $T \to o$ $m(T) \to n_+ - n_-$. In particular we are interested for values of T, $T = T_m$, such that $m(T_m) = 0$.

6. Various thermodynamic functions like the free energy F, the energy E, the entropy S and the specific heat C as functions of the temperature.

7. The magnetic susceptibility χ as a function of the temperature T.

The behavior of the system can be summarized in a phase diagram where the T - U (V is assumed constant) plane is separated in various regions by the lines $T_c = T_c(U)$, $T_o = T_o(U)$, $T_{MI} = T_{MI}(U)$,

$T_m = T_m(U)$.

IV. METHODS OF CALCULATIONS

In order to carry out the program outlined above we should be able to calculate partially and fully averaged density of states in a random tight binding Hamiltonian even when some order is present ($P_{B/A} \neq .5$). This problem has been examined by Licciardello and Economou[9] through an extension of the so-called coherent potential approximation.[10] White and Economou have recently refined and improved the method to the extent that no appreciable error is introduced at this stage.

One needs also $S_{latt}(P_{B/A})$. This function is quite trivial for 1-D systems; it is also known (from the Onsager solution for the Ising model) in 2-D systems. There is no exact expression for 3-D systems. However, its behavior around the points $P_{B/A} = .5$, $P_{B/A} = 0$ or 1 is rigorously known. Also the behavior around the critical points p_c is known to a high degree of accuracy.[11]

The straightforward way to set up the calculation is to start with a given U, V and $P_{B/A}$, solve the self-consistency equations which will determine the effective Hamiltonian and then proceed with the minimization of the free energy. The procedure can be significantly simplified numerically if one starts with the quantity $\delta \equiv U(n_+ - n_-)/ZV$ and $P_{B/A}$. One then can find the partial density of states and calculate n_+, n_- and consequently U itself. Z is the coordination number of the lattice.

The mobility $\mu(E)$ can be calculated from the relation $\mu(E) = \sigma(E)/e\rho(E)$, where the energy dependent conductivity $\sigma(E)$ has been studied within the coherent potential approximation by Velicky.[12] The results can be improved if one combines Velicky's approach with the localization theory method[8] by putting $\mu(E) = 0$ if the eigenstates at E are localized and $\mu(E) = \sigma(E)/e\rho(E)$ if the eigenstates at E are extended; the quantity $\sigma(E)$ can be calculated following Velicky's theory.

V. PRELIMINARY RESULTS AND DISCUSSION

I report here briefly some preliminary results obtained in a two-dimensional square lattice. A detailed presentation of our results will appear elsewhere.[13,14]

The most significant result of the present scheme is that a first order phase transition is predicted. Mathematically this occurs because the function $S_{latt}(P_{B/A})$ has an infinite slope at

$p_{B/A} = p_c$. At very low temperatures the minimum of the free energy occurs for $p_{B/A} > p_c$ while at high temperatures for $p_{B/A} < p_c$. Since the slope at p_c is infinite the minimum will never occur at the critical value. Thus the behavior of the system is as follows: For $T = 0$ $p_{B/A} = 1$ and the system behaves as an antiferromagnetic insulator. As the temperature is raised $p_{B/A}$ remains larger than p_c so that long range magnetic order is exhibited. At a critical temperature T_c the quantity $p_{B/A}$ changes discontinuously from a value above p_c to a value below p_c so that long range magnetic order disappears. This phase transition is of first order since it is accompanied by a discontinuity in $p_{B/A}$ and consequently a discontinuity is the thermodynamic quantities like E, F, S, etc. Despite the destruction of long range order at T_c short range order remains for $T_c<T<T_o$. At T_o the minimum occurs at $p_{B/A} = .5$ and the short range disappears as well.

It seems that the present way of incorporating magnetic order in the system is not adequate for very low temperatures. The reason is that at very low temperature the thermodynamic quantities are dominated by spin waves which are absent from the Ising type of magnetic interaction that we have incorporated in our scheme.

The present approach predicts also a region of metallic behavior accompanied by the existence of local moments. This regime seems to be extremely important for explaining the appearance of negative magnetoresistance in impurity bands. Toyozawa[15] has developed a many body theory in order to account for this effect. He treats the electrons on the one hand as localized in order to produce the local moments and on the other hand as itinerant being scattered by the local moments in order to account for the metallic conductivity. In the present method the electrons naturally produce local moments while they remain itinerant. Their scattering from the local moments is already included in the effective one-electron Hamiltonian and will appear in the function $\sigma(E)$.

Another attractive feature of the present method is the possibility of having a rather high $\rho(E)$ while $\mu(E)$ [or $\sigma(E)$] is very low.[16] This feature appears in an equivalent form in many body treatments where a high density of electronic states is associated with a low density of carrier states.[17,1]

A metal-insulator transition is predicted, although not a sharp one. The reason is that at $T = 0$, where the metallic and insulating behaviors are clearly distinguishable, the system behaves as an insulator, at least for certain simple lattices.[5] At finite temperatures the metal-insulator transition can be conventionally defined by the equation $|E_c - E_F| = k_B T_{MI}$. At high temperatures ($k_B T \sim V$) the values of U for which the metal-insulator transition occurs is of the order of the bandwidth.

VI. REFERENCES

1. N. F. Mott in "Metal-Insulator Transitions," Taylor and Francis, Ltd., London, 1974.
2. See e.g. I. F. Schegolev, Phys. Stat. Solidi A12, 9 (1972); V.K.S. Shante, C. M. Varma, and A. N. Bloch, Phys. Rev. B8, 4885 (1973); A. J. Epstein, S. Etemad, A. F. Garito and A. J. Heeger, Phys. Rev. B5, 952 (1972); L. B. Coleman, J. A. Cohen, A. F. Garito, and A. J. Heeger, Phys. Rev. B7, 2122 (1973).
3. J. Hubbard, Proc. Roy. Soc. A281, 401 (1964).
4. E. H. Lieb and F. Y. Wu, Phys. Rev. Lett. 20, 1445 (1968).
5. M. Cyrot and P. Lacour-Gayet, Sol. St. Comm. 11, 1767 (1972); P. Lacour-Gayet and M. Cyrot, J. Phys. C7, 400 (1974).
6. D. C. Licciardello and E. N. Economou, in preparation.
7. M. Plischke, Sol. Stat. Comm. 13, 393 (1973).
8. E. N. Economou and M. H. Cohen, Phys. Rev. B5, 2931 (1972); D. C. Licciardello and E. N. Economou, Phys. Rev. B11,3697(1975)
9. D. C. Licciardello and E. N. Economou, Sol. Stat. Comm. 12, 1275 (1973).
10. For a review see R. J. Elliot, J. A. Krumhansl, P. L. Leath, Rev. Mod. Phys. 46, 437 (1974).
11. See, e.g. C. J. Thompson "Mathematical Statistical Mechanics" MacMillan Co., New York, 1972, Chapters 5 and 6.
12. B. Velicky, Phys. Rev. 184, 614 (1969).
13. R. Demarco, C. White, E. N. Economou, in preparation.
14. C. White, E. N. Economou, in preparation.
15. Y. Toyozawa, J. Phys. Soc. Jap 17, 986 (1962); see also N. Ohata and R. Kubo, J. Phys. Soc. Jap 28, 1402 (1970); A. V. Matveenko, S. S. Shalyt, M. L. Shubnikov and V. S. Vekshina, Sov. Phys.-Semic. 5, 949 (1971).
16. E. N. Economou, D. C. Licciardello and K. L. Ngai, Contr. to "Phase Transitions - 1973" ed. by H. K. Henisch, R. Roy and L. E. Cross, Pergamon Press, 1973.
17. W. F. Brinkman and T. M. Rice, Phys. Rev. B2, 1324 (1970); 4302 (1970).

OPTICAL DETERMINATION OF HOT CARRIER DISTRIBUTION FUNCTIONS

G. Bauer

I. Physikalisches Institut, RWTH Aachen

Aachen, Germany

I. Introduction

In dealing with a hot carrier situation, the most important problem to be solved both experimentally and theoretically is the determination of the carrier distribution function (1). The shape of the distribution function, e.g. its energy (E) and wavevector ($\underline{k}$) dependence directly reflects the interactions of the carriers with the lattice. A transport property like the drift velocity always represents an integral average weighted by the distribution function (DF) and thus does not more contain the full information which is buried in the distribution function.

However by employing optical methods some information on the distribution function f(E) can be obtained. In a direct optical transition at a well defined energy $E=h\nu$ between an initial state E_1 and a final state $E_2(E_2-E_1=h\nu)$, the transition rato depends of course on the occupancy of both and thus on $f(E_1)$ and $f(E_2)$. By varying $h\nu$ and by knowing one of them either $f(E_1)$ or $f(E_2)$, the other DF can be probed if well defined states E_1 and E_2 exist.

Until now three main methods have been employed to get information on this subject:

a) optical absorption measurements
b) inelastic light scattering
c) emission measurements

In every case the knowledge of the exact bandstructure is a necessary prerequisite : in a) and c) the $E(k)$ dependence of the energy bands involved has to be known accurately, in b) in addition the frequency shift of the scattered light.

The change of the distribution function away from a thermal equilibrium one (Fermidistribution) can be obtained either by applying d.c. electric fields to a semiconductor sample, by microwave fields or at still higher frequencies, by using the radiation of an intense laser.

There is however a basic difference in the experimental methods : if the field induced change of the DF due to a static electric field is to be analyzed, the intensity of the probing light has to be sufficiently small in order to observe a DF unaffected by the probing procedure.

If the light itself is used to create a hot carrier distribution, of course the intensity of it determines the shape of the DF . Until now the latter case was only studied in connection with emission measurements.

II. Optical Absorption Measurements

One might think that a convenient way to study the field induced change of $f(E)$ of carriers e.g. in the conduction band is a measurement of the electronic fundamental absorption in direct gap materials. However, if the Fermienergy lies between the conduction (CB) and the valence band (VB) , the increase of the absorption constant with E near the gap energy E_g is so steep, that a change of $f(E)$ with electric field F is masked. Only if the Fermienergy E_F lies within e.g. the conduction band then the electronic fundamental absorption changes sufficiently by F to be observable. In this situation, for a standard band model the absorption constant $K(\hbar\omega)$ would be proportional to

$$K(\hbar\omega) \propto 1 - f_e(E) \tag{1}$$

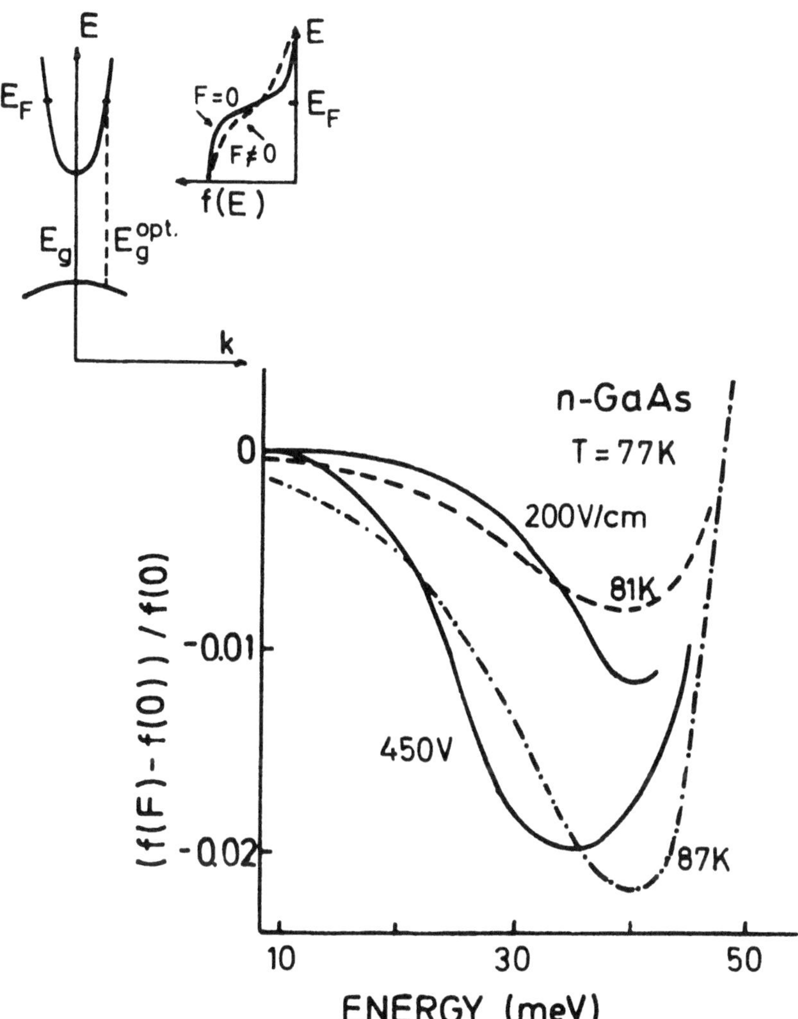

Fig. 1 : Determination of hot carrier DF from optical absorption in degenerate n-GaAs (after [2]).

where $\hbar\omega = E_g+E+E'$, and $f_e(E)$ denotes $f(E)$ in the CB, E is counted from the conduction band edge and E' from the valence band edge.
It is assumed that the valence band is completely filled with electrons. By comparing $K(\hbar\omega)$ under thermal equilibrium conditions and $K_F(\hbar\omega)$ in the presence of an electric field, the following relation results

$$1 - f_e(E,F) = K_F(\hbar\omega)/K_{th}(\hbar\omega) \, . \, (1-f_e(E,o)) \quad (1a)$$

where $f_e(E,o) = 1/(\exp(E-E_F/kT)+1)$. Thus by measuring K_{th} and K_F , $f_e(E,F)$ can be obtained. Of course it is assumed that the transition probabilities are not influenced by the electric field in making this comparison. Figure 1 shows results of such measurements in n-GaAs with $n=1X10^{18}cm^{-3}$ by Jantsch and Heinrich [2] . The experimentally deduced distribution functions are compared with Fermidistributions in which the lattice temperature T was replaced by an electron temperature T_e . As can be seen, the electron temperature model does not describe the observed distribution functions.

Unfortunately there are however several problems in the analysis of these data. In high doped materials, band tails exist, cousing a non-zero density of states at energies $E<0$. These tails cause uncertainities in the energy scale of the abscissa beside the limited experimental resolution of about 5meV . The change of E_g due to the presence of fields of the order of 500 V/cm is negligible [3] . In addition problems with shift of E_g smaller energies at high carrier concentrations occur [4] .

A less complicated method to gain information on DF should involve interband transitions which do not lead to such huge absorption constants as at the fundamental electronic gap. The most thoroughly investigated method fulfilling this requirement is the study of intervalence-band transitions [5-9] . In diamond and zinc-blende-type semiconductors the uppermost valence band consists of three doubly degenerate bands, two of these are degenerate at k=o (Fig.2a) and the third is split-off by spin-orbit-interaction. The splitting Δ may reach from

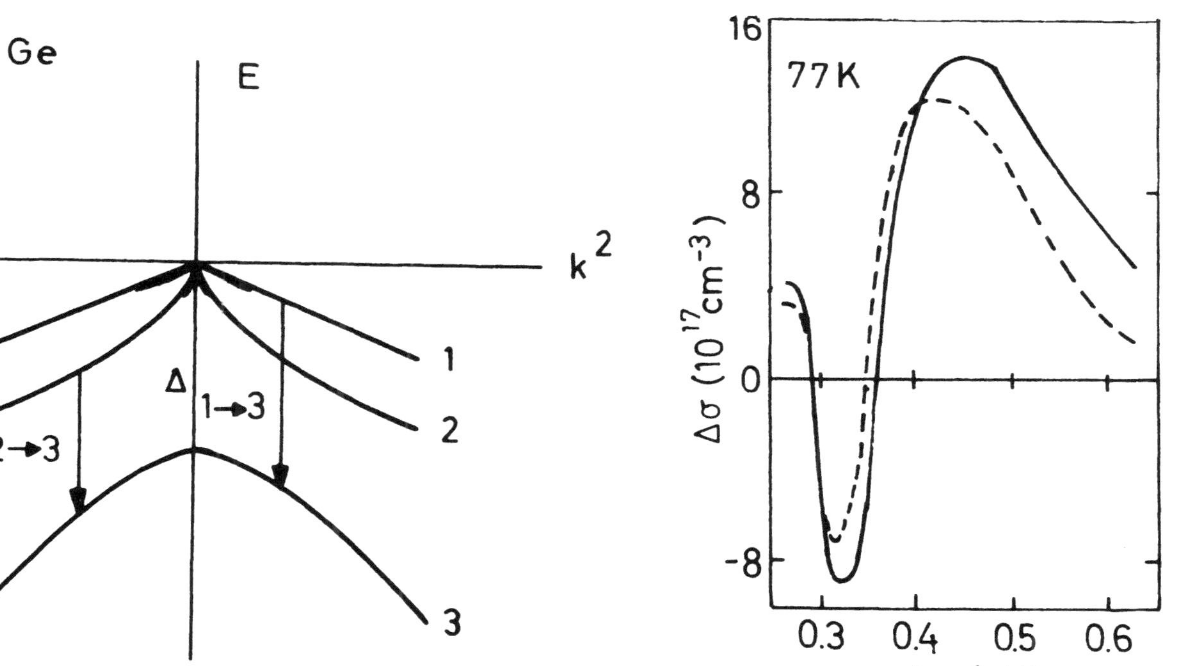

Fig. 2 : a) valence band structure of p-Ge
b) change of absorption $\Delta\sigma = K/p$ induced by F=1710V/cm and F=810V/cm (after [5]).

0.04 eV (Si) to about 0.9 eV (InSb). If $\Delta > E_g$, than the fundamental electronic transitions from VB to CB will make it impossible to observe intervalence-band transitions between the bands $1 \rightarrow 3$, $2 \rightarrow 3$. On the other band if Δ is small or even comparable to kT, then band 3 will already be populated with holes thermally, or at least by carrier transfer from the bands 1 or 2 in moderate electric field. Thus in order to observe and to analyze intervalence-band transitions the following inequality shoult be fulfilled

$$kT \, , \, \langle E \rangle \ll \Delta < E_g \tag{2}$$

($\langle E \rangle$ denotes the mean carrier energy in the presence of an electric field).
In Fig. 2b the field induced change of the absorption constant $\Delta\sigma$ (=K/p, p... hole concentration) is shown. For photon energies higher than 0.3 eV almost only $1 \rightarrow 3$ transitions contribute to the observed absorption.

As expected, the electric field causes a descrease of K at low carrier energies and an increase at higher ones. The number of carriers with higher energies increases at the expense of the number with lower energies. In order to evaluate the distribution function in band 1 again it was assumed that a direct proportionality between $K(\hbar\omega)$ and $f_1(E)$ exist : $K(\hbar\omega) = C \cdot f_1(E)$. Band 3 is not populated with holes. If the proportiinality function $C(\hbar\omega)$ is field independent, $f_1(E,F)$ can be determined as above from a comparison with measurements under thermal equilibrium. Fig. 3 shows results of experiments of Pinson and Bray [5] compared with calculations of Budd [10]. The calculated kink of the distribution function near the optical phonon energy ($\hbar\omega$ = 0.037 eV) is smeared out in the curves deduced from the experiments. However calculated curves based on the T_e model do not fit at all.

In the analysis leading to the DF shown in Fig. 3, however, an assumption concerning the connection between the photon energy and a particular heavy hole energy E_1 is made. Since the bands 1,2,3 are warped, the relationship between E_1 and $\hbar\omega$ ($\hbar\omega = E_3 - E_1$) is not as

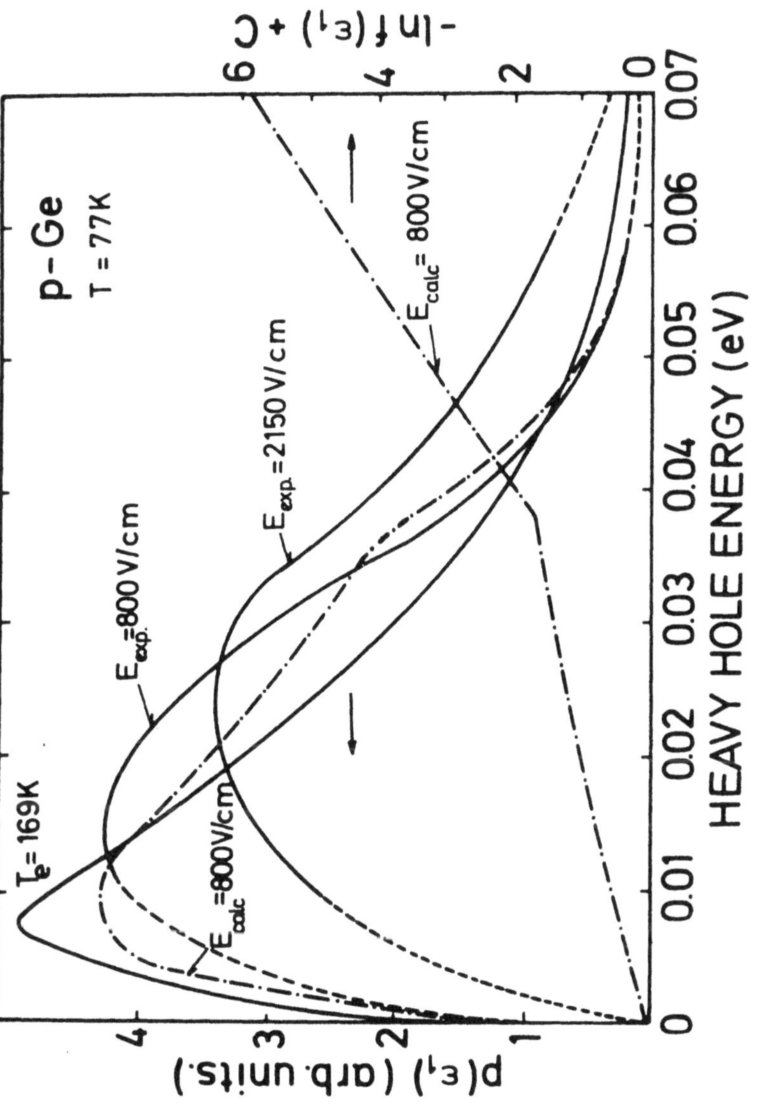

Fig. 3 : occupancy (p(E)dE=f(E)·N(E)dE) of heavy holes as a function of energy. Experimental data from Ref. [5] , calculated data from Ref. [10].

close as necessary for an unambiguous determination of $f_1(E)$. Due to the band warping light with a certain energy $\hbar\omega$ causes at different states $\underline{k}$ in the Brillouin zone transitions from $1 \rightarrow 3$. Thus $f_1(\underline{E})$ cannot be determined directly. Since the absorption constant for the direct $1 \rightarrow 3 (i \rightarrow j)$ transitions is given by

$$K \propto \frac{1}{\hbar\omega} \int d^3k \, |\langle j|\underline{a}\cdot\underline{p}|i\rangle|^2 \delta(E_j(\underline{k}) - E_i(\underline{k}) - \hbar\omega) \times$$

$$\times \, (f_i(\underline{k}) - f_j(\underline{k})) \qquad (3)$$

where $\underline{a}$ denotes the polarization vector of the radiation, a parametrized model for $f_i(\underline{k})$ has to be made in order to fit the observed $K(\hbar\omega)$ [9]. No analytical expression for $f_i(\underline{k})$ can be given.

Since $f_i(\underline{k})$ will generally be anisotropic in the presence of an electric field Eq. 4 offers however a possibility to get information on this property : since the matrix elements depend on the polarization vector of the light $\underline{a}$, the absorption coefficient K will be different for $F||a$ and $F\perp a$. Thus also some information on the anisotropy of $f_i(\underline{k})$ is got [6-7] .

Although some III-V compounds like p-Ga As would also satisfy the requirements of Eq. 3, until now only in p-Ge distribution functions have been determined.

III. Inelastic Light Scattering

Inelastic scattering of light in a semiconductor is not only caused by elementary excitations such as phonons or plasmon. Beside the collective phenomena also single particle excitations contribute to light scattering [11] .

Since the quasifree carriers in the bands contribute to the polarizability, a change of their properties due to some external perturbations results in changes of the

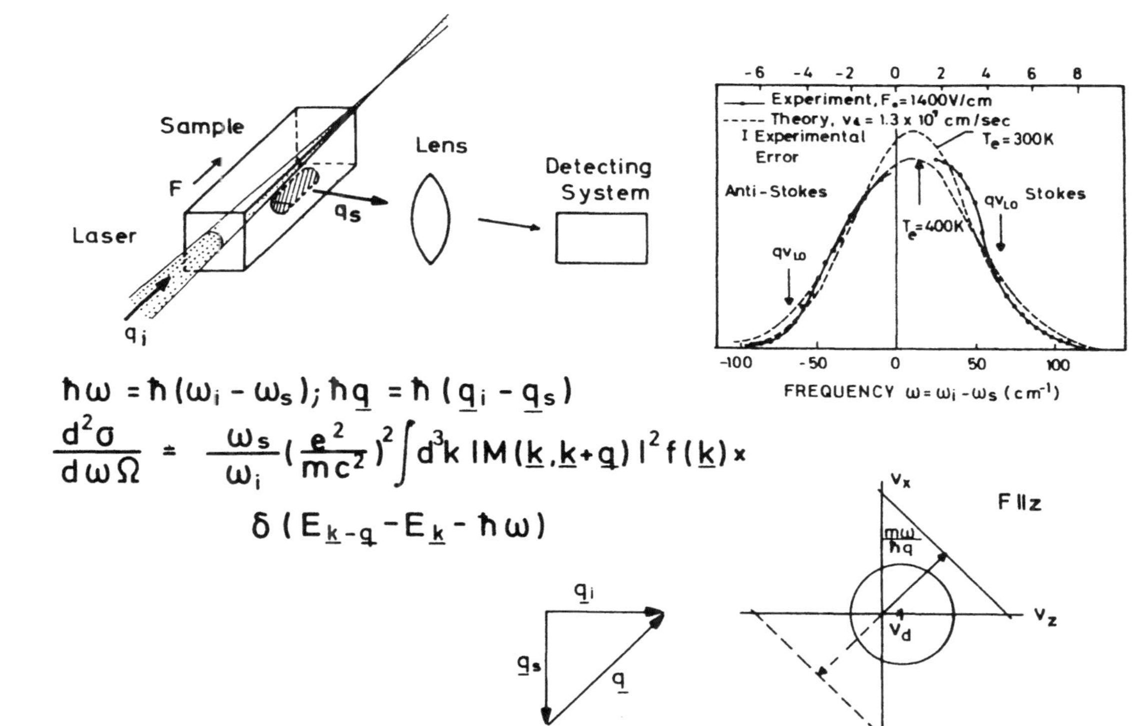

Fig. 4 : Experimental arrangement, experimental results and geometry used for inelastic light scattering in n-GaAs at T≈25K . Data from Ref. [13] .

scattered spectrum. If the momentum transfer from the light $\hbar \underline{q}$ to an electron is large compared to $\hbar \underline{q}_D$ ($\underline{q}_D$ being the screening wave vector of the electron gas) scattering due to single particle excitations even is predominant over the collective modes.

In the scattering process radiation of energy $\hbar\omega_i$ and wave vector $\underline{q}_i$ is scattered by the free carriers into sate states with frequency ω_s and wave vector $\underline{q}_s$. The transferred energy and momentum are given by

$$\hbar\omega = \hbar(\omega_i - \omega_s) \tag{4}$$

$$\hbar\underline{q} = \hbar(\underline{q}_i - \underline{q}_s) \tag{5}$$

For scattering caused primarily be charge density fluctuations the scattering cross section is given by[11]

$$\frac{d^2\sigma}{d\omega\, d\Omega} = \frac{\omega_s}{\omega_i}\left(\frac{e^2}{mc^2}\right)^2 \int d^3k \,|\, M(\underline{k},\underline{k}+\underline{q})\,|^2\, f(\underline{k}) \times$$

$$\times\ \delta\,(E_{\underline{k}+\underline{q}} - E_{\underline{k}} - \hbar\omega) \tag{6}$$

It is assumed that the population of the final states can be neglected. If the transition matrix element is a slowly varying function of $\underline{k}$ Eq.6 can be interpreted in a simple way geometrically [12] as can be seen from Fig.4 : if the q^2 term in the δ-function of Eq.6 is neglected than this δ function is given by

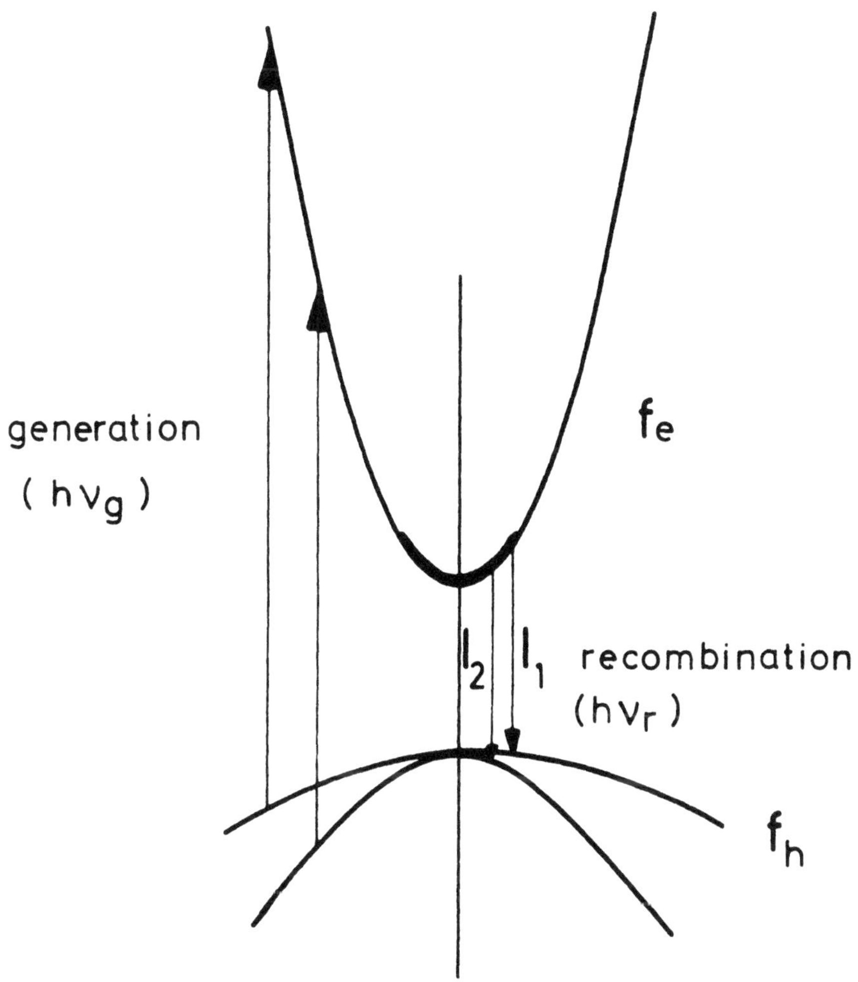

Fig. 5 : Generations of electron-hole pairs and recombination.

$$\delta(\underline{k}.\underline{q} - m\omega/\hbar)$$

Thus the contribution of f(k) to the integral comes from a plane perpendicular to $\underline{q}$ at a distance $m.\omega/\hbar q$ from the origin. Thus this method is in principle extremely useful for studying asymmetries of the non-equilibrium distribution function.

The only experiment until now [13] used only a 90° geometry (Fig. 4) . Thus information on the anisotropy of f(k) could not be obtained, unfortunately.

IV. Emission measurements

The recombination radiation of hot electrons with donors, acceptors, and with holes yields in principle also information on the distribution function of the electrons. From the spectral intensity distribution of the luminescence radiation the occupancy of the initial states can be calculated if again the bandstructure and-or the exact position of donors or acceptors known.

In this section we distinguish between different kinds of producing a hot electron distribution either by d.c. fields or by photoexcitation.

a) dc electric field excitation

In order to observe recombination radiation a small amount of photoexcited electrons is produced in an n-type semiconductor by transferring electrons from the filled valence band to the conduction band. Compared to the energy range where a reasonable occupancy of electrons is possible in the CB due to the DF (usually < 100meV) the photoexcited electrons (see Fig.5) have usually much higher energies. By subsequent emission of LO phonons these carriers loose very rapidly their initial energy ($\tau_{e,LO} \lesssim 10^{-12}$s) and reach a quasiequilibrium with the distribution of the carriers determined by the applied electric field. The luminescence from the subsequent electron-hole recombination thus probes the electron distribution.

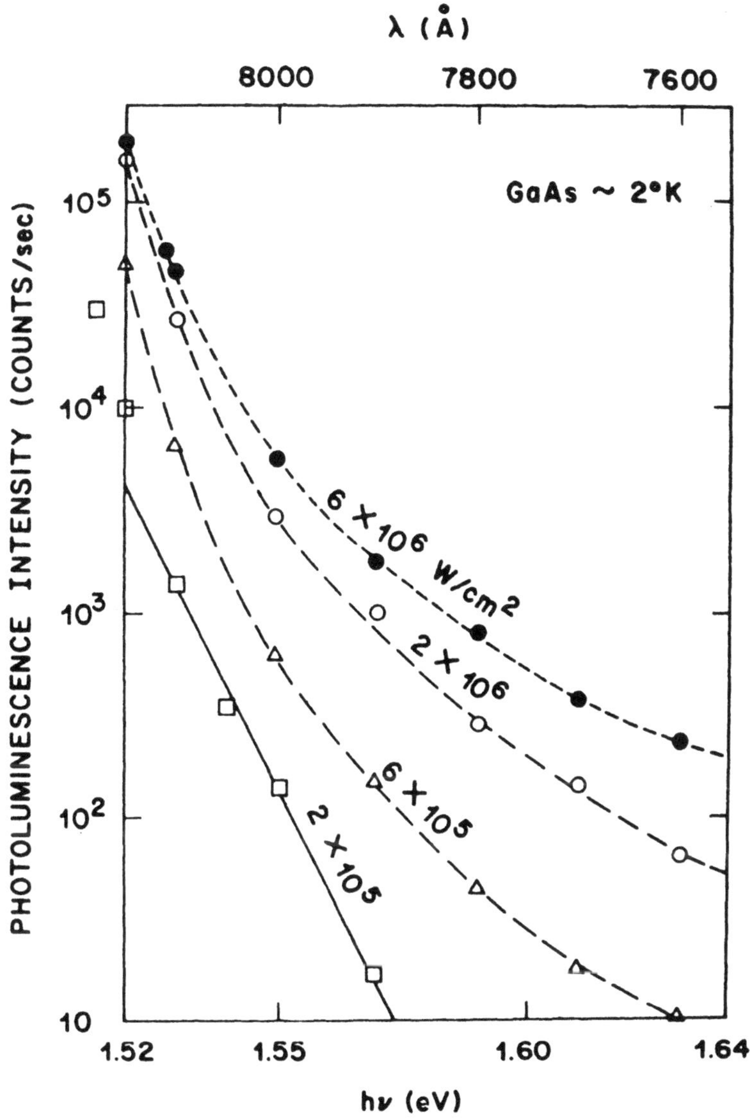

Fig. 6 : Photoluminescence of laser-excited hot carriers in GaAs as a function of light intensity. Data from Ref. [17].

The most detailed experimental data published until now [14] deal with n-GaAs with fields up to 2000V/cm at 77 and 200K. The determination of the DF from the intensity dependence of the luminescence spectrum is however not quite straight forward. At a certain energy $h\nu$ (Fig. 5) in materials like GaAs transitions of the electrons to the heavy as well as to the light hole band occur. In addition assumptions concerning the hole distribution have to be made. The experimentally derived resolution was not good enough to distinguish between the results of an electron temperature model and a more elaborate model calculation based on an iterative solution of the Boltzmann transport equation although information up to 50 meV in the CB was obtained. This energy range extends beyond the LO phonon energy.

Recombination radiation in the far infrared region of the electromagnetic spectrum was observed in n-Ge and N-GaAs at liquid He temperatures [15] and electric fields of a few V/cm. Impact ionized carriers recombine with shallow donors by emitting radiation in the spectral range below 15 cm^{-1}. The spectral spread of the recombination radiation yields information on the electron DF up to 8meV in the CB.

b) Photoexcitation

Already in 1969 Shah and Leite [16] analyzed radiative recombination from photoexcited hot carriers in GaAs. Using an Ar-laser, electron-hole pairs were created deep in both the CB as well as the VB. Observed recombination radiation extended to about 100 meV beyond the gap energy and from the dependence of the emitted photoluminescence on energy it was concluded that the hot carriers have a Maxwellian distribution with $T_e > T$. By increasing the excitation intensity of the laser beyond $2\times10^5 W/cm^2$, however, the results could no longer be interpreted in terms of an electron temperature model [17] (see Fig.6). This is a somewhat puzzling feature as one would expect just the contrary : at high excitation intensities carrier-carrier scattering should eventually prodominate over all other scattering mechanisms, establishing thereby a Maxwellian DF. Since the excitation energy is high compared to E_g, at high excitation an appreciable number of optical phonons will be emitted from the photoexcited carriers. These phonons will not immediately thermalize into the "phonon bath" but establish a non-equilibrium phonon distribution too (Fig.7). Depending

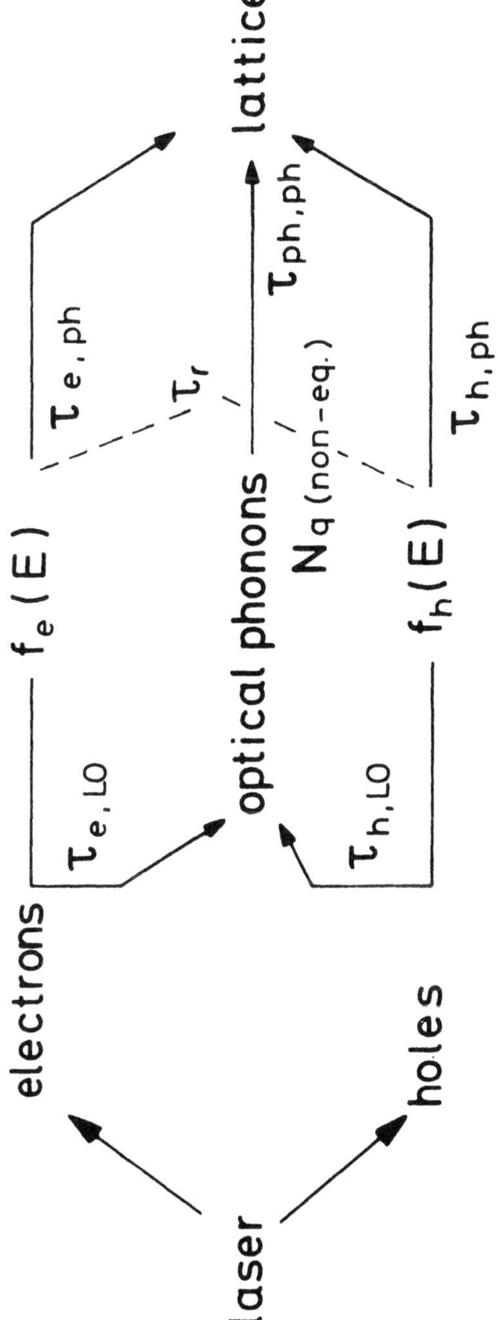

Fig. 7 : Schematic diagram of processes following electron hole production.

on the magnitude of the phonon-phonon relaxation τ_{ph-ph}, this nonequilibrium phonon distribution may be observable directly in Raman scattering experiments [18-19]. Indeed it was found that photoexcitation produces with increasing laser intensity a shift of the ratio of Stokes (S) to Antistokes (A) scattering. Whereas the ratio S/A decreases significantly for the LO phonon, giving evidence for LO phonon emission as the laser intensity is increased, it decreases for the TO phonon peaks only slightly. This indicates, that the lattice temperature stays nearly constant whereas the number of LO phonons increased drastically.

Thus a complicated situation is found in a photoexcitation experiment (Fig.7). After the initial excitation of electronhole pairs by the laser radiation, these carriers loose energy by LO emission ($\tau_{e,LO}$ and $\tau_{h,LO}$) establishing a non-equilibrium phonon distribution and at least two hot carrier distribution functions. The relaxation times $\tau_{e,ph}$ and $\tau_{p,ph}$ involve then absorption and emission of all kinds of phonons. These times are substantially smaller than the radiative recombination time τ_r.

First attempts to discriminate between hot phonon and hot carrier DF from the observed photoluminescence have been performed by Motisuke et al [20] in experiments with CdS.

The situation is somewhat simpler if not CB-VB recombination is analysed but conduction band-acceptor luminescence. Then the spectral dependence of the luminescence intensity $I(\hbar\omega)$ directly reflects the electron DF since:

$$I(\hbar\omega) = f(E).N(E) . |M_{c,A}(\underline{k})|^2$$

where $E = \hbar\omega - E_g + E_A$. If the energy dependence of the dipole matrix element M is known and also the density of states $N(E)$ then $f(E)$ is obtained. In an elegant experiment Ulbrich [21] was able to get with this method even information of the time dependent change of the DF in GaAs after irradiation with 0.2ns long laser pulses. The results are shown in Fig.8. Beside electron-accep-

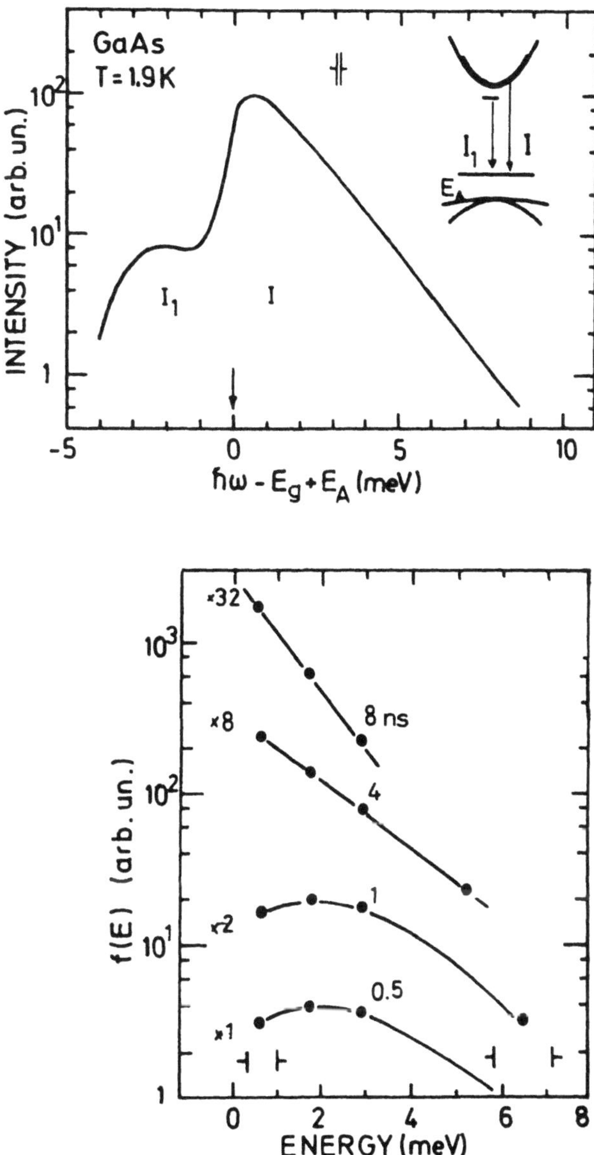

Fig. 8 : Upper half : photoluminescence from conduction band-acceptor and donor-acceptor recombination in GaAs. Lower half : f(E) as a function of time and E . Curves are shifted vertically by factors indicated. After Ref. [21].

tor recombination the shoulder at the low energy side is caused by radiative transitions of carriers between donor and acceptor states.

c) Emission in quantizing magnetic fields

In the presence of a high magnetic field B, the energy bands in an effective mass approximation are split into Landau levels according to

$$E(N,s,k_z) = (N+1/2)\hbar\omega_c + \frac{\hbar^2 k_z^2}{2m} + g.\mu_B.B.s \qquad (7)$$

where $N=0,1,2...$, $\omega_c=\frac{e}{m}.B$, g is the effective g-factor, μ_B the Bohr magneton $(\mu_B=e\hbar/2mc)$. Also the density of states function $N(E)$ is altered according to

$$N(E) \propto \frac{eB}{\hbar} . \left(\frac{2m}{\hbar^2}\right)^{1/2} \sum_{N=o}^{N=m} (E-[(N+1/2)\ \omega_c+g\mu_B Bs])^{-1/2} \qquad (8)$$

This density of states exhibits singularities whenever in Eq.7 $k_z=0$. In a real situation however, due to a finite scattering time in the Landau levels, these singularities will be somewhat smeared out.

By applying moderate electric fields to semiconductors like n-InSb, [22] or n-GaAs [23] at low temperatures impact ionization of frozen out electrons from the donor levels occurs. The electrons can gain energy enough to populate higher lying Landau levels (e.g. $1^+, 1^-$, see Fig.9). A small part of the electrons in these higher Landau levels will loose their energy by spontaneous transitions ($1^+ \rightarrow 0^+$, $1^- \rightarrow 0^-$) by emitting radiation whereas most of them emit phonons. The spontaneous emission intensity due to Landau level transitions probes the occupancy and thus the distribution function. In fig.9 emission spectra as a function of the magnetic field strength

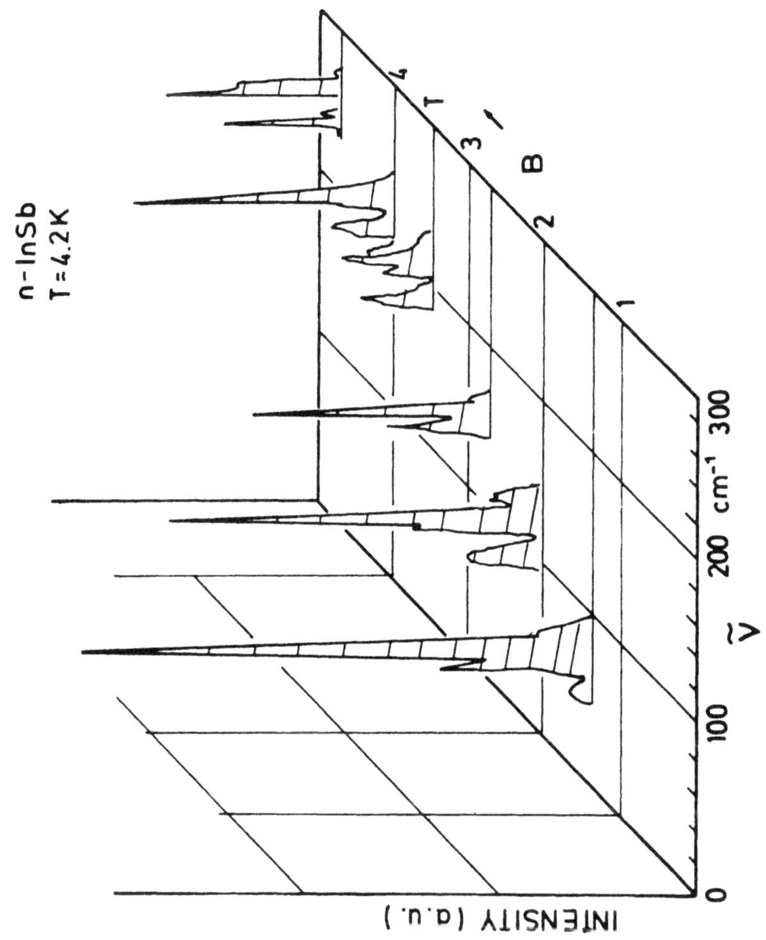

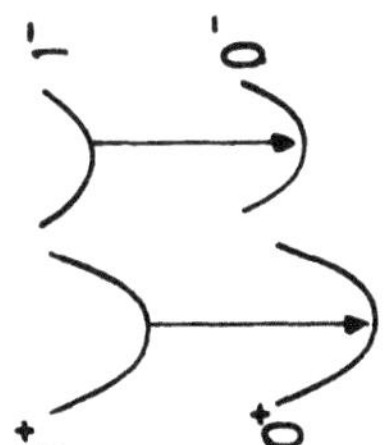

Fig. 9 : Inset: Landau level transitions $1^+ \to 0^+$ and $1^- \to 0^-$ of hot electrons. Variation of intensity and spectral position of Landau emission with magnetic field.

are shown for various fields in n-InSb. In Fig.10 the magnetic field dependence of this intensity is shown for three different electric fields. The most prominent feature of these curves is the existence of Landua emission beyond the LO phonon energy [24] . This feature was unexpected, since model calculations of Yåmada [25] had shown, that a more or less strong cut-off of the hot electron DF at the LO phonon energy should occur.

The peak positions have to be calculated, considering the nonparabolicity of n-InSb, for the $1^+ \to 0^+$ and $1^- \to 0^-$ transitions according to :

$$E(N,s,k_z) = -\frac{E_g}{2} + \left[\left(\frac{E_g}{2}\right)^2 + E_g \cdot D(N,s,k_z) \right]^{1/2}$$

$$D(N,s,k_z) = \hbar\omega_c (N+1/2) + \frac{\hbar^2 k_z^2}{2m} + s.g.\mu_B.B \quad (9)$$

where here m denotes the band edge effective mass. In order to derive from the luminescence data the carrier DF , the emission rate for spontaneous transitions in the electric dipole approximation [26] can be used. Using as a first step a δ-function approximation for the density of states, the transitionrate is given by

$$R = \frac{n_r e^2 \mu_o \omega_c^3}{3\pi c\hbar} |Mi_{ij}|^2 f(E_i)\,(1-f(E_j)) \qquad (10)$$

where $|M_{i,j}|^2$ is the transition probability from $i \to j$ and n_r the refractive index. In our experiment i is characterized by $(1,k_z,\pm 1/2)$ and j by $(0,k_z,\pm 1/2)$. In a reasonable analysis however one introduces some broading in the density of states equation (8) . Since the amount of broading is determined by the scattering mechanisms a problem of self-consistency exists. A similar situation in cyclotron resonance absorption

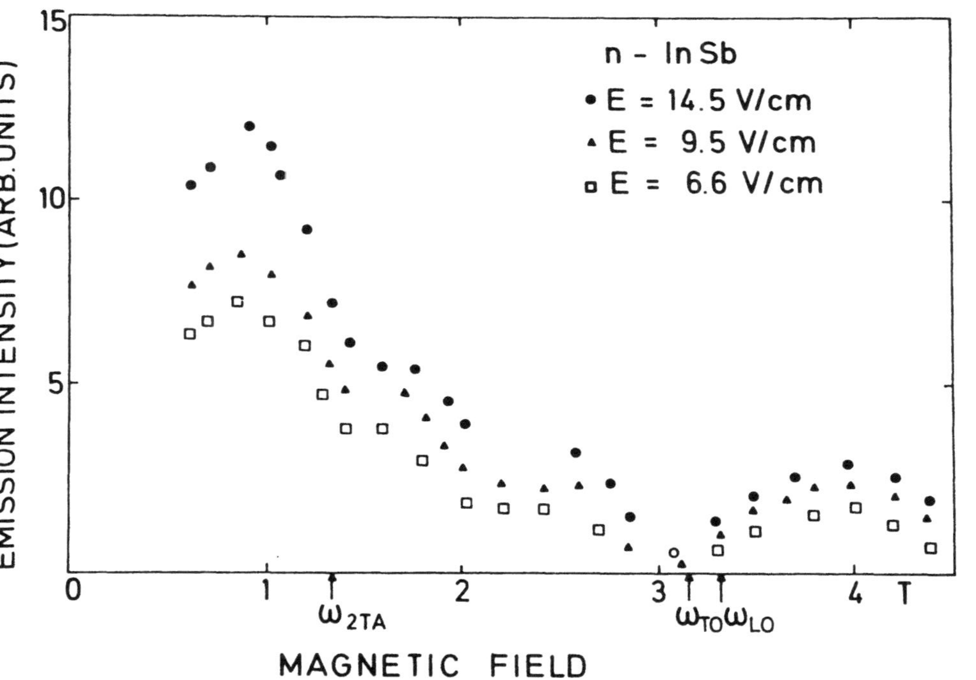

Fig. 10 : Spontaneous Landau emission intensity vs magnetic field. Parameter of the data ;s the electric field. Arrows indicate magnetic field positons where 1^+ 0^+ separation equals 2TA(X) phonon energy, and the TO and LO phonon energy, respectively.

$(0^+ \rightarrow 1^+, 0^- \rightarrow 1^-)$ has until now not been solved quantitatively as far as the problem of level broading, which determines the line width, is considered. Thus, in principle the distribution function could be determined from Landau emission measurements, however it depends on the further progress in dealing with the problem of level broadening.

V. Conclusions

The probing of a hot carrier distribution function by optical methods such as absorption, inelastic light scattering, and emission spectroscopy presents a difficult experimental problem. In order to avoid lattice heating, the excitation has always to be performed in pulses of about or below 1µs duration. Thus all optical experiments have to overcome this requirement of time resolution.

Until now, data on a few materials like Ge , GaAs , and InSb have been published. Interesting phenomena like non-equilibrium phonon distributions have been directly observed. However, nearly all experiments do not give directly a hot carrier distribution function and thus one should be very careful in deducing the DF from the observed optical phenomena.

Acknowledgements :

I thank Prof. P. Grosse for discussions. The experiments reported in part IV c were carried out in collaboration with E. Gornik and M. Overhamm at the RWTH Aachen. The technical assistance of R. Siedling and R. Kohnen is acknowledged. Part of this work was supported by Deutsche Forschungsgemeinschaft, Bonn.

★ Seminar presented at the NATO Advanced Study Institute on Linear and Nonlinear Electronic Transport in Solids, Antwerpen July 21 - August 2, 1975.

References

1. For a comprehensive review on high field transport in semiconductors see : E.M. Conwell : in Solid State Physics, Suppl. 9, ed. F. Seitz, D. Turnbull, H. Ehrenreich (Academic Press N.Y.) 1967.
   Theoretical methods to obtain high field distribution functions are reviewed in : W. Fawcett : in Electrons in Crystalline Solids (IAEA, Vienna) 1973, p. 531.
   See also : J.T. Devreese, R. Evrard, Lectures at the present Institute.
2. W. Jantsch and H. Heirich, Solid State Commun. 13, 715 (1973) .
3. J.C. McGroddy and O. Christensen, Bull. Am. Phys. Soc. Ser. II, 17, 325 (1972).
4. J.C. Hwang, J. Appl. Phys. 41, 2668 (1970).
5. W.E. Pinson, and R. Bray, Phys. Rev. 136, A 1449 (1964).
6. A.C. Baynham and E.G.S. Paige, Phys Letters 6, 7 (1963).
7. R. Bray and W.E. Pinson, Phys. Rev. Letters 11, 502 (1963).
8. M.A. Vasileva, L.E. Vorob'ev, V.I. Stafeev, Sov. Phys. Semiconductors 1, 273 (1967).
9. O. Christensen, Phys. Rev. B7, 763 (1973).
10. H.F. Budd, Phys. Rev. 158, 798 (1967).
11. P.A. Wolff, in Light Scattering Spectra of Solids (Springer Verlag N.Y.) 1969, p.273.
12. D. Healey and T.P. McLean, Phys. Letters 29A, 607 (1969).
13. A. Mooradian and A.L. McWorther in : Proc. Int. Conf. Physics of Semiconductors, Cambridge, Mass. (U.S.A.E.C.) 1970, p. 380.
    A. Mooradian : in : Laser Handbook, ed. F.T. Arecchi and E.O. Schulz-Dubois (North Holland) 1972, p.1409.
14. P.D. Southgate, D.S. Hall, and A.B. Dreeben J. Appl. Phys. 42, 2868 (1971).
15. S.R. Thomas and H.Y. Fan, Phys. Rev. B9, 4295 (1974).
16. J. Shah and R.C.C. Leite, Phys. Rev. Letters 24, 1304 (1969).
17. J. Shah Phys. Rev. B10, 3697 (1974).
    see also : J. Shah, Phys. Rev. B10, 562 (1974).
    J. Shah, R.F. Leheny, and W.F. Brinkman , Phys. Rev. B10, 659 (1974).
18. J. Shah, R.C.C.Leite, and J.F. Scott, Solid State Commun. 8, 1089 (1970).
19. I.B. Levinson, Sov. Phys. Semiconductors 7, 1121 (1974).

20. P. Motisuke, C.A. Argüello, and R.C.C. Leite, Solid State Commun. 16, 763 (1975).
21. R. Ulbrich, Phys. Rev. B8, 5719 (1973).
22. E. Gornik, Phys. Rev. Letters 29, 595 (1972), Optics and Laser Thechnology 121 (1975).
23. J. Waldman, T.S. Chang, H. Fetterman, G. Stillman, C. Wolffe, Conference Digest (Conf. on Submillimeter waves and their applications) Atlanta 1974.
24. G. Bauer, M. Overhamm, P. Grosse, E. Gornik, W. Müller, H.W. Pötzl, to be published.
25. E. Yamada, Proc. Inf. Conf. Phys. Semiconductors Warsaw (P.W.N. Sientific Publishers) Warsaw 1972.
26. F. Stern, in Solid State Physics (ed. F. Seitz and D. Turnbull, Academic N.Y.) 1963 p. 370.

Part V
Special Seminar

SOFTONS, SOFTARONS AND BISOFTARONS IN AMORPHOUS SOLIDS*

E. N. Economou

Department of Physics, University of Virginia
Charlottesville, Virginia 22901, USA

and

K. L. Ngai and T. L. Reinecke

Naval Research Laboratory
Washington, D.C. 20375, USA

I. INTRODUCTION

Many amorphous solids as well as other non-crystalline substances exhibit[1] an anomalously high specific heat which seems to vary linearly with temperature T at very low T. This implies a finite density of states down to energies comparable to kT, where T is the lowest temperature in these experiments. Anderson et al.[2] and Phillips[3] have attributed this high density of states to the existence, locally, of another configuration having about the same energy as the ground state. In other words the potential energy of the system as a function of some appropriate local atomic coordinates is assumed to exhibit a double minimum form instead of the single minimum of the crystalline solids. If the potential barrier between the two minima is not too high, tunneling from one minimum to the other is possible, and it can account for the linear specific heat. This picture has been further supported by experiments on thermal conductivity,[1] ultrasonic attenuation,[4] sound velocity[5] and nuclear spin lattice relaxation.[6,7] Note that the existence

*Work supported in part by NSF Grant No. GH-37264

of the double minimum structure is associated with the presence of a soft (≡ easily deformable) chemical arrangement of comparable energy with an alternative local chemical bonding. The existence of such conditions is favored both by randomness and also by a low coordination number as in the case of chalcogenide glasses. There, the two minima are probably associated with two different local arrangements of the bonds;[8,9] independent evidence for such a local rearrangement of the bonds was provided by Raman studies.[9] The photostructural changes exhibited by chalcogenide glasses,[10] that we shall discuss in some detail later, is another indication of the softness of their structure as well as the existence of rearrangeable bond configurations with low energy differences. A convenient way to represent the two-minima structure is by introducing local fictitious spin 1/2 variables such that the lowest minimum will correspond to $\sigma_z = -1$ and the upper minimum to $\sigma_z = 1$. In the remainder of this work we call softon each local fictitious spin 1/2 variable representing the two local configurations of minimum energy.

The concept of a softon may be useful in describing in a continuous way the transition from a crystalline solid to a liquid (through intermediate amorphous states) as a gradual increase of the softon concentration from zero (in the crystalline state) to a maximum value of the order of atomic concentration (in the liquid state).

In recent years electronic properties in amorphous semiconductors have also been examined in detail.[12] In order to interpret experimental results the following concepts have been utilized: localized states due to static disorder[11,13] polaron states,[14] and paired electronic states (of opposite spins), where the attraction is provided by phonon exchange.[15,16] To the best of our knowledge no one has considered the role of softons in determining the electronic properties of amorphous semiconductors. In the present work we address ourselves to this question and we report several preliminary results.

The interaction between the softon and electrons is treated here with a simplified but physically reasonable Hamiltonian to be introduced in the next section. Our main results are: (a) A softon can localize the electron hole in its vicinity creating a composite state to be called softaron (in analogy with the polaron state); (b) Softons can mediate a substantial el-el (hole-hole) attraction (analogous to the phonon mediated el-el attraction in superconductivity) which may overcome the Coulomb el-el (hole-hole) repulsion and create an el-el (hole-hole) localized pair to be called bi-softaron. Thus the electron-softon interaction produces all the fundamental concepts which are thought to be necessary to account for the experimental data in a-semiconductors. This, of course, does not imply that softons provide a complete picture which can explain all the experimental observations. It rather means that softons and electron-softon interactions give an alternative way for analyzing electronic properties.

It is worth pointing out that the electron-softon interaction model unites the somewhat different concepts and approaches used in studying on the one hand electronic properties (e.g. localization, states in the gap, optical properties, etc.) and on the other hand the universal properties believed to reflect the structure of amorphous semiconductors (e.g. linear specific heat, thermal conductivity, saturation of ultrasonic attenuation, etc.). Further as we shall see in Section III the softaron model provides a transparent interpretation of the physical origin of photostructural changes,[10] photodarkening effect,[17] the Urbach tail,[18,19] photoinduced ESR centers[20] and midgap absorption[20] in chalcogenide glasses. Hence the present model has the merit of being simple and elegant and its predictions are broad in scope.

For a-Si, and a-Ge with their highly coordinated 3-D strong structure we expect a few, if any, softons to be present; therefore the conventional pictures emphasizing the role of static disorder and phonons may be adequate for interpretation of most but possibly not all of the experimental data. On the other hand, in chalcogenide glasses (e.g. As_2S_3, As_2Se_3) exhibiting low coordinated, 2-D soft structure, it is expected that the concentration of softons is rather high, probably around 10^{19} per cc; therefore the softon based picture proposed here may be most appropriate for the electronic properties of these materials.

II. THEORY

A. The Model

We consider first the case of a single softon interacting with electrons. The Hamiltonian is assumed to have the following form

$$H = H_{el} + H_s + H_{el-s} \tag{1}$$

where H_{el} is a Hubbard Hamiltonian,

$$H_{el} = \sum_{is} \varepsilon \hat{n}_{is} + \sum_{ijs}{}' V_{ij} a^+_{is} a_{js} + U \sum_i \hat{n}_{i\uparrow} \hat{n}_{i\downarrow} \tag{2}$$

the softon part is

$$H_s = \omega \sigma_{zo} \tag{3}$$

and the electron-softon interaction has the general form

$$H_{el-s} = t\cdot f_1(\hat{n}_{o\uparrow},\hat{n}_{o\downarrow})\sigma_{xo} + \tau f_2(\hat{n}_{o\uparrow},\hat{n}_{o\downarrow})\sigma_{zo}. \tag{4}$$

In Eq.(1) $\hat{n}_{is} = a^+_{is}a_{is}$ is the number operator for electrons in a local Wannier state $|i\rangle$ centered at lattice site i, with spin s (s up or down); σ_{zo} is the fictitious spin 1/2 Pauli operator representing the softon site 0; for $\sigma_{zo} = -1$ we obtain the lower softon state of energy $-\omega$ and for $\sigma_{zo} = 1$ we obtain the higher softon state of energy ω. The term H_{el-s} represents the effects of electrons on the softon; the first term describes the flipping of the softon due to the presence of electron(s) at site 0; the second term represents the change of the softon energy $\pm\omega$ due to the presence of electron(s) at site 0. The functions f_1, f_2 satisfy the conditions $f_1(0,0) = f_2(0,0) = 0$. Their simplest form is

$$f_1(n_{o\uparrow},n_{o\downarrow}) = f_2(\hat{n}_{o\uparrow},\hat{n}_{o\downarrow}) = \hat{n}_{o\uparrow}+ \hat{n}_{o\downarrow} \quad . \tag{5}$$

It is conceivable however that f_1, f_2 depend non-linearly on their arguments. For example, if the electron-softon interaction is electrostatic in nature $f_1 = f_2 = (\hat{n}_{o\uparrow} + \hat{n}_{o\downarrow})^2$; f_1 may even depend in a more complicated way on $n_{o\uparrow}$, $n_{o\downarrow}$, if one takes into account the effect of the electrons on the barrier between the two minima defining the softon.

B. Filled Band; Electron Hole Similarity

Consider first the case of a completely filled valence band; then $n_{i\uparrow} = n_{i\downarrow} = 1$ and electrons decouple from the softon which is described by a renormalized Hamiltonian $H'_s \equiv \omega'\sigma'_{zo} = tf_1(1,1)\sigma_{xo} + \left(\omega + \tau f_2(1,1)\right)\sigma_{zo}$.

If some excess electrons were introduced in the conduction band their Hamiltonian H will be of the form $H_c = H_{c,el}+ H'_s + H'_{c,el-s}$ where the subscript c denotes the conduction band and $H'_{c,el-s}$ has the same form as in Eq.(4) with σ_{xo} σ_{zo} replaced by σ'_{xo} σ'_{zo} and the parameters having values appropriate for electrons in the conduction band.

If few electrons were missing from a filled valence band, one may introduce in Eqs.(1)-(4) hole operators $b^+_{is} = a_{is}$, $b_{is} = a^+_{is}$ $b^+_{is}b_{is} = 1 - n_{is}$ and reexpress the original Hamiltonian in terms of b's and the renormalized softon variables σ'_{zo}, σ'_{xo}. Defining the hole energy as minus the missing electron energy and omitting an unimportant constant term, the softon-hole Hamiltonian looks like $H_v = H_{v,h}+ H'_s + H'_{v,h-s}$, where the subscript v denotes the valence band, $H_{v,h}$ has the form (in hole operators) shown in Eq.(2), and $H'_{v,h-s}$ has the same form as in Eq.(4) with σ_{xo}, σ_{zo} replaced by σ'_{xo} σ'_{zo}, hole operators replacing the electron operators, and the parameters t, τ, f_1, f_2 replaced by t_h, τ_h, f_{1h}, f_{2h}; the latter can be easily expressed in terms of the former.

Thus both for a few electrons in the conduction band or a few holes in the valence band the Hamiltonian has the same form as in Eqs.(1)-(4). It is enough therefore to examine Hamiltonian (1) in the case of a few electrons present.

C. Single Softon-single Electron

In this case the electron spin index can be dropped $\hat{n}_{i\uparrow}\hat{n}_{i\downarrow}=0$, and $f_1 = f_2 = \hat{n}_o$. We define $H_o \equiv H_{el} + H_s + \tau\hat{n}_o\sigma_{zo}$ and $V \equiv t\hat{n}_o\sigma_{xo}$, then $H = H_o + V$; we consider the Green operators $\hat{G}_o(E) \equiv (E-\hat{H}_o)^{-1}$ and $G(E)$ $(E - H)^{-1} = (E - H_o - V)^{-1}$. One can easily show that

$$G = G_o + G_oVG_o + G_oVG_oVG_o + \ldots \tag{6}$$

Introducing a 2 x 2 matrix notation for the softon pseudospin space, one can sum Eq.(6) and recast it in the form

$$\langle i|\tilde{G}|j\rangle = \langle i|\tilde{G}_o|j\rangle + \langle i|\tilde{G}_o|o\rangle\tilde{t}(1-\langle o|\tilde{G}_o|o\rangle\tilde{t})^{-1} \langle o|\tilde{G}_o|j\rangle \tag{7}$$

where

$$\langle i|G_o|j\rangle = \begin{pmatrix} \langle i|G_o^{\uparrow\uparrow}|j\rangle & o \\ o & \langle i|G_o^{\downarrow\downarrow}|j\rangle \end{pmatrix}, \tag{8a}$$

$$\tilde{t} = \begin{pmatrix} o & t \\ t & o \end{pmatrix} \tag{8b}$$

$$\langle i|G_o^{\uparrow\uparrow}(E)|j\rangle \equiv \langle i|(E - H_{el} - \omega - \tau\hat{n}_o)^{-1}|j\rangle \tag{8c}$$

$$\langle i|G_o^{\downarrow\downarrow}(E)|j\rangle = \langle i|(E - H_{el} + \omega + \tau\hat{n}_o)^{-1}|j\rangle \tag{8d}$$

We call $\langle o|G_o^{\uparrow\uparrow}(E)|o\rangle \equiv f_-(E)$ and $\langle o|G_o^{\downarrow\downarrow}(E)|o\rangle \equiv f_+(E)$. From Eq.(7) we then obtain

$$\langle o|\tilde{G}|o\rangle = \begin{pmatrix} f_- & o \\ o & f_+ \end{pmatrix} + \frac{tf_+f_-}{1 - t^2 f_+ f_-} \begin{pmatrix} tf_- & 1 \\ 1 & tf_+ \end{pmatrix} \tag{9}$$

The poles of $\langle o|\tilde{G}|o\rangle$ give the eigenenergies of the bound eigenstates of the system (softarons), while the branch cuts correspond to the continuous spectrum. The latter coincides with the region where either or both f_+, f_- possess an imaginary part. The poles of $\langle o|\tilde{G}|o\rangle$ are given by

$$1/t^2 = f_+(E)f_-(E) \tag{10}$$

From Eqs. (8c) (8d) one sees that $f_+(E)$ is the 0,0 matrix element of a tight binding Hamiltonian with E replaced by $E + \omega$ and with a single impurity $-\tau$ present at the 0 lattice site. Similarly $f_-(E)$ is the 0,0 matrix element of the same tight binding Hamiltonian with E replaced by $E-\omega$ and a single impurity τ present at the 0 lattice site. In Fig. 1, f_+, f_-, f_+f_- are plotted schematically vs E for some particular choice of the parameters ($|\tau|$ large enough to create a bound state in f_+ or f_- and ω such that the bound state in Fig. 1b' lies above $E_c-\omega$). In the case where $\tau>o$ Eq. (10) is satisfied for $E = E_b$, where E_b is the eigenvalue of the softaron. On the other hand in the case $\tau<o$ Eq. (10) may not be satisfied as shown in Fig. 1c'; still, however, for $E = E_R$, $\langle o|\tilde{G}|o$ will exhibit a peak associated with a resonance softaron state, i.e. an extended softaron state where a substantial part of the electronic eigenfunction is localized in the vicinity of the softon. Note that the resonance "eigenenergy" E_R lies in the gap ($E_R<E_c$) but above $E_c-\omega$ as it should. If t were zero this resonance electronic state would reduce to a static impurity [21] localized state shown as the pole in Fig. 1b'; the softon, however, mixes this state with band states reducing it to an extended resonance state which will make a small but non-zero contribution to the DC conductivity. Note that large values of t and τ, and small values of ω favor the formation of a softaron (which by definition is a bound electronic state).

Explicit expressions for the softaron eigenstates in terms of $\langle o|\tilde{G}_o|j\rangle$ have been obtained by using a generalization of the Koster-Slater method.[21]

D. Single Softon-two Electrons (Holes)

We consider next the case of a single softon interacting with two electrons. The Hamiltonian is again given by Eqs. (1-4). We were not able to solve this problem exactly. Instead we have used a variational technique, analogous to the solution of the ground state of the helium atom problem. The two electrons of

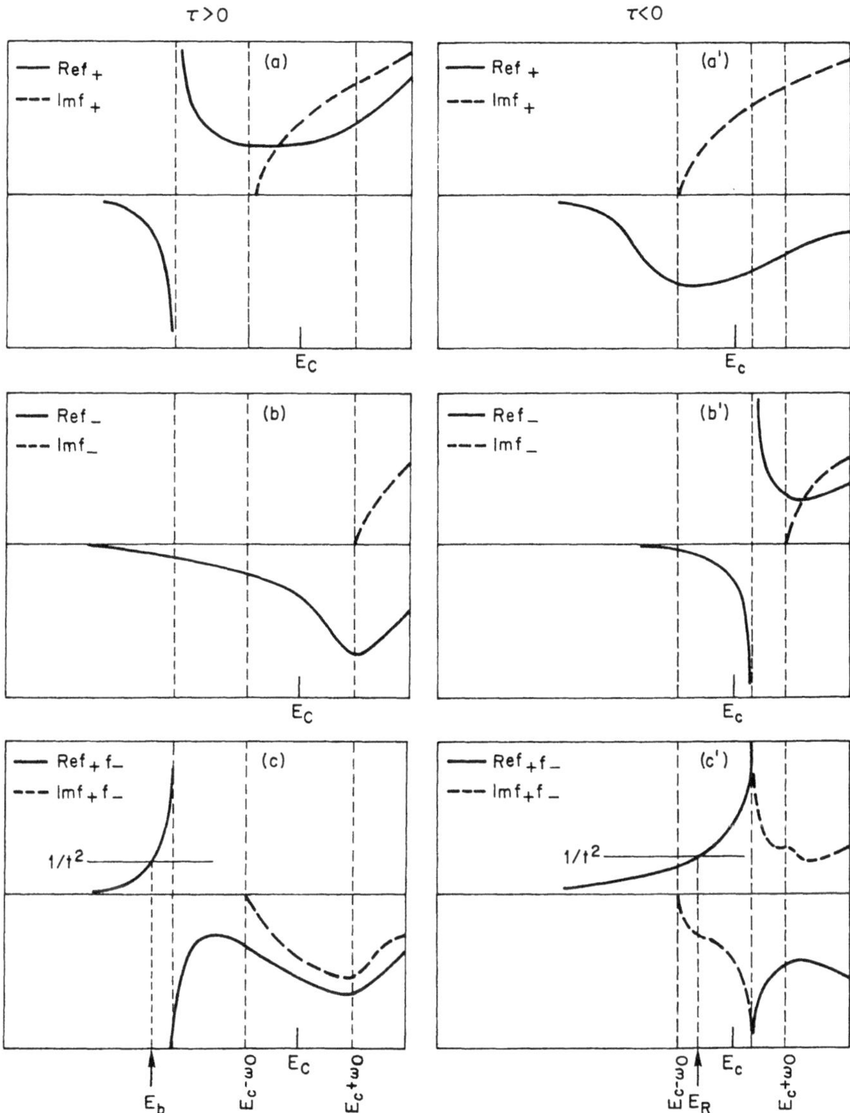

Fig. 1: Schematic plot of $f_+(E)$, $f_-(E)$, $f_+(E)f_-(E)$ vs E for positive and negative τ(see text). The eigenenergy E_b (c) and the resonance energy E_R (c') are indicated (see text). The former is a solution of the equation $f_+(E)f_-(E) = 1/t^2$ while the latter satisfies the equation $\mathrm{Re}f_+(E)f_-(E) = 1/t^2$, and makes $\mathrm{Im}f_+(E)f_-(E)$ small.

opposite spin are assumed to have wavefunctions of the form obtained from the solution of the softaron problem described in Section (II.C) except that the parameters that describe (1) the spatial extent of the wavefunction and (2) the ratio of its up and down softon pseudospin components are treated as variational parameters. By minimizing the energy with respect to these variational parameters, we obtain an approximate variational eigenfunction for the two electrons (holes)-softon system as well as the eigenenergy. In general an effective electron-electron (hole-hole) attraction (mediated by the softon) results; this attraction may overcome the direct Coulomb repulsion (due to the Hubbard term) in which case the two electrons (holes) form a localized pair state called a bisoftaron; non-linear forms of f_1 and f_2 substantially enhance the binding energy of the bisoftaron. On the other hand if t is small and f_1 and f_2 are linear as in Eq.(5) the effective electron-electron attraction is weak compared with Coulomb repulsion, and the two electrons repel each other. In analogy with softaron resonance states, bisoftaron resonance states may exist although the variational approach is not appropriate for examination of this question.

Nonlinear forms of f_1 and f_2 are quite plausible and can arise from electrostatic interaction of electrons with softons or from the lowering of the energy barrier that defines the softon in the presence of the electrons (holes). We have examined several nonlinear models for f_1 and f_2 and calculated variationally (numerically) the bisoftaron eigenfunction and eigenenergy for (1) the one dimensional lattice of equally spaced atoms and (2) the Bethe lattice. In both cases, the criteria for the formation of bisoftaron states and their relation with the softaron states are obtained. Detailed calculations will be presented elsewhere. The important conclusion from these computations for our present purpose is that bisoftaron states of two electrons or two holes can exist in soft structures.

E. More Complicated Cases

Since both electrons and holes are present in an a-semiconductor one should examine the question of an electron-hole pair in the presence of a softon. The effective, softon mediated, electron-hole interaction is not always attractive. The reason is that the electron and the hole parameter are in general different and consequently the softon is affected differently in the presence of a hole. The simultaneous action of electron and hole forces the softon to a quite new state which tends to be of higher energy. Thus one expects electron-hole repulsion more often than electron-hole attraction. Note that if electron-hole attraction is achieved the energy of the system will be substantially lowered if one brings a pair of electrons and a pair of holes at the vicinity of the same softon: We will refer to this possibility in the next section.

In a realistic case we have a finite concentration of local softons (at least two to three orders of magnitude less than the atomic concentration). Each of these softons is characterized by its own parameters, ω, τ, t, which should be viewed as random variables, which may be statistically correlated; further the softon may be associated with more than one lattice site. For low concentration of softons each one can be treated independently of the rest; as the concentration increases the overlap of the softaron eigenstates becomes significant and one should treat the problem as the impurity bands by using a generalized Anderson's Hamiltonian for random lattices.[22] In the present paper we assume that the softon concentration is low enough as to avoid this complication.

III. APPLICATIONS TO AMORPHOUS SEMICONDUCTORS

A. Ground State, States in the Gap

In Fig. 2 we present schematically a density of states diagram for an amorphous semiconductor on the basis of electron (hole)-softon interaction as developed in the previous section. Note that the lower part of the diagram is a hole-density of states plot. This means that, if electrons are removed from the filled valence band, holes are created initially below E_V. These holes interact with the softons and fall to lower energy softaron states. The lowest energy hole states are obtained if a hole-pair is localized in the vicinity of an appropriate softon. After these (hole) bisoftaron states are occupied by hole pairs, (hole) softaron states start being (singly) occupied by holes. Even higher in energy lie the resonance (hole) softaron states. Similar things happen to electrons below the bottom of the conduction band. The energy overlap of (electron) and (hole) bisoftaron states shown in Fig. 2 is very significant. It simply means that the total energy of the system is lowered if two electrons are removed from the valence band and placed in an (electron) bisoftaron state while the two holes left behind arrange themselves in a (hole) bisoftaron state. One can distinguish two cases: these two bisoftarons are either associated with the same softon or with two different ones. In the first case the rearranged state is neutral, since simply two electrons from the valence band (probably the lone pair for chalcogenide glasses in the vicinity of the softon) are removed and placed in the vicinity of the same softon but in an antibonding type of compound state. The total energy is lowered because of interaction with the softon. In the second case the vicinity of one

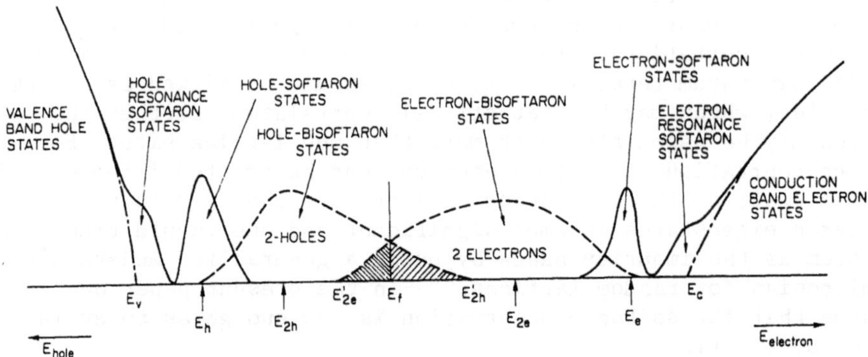

Fig. 2. Density of states for single electron and single hole (solid lines) as well as bound pairs of two electrons or two holes (dashed lines). The Fermi level, E_F, the top of the valence band, E_v, and the bottom of the conduction band, E_c, (in the absence of el-softon interaction) are indicated.

softon is left positively charged while the vicinity of the second softon is negatively charged. We expect that both cases are realized although we think that the first one is more probable. The Fermi level E_F is determined from the requirement that the total number of (electron)bisoftaron states from E'_{2e} to E_F is equal to the total number of (hole) bisoftaron states from E'_{2h} to E_F. One can summarize the ground state picture as follows: Some of the softons (those associated with the shaded areas in Fig. 2) manage to lower the total energy of the system, thereby being stabilized, either by localizing an extra electron (hole) pair in their vicinity or by moving an electron pair from a localized lone pair state to a localized antibonding state. For the rest of the softons such processes will increase the total energy and consequently they are not realized in the ground state. Thus the rest of the softons remain in their ground state decoupled from the electronic degrees of freedom. The Fermi level is pinned as shown in Fig. 2.

B. Physical Consequences

The picture presented above which was derived from the electron (or hole)-softon interaction is consistent with several experimental data:

The pinning of the Fermi level, the absence of paramagnetism (since only paired states are occupied), an optical gap E_g roughly equal to $E_c - E_v$, since the electrons initially are below E_v and can essentially be excited to states in the vicinity of E_c; also if the bisoftaron states just below E_F are excited the optical gap will be equal[15] to $(E_c - E_F) + (E_e - E_F) \approx E_c - E_v$. (See Fig. 2)

Two linear contributions to the specific heat are predicted from the present picture. One is associated with the decoupled (in the ground state) softons; the other with occupation of bisoftaron states just above the Fermi level. It is expected that the first contribution will interact strongly with phonons while the second one, being entirely electronic in nature, will have a much weaker interaction with phonons. This seems consistent with recent experimental evidence[23] pointing to the existence of two contributions to the specific heat, one being the usual two level system (softon) strongly affecting thermal conductivity and the other weakly influencing the thermal conductivity.

If the material is illuminated by photons some of the (hole) and (electron) softaron states will be occupied. Thus an illuminated material should exhibit ESR coming from localized holes (softaron states at E_h) and localized electrons (softaron states at E_e). The concentration of these occupied softaron states is expected to be very low since only a small fraction of the softons will be singly occupied, the rest being either doubly occupied or unoccupied. This feature is consistent with recent experimental data[20] on photo-induced states in the gap detected by ESR.

After illumination one expects absorption for energies well below the gap since the singly occupied softaron states can be excited either by removing the electron from the (electron) softaron to the conduction band or by removing the hole from the (hole) softaron to the valence band. This property has been observed experimentally[20] as photo-induced "midgap" absorption.

Amorphous chalcogenide semiconductors and their alloys undergo changes in optical and other properties when irradiated with band gap light. Detectable structural changes (photo-structural[10] transformations) often occur along with such changes in optical properties as shifts of the absorption edge energy[17] (photodarkening) and modified values of the refractive index.[10] The softon, softaron and bisoftaron model introduced here is also found to be very useful in explaining the physical mechanisms behind these interesting effects. Similarly we can also give an explanation for the Urbach tail[18,19] which is the exponential absorption edge in amorphous solids. Before proceeding to discuss these topics, it is useful to recall that the ground state of a glass before optical excitation consists of occupied bisoftaron states associated with a certain fraction of the softons, while the rest of the softons can be considered to be decoupled from the electronic degrees of freedom.

Photostructural Transformations: When a glass is illuminated with band gap radiation, electrons and holes are created by excitation from the valence band. These excess electrons and holes can then interact with the softons with the Hamiltonian given by Eqs. (1-4), forming electron and hole softaron eigenstates and resonances and in the process changing the pseudospin from its initial state to a final softon state which is part of the softaron. These softaron states are metastable and favor bisoftaron formation. On the other hand, the radiation also excites the already occupied bisoftarons in the energy gap, and the associated softons are relaxed. Both the creation of new bisoftarons and the destruction of occupied bisoftarons lead to softon renormalizations, and hence the structure of the glass is altered locally. These processes leading to local alterations of structure are believed to be reversible. If on the other hand the external radiation is sufficiently intense so that many softaraons are created and many bisoftarons created and destroyed, then several of the associated softons may undergo a collective transformation and the concomitant structure change will involve a large number of atoms or bonds. In the language of our softaron Hamiltonian, Eqs. (1-4), we can alternatively say that the term $t \cdot f_1 \sigma_x$ increases with increasing number of electrons and holes excited at bonds within the extent of the softon. Therefore even a softon which has a high barrier and is difficult to flip would do so when the softon has many of its member atoms excited by external radiation. When such a "difficult" softon flips, we believe that the structural change is irreversible.

Photodarkening Effect: After reversible photostructural transformation in several chalcogenide glasses, the absorption edge of the glasses are often observed to have shifted to longer wavelengths. Photon absorption and consequent excitation of the electrons-softon system obey the Franck-Condon Principle, in the sense that electron velocities are large compared with those of softon flip, and in the time duration of an electronic transition, the softon cannot significantly have its configuration changed. The physics can best be described in terms of a diagram giving the energy of the electron-softon system as a function of the softon configuration-coordinate as in Fig. 3. In this figure, the lower curve represents the softon decoupled from the valence band electrons before the band gap transition. The upper curve represents the electron and/or hole now interacting with the softon by means of the $\tau f_2(n_e, n_h)\sigma_z$ term. The off-diagonal term $tf_1(n_e,n_h)\sigma_x$ is not operative in the duration of the electronic transition because of the Franck-Condon Principle. The optical excitation labelled E_{g1} is vertical, with the initial state on the lower curve and the final state on the upper curve having the same softon configuration coordinate. This transition with energy E_{g1} contributes to the band edge absorption. After a glass has been illuminated so as to transform photostructurally, the softon coordinate is shifted to position 2 which defines the ground state of the structurally transformed glass. A band edge absorption measurement then will give the vertical transition E_{g2}. In general E_{g2} will be different from E_{g1} and hence a shift of the absorption edge is seen. Whether this is a red or blue shift depends on the preparation method and annealing procedure of the original glass.

The shift in refractive index can also be readily understood in terms of the softon model. It has been suggested from experimental evidence by von Schickfus et al.[24] that softons contribute to the refractive index of the glass. Hence, after a photostructural change has occurred, the coordinates of the softon are displaced (Fig. 3) and the contributions of the softon to the refractive index is altered. In addition, the shift of the absorption, which is a consequence of softon coordinate displacement, implies that the interband contribution to the refractive index is also modified.

Urbach Tail: The exponential tail of the absorption edge is usually referred to as the Urbach tail and is empirically described by the relation

$$\alpha(\omega)\sim\exp[\sigma\left(\hbar\omega - E_g(T)\right)/kT^*], \quad \hbar\omega<E_g(T) \tag{11}$$

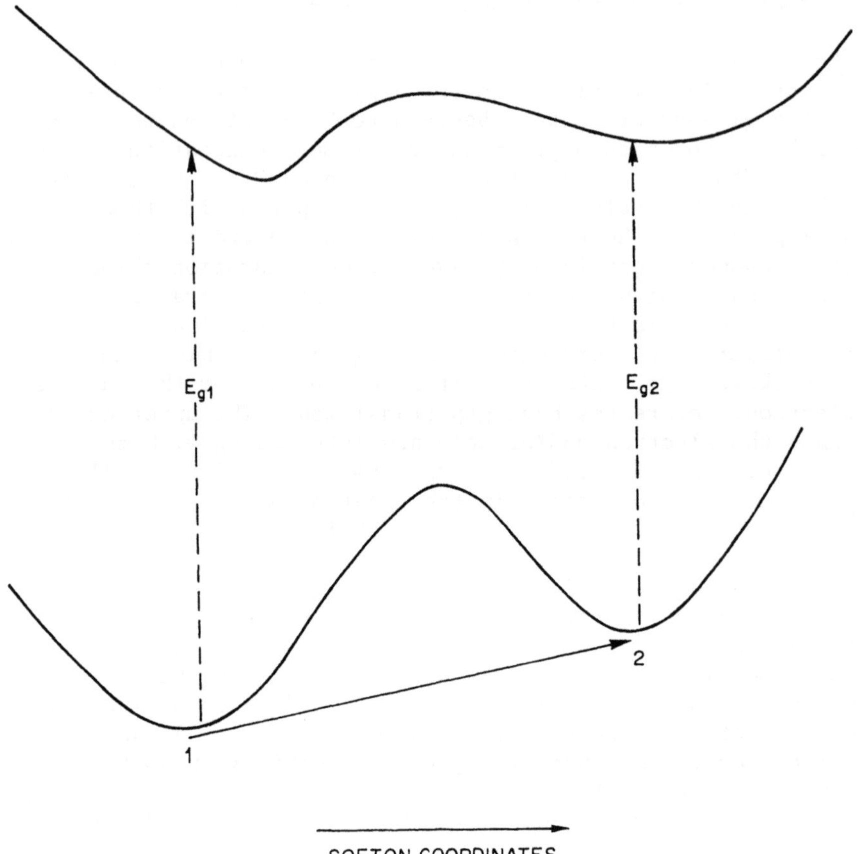

Fig. 3 The solid line with an arrow indicates either a photo-structural transformation (in the photodarkening case) or a thermal excitation (in the high temperature Urbach tail case).

where α is the absorption constant of order unity, $E_g(T)$ is the temperature dependent gap, and T^* an effective temperature. The effective temperature T^* is approximately constant at low temperature and proportional to T at high temperature. Discussions of the Urbach tail can be found in recent reviews by Dow[18] and Tauc.[19] Urbach tails occur in some crystalline ionic solids and in amorphous solids. We shall limit ourselves to the chalcogenide glasses in applying the softon-softaron approach to the Urbach tail. This is especially appropriate because crystalline As_2Se_3 does not exhibit an Urbach tail while amorphous As_2Se_3 does.

At low temperatures the Urbach tail is treated in the softon approach by generalizing the softaron Hamiltonian, Eqs. (1-4), from one site to multiple sites for softons of finite extent. Instead of Eq.(4) we have then

$$H_{el-s} = \sigma_x \sum_i t_i f_1(\hat{n}_{ei}, \hat{n}_{hi}) + \sigma_z \sum_i \tau_i f_2(\hat{n}_{ei}, \hat{n}_{hi}). \tag{12}$$

Here the softon (σ_z, σ_x) acts on a number of sites, hence the summation over i. The electron-hole pair excited by photon absorption appears in f_1 and f_2. The interaction (Hamiltonian) H_{el-s} in Eq. (12) can be interpreted as "microfields" acting on the electron-hole pair, with the origin of the microfields due to softons. Having in mind this interpretation, one can use an approach similar to that of Dow and Redfield[25] giving $\alpha(\omega)$ as in Eq. (11) with T^* proportional to the "microfield" strength.

At high temperature an additional process begins to compete. Any softon which is decoupled from the valence electrons can be thermally excited to its other softon coordinates with higher energy ΔE (Fig. 3). This is analogous to the photostructural transformation escept that the thermal excitation probability is proportional to $e^{-\beta\Delta E}$. Then the Urbach tail can be interpreted in terms of Franck-Condon principle as follows: During thermal excitation of the softon ($1 \rightarrow 2$ in Fig. 3), an optical transition to the upper curve occurs. If the overall shift in energy from the edge in this vertical transition bears a simple relationship to ΔE, an exponential absorption edge of the form shown in Eq. (11) with $T^* = T$, the temperature, is obtained.[26]

IV. CONCLUDING REMARKS

In these lectures we have introduced a novel approach to the electronic states in the gap of amorphous semiconductors and insulators. The salient feature of the approach is the recognition that structural disorder modes in glasses, which manifest themselves explicitly in several nonelectronic physical properties, e.g., the linear T-dependence of the specific heat, also play a central role in determining the electronic states in the energy gap. This is true in all glasses whose structure is "soft" and thus it allows an appreciable number of these two level structural disorder modes to exist. This class of structural disorder modes is collectively called softons. They can capture an electron (or hole) forming a softaron state; they can even bound two electrons (or holes) in their vicinity creating a bisoftaron state.

It has been shown that the foregoing "softon-softaron" model accounts in a natural way for a wide variety of features considered essential in understanding the electronic properties of amorphous semiconductors. These results include: (i) the pinning of the Fermi energy near midgap due to the overlap of bisoftaron hole and electron states (ii) the diamagnetism of most amorphous semiconductors due to opposite spin electron-electron and hole-hole pairing in the ground state, (iii) the appearance of photo-induced paramagnetic centers in the gap and midgap absorption by excitation of bisoftarons into softaron states, and (iv) the photostructural and photodarkening effects and the associated Urbach tail in the optical absorption as due to modifications in softons and associated electron states upon photo-excitation.

The model developed here produces simultaneously single electron localized states and bound electron pairs of the type proposed by Anderson[15] as well as hole localized states and bound hole pairs of the type proposed by Street and Mott[16] and Mott, Davis and Street.[16] However, the present model goes beyond the work by these authors in that it has the following desirable features: (i) it employs a single, novel, plausible mechanism (softons) known to be appropriate for amorphous materials for the electron (hole) localization, and electron-electron (hole-hole) attraction, (ii) it provides a derivation of the localized and resonant electronic states in a unified and natural way from a single Hamiltonian which involves a single, simple electron-softon interaction and treats electrons and holes symmetrically (iii) it accounts for the variety of experimental results obtained in particular materials and (iv) it directly links electronic properties to structural features, thereby providing new insight into photostructural and photodarkening effects which involve both electronic and structural aspects. It should be pointed out, however, that both softon derived states and conventional (e.g. dangling band defect[16]) states may be present simultaneously in some materials.

ACKNOWLEDGEMENTS

We would like to thank S. G. Bishop, H. Fritzsche, U. Strom and P. C. Taylor for informative discussions.

REFERENCES

1. R. C. Zeller and R. O. Pohl, Phys. Rev. B4, 2029 (1971); R. B. Stephens Phys. Rev. B8, 2896 (1973) and Thesis, Cornell University, 1974; M. P. Zaitlin and A. C. Anderson, Phys. Rev. Lett. 33, 1155 (1974); J. C. Lasjaumas, D.

Thoulouze and F. Pernot, Solid St. Comm. 14, 957 (1974).

2. P. W. Anderson, B. I. Halperin and C. M. Varma, Phil. Mag. 25, 1 (1972).

3. W. A. Phillips, J. Low Temp. Phys. 7, 351 (1972).

4. S. Hunklinger, W. Arnold, S. Stein, R. Nava, and K. Dransfeld, Phys. Lett. 42A, 253 (1972); B. Golding, J. E. Graebner, B. I. Halperin, and R. J. Schutz, Phys. Rev. Lett. 30, 223 (1973); W. Arnold, S. Hunklinger, S. Stein and K. Dransfeld, J. Non-Cryst. Solids, 14, 192 (1974).

5. L. Piché, R. Maynard, S. Hunklinger and J. Jäckle, Phys. Rev. Lett. 32, 1426 (1974).

6. J. Szeftel and H. Alloul, Phys. Rev. Lett. 34, 667 (1975).

7. M. Rubinstein, H. A. Resing, T. L. Reinecke and K. L. Ngai, Phys. Rev. Lett. 34, 1444 (1975); T. L. Reinecke and K. L. Ngai, Phys. Rev. B 12, 3476 (1975).

8. E. Finkman, A. P. DeFonzo, and J. Tauc, in "Amorphous and Liquid Semiconductors," edited by J. Stuke and W. Brenig (Taylor and Francis, London, 1974) p. 1275.

9. G. N. Papatheodorou and S. A. Solin, Phys. Rev. issue (1976).

10. J. P. deNeufville chapter in "Optical Properties of Solids - Recent Developments" edited by E. O. Seraphin, North Holland Press, 1975. H. Fritzsche, in "Electronic and Structural Properties of Amorphous Semiconductors," edited by P. G. LeComber and J. Mort (Academic Press, 1973), p. 575.

11. N. F. Mott and E. A. Davis, "Electronic Processes in Non-Crystalline Materials" (Clarendon Press, Oxford, England, 1971).

12. See, for example, "Amorphous and Liquid Semiconductors" edited by J. Stuke and W. Brenig (Taylor and Francis, London, 1974).

13. M. H. Cohen, H. Fritzsche, and S. R. Ovshinsky, Phys. Rev. Lett. 22, 1065 (1969).

14. D. Emin, in "Electronic and Structural Properties of Amorphous Semiconductors," edited by P. G. LeComber and J. Mort (Academic Press, 1973), p. 261.

15. P. W. Anderson, Phys. Rev. Lett. 34, 953 (1975).

16. R. A. Street and N. F. Mott, Phys. Rev. Lett. 35, 1293 (1975); N. F. Mott, E. A. Davis and R. A. Street, to be published.

17. R. Chang, Mat. Res. Bull. 2, 145 (1967); K. Tanaka and M. Kikuchi, "Amorphous and Liquid Semiconductors, edited by J. Stuke and W. Brenig (Taylor and Francis, London, 1974) p. 439.

18. J. D. Dow, Comments on Solid State Physics 4B, 35 (1972).

19. J. Tauc in "Amorphous and Liquid Semiconductors" edited by J. Tauc (Plenum 1974), p. 159.

20. S. G. Bishop, U. Strom and P. C. Taylor, Phys. Rev. Lett. 34 1346 (1975); S. G. Bishop, U. Strom and P. C. Taylor, Solid St. Comm. (to appear).

21. G. F. Koster and J. C. Slater, Phys. Rev. 95, 1167 (1954); J. Callaway, J. Math. Phys. 5, 783 (1964).

22. See e.g., D. C. Licciardello and E. N. Economou, Phys. Rev. B11, 3697 (1975).

23. H. S. Hunklinger, L. Piché, J. C. Lasjaunias and K. Dransfeld, J. Phys. (London) C8, L423 (1975).

24. M. von Schickfus, S. Hunklinger, and L. Piché, Phys. Rev. Lett. 35, 876 (1975).

25. J. D. Dow and D. Redfield, Phys. Rev. B1, 3358 (1970).

26. H. Sumi and Y. Toyozawa, J. Phys. Soc. Jap. 31, 342 (1971).

LIST OF LECTURERS

I. G. Austin, Department of Physics, University of Sheffield, Sheffield S3 7RH, Great Britain

P. N. Butcher, Department of Physics, University of Warwick, Coventry, Warwickshire CV4 7AL, Great Britain

J. T. Devreese, Leerstoel voor Toegepaste Wiskunde voor de Wetenschappen, Rijksuniversitair Centrum Antwerpen, B-2020 Antwerpen, Belgium; and Departement Natuurkunde, Universitaire Instelling Antwerpen, B-2610 Wilrijk, Belgium

D. Emin, Sandia Laboratories, Albuquerque, New Mexico 87115, U.S.A.

R. Evrard, Institut de Physique, Université de Liège, Sart Tilman par Liège I, Belgium

D. Langreth, Department of Physics, Rutgers University, New Brunswick, New Jersey 08903, U.S.A.

N. H. March, Department of Physics, Imperial College, London SW7 2BZ, Great Britian

N. Mott, Cavendish Laboratory, University of Cambridge, Cambridge CB3 OHE, Great Britain

P. Nagels, Rijksuniversitair Centrum Antwerpen, B-2020 Antwerpen, Belgium; and Departement Materiaal Onderzoek, S.C.K., B-2400 Mol, Belgium

K. L. Ngai, Naval Research Laboratories, Washington, D. C. 20375 U.S.A.

G. Papadopoulos, University of Leeds, Leeds LS2 9JT, Great Britain; and E. S. I. S., Universitaire Instelling Antwerpen, B-2610 Wilrijk, Belgium

K. S. Singwi, Department of Physics, Northwestern University, Evanston, Illinois 60201, U.S.A.

K. K. Thornber, Bell Laboratories, Murray Hill , New Jersey 07974, U.S.A.

M. Tosi, Istituto di Fisica "Guglielmo Marconi," Università degli Studi Roma, Roma, Italy

N. Tzoar, Department of Physics, City University of New York, New York, N. Y. 10031, U.S.A.

W. Zawadzki, Instytut Fizyki Polskiej, Akademii Nauk, Warszawa, Poland

AUTHOR INDEX

SUBJECT INDEX

MIX
Papier aus verantwortungsvollen Quellen
Paper from responsible sources
FSC® C105338

If you have any concerns about our products,
you can contact us on
ProductSafety@springernature.com

In case Publisher is established outside the EU,
the EU authorized representative is:
Springer Nature Customer Service Center GmbH
Europaplatz 3, 69115 Heidelberg, Germany

Printed by Libri Plureos GmbH
in Hamburg, Germany